VOLUME SEVENTY SEVEN

ADVANCES IN INORGANIC CHEMISTRY

Catalysis in Biomass Conversion

ADVISORY BOARD

VOLUME SEVENTY SEVEN

ADVANCES IN INORGANIC CHEMISTRY

Catalysis in Biomass Conversion

Edited by

PETER C. FORD

Department of Chemistry and Biochemistry,
University of California Santa Barbara, CA, United States

RUDI van ELDIK

University of Erlangen–Nuremberg, Germany
N. Copernicus University, Torun, Poland
Jagiellonian University, Krakow, Poland

ELSEVIER

ACADEMIC PRESS

An imprint of Elsevier

Academic Press is an imprint of Elsevier
50 Hampshire Street, 5th Floor, Cambridge, MA 02139, United States
525 B Street, Suite 1650, San Diego, CA 92101, United States
The Boulevard, Langford Lane, Kidlington, Oxford, OX5 1GB, United Kingdom
125 London Wall, London, EC2Y 5AS, United Kingdom

First edition 2021

ISBN: 978-0-323-85058-2
ISSN: 0898-8838

For information on all Academic Press publications
visit our website at https://www.elsevier.com/books-and-journals

Publisher: Zoe Kruze
Acquisitions Editor: Sam Mahfoudh
Developmental Editor: Federico Paulo Mendoza
Production Project Manager: Denny Mansingh
Cover Designer: Alan Studholme

Typeset by SPi Global, India

Contents

Contributors

Mahdi M. Abu-Omar
Department of Chemistry and Biochemistry; Department of Chemical Engineering, University of California, Santa Barbara, CA, United States

Wouter Arts
Centre for Sustainable Catalysis and Engineering, KU Leuven, Leuven, Belgium

Alina M. Balu
Departamento de Química Orgánica, Universidad de Córdoba, Cordoba, Spain

Jacob A. Barrett
Department of Chemistry and Biochemistry, University of California, San Diego, CA, United States

Katalin Barta
Institute for Chemistry, Organic and Bioorganic Chemistry, University of Graz, Graz, Austria

Mauricio Boscolo
Sao Paulo State University (UNESP), Chemistry and Environmental Sciences Department (IBILCE), Sao Jose do Rio Preto, SP, Brazil

Zhenweng Cao
Max-Planck-Institut für Kohlenforschung, Mülheim an der Ruhr, Germany; Qingdao Institute of Bioenergy and Bioprocess Technology, Chinese Academy of Sciences, Qingdao, China

Yanbin Cui
Department of Chemistry, University of Wisconsin-Madison, Madison, WI, United States

Mario De Bruyn
Institute for Chemistry, Organic and Bioorganic Chemistry, University of Graz, Graz, Austria

Michael Dierks
Max-Planck-Institut für Kohlenforschung, Mülheim an der Ruhr, Germany

Shannon L. Goes
Department of Chemistry, University of Wisconsin-Madison, Madison, WI, United States

István T. Horváth
Department of Chemical and Environmental Process Engineering, Budapest University of Technology and Economics, Budapest, Hungary

Baoyuan Liu
Department of Chemistry and Biochemistry, University of California, Santa Barbara, CA, United States

Rafael Luque
Departamento de Química Orgánica, Universidad de Córdoba, Cordoba, Spain; Peoples Friendship University of Russia (RUDN University), Moscow, Russian Federation

Gustavo Metzker
Sao Paulo State University (UNESP), Chemistry and Environmental Sciences Department (IBILCE), Sao Jose do Rio Preto, SP, Brazil

László T. Mika
Department of Chemical and Environmental Process Engineering, Budapest University of Technology and Economics, Budapest, Hungary

Roberto Rinaldi
Imperial College London, Department of Chemical Engineering, South Kensington Campus, London, United Kingdom

Daily Rodríguez-Padrón
Departamento de Química Orgánica, Universidad de Córdoba, Cordoba, Spain

Antonio A. Romero
Departamento de Química Orgánica, Universidad de Córdoba, Cordoba, Spain

Dieter Ruijten
Centre for Sustainable Catalysis and Engineering, KU Leuven, Leuven, Belgium

Bert Sels
Centre for Sustainable Catalysis and Engineering, KU Leuven, Leuven, Belgium

Shannon S. Stahl
Department of Chemistry; Wisconsin Energy Institute, University of Wisconsin–Madison, Madison, WI, United States

Zhuohua Sun
Beijing Advanced Innovation Center for Tree Breeding by Molecular Design, Beijing Key Laboratory of Lignocellulosic Chemistry, Beijing Forestry University, Beijing, PR China

Laura Trullemans
Centre for Sustainable Catalysis and Engineering, KU Leuven, Leuven, Belgium

Korneel Van Aelst
Centre for Sustainable Catalysis and Engineering, KU Leuven, Leuven, Belgium

Jorge Andrés Mora Vargas
Sao Paulo State University (UNESP), Chemistry and Environmental Sciences Department (IBILCE), Sao Jose do Rio Preto, SP, Brazil; Universidad Santiago de Cali (USC), Facultad de Ciencias Basicas, Santiago de Cali, Colombia

Feng Wang
State Key Laboratory of Catalysis, Dalian National Laboratory for Clean Energy, Dalian Institute of Chemical Physics, Chinese Academy of Sciences, Dalian, China

Chaofeng Zhang
State Key Laboratory of Catalysis, Dalian National Laboratory for Clean Energy, Dalian Institute of Chemical Physics, Chinese Academy of Sciences, Dalian, China

Preface

The idea of a thematic volume of *Advances in Inorganic Chemistry* devoted to "Catalysis in Biomass Conversion" was initiated in discussions between AIC series editor Rudi van Eldik and volume editor Peter C. Ford in March 2020. This, as everyone will remember, was in the early stages of the world-wide pandemic. So, with a deep sigh of relief and much gratitude to the contributors, we are pleased to be able to present a volume on this topic consisting of 10 chapters covering most aspects of biomass conversion to chemicals and fuel.

Biomass conversion has long been of interest to a relatively small cohort of chemists, many of whom were residents in departments devoted to forestry or paper manufacture. However, the increasing concern about climate change related to the CO_2 burden in the atmosphere has stimulated growing interest in this topic among chemists and chemical engineers. In principle, the fuels and/or chemicals derived from biomass can be carbon-neutral. One can easily imagine that much of the world's energy requirements can be addressed with sustainable sources such as wind turbines and photovoltaics or low carbon intensity energy sources such as nuclear reactors. However, for the foreseeable future, there will continue to be major requirements for renewable liquid aviation and marine fuels and for the organic feedstocks to prepare the multitude of materials and chemicals that modern society demands.

The intellectual challenges of biomass conversion are substantial, and require a paradigm shift, since the chemical establishment over the past century has focused its attention on fossil carbon feedstocks. Petroleum, for example, has high thermal stability, can be volatized and has low functionality. By comparison, lignocellulose, by far the most plentiful form of biomass, has low thermal stability, low volatility, and high functionality. Procedures designed to upgrade a simple hydrocarbon to a higher value chemical, simply do not work with a multifunctional substrate. Furthermore, dry lignocellulose is a complex mixture of three major components, the carbohydrates cellulose and hemicelluloses, plus lignin. These have first to be separated and then each be treated with different chemical procedures. To add further complications, the percentage compositions of lignocellulose vary from species to species and even within a species.

Lignin, which comprises 15%–30% of the woody biomass dry weight is particularly complex given that it is formed from the radical catalyzed polymerizations of several propyl phenol derivatives. It has the highest relative carbon content of the three major components of woody biomass and is a potential source of electron-rich aromatic compounds. Nonetheless, owing to its complexity, heterogeneity, and recalcitrant nature, lignin remains largely underutilized. It is a by-product of paper manufacture and second-generation bioethanol refineries, and is mostly burned to produce low-grade heat. However, the economic success of large-scale commercialized biorefining will depend on the integrated and efficient utilization of all the material and energy stored in lignocellulose including lignin.

Despite these challenges, biomass utilization presents some real opportunities. If the goal were simply to prepare chemicals and fuels as drop-in replacements for those derived from petroleum, real opportunities would be lost. One should be able to build new materials from biomass-derived substrates that offer key functional advantages over current technology, including biodegradability. In addition, biomass is widely distributed and expensive to transport, therefore integrated biorefineries will need to locate in communities proximal to the districts that produce the forest or agricultural biomass feedstocks. The impact would be a redistribution of industrial activity that may have major socio-economic consequences.

The Ford group's research adventures in biomass conversion began about 15 years ago with a commitment to shift the focus of catalysis studies away from petrochemical processes such as carbonylations to the chemistry of renewable feedstocks. We began by examining the dehydrogenation of polyols with homogeneous catalysts as a possible entry into the conversion of carbohydrates to syngas. In parallel studies, we developed solid bases for the conversion of triglyceride fats into biodiesel.[1] These solid bases were prepared by calcining transition metal-doped hydrotalcites (Mg: Al double layered hydroxides) into porous metal oxides (PMOs). Naively, we thought that these solid bases would also prove effective for disassembling the biopolymers present in lignocellulose. Perhaps surprisingly, we were successful in converting lignin and even lignocellulose to organic liquids, the most effective catalyst being Cu-PMO in super critical methanol.[2] This turned out to be serendipity, since our initial premise was insufficient. Much of the disassembly activity seen was the result of hydrogenolysis and hydrodeoxygenation reactions catalyzed by the copper centers. The hydrogen needed for these processes was produced by the copper reforming of the methanol.[3] Our subsequent efforts in this area have focused on improving selectivity since the plethora of products formed in the original process makes separation impractical.

The 19th Century was the century of coal while the 20th Century was the century of oil and natural gas. It is inevitable that the 21st Century will be the century of renewable energy and feedstocks. From the comments above, it should be obvious that catalytic biomass conversion should play a significant role in this societal transformation. This is a very broad topic given the marked diversity of biomass-based resources of this substrate, but real progress is being made in addressing the key challenges. This volume has brought together highly respected experts on biomass conversion to discuss their approaches to upgrading key components to the renewable fuels and chemicals for societal needs in the future.

In Chapter 1, *Mika* and *Horváth* (Budapest University of Technology and Economics) discuss transition metal homogeneous catalysts for the transformations of levulinic acid (LA) into higher value compounds, principally gamma-valerolactone. LA is a platform chemical produced from a wide range of carbohydrate containing biomass, including agricultural residues and food wastes.

Chapter 2 by *Rodríguez-Padrón,* Balu, Romero, and Luque** (Universidad de Córdoba) describe both conventional (heating, continuous flow, etc.) strategies, and less conventional (microwave, ultrasound, photochemical) technologies directed toward upgrading biomass residues and biomass derived platform molecules. Their catalysts are largely "benign by design" nanomaterials.

In Chapter 3, *De Bruyn, Sun, and Barta** (University of Graz) describe these researchers' application of copper-doped porous metal oxide catalysts having different compositions including other transition metal dopants. The catalyzed processes include lignin depolymerization, conversions of cellulose and other carbohydrate-derived platform chemicals, and upgrading pyrolysis oils and ethanol among other topics.

Chapter 4 by *Cui, Goes, and Stahl** (University of Wisconsin) is a general overview of lignin depolymerization methods. This is followed by discussion of oxidative-depolymerization methods with a focus on a two-step process, where specific lignin alcohol groups are first oxidized followed by oxidative, reductive, or redox-neutral steps that cleave the resulting polymer into aromatic monomers and oligomers.

Chapter 5 by *Liu and Abu-Omar** (University of California, Santa Barbara) contrasts techniques for preparing technical lignin and reviews the use of heterogeneous transition metal catalysts for lignin valorization into chemicals and fuels. In addition, it offers specific examples of how lignin derived monomers might be utilized in preparing new materials.

In Chapter 6, *Zhang and Wang** (State Key Laboratory of Catalysis, Dalian Institute of Chemical Physics, Chinese Academy of Sciences) describe nickel catalyzed depolymerization of protolignin and technical lignin. They highlight pretreatments involving adjacent functional group modification that decreases bond dissociation energies of targeted bonds and summarize possible applications of lignin monomers to prepare high-value synthetic building blocks.

Chapter 7 by *Dierks, Cao, and Rinaldi** (Imperial College London) provides a critical analysis of the sustainable manufacture of advanced biofuels as catalyzed by metal phosphides. Emphasis is placed on optimizing activity, product distribution, product selectivity by hydrodeoxygenation reactions with various biomass substrates as catalyzed by nickel phosphides.

Chapter 8 by *Arts, Ruijten, Van Aelst, Trullemans, and Sels** (Centre for Sustainable Catalysis and Engineering, KU Leuven) is a *tour de force* discussion of reductive catalytic fractionation technology for lignin depolymerization and the downstream valorization of the resulting phenolic monomers. This discussion includes offering a plan for future biorefineries based on this technology, and techno-economic and life cycle assessments of such biorefinery production of phenol and propylene from such monomers. These are concepts that are crucial to the successful implementation of biomass as an economically viable substitute for non-renewable fossil fuels.

In Chapter 9 *Barrett** (University of California, San Diego) describes strategies for conversion of lignin models and organosolv poplar lignin, and for the condensation of ethanol using copper-doped porous metal oxide catalysts, with the goal of improving the selectivity of product streams in these systems. He also describes studies of the structural changes that these catalysts undergo under the reaction conditions employed.

In Chapter 10, *Metzker,* Mora Vargas, and Boscolo** (Sao Paulo State University, Sao Jose do Rio Preto) discuss various catalytic methodologies for converting ethanol to value-added products. Their particular focus is on the efforts in Brazil to utilize ethanol as a chemical feedstock in addition to its considerable use as transportation fuel in that country.

As a final note, the editors want to thank again all of the authors for their contributions to this volume under the unusual circumstance brought on by a world-wide pandemic.

Peter C. Ford
Co-Editor of this volume
Distinguished Professor of Chemistry
University of California Santa Barbara, United States

Rudi van Eldik
Editor of Advances in Inorganic Chemistry
Emeritus Professor of Inorganic Chemistry
University of Erlangen–Nuremberg, Germany

Research Professor of Inorganic Chemistry
N. Copernicus University, Torun, Poland

Research Professor of Inorganic Chemistry
Jagiellonian University, Krakow, Poland

References

1. Macala, G. S.; Robertson, A. W.; Johnson, C. L.; Day, Z. B.; Lewis, R. S.; White, M. G.; Iretskii, A. V.; Ford, P. C. *Catal. Lett.* **2008**, *122*, 205–209.
2. Macala, G. S.; Matson, T. D.; Johnson, C. L.; Lewis, R. S.; Iretskii, A. V.; Ford, P. C. *ChemSusChem* **2009**, *2*, 215–217.
3. Barta, K.; Ford, P. C. *Acc. Chem. Res.* **2014**, *47*, 1503–1512.

CHAPTER ONE

Homogeneous transition metal catalyzed conversion of levulinic acid to gamma-valerolactone

László T. Mika* and István T. Horváth*
Department of Chemical and Environmental Process Engineering, Budapest University of Technology and Economics, Budapest, Hungary
*Corresponding authors: e-mail address: laszlo.t.mika@edu.bme.hu; istvan.t.horvath@edu.bme.hu

Contents

Abstract

Levulinic acid (LA) is one of the most studied platform chemicals produced from a wide range of carbohydrates containing biomass including agricultural residues and food wastes. The reduction of the carbonyl group of LA leads to the transient formation of 4-hydroxyvaleric acid (4-HVA), which readily undergoes facile dehydration to give gamma-valerolactone (GVL) in very high yields. The homogeneous catalytic conversion of LA to GVL via 4-HVA can be achieved by transition metal complexes by using molecular hydrogen. Alternatively, the transfer hydrogenation of LA with formic acid or sodium formate or isopropanol can be used for the transformation of LA to GVL. Although no catalytic systems have been reported to yield 4-HVA selectively, the formation of chiral GVL by the asymmetric reduction of LA confirms its transient formation. The application of transition metal complexes as catalysts for the hydrogenation and transfer-hydrogenation of LA are reviewed including the use of chiral ligands for the asymmetric catalytic reductions of LA.

Advances in Inorganic Chemistry, Volume 77
ISSN 0898-8838
https://doi.org/10.1016/bs.adioch.2021.02.004

Abbreviations

AA	acetic acid
AAL	alfa-angelica lactone
BAL	beta-angelica lactone
EL	ethyl levulinate
FA	formic acid
FF	furfural
FFA	furfuryl alcohol
GAL	gamma-angelica lactone
GVL	gamma-valerolactone
5-HMF	5-hydroxymethylfurfural
4-HVA	4-hydroxyvaleric acid
IPC	initial platform chemicals
LA	levulinic acid
ML	methyl levulinate
MNP	magnetic nanoparticle
NHC	N-heterocyclic carbene
pseudo-LA	pseudo-levulinic acid
TOF	turnover frequency
TON	turnover number

1. Introduction

The conversion of biomass including agricultural residues and food wastes to chemicals is a rapidly expanding approach for the gradual replacement of fossil resources for the chemical enterprise.[1,2] Among the various components of biomass, the utilization of polymers of carbohydrates has opened the door for the production of various platform chemicals.[3] The dehydration of hemicellulose or cellulose leads to the formation of furfural (FF)[4] and 5-hydroxymethylfurfural (5-HMF),[5] respectively, which were defined as initial platform chemicals (IPC)[6] (Scheme 1). Similarly, chitin can be converted to 5-HMF, acetic acid (AA) and ammonia.[7] The reduction of FF results in the formation of furfuryl alcohol (FFA).[8] While the hydration of 5-HMF leads to the formation of equimolar levulinic acid (LA)[9] and formic acid (FA),[10] the hydroisomerization of FFA also results in the formation of LA.[11] It should be emphasized, that FA can be used for the transfer hydrogenation of FF to FFA.[12]

Theoretically, the conversion of cellulose and chitin could offer a great opportunity for complete conservation of all C-, H-, and O-atoms of cellulose[12] and all C-, H-, O-, and N-atoms of chitin (Table 1), since the carbon-economy of the formation of AA, LA and FA is 100%. The

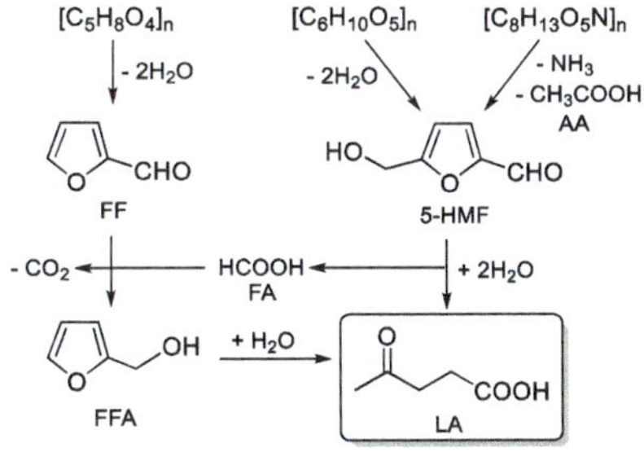

Scheme 1 Conversion of poly-carbohydrates to AA, ammonia, FA and LA.

Table 1 Conservation of the C-, H-, and O-atoms of cellulose, hemicellulose, and chitin.[a]

Carbohydrates	Required co-substrate	Product(s)	Side-product(s)	Atom-economy C (%)	Atom-economy H (%)	Atom-economy O (%)
Cellulose $[C_6H_{10}O_5]_n$	**None**	n 5-HMF	2n H_2O	100	60	60
		n LA + n FA	**none**	**100**	**100**	**100**
		n LA	n CO_2, n H_2	83.3	80	60
		n 4-HVA	2n CO_2	83.3	100	60
		n GVL	n CO_2, n H_2O	83.3	80	40
		n BuOH	2n CO_2	66.6	100	20
	n H_2O	2n EtOH	2n CO_2	66.6	100	33.3
Hemicellulose $[C_5H_8O_4]_n$	None	n FF	2 H_2O	100	50	50
	n H_2	n FFA	2 H_2O	100	60	50
		n LA	H_2O	100	80	75
Chitin $[C_8H_{13}O_5N]_n$	**None**	**n 5-HMF + n AA + n NH_3**	**none**	**100**	**100**	**100**
	2n H_2O	**n LA + n FA + n AA + n NH_3**	**none**	**100**	**100**	**100**
		n LA + n AA + n NH_3	n CO_2, n H_2	87.5	88.2	71.4
		n 4-HVA + n AA + n NH_3	n CO_2	87.5	100	71.4
		n GVL + n AA + n NH_3	n CO_2, n H_2O	87.5	88.2	57.1

[a] 5-HMF: 5-(hydroxymethyl)-2-furaldehyde ($C_6H_6O_3$), LA: levulinic acid ($C_5H_8O_3$), FA: formic acid ($C_1H_2O_2$), 4-HVA: 4-hydroxyvaleric acid ($C_5H_{10}O_3$), BuOH: butanol ($C_4H_{10}O$), EtOH: ethanol (C_2H_6O), GVL: gamma-valerolactone ($C_5H_8O_2$), FF: furfural ($C_5H_4O_2$), FFA: furfuryl alcohol ($C_5H_6O_2$), AA: acetic acid ($C_2H_4O_2$).

[a]5-HMF, 5-(hydroxymethyl)-2-furaldehyde ($C_6H_6O_3$); LA, levulinic acid ($C_5H_8O_3$); FA, formic acid ($C_1H_2O_2$); 4-HVA, 4-hydroxyvaleric acid ($C_5H_{10}O_3$); BuOH, butanol ($C_4H_{10}O$); EtOH, ethanol (C_2H_6O); GVL, gamma-valerolactone ($C_5H_8O_2$); FF, furfural ($C_5H_4O_2$); FFA, furfuryl alcohol ($C_5H_6O_2$); AA, acetic acid ($C_2H_4O_2$).

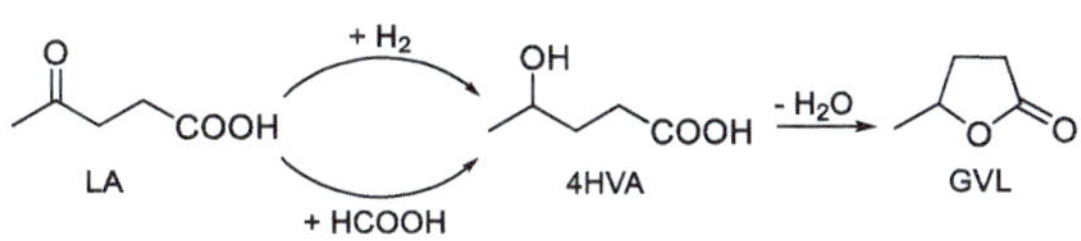

Scheme 2 Conversion of LA to GVL *via* the transient formation of 4-HVA.

carbon-economy of the conversion of cellulose to butanol or ethanol is only 66.6%, due to the formation of carbon dioxide.[12]

The product of the catalytic hydrogenation of the carbonyl group of LA is 4-hydroxyvaleric acid (4-HVA), which can be readily dehydrated to gamma-valerolactone (GVL) (Scheme 2).

GVL has been considered as a sustainable liquid for the production of energy and carbon-chemicals.[13] It should be noted that GVL does not react with water up to 60 °C for several weeks. At higher temperatures the equilibrium between GVL, water and 4-HVA can be reached in a few days. Aqueous acids or bases can catalyze the establishment of the equilibrium between GVL, water and 4-HVA.[14] The presence of 4-HVA up to 4 mol% could be worrisome as it could be an excellent competing bi- or even a tridentate ligand for transition metals. Although no catalytic system was reported to yield 4-HVA selectively, the isolation of chiral GVL by the asymmetric hydrogenation of LA confirmed the transient formation of chiral 4-HVA.[15]

The molecular hydrogen can be replaced with FA (Scheme 2) or sodium formate or isopropyl alcohol. The utilization of FA in the catalytic transfer-hydrogenation of LA is particularly attractive as FA is the stoichiometric co-product with LA in the hydration of 5-HMF.[10] Thus, the catalytic transfer-hydrogenation of LA with FA offers a one-pot approach for the production of GVL from cellulose or chitin with carbon-economy higher than 80%.

The isomerization of LA to pseudo-LA (or 5-hydroxy-gamma-valerolactone) at higher temperatures, opens the way for the formation of alfa-angelica lactone (AAL) or gamma-angelica lactone (GAL), which can be isomerized to beta-angelica lactone (BAL) (Scheme 3).[16] The catalytic hydrogenation of these angelica lactones results in the formation of GVL. Because of the required reaction temperatures are high, the use of heterogeneous transition metal catalysts has been preferred.

The application of transition metal complexes as catalysts for the conversion of LA to GVL via 4-HVA by hydrogenation and transfer-hydrogenation of LA are reviewed, including the use of chiral ligands for the asymmetric catalytic reductions of LA. The most accepted reaction mechanisms are included and some of the challenges are outlined.

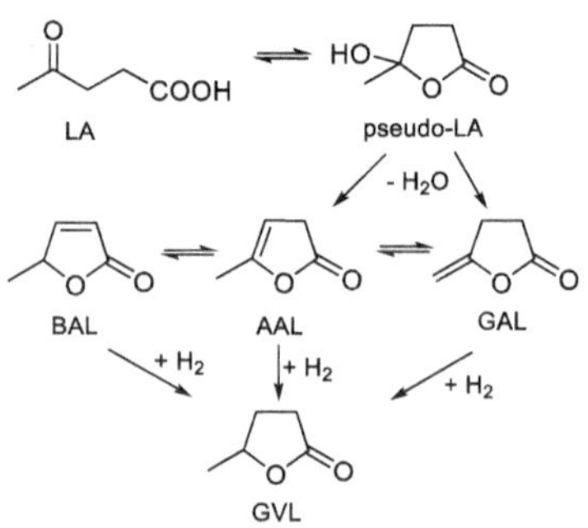

Scheme 3 Conversion of LA to GVL via pseudo-LA and α- or β- or γ-angelica lactone.

2. Homogeneous catalytic hydrogenation

The rapidly increasing interests in the chemistry of GVL have resulted in the development of transition metal catalysts for the conversion of LA to GVL. Transition metal complexes of mono-, bi-, and multidentate phosphines (Scheme 4) as well as nitrogen containing ligands (Scheme 5) have been developed, some of which exhibit high stability and catalytic activity. The ruthenium- (Table 2, #1–38), iridium- (Table 2, #39–42) and palladium-based (Table 2, #43) catalysts can be used either in an appropriate solvent or under solvent-free conditions.

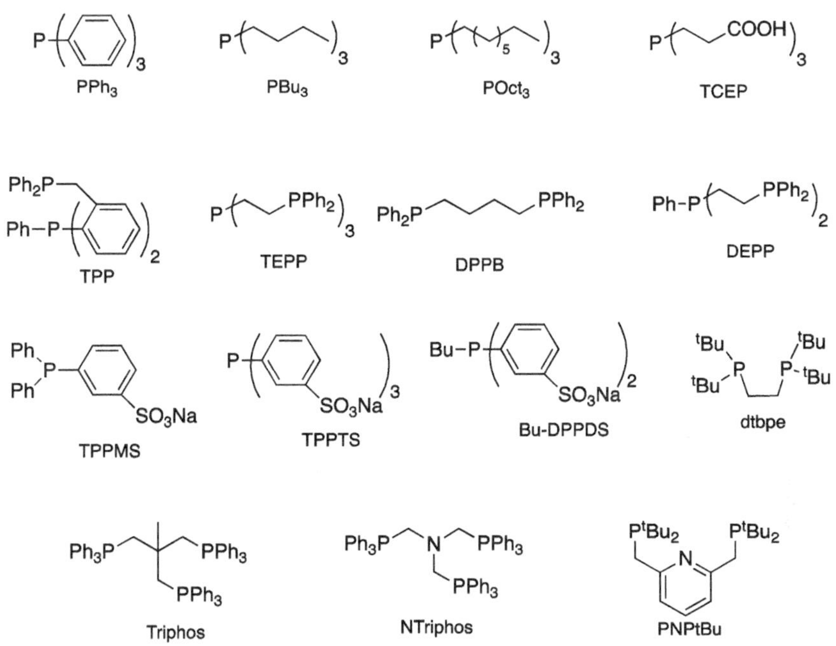

Scheme 4 Phosphine ligands used for the hydrogenation of LA.

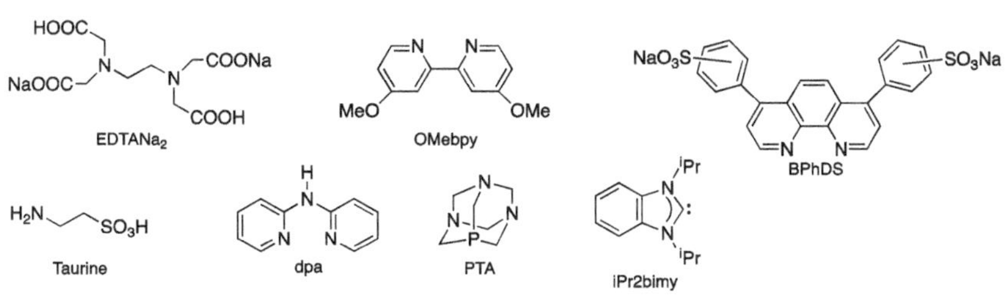

Scheme 5 Nitrogen containing ligands used for the hydrogenation of LA.

Table 2 Homogeneous catalytic conversion of LA to GVL using molecular hydrogen.

#	Substrate	Solvent	Substrate concentr.	Catalyst system	Catalyst concentr.	Additive	p (bar)	T (°C)	t (h)	TON (−)	TOF (h^{-1})	$X_{subst.}$ (%)	S_{GVL} (%)	Y_{GVL} (%)	Ref.
1	LA	Water[a]	0.02 M	$HRuCl(TPPMS)_3$	5×10^{-4} M	–	0.81	60	3	40	13	>99.9	100	>99.9	17
2	LA	Water[b]	0.05 M	$HRu(OAc)(TPPMS)_3$	0.001 M	–	0.81	60	2	50	25	>99.9	100	>99.9	17
3	LA	Water	0.66 M	$Ru(acac)_3$/TPPTS	1.8×10^{-3} M	–	70	140	12	600	50	>99.9	100	>99.9[c]	18
4	LA	Water/ CH_2Cl_2[d]	0.6 M	$RuCl_3$/TPPTS	1 mol%	–	45	90	1.3	100	77	>99.9	100	>99.9	19
5	LA	Water	0.66 M	$Ru(acac)_3$/PTA	0.0011 M	–	50	140	5	18	3.6	>99.9	100	>99.9	20
6	LA	Water	0.66 M	$Ru(acac)_3$/TPPMS	0.0011 M	–	50	140	5	582	113	94	94	88	20
7	LA	Water	0.66 M	$Ru(acac)_3$/TPPTS	0.0011 M	–	50	140	5	594	128	99	97	96	20
8	LA	–	–	$Ru(acac)_3/PBu_3$	0.06 mol%	NH_4PF_6	100	135	8	1559	65	>99.9	100	>99.9	18
9	LA	–	–	$Ru(acac)_3/POct_3$	0.1 mol%	NH_4PF_6	100	160	18	1000	56	>99.9	100	>99.9	21
10	LA	–	–	$Ru(acac)_3$/DPPB	0.1 mol%	NH_4PF_6	100	160	18	1000	56	>99.9	89	89	21
11	LA	–	–	$Ru(acac)_3$/Triphos	0.1 mol%	pTsOH	100	160	18	1000	56	>99.9	58	58[f]	21
12	LA	–	–	$RuH_2(CO)$(Triphos)	0.1 mol%	–	100	160	18	1000	56	>99.9	90	90	22
13	LA	–	–	$Ru(acac)_3$/DEPP	0.001 mol%	pTsOH	80	140	96	59,265	617	na	na	59	23
14	ML	–	–	$Ru(acac)_3$/DEPP	0.001 mol%	pTsOH	80	140	96	66,000	694	na	na	67	23
15	LA	–	–	$Ru(acac)_3$/TEPP	0.001 mol%	$Sc(OTf)_3$	80	140	96	73,142	762	na	na	73	23

16	ML	–	–	$Ru(acac)_3$/TEPP	0.001 mol%	$Sc(OTf)_3$	80	140	96	61,866	644	na	na	62	23
17	LA	–	–	$Ru(acac)_3$/TPP	0.001 mol%	pTsOH	80	140	96	12,000	125	na	na	12	23
18	ML	–	–	$Ru(acac)_3$/TPP	0.001 mol%	pTsOH	80	140	168	75,855	451	na	na	76	23
19	LA	–	–	$Ru(acac)_3$/TPPTS	0.016 mol%	–	100	140	1.8	1456	63	23	100	23	24
20	LA	–	–	$Ru(acac)_3$/Bu-DPPDS	0.016 mol%	–	100	140	1.8	6370	3450	>99.9	100	>99.9	24
21	LA	–	–	$Ru(acac)_3$/DPPB	0.01 mol%	–	50	160	0.6	12,740	21,233	>99.9	100	>99.9	25
22	LA	Water[e]	0.3 M	$RuCl_3$/$EDTANa_2$	0.004 mol%	–	40	140	1	970	970	85	99.5	84.5[f]	26
23	LA	Water[e]	3 M	$RuCl_3$/BPhDS	0.03 mol%	–	80	140	1	3000	3000	>99.9	100	>99.9	26
24	LA	Water[e]	3 M	$RuCl_3$/taurine	0.03 mol%	–	80	140	1	2509	2509	85	98.4	83.64[f]	26
25	LA	Water	0.43	$RuCl_2(PPh_3)_2$	0.1 mol%	–	12	130	2.6	137	52	na	na	14	27
26	LA	Water	0.43	$[RuCl_2(pcy)]_2$	0.1 mol%	–	12	130	2.6	995	373	>99.9	100	>99	27
27	LA	Water	0.43	$[RuCl_2(pcy)(iPr2bimy)]_2$	0.1 mol%	–	12	130	2.6	999	374	>99.9	100	>99	27
28	LA	Toluene	1.4 M	$RuCl_2(PPh_3)_2$	0.5 mol%	–	12	180	24	200	8.3	>99.9	100	99	28
29	LA	Toluene	1.4 M	$RuH_2(PPh_3)_2$	0.5 mol%	–	12	180	24	116	4.8	58	100	58	28
30	LA	Toluene	1.4 M	$RuCl(PPh_3)_4$	0.5 mol%	–	12	180	24	8	<1	8	100	4	28
31	ML	THF	4 M	$Ru(acac)_3$/TPP	0.025 mol%	pTsOH	50	140	6	3050	508	na	na	76	23
32	LA	THF	0.5 M	$RuH_2(CO)(NTriphos)$	0.5 mol%	NH_4PF_6	65	150	25	200	8	>99.9	95	95	29

Continued

Table 2 Homogeneous catalytic conversion of LA to GVL using molecular hydrogen.—cont'd

#	Substrate	Solvent	Substrate concentr.	Catalyst system	Catalyst concentr.	Additive	p (bar)	T (°C)	t (h)	TON (–)	TOF (h^{-1})	$X_{subst.}$ (%)	S_{GVL} (%)	Y_{GVL} (%)	Ref.
33	LA	THF	0.5 M	$RuH_2(CO)$(Triphos)	0.5 mol%	NH_4PF_6	65	150	25	200	8	>99.9	70	70	29
34	LA	THF	0.5 M	$Ru(acac)_3$/Triphos	0.5 mol%	NH_4PF_6	65	150	25	200	8	>99.9	73	73	29
35	LA	–	0.2 M	$RuCl_2(PPh_3)_3$	0.1 mol%	–	40	150	6	764	127	78	93	72	30
36	LA	–	0.2 M	$RuHCl(CO)(PPh_3)_3$	0.1 mol%	–	40	150	6	980	163	98	98	96	30
37	LA	Water:CO_2	0.2 M	$RuCl_2(PPh_3)_3$	0.1 mol%	–	130	150	6	911	151	93	93	86	30
38	LA	Toluene	1 M	Shvo catalyst (Ar=Ph)[g]	0.01 M	–	50	100	16	100	6	na	na	99	31
39	LA	EtOH	0.75 M	$[Ir(COE)_2(Cl)]_2$/PNPtBu	0.5 mol%	KOH	50	100	15	960	64	96	99	96	32
40	LA	EtOH	0.75 M	IrH_3PNPtBu	0.5 mol%	KOH	100	100	48	71,000	1480	72	99	71	32
41	LA	Water	1 M	(Cp*)Ir(OMebpy)	0.01 mol%	–	10	120	4	980	240	99	98	98	33
42	LA	Water	2 M	$[Cp^*Ir(dpa)(OSO_3)]$	0.001 mol%	–	5	130	32	97,000	3031	98	100	100	34
43	LA	Water	0.22 M	$Pd(DTBPE)Cl_2$	0.1 mol%	–	5	80	5	1000	200	>99.9	>99.9	>99.9	35

[a] pH = 1.
[b] pH = 4.8.
[c] Isolated yield of GVL was 95%.
[d] Biphasic system.
[e] pH = 2.4–3.8 depending on ligand.
[f] By-product: 1,4-PDO or 2-MeTHF.
[g] Structure presented on Scheme 8.
The structures of the ligands are presented in Schemes 4 and 5.

The water-soluble phosphine-modified Ru-catalysts were pioneered by Joó et al.[17] For selective reduction of C=O group of LA, HRuCl $(TPPMS)_3$ gave full conversion in water at pH 1 and 60 °C with a TOF of 13 h^{-1}. The activity can be doubled by the use of HRu(OAc) $(TPPMS)_3$ (Table 2, # 1, 2). Later on, Horváth et al. applied a Ru/ TPPTS catalyst exhibiting quantitative formation of GVL in water under 70 bar at 140 °C.[18] To facilitate an easier GVL separation, Heeres developed a CH_2Cl_2:H_2O = 4:1 (v/v) biphasic hydrogenation system, which operated under milder conditions (45 bar, 90 °C) and resulted in quantitative yield of GVL for 1.3 h (Table 2, #3).[19] After catalyst recycling, only 55% conversion was observed in the second run. A detailed kinetic model showed a first order dependence for LA (Table 2, #4). Kühn et al. discussed the effect of the structures of the water-soluble ligands on hydrogenation efficiency under aqueous conditions by using Ru(III) precursors. It was revealed that ligands having less bulky substituents such as PTA or TCEP gave poor activities (TOF ⟨20 h^{-1}) and the use of more sterically demanding ones resulted in satisfactory TOFs from 30 to 128 h^{-1}. Excellent results were obtained by TPPMS- and TPPTS-containing catalysts (Table 2, #5–7).[20] The first solvent-free hydrogenation of LA that utilized Ru/PBu_3 catalysts in the presence of NH_4PF_6 as an additive, was shown by Horváth et al. (Table 2, # 8).[18] The in situ high-pressure IR (Fig. 1) verified the complete conversion of LA under 100 bar H_2 at 135 °C.

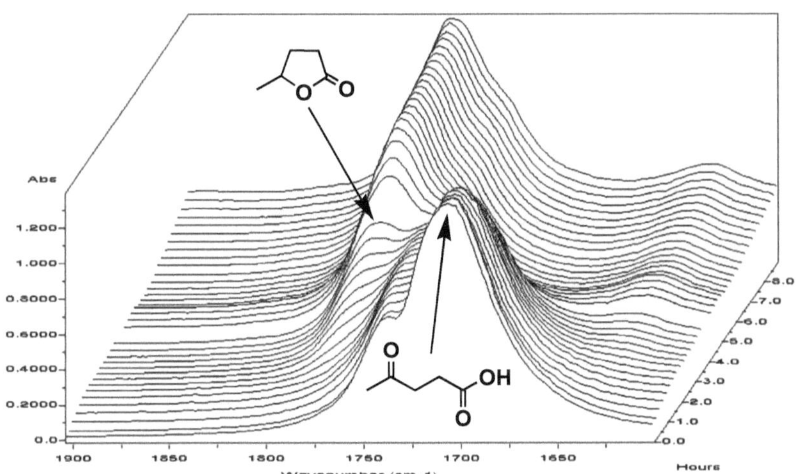

Fig. 1 In situ IR of the conversion of LA to GVL.[18]

Following this study, Leitner and co-workers evaluated the effect of monodentate ($POct_3$) and chelating phosphines, i.e. bidentate DPPB and tridentate Triphos as well as various acidic additives on the transformation of LA.[21] The catalyst generated in situ from $Ru(acac)_3$ and $POct_3$ in the presence of NH_4PF_6, gave GVL in 99% yield at 160 °C under neat conditions. However, ca. 2-times higher catalyst loading, and reaction time were necessary to obtain a comparable conversion to that of Ru/PBu_3 (Table 2, # 8–9). Under similar conditions, the use of the bidentate DPPB and Triphos ligands in the presence of NH_4PF_6, resulted in >99.9% conversion and yields of 89% and 8% GVL, respectively. It was shown that depending on the desired product (GVL, 2-methyltetrahydrofurane (2MeTHF) or 1,4-pentanediol (1,4-PDO)), the Ru/Triphos system could easily be fine-tuned by the applied additive. By the use of pTsOH, 58% selectivity toward GVL was observed maintaining high conversion (>99%) under 100 bar at 160 °C (Table 2, #11). In the absence of any additive, 1,4-PDO was formed with a yield of 95%. When the $RuH_2(CO)(Triphos)$ complex was applied as the catalyst precursor, GVL was obtained with a yield of 90% (Table 2, #12).[22] On the basis of in situ NMR measurements, which have established the coordination of LA to the Ru center (Scheme 6), a $[(Triphos)RuH]^+$ complex was proposed as a catalytically active species.

Beller and co-workers evaluated the commercially available DEPP, TEPP and a recently designed Triphos analogue TPP ligand, for neat conversion of both LA and ML under mild conditions (80 bar, 140 °C).[23] It was shown that all the tested tridentate ligand-modified catalytic systems, exhibited excellent

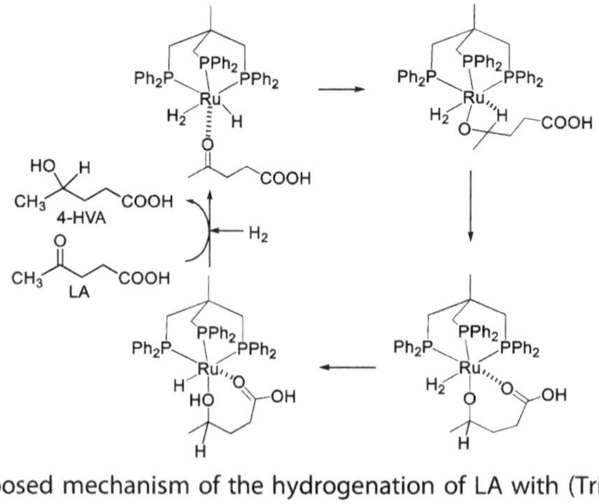

Scheme 6 Proposed mechanism of the hydrogenation of LA with $(Triphos)Ru(CO)H_2$.

activities and selectivities with TOF of 125–762 h^{-1}. However, the acidic additive such as pTsOH or $Sc(OTf)_3$ and very long reaction times (up to 168 h) were required. No significant changes between the conversion of LA and ML were shown (Table 2, #13–18) as well as use of THF as solvent did not enhance the activity (cf. Table 2, #31). Although very high TON were achieved, the TOF still remained moderate. These results clearly demonstrated that while high conversions and selectivities could be achieved in water, the efficient neat transformation required the application of acidic auxiliary material. From a technological point of view, the elimination of a solvent as well as any additive(s) could ease the product separation and even catalyst recycling. To investigate the beneficial combination of the high selectivities of sulfonated phosphines and neat conditions established by the use of strongly electron-donating tertiary phosphines, a series of water-soluble mixed phosphines ($R_nP(C_6H_4\text{-}m\text{-}SO_3Na)_{3\text{-}n}$ ($n=1$ or 2; R = Me, Pr, *i*Pr, Bu, Cp)) bearing one or two alkyl groups, were investigated for Ru-catalyzed reduction of LA by Mika et al.[24] Since the precursor $Ru(acac)_3$ and the ligands were readily dissolved in LA, the reactions could be performed in neat LA even in the absence of additives. Screening experiments established that by applying a catalyst in situ formed from $Ru(acac)_3$ and Bu-DPPDS, which has a slightly higher electron-donating ability ($\chi_{(CO)Tolman} = 12.5\ cm^{-1}$) compared to TPPTS ($\chi_{(CO)Tolman} = 15.3\ cm^{-1}$), quantitative conversion of LA to GVL could be achieved under 100 bar at 140 °C. The TOF value of 3530 h^{-1} was 4-times higher than that obtained for TPPTS (TOF = 800 h^{-1}). In addition, the catalyst loading could be reduced to 0.016 mol% as well as the typically used inert atmosphere was not necessary. Noteworthy, the Ru/Bu-DPPDS catalyst could be separated by vacuum distillation and recycled for six consecutive runs without any decrease in the yield of GVL. The system was further improved by the use of analogous chelating type bidentate $Ph_2P(CH_2)_nPPh_2$ ($n = 1$–3) ligands.[25] The variation of the methylene spacers between the phosphorus atoms showed that BDPPP provided the highest TOF value of 21,233 h^{-1} (Table 2, #19–21). So far, it is the highest TOF value reported for selective hydrogenation of LA. The catalyst could be recovered by vacuum distillation and recycled for 10 consecutive runs at 140 °C exhibiting no activity decay in this case, as well.

Papadogianakis et al. have recently examined a wide range of water-soluble mono- and multidentate N-containing ligands such as $EDTANa_2$, taurine or BPhDS for Ru-catalyzed reduction of LA. From the series of 26 tested ligands, the Ru/BPhDS exhibited the highest activity with full conversion of LA and TOF of 3000 h^{-1}.[26] When the taurine-modified system was investigated, no significant effect of additive (NaI, $AlCl_3$, NaH_2PO_4) on product yield was observed. By comparing different Ru precursors, the following order was revealed: $RuCl_3 \cdot 3H_2O \approx Ru(NO)(OAc)_3 > Ru(NO)(NO_3)_3 > Ru(acac)_3 > [Ru(NO)]_2(SO_4)_3 > RuO_2 \cdot H_2O$ to give TOF's of 2000, 1620, 1440, 890 and 770 h^{-1} in water, respectively. The well-known $RuCl_2(PPh_3)_2$ and $[RuCl_2(pcy)]_2$ catalyst converted LA to GVL with low TON of 51 and 373 h^{-1}, respectively (Table 2, #25, 26).[27] In addition to the phosphine-modified catalyst discussed above, Stubbs demonstrated the Ru–NHC (ruthenium N-heterocyclic carbene) complex catalyzed conversion of LA in water under 12 bar at 160 °C. However, the TEM analysis established the significant aggregation of Ru nanoparticles, of which catalytic activity was associated to their interaction with LA.

Beyond neat and aqueous systems, some conventional solvents (toluene or THF) and environmentally benign super critical CO_2 ($scCO_2$) were also examined as reaction media for Ru-catalyzed hydrogenation of LA. Earlier, Ikariya and co-workers examined the reaction of LA with $RuH_2(PPh_3)_4$ complexes in toluene. They found that both chloro and hydrido Ru–PPh_3 complexes were able to hydrogenate the C=O bond giving GVL with moderate (58%) to excellent (99%) yields (Table 2, #28–30).[28] Miller et al. applied the $RuH_2(CO)(NTriphos)/NH_4PF_6$ system in THF, resulting in comparable yields to that obtained by $RuH_2(CO)(Triphos)$ under neat conditions (Table 2, #12, 32). Both in situ generated and pre-prepared catalyst gave similar yields of GVL (Table 2, #33, 34).[29] Zhao et al. tested $scCO_2$ as reaction media for Ru-catalyzed reduction of LA. The use of $RuCl_2(PPh_3)_3$ catalyst yielded 72% GVL in the absence of co-solvent under 40 bar H_2 at 150 °C. In water and water/CO_2 (90 bar), 41% and 47% conversions were achieved, respectively (Table 2, #35–37). When neat CO_2 was used as initial reaction medium, the conversion increased up to 93% with the same selectivity. By introducing $RuHCl(CO)(PPh_3)_3$ as catalyst, 96% product yield was obtained under neat conditions. They proposed that $RuCl_2(PPh_3)_3$ converted to $RuHCl(CO)(PPh_3)_3$ under acidic conditions, which formed from CO_2 and water as a by-product of 4-HVA lactonization.[30] Wildeman reported a test for the Shvo-type catalyst assisted conversion of LA to GVL giving GVL yield over 99% under 50 bar H_2.[31]

Iridium complexes of tridentate princer ligand as well as bipyridyl-based half-sandwich pentamethyl cyclopentadienyl complexes, have been used for hydrogenation of LA (Scheme 7.)

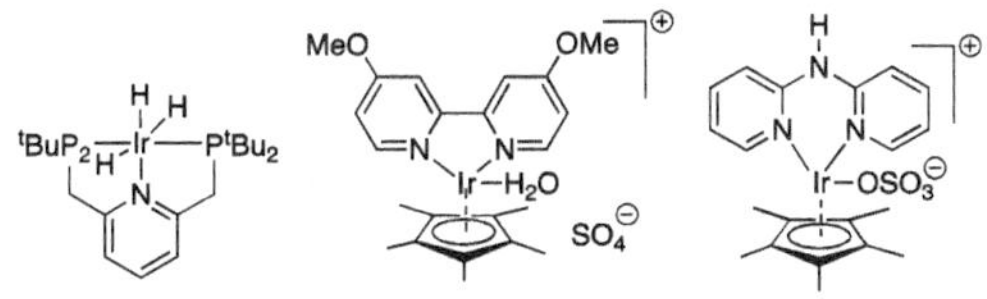

Scheme 7 Homogeneous Ir-based hydrogenation catalyst.

The iridium trihydride complexes formed in situ from $[Ir(COE)_2(Cl)]_2$ and tridentate princer type ligands, were reported for hydrogenation of LA under 50 bar H_2 at 100 °C (Scheme 7, complex **a**). The catalyst activity was significantly enhanced by incorporation of electron-donating substituents (*t*Bu and *i*Pr) in the P-ligand, as well as the reaction in EtOH afforded the highest yield.[32] It was shown that Ir-trihydride complexes gave better performance than the Ir–Cl complexes. Although the TON of 71,000 obtained by using rather low catalyst loading, was outstanding, the TOF still remained average (Table 2, #40). Unexpectedly, no information on the possible *in situ* formation of EL was provided. Iridium bipyridyl-based half-sandwich complexes (Scheme 7, complexes **b** and **c**) was also found to be efficient hydrogenation catalyst operating under mild conditions (10 bar H_2, 120 °C) for this purpose. A wide range of bipyridyl ligands bearing electron-donating groups in sterically unhindered positions, were screened to find the best ligand structure. It was revealed that the OMebpy modified IrCp* (Scheme 7, complexes **b**) gave the best activity yielding GVL in 98%.[33] Fischmeister et al. reported even more active iridium complexes for the reduction of LA, from which the [Cp*Ir(dpa)(OSO_3)] (Scheme 7, complex **c**) resulted in almost quantitative formation of GVL at 130 °C with TON of 97,000.[34]

From the series of Pd-based catalyst, the 1,2-(bis-di-tert-butylphosphino) ethane (DTBPE) ligand-modified Pd(II) system exhibited the highest activity and selectivity. Under aqueous conditions, complete conversion of LA was detected under 5 bar H_2 at 80 °C (Table 2, #43).[35]

3. Homogeneous catalytic transfer-hydrogenation

Beyond the conventional hydrogenation processes the transfer-hydrogenation reaction using a hydrogen carrier species such as FA, iPrOH, etc., could provide a safer alternative for GVL production. Because FA is produced along with LA in equimolar amounts during the hydration of the biomass based 5-HMF, this approach could result in a hydrogen-free protocol for GVL production. Accordingly, several groups have investigated the Ru-, Ir-, Pd-, and Fe-based homogeneous catalyst systems (Table 3), of which Ru-catalysts have been the most studied. Horváth and co-workers utilized $[(Cp^{*})Ru(bpy)(H_2O)][SO_4]$ catalyst (Scheme 8, complex **a**) under aqueous conditions, which gave 25% of GVL and 25% of 1,4-PDO from LA at 70 °C (Table 3, #1).[18]

The Shvo's type catalysts $\{[2,5\text{-}Ph_2\text{–}3,4\text{-}Ar_2(\eta^5\text{-}C_5O)]_2H\}Ru_2(CO)_4(\mu\text{-}H)$ (Ar=Ph, 4-MePh, 4-MeOPh) (Scheme 8, complex **b**) were introduced for the GVL production by the same group, which yielded GVL quantitatively from the initial mixture of LA to FA (1:2) at 95 °C. The in situ analysis established the conversion of 4-MeOPh-type Shvo's precursor to a mononuclear hydride species $[2,5\text{-}Ph_2\text{-}3,4\text{-}(4\text{-}MeOPh)_2(\eta^5\text{-}C_5OH)]Ru(CO)_2H$, which was proposed to be a key species of the catalytic cycle (Scheme 9).[36] While the hydride ligand is transferred to the carbon atom of the carbonyl group, the proton of the hydroxyl group is transferred to the oxygen atom of the carbonyl group to from 4-HVA. The elimination of 4-HVA from the Ru-center results in the $16e^-$ $\{[2,5\text{-}Ph_2\text{–}3,4\text{-}(4\text{-}MeOPh)_2(\eta^4\text{-}C_4C{=}O)]Ru(CO)_2\}$ species, which activates the FA. After the transfer of the proton of the coordinated FA to the carbonyl group of the cyclopentadienone ligand, the elimination of CO_2 from the coordinated formate results in the key species $\{[2,5\text{-}Ph_2\text{-}3,4\text{-}(4\text{-}MeOPh)_2(\eta^5\text{-}C_5OH)]Ru(CO)_2H\}$. The conversion was completed by the dehydration of 4-HVA to give GVL (Table 3, #3).

It is noteworthy that the Shvo catalyst was successfully recycled for four consecutive runs without loss of activity. This could compensate the cost of the catalyst as well as it could be introduced to the crude fructose dehydration mixture, containing LA and FA, yielding GVL with ca. 25%.[37] In toluene, the Shvo-type catalyst showed lower activity. Similar effect was observed by replacing the H-source FA to iPrOH (Table 3, #4, 5).[31] Fu et al. have applied a simple mixture of $RuCl_3$ and PPh_3 with a ratio of 1:3 for the reduction of LA using FA as a H-source with a ratio of 1:1 at 150 °C. The yield of GVL was 93% and 94% in the presence of pyridine and NEt_3, respectively (Table 3, #2).[38] Makhubela developed a robust Ru(II)–Zn(II) bimetallic and recyclable catalyst for conversion of LA yielding GVL of 92% at 120 °C (Table 3, #6).[39]

From the group of d^9- and d^{10}-transition metals, both Ir- and Pd-based catalysts were active in the LA to GVL conversion (Table 3, #7–9). The $(Cp^{*})Ir(OMebpy)$ (Scheme 7, complex **b**) gave excellent (99%) yield of GVL by using 2 equiv. of FA with respect to LA at 120 °C for 4 h.

Table 3 Homogeneous transfer-hydrogenation of LA and EL to produce GVL.

#	Substrate	Solvent	Substrate concentr.	Catalyst system	Catalyst concentration	H-Source	Additive	T (°C)	t (h)	TON (−)	TOF (h^{-1})	$X_{subst.}$ (%)	S_{GVL} (%)	Y_{GVL} (%)	Ref.
1[a]	LA	Water	0.02 M	$[(Cp^*)Ru(bpy)(H_2O)][SO_4]$	3×10^{-4} M	HCO_2Na	HNO_3	70	18	254	14	50	25	12.5	18
2	LA	LA/FA	0.5 mol%	$RuCl_3/PPh_3$	0.01 M	FA	pyridine	150	12	1000	83	na	na	93	38
3	LA	LA/FA	33 mol%	Shvo catalyst (Ar=Ph)	0.01 mol%	FA	–	95	6	1065	177	>99.9	>99.9	>99.9	36
4	LA	Toluene	1 M	Shvo catalyst (Ar=Ph)	0.01 M	FA	–	100	20	95	5	na	na	95	31
5	LA	LA/iPrOH	75 mol%	Shvo catalyst (Ar=Ph)	0.01 M	iPrOH	–	80	16	100	6	>99.9	>99.9	>99.9	31
6[b]	LA	LA/FA	50 mol%	$[Ru_4Cl_8(pcy)_4(L1)_4Zn_2]$	0.1 mol%	FA	Et_3N	120	4	920	230	92	>99.9	92	39
7	LA	Water	1 M	$(Cp^*)Ir(OMebpy)$	0.01 M	FA	–	120	4	990	248	99	>99.9	99	33
8	LA	Water	1 M	$[Cp^*Ir(dpa)(OSO_3)]$	0.1 mol%	FA	–	110	16	100	62	>99.9	>99.9	>99.9	34
9	LA	Water	0.67 M	$Pd(DTBPE)Cl_2$	0.1 mol%	FA	Et_3N	100	1	1000	1000	>99.9	>99.9	>99.9	35
10	EL	1,4-Dioxane	1 M	$Fe(OTf)_2/TEPP$	5 mol%	FA	–	140	24	25	1	>99.9	>99.9	>99.9	40
11	EL	iPrOH/EL	2 M	Casey-type catalyst	1 mol%	iPrOH	$NaHCO_3$	100	19	100	5	96	99	95	41
12	LA	Toluene	0.17 M	Casey-type catalyst	0.006 M	iPrOH	–	80	20	na	na	na	na	38	31
13	LA	LA/FA	0.1 M	$[Fe_3(CO)_{12}]$	4 mol%	FA	imidazole	180	15	23	1.5	na	na	92	42

[a]pH = 4, by-product: 1,4-PDO.
[b]L1 = $Ph_2P(4\text{-}C_6H_4\text{-}COOH)$.

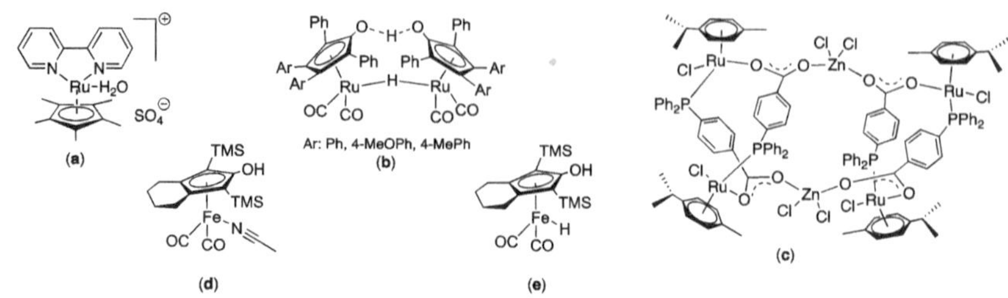

Scheme 8 Homogeneous transfer-hydrogenation catalyst precursors.

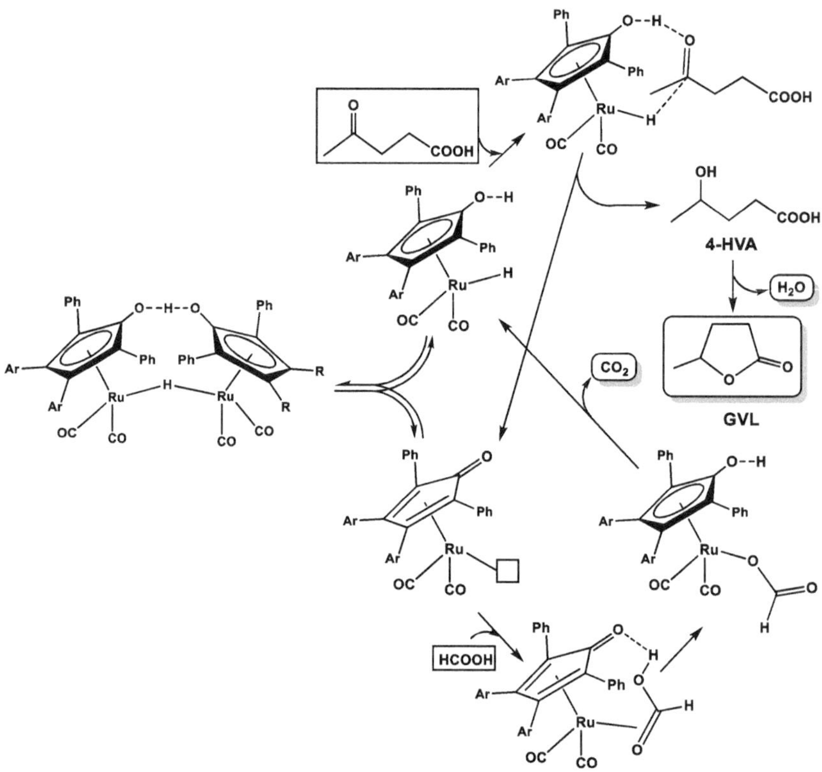

Scheme 9 Proposed mechanism of the homogeneous transfer-hydrogenation of LA with FA with Shvo's precursor {[2,5-Ph_2-3,4-Ar_2(η^5-C_5O)]$_2$H}$Ru_2(CO)_4$(μ-H), Ar = 4-MeOPh.

However, at room temperature, 10 times higher catalyst loading and 24 h were necessary to achieve high (94%) product yield. The reusability test revealed that the catalyst could be recovered for five consecutive runs without significant loss in its activity (Table 3, #7).[33] Similarly, high activity was reported for [Cp*Ir(dpa)(OSO$_3$)] (Scheme 7, complex **c**) in the presence of 2 equiv. of FA (Table 3, #8).[34] García investigated several Pd-diphosphine complexes of the type [Pd(*PP*)X$_2$] (X=Cl or Me) for the LA to GVL conversion in the presence of FA as H-source. The best-performing catalyst was the 1,2-(bis-di-tert-butylphosphino)ethane (DTBPE) modified one, which converted LA to GVL with high rate exhibiting high TOF of 1000 h^{-1} in water at 100 °C (Table 3, #9). In non-aqueous solvents, e.g., 1,4-dioxane or THF, the reaction of the precursor Pd(DTBPE)Cl$_2$ with FA led to the formation of catalytically inactive [(DTBPE)$_2$Pd$_2$(μ-H)(μ-CO)]$^+$ species.[35]

Since the cost of the Ru and Ir catalyst could be a crucial factor for larger scale industrial applications, attempts were made to introduce an iron-based system for GVL production. Wildeman compared the Ru-based Shvo-type catalyst (Ar=Ph) with Fe-based Casey's catalyst (Scheme 8, complex **d**) showing that the latter exhibited significantly lower activities under the same conditins.[31] Fu et al. have demonstrated that a catalyst formed in situ from Fe(OTf)$_2$ and TEPP ligand gave the most effective system generating GVL from EL in 98% yield (Table 3, #10).[40] Although, it seems to be an attractive method, the high catalyst loading and the expensive ligand should be considered. Other phosphines, i.e., DPPB, PPh$_3$ or Triphos, gave negligible activity. The same group utilized Casey's catalyst (Scheme 8, complex **e**) for conversion of EL in iPrOH under basic conditions and at lower temperature (100 °C). Although the catalyst loading could be reduced to 1 mol%, the recycling was unsuccessful (Table 3, #11).[41] The activity of the same catalyst system in toluene gave negligible activity (Table 3, #12).[31] The cheap and readily available [Fe$_3$(CO)$_{12}$] precursor was used for neat conversion of LA and FA at 180 °C. The reaction gave 92% product yield, but the catalyst and additive concentrations were rather high (Table 3, #13).[42] The addition of PPh$_3$ to the system resulted in the formation of Fe(CO)$_3$(PPh$_3$)$_2$ and a reduced activity. It was also shown that no [Fe$_3$(CO)$_{12}$] was present indicating that the precursor converted into unknown active catalyst species.

4. Homogeneous asymmetric catalytic hydrogenation

The synthesis of GVL in the presence of non-chiral catalysts, reviewed in Section 3, resulted in racemic mixtures. Although major applications of GVL such as solvent or even fuel do not require one specific enantiomer, its enantiomers could be excellent building blocks for the synthesis of chiral compounds or chiral solvents (Scheme 10).[43] Among the d^8–d^9 transition metals, the chiral chelating phosphine-modified Ru systems showed the highest activity and enantioselectivity for the asymmetric reduction of LA or its esters (Scheme 11). Typically used chiral phosphines for this purpose are summarized in Schemes 12 and 13.

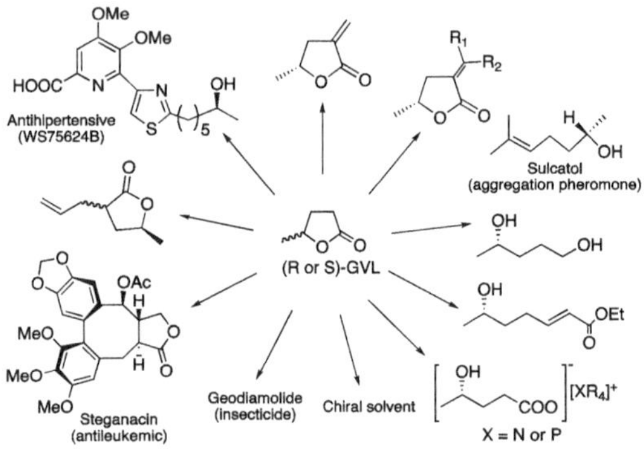

Scheme 10 Utilization of chiral GVL.

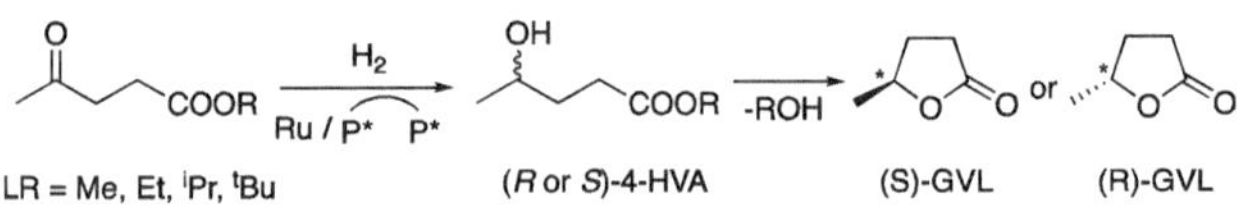

Scheme 11 Synthesis of chiral GVL from LA or its esters.

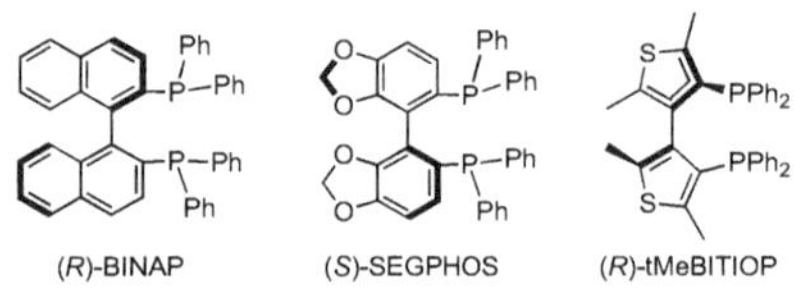

Scheme 12 Chiral chelating phosphines used for GVL production.

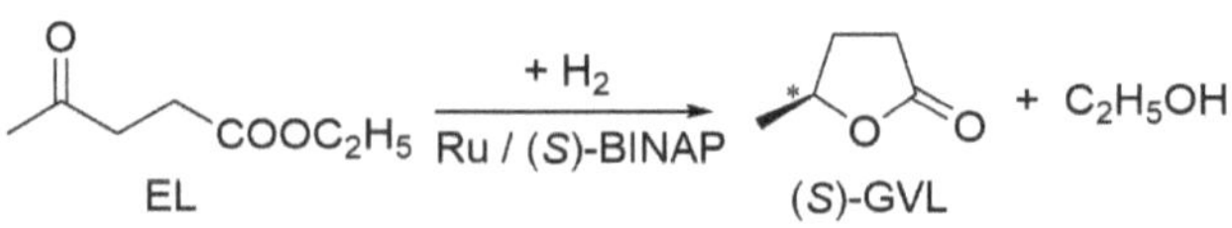

Scheme 13 Conversion of EL to (S)-GVL.

The first synthesis of enantiopure GVL from EL by using Ru/BINAP catalyst was reported by Noyori as early as 1990 (Scheme 13). The (*S*)-GVL was isolated from EtOH with high yield (96%) and *ee* (99.5%) (Table 4, #1).[44] Later on Saito et al. reported the SEGPHOS ligand, which enforces a smaller dihedral angle at the Ru-center and exhibited higher activity and selectivity for reduction of ketones. GVL was also obtained from EL with excellent *ee* (99%); however, 5-times shorter reaction time was necessary (Table 4, #2).[45] Sannicolo utilized tMeBITIOP ligand-modified system for conversion of EL in EtOH. The electron rich nature and lower dihedral angle were proposed to be responsible for high selectivity (*ee*=98%) (Table 4, #3).[46] Vinogradov and co-workers investigated in detail the Ru–BINAP–HCl catalyst system for reduction of levulinic acid and its esters (Table 4, #4, 5). LA was converted to (*S*)-GVL by the use of catalyst in situ formed from [Ru(Me-allyl)$_2$(COD)], (*S*)-BINAP and HCl resulting in (*S*)-GVL with high *ee* (98%), but yield (66%) was moderate under 60 bar H_2 at 60 °C. Significantly higher yields (95%) were obtained by using EL. It was thought that the LA first esterized by the solvent EtOH and then hydrogenated to the (*S*)-4-hydroxy ethyl levulinate. The effect of the ester substituents (Me, Et, iPr, tBu) as the steric bulk of the alkoxy moiety on the rate of conversion was demonstrated. It was shown that other diphosphines, i.e., DuPHOS or MeBPE, gave significantly lower activities.[47] By the use of cheaper $RuCl_3$ precursor and (*R*)-BINAP ligand also gave high yields from ML under similar conditions (Table 4, #6).[48] The same group performed a detailed kinetic study for this system revealing that the reaction order changed from first to zero order by increasing the initial concentration of ML and HCl. A detailed mechanism based on kinetic and ^{31}P NMR studies was proposed. The intramolecular hydride transfer step in the coordinated ML containing [(BINAP)(H)Ru(ML)(μ-Cl_2)Ru(BINAP)Cl_2]$^-$ complex was assumed as rate-limiting step of the catalytic cycle.[49] Mika et al. reported the first solvent-free asymmetric reduction of LA to GVL by using the Ru(acac)$_3$/(*R*)-BINAP system operated under 60 bar H_2 at 140 °C. Although complete conversion was achieved, the enantioselectivity was only 23% (Table 4, #7).[15] When (*S*)-[(RuCl(SEGPHOS))$_2$(*μ*-Cl)$_3$][NH_2Me_2] (Fig. 2) was used, 82% of *ee* was detected in MeOH under 60 bar at 150 °C (Table 4, #8). By the use of higher aliphatic alcohols (EtOH, iPrOH, BuOH) significant decrease in *ee* was shown. It should be noted that the Ru/SEGPHOS system was able

Table 4 Ru-Catalyzed Asymmetric hydrogenation of LA and it esters.

#	Substrate	Solvent	Catalyst	S/C	p (bar)	T (°C)	t (h)	TON	TOF	$X_{subst.}$ (%)	Y_{GVL} (%)	ee (%)	Ref.
1	EL	EtOH	$Ru(OAc)_2$/(*S*)-BINAP	1000	100	25	110	960	9	96	96	99	44
2	EL	EtOH	Ru/SEGPHOS	1000	50	50	20	1000	50	>99.9	>99.9	99	45
3	EL	EtOH	$[Ru(tMeBTIOP)][OTf]_2$	500	100	45	76	500	7	>99.9	>99.9	98	46
4	LA	EtOH	Ru/BINAP/HCl[a]	200	60	60	5	138	28	69	66	98	47
5	EL	EtOH	Ru/BINAP/HCl[a]	200	60	60	5	100	40	>99.9	95	99	47
6	ML	MeOH	$RuCl_3$/(*R*)-BINAP	200	65	60	6	190	32	>99.9	96	99	48
7	LA	–	$Ru(acac)_3$/(*R*)-BINAP	6314	60	140	12	6314	526	>99.9	>99.9	23	15
8	LA	MeOH	(*S*)-$[(RuCl(SEGPHOS))_2\ (\mu\text{-}Cl)_3]$ $[NH_2Me_2]$	2400	60	150	20	2400	120	>99.9	>99.9	82	15

[a]HCl/Ru = 10.

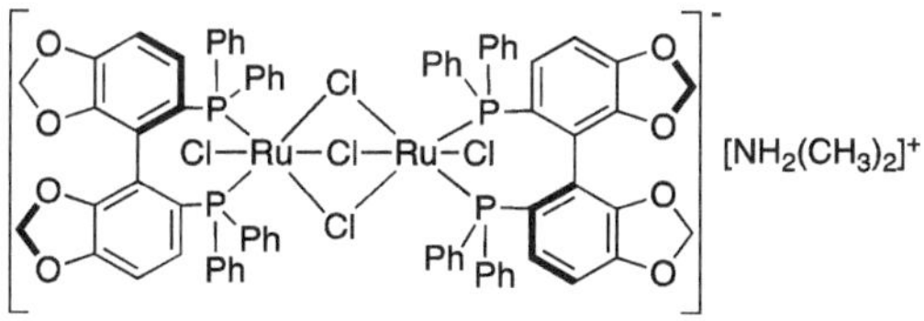

Fig. 2 The structure of (*S*)-[(RuCl(SEGPHOS))$_2$(μ-Cl)$_3$][NH$_2$Me$_2$].

to convert biomass-originated (non-purified) LA without decrease in its selectivity.[15]

5. Immobilized homogeneous transition metal catalysts

Immobilization of homogeneous catalytic systems is an attractive method to combine the high product yields of homogeneous catalysis with facile catalyst separation of heterogeneous catalytic systems.[50] The attachment of ligands of homogeneous catalysts to the surface of inorganic or organic solid supports via covalent bonds, is a typical method to prepare immobilized catalysts, which could be used in continuous and batch reactors. In the latter case, simple filtration of the reaction mixture is enough for separation and catalyst recycling. The immobilization of the dimeric form of the Shvo's catalyst has been demonstrated by using the sol-gel method[51] and simply adding it to silica.[52] A triethoxysilylpropoxy-functionalized monomeric Shvo's catalyst precursor {3,4-[p-$(EtO)_3Si(CH_2)_3OPh]_2$-2,5-$Ph_2(\eta^4$-$C_4C{=}O)$}$Ru(CO)_3$, was immobilized to silica by using covalent anchoring (Scheme 14) and the sol-gel methods.[53] The homogeneous and immobilized transfer hydrogenation catalysts were used for the conversion of LA and FA to GVL.

The immobilized catalysts prepared by the sol-gel method showed higher activity than the covalently grafted catalysts. Hot filtration tests showed no leaching of the immobilized Shvo's catalysts, opening the door to facile catalyst recycling.

It should be noted that the production of LA and FA from carbohydrates via 5-HMF is typically accompanied by the formation of insoluble solid side-products such as humins,[54] which makes the recycling of the silica supported Shvo's catalysts practically impossible. The immobilization of a catalyst to magnetic nanoparticles (MNPs) could offer easy catalyst separation from liquids and solids.[55] Therefore, the functionalized Shvo's catalyst precursor {3,4-[p-$(EtO)_3Si(CH_2)_3OPh]_2$-2,5-$Ph_2(\eta^4$-$C_4C{=}O)$}$Ru(CO)_3$

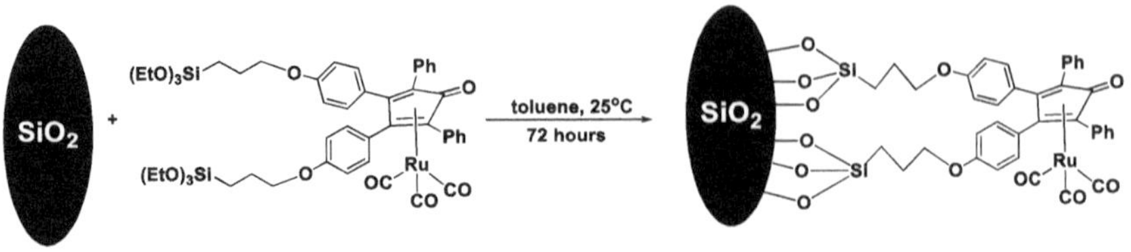

Scheme 14 Immobilization of {3,4-[p-$(EtO)_3Si(CH_2)_3OPh]_2$-2,5-$Ph_2(\eta^4$-$C_4C{=}O)$}$Ru(CO)_3$ to silica.

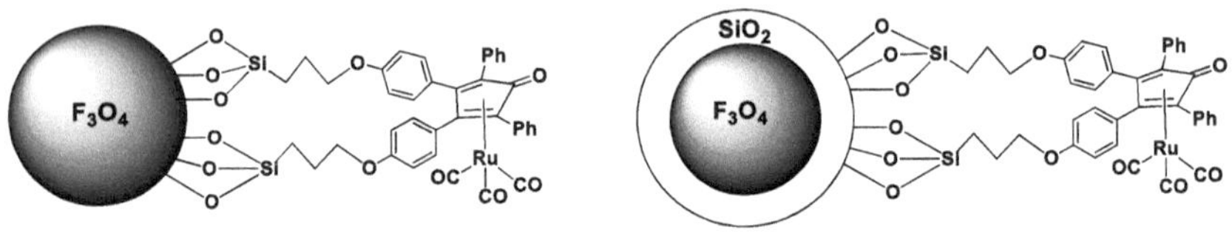

Scheme 15 Immobilized catalysts Fe_3O_4@Ru and Fe_3O_4@SiO_2@Ru.

was attached to magnetite and the resulting Fe_3O_4@Ru catalyst (Scheme 15) was separated from the solution by using an external magnet.[56]

Because of significant leaching of iron during the test reactions with Fe_3O_4@Ru, the magnetite was covered by one, two and three independently added silica coatings to give Fe_3O_4@SiO_2, Fe_3O_4@SiO_2@SiO_2, and Fe_3O_4@SiO_2@SiO_2@SiO_2. Attachment of the Shvo's catalyst precursor {3,4-[p-$(EtO)_3Si(CH_2)_3OPh]_2$-2,5-$Ph_2(\eta^4$-$C_4C{=}O)$}$Ru(CO)_3$ resulted in Fe_3O_4@SiO_2@Ru, shown in Scheme 15, Fe_3O_4@SiO_2@SiO_2@Ru, and Fe_3O_4@SiO_2@SiO_2@SiO_2@Ru. The catalysts on the silica coated MNPs exhibited higher activity than that of Fe_3O_4@Ru. There were no significant differences in TONs, TOFs and yields of GVL using these catalysts. Leaching tests have established that increasing the number of independent silica coatings on the surface of magnetite significantly decreased the leaching of iron. Fe_3O_4@SiO_2@Ru was reused several times without significant loss of the catalytic activity. Hot filtration tests showed that the catalytic activity was due to the supported ruthenium catalyst and not from some active ruthenium species leached from the solid support to the solution under the reaction conditions. Finally, these catalysts can be separated from the reaction mixture by using an external magnet.

6. Conclusions

The homogeneous transition metal catalyzed conversion of LA to GVL by hydrogenation or transfer-hydrogenation is a well-established method and part of the valorization of biomass including agricultural residues and food wastes. The most frequently used catalysts are based primarily on ruthenium and a lesser extent on iridium, some of which have shown spectacular TONs and high yields. Although environmentally friendly catalysts based on iron were demonstrated, further improvements and the inclusion of other biometals should be developed. More selective and robust chiral catalyst should be invented for the asymmetric hydrogenation of LA under neat conditions to produce chiral GVL. Immobilization has been shown and further improvements required for the minimalization for leaching and long term recyclability.

Acknowledgments

This work was funded by National Research, Development and Innovation Office – NKFIH (KH 129508) and Higher Education Excellence Program of the Ministry of Human Capacities in the frame of biotechnology research area of Budapest University of Technology and Economics (BMEFIKP-BIO). We are grateful for the support of the Technology Support Program of the Innovation and Technology Fund of the Government of the Hong Kong SAR (ITS/079/13). Any opinions, findings, conclusions or recommendations expressed in this material/event (or by members of the project team) do not reflect the views of the Government of the Hong Kong Special Administrative Region, the Innovation and Technology Commission or the Panel of Assessors for the Innovation and Technology Support Program of the Innovation and Technology Fund. We also thank the Environment and Conservation Fund (ECF/31/2014) for partial financial support.

References

1. (a) Clark, J. H.; Deswarte, F., Eds. *Introduction to Chemicals from Biomass*; 2nd ed.; John Wiley & Sons, Ltd.: Chichester, UK, 2015; (b) Kuila, A.; Mukhopadhyay, M., Eds. *Biorefinery Production Technologies for Chemicals and Energy*; John Wiley and Sons: Hoboken, NJ, USA, 2020.
2. Tuck, C. O.; Perez, E.; Horváth, I. T.; Sheldon, R. A.; Poliakoff, M. *Science* **2012**, *337*, 695–699.
3. Werpy, T.; Petersen, G. *Top Value-Added Chemicals from Biomass; NREL/TP-10-35523*; National Renewable Energy Laboratory: Golden, CO, USA, 2004.
4. (a) Gomez Bernal, H.; Bernazzani, L.; Galletti, R. *Green Chem.* **2014**, *16*, 3734–3740; (b) Qing, Q.; Guo, Q.; Zhou, L.; Wan, Y.; Xu, Y.; Ji, H.; Gao, X.; Zhang, Y. *Bioresour. Technol.* **2017**, *226*, 247–254.
5. (a) Yu, I. K. M.; Tsang, D. C. W. *Bioresour. Technol.* **2017**, *238*, 716–732; (b) Cao, L.; Yu, I. K. M.; Chen, S. S.; Tsang, D. C. W.; Wang, L.; Xiong, X.; Zhang, S.; Ok, Y. S.; Kwon, E. E.; Song, H.; Poon, C. S. *Bioresour. Technol.* **2018**, *252*, 76–82.
6. Mika, L. T.; Cséfalvay, E.; Németh, Á. *Chem. Rev.* **2018**, *118*, 505–613.
7. Omari, K. W.; Besaw, J. E.; Kerton, F. M. *Green Chem.* **2012**, *14*, 1480–1487.
8. Srivastava, S.; Solanki, N.; Mohanty, P.; Shah, K. A.; Parikh, J. K.; Dalai, A. K. *Catal. Lett.* **2015**, *145*, 816–823.
9. Pileidis, F. D.; Titirici, M. M. *ChemSusChem* **2016**, *9*, 562–582.
10. Horvat, J.; Klaic, B.; Metelko, B.; Sunjic, V. *Tetrahedron Lett.* **1985**, *26*, 2111–2114.

11. Maldonado, G. M. G.; Assary, R. S.; Dumesic, J.; Curtiss, L. A. *Energ. Environ. Sci.* **2012**, *5*, 6981–6989.
12. Lui, M. Y.; Wong, Y. Y.; Choi, W.-T.; Mui, Y. F.; Qi, L.; Horváth, I. T. *ACS Sustain. Chem. Eng.* **2019**, 7, 17799–17807.
13. Horváth, I. T.; Mehdi, H.; Fábos, V.; Boda, L.; Mika, L. T. *Green Chem.* **2008**, *10*, 238–242.
14. Wong, Y. Y.; Choi, W. T.; Lui, M. Y.; Fridrich, B.; Horváth, A. K.; Mika, L. T.; Horváth, I. T. *Struct. Chem.* **2017**, *28*, 423–429.
15. Tukacs, J. M.; Fridrich, B.; Dibó, G.; Székely, E.; Mika, L. T. *Green Chem.* **2015**, *17*, 5189–5195.
16. Mamun, O.; Saleheen, M.; Bond, J. Q.; Heyden, A. *J. Phys. Chem. C* **2017**, *121*, 18746–18761.
17. Joó, F.; Tóth, Z.; Beck, M. T. *Inorg. Chim. Acta* **1977**, *25*, L61–L62.
18. Mehdi, H.; Fábos, V.; Tuba, R.; Bodor, A.; Mika, L. T.; Horváth, I. T. *Top. Catal.* **2008**, *48*, 49–54.
19. Chalid, M.; Broekhuis, A. A.; Heeres, H. J. *J. Mol. Catal. A: Chem.* **2011**, *341*, 14–21.
20. Delhomme, C.; Schaper, L.-A.; Zhang-Preße, M.; Raudaschl-Sieber, G.; Weuster-Botz, D.; Kühn, F. E. *J. Organomet. Chem.* **2013**, *724*, 297–299.
21. Geilen, F. M. A.; Engendahl, B.; Harwardt, A.; Marquardt, W.; Klankermayer, J.; Leitner, W. *Angew. Chem. Int. Ed.* **2010**, *49*, 5510–5514.
22. Geilen, F. M. A.; Engendahl, B.; Hölscher, M.; Klankermayer, J.; Leitner, W. *J. Am. Chem. Soc.* **2011**, *133*, 14349–14358.
23. Dutta Chowdhury, A.; Jackstell, R.; Beller, M. *ChemCatChem* **2014**, *6*, 3360–3365.
24. Tukacs, J. M.; Király, D.; Strádi, A.; Novodarszki, G.; Eke, Z.; Dibó, G.; Kégl, T.; Mika, L. T. *Green Chem.* **2012**, *14*, 2057–2059.
25. Tukacs, J. M.; Novák, M.; Dibó, G.; Mika, L. T. *Catal. Sci. Technol.* **2014**, *4*, 2908–2912.
26. Moustani, C.; Anagnostopoulou, E.; Krommyda, K.; Panopoulou, C.; Koukoulakis, K. G.; Bakeas, E. B.; Papadogianakis, G. *Appl. Catal. Environ.* **2018**, *238*, 82–92.
27. Tay, B. Y.; Wang, C.; Phua, P. H.; Stubbs, L. P.; Huynh, H. V. *Dalton Trans.* **2016**, *45*, 3558–3563.
28. Osakada, K.; Ikariya, T.; Yoshikawa, S. *J. Organomet. Chem.* **1982**, *231*, 79–90.
29. Phanopoulos, A.; White, A. J. P.; Long, N. J.; Miller, P. W. *ACS Catal.* **2015**, *5*, 2500–2512.
30. Yang, W.; Cheng, H.; Zhang, B.; Li, Y.; Li, Y.; Liu, T.; Lan, M.; Yu, Y.; Zhang, C.; Lin, W.; Fujita, S.; Arai, M.; Zhao, F. *Green Chem.* **2016**, *18*, 3370–3377.
31. Van Slagmaat, C. A. M. R.; Wildeman, S. M. A. D. A. *Eur. J. Inorg. Chem.* **2018**, *6*, 694–702.
32. Li, W.; Xie, J.-H.; Lin, H.; Zhou, Q.-L. *Green Chem.* **2012**, *14*, 2388–2390.
33. Deng, J.; Wang, Y.; Pan, T.; Xu, Q.; Guo, Q.-X.; Fu, Y. *ChemSusChem* **2013**, *6*, 1163–1167.
34. Wang, S.; Huang, H.; Dorcet, V.; Roisnel, T.; Bruneau, C.; Fischmeister, C. *Organometallics* **2017**, *36*, 3152–3162.
35. Ortiz-Cervantes, C.; Flores-Alamo, M.; García, J. J. *ACS Catal.* **2015**, *5*, 1424–1431.
36. Fábos, V.; Mika, L. T.; Horváth, I. T. *Organometallics* **2014**, *33*, 181–187.
37. Qi, L.; Horváth, I. T. *ACS Catal.* **2012**, *2*, 2247–2249.
38. Deng, L.; Li, J.; Lai, D.-M.; Fu, Y.; Guo, Q.-X. *Angew. Chem. Int. Ed.* **2009**, *48*, 6529–6532.
39. Amenuvor, G.; Darkwa, J.; Makhubela, B. C. E. *Catal. Sci. Technol.* **2018**, *8*, 2370–2380.
40. Fu, M.-C.; Shang, R.; Huang, Z.; Fu, Y. *Synlett* **2014**, *25*, 2748–2752.
41. Dai, N.; Shang, R.; Fu, M.; Fu, Y. *Chin. J. Chem.* **2015**, *33*, 405–408.

42. Metzker, G.; Burtoloso, A. C. B. *Chem. Commun.* **2015**, *51*, 14199–14202.
43. Fridrich, B.; Tukacs, J. M.; Mika, L. T. In *Advances in Asymmetric Autocatalysis and Related Topics*; Palyi, G., Kurdi, K., Zucchi, C., Eds.; Elsevier Academic Press: Amsterdam, 2017; pp. 223–240. 408 p.
44. Ohkuma, T.; Kitamura, M.; Noyori, R. *Tetrahedron Lett.* **1990**, *31*, 5509–5512.
45. Saito, T.; Yokozawa, T.; Ishizaki, T.; Moroi, T.; Sayo, N.; Miura, T.; Kumobayashi, H. *Adv. Synth. Catal.* **2001**, *343*, 264–267.
46. Benincori, T.; Rizzo, S.; Pilati, T.; Ponti, A.; Sada, M.; Pagliarini, E.; Ratti, S.; Giuseppe, C.; de Ferra, L.; Sannicolò, F. *Tetrahedron: Asymmetry* **2004**, *15*, 2289–2297.
47. Starodubtseva, E. V.; Turova, O. V.; Vinogradov, M. G.; Gorshkova, L. S.; Ferapontov, V. A. *Russ. Chem. Bull.* **2005**, *54*, 2374–2378.
48. Starodubtseva, E. V.; Turova, O. V.; Vinogradov, M. G.; Gorshkova, L. S.; Ferapontov, V. A.; Struchkova, M. I. *Tetrahedron* **2008**, *64*, 11713–11717.
49. Turova, O. V.; Starodubtseva, E. V.; Vinogradov, M. G.; Ferapontov, V. A. *J. Mol. Catal. A: Chem.* **2009**, *311*, 61–65.
50. Collis, A. E. C.; Horváth, I. T. *Catal. Sci. Technol.* **2011**, *1*, 912–919.
51. Choi, J. H.; Kim, N.; Shin, Y. J.; Park, J. H.; Park, J. *Tetrahedron Lett.* **2004**, *45*, 4607–4610.
52. Hanna, D. G.; Shylesh, S.; Parada, P. A.; Bell, A. T. *J. Catal.* **2014**, *31*, 52–58.
53. He, D.; Horváth, I. T. *J. Organomet. Chem.* **2017**, *847*, 263–269.
54. van Zandvoort, I.; Wang, Y.; Rasrendra, C. B.; van Eck, E. R. H.; Bruijnincx, P. C. A.; Heeres, H. J.; Weckhuysen, B. M. *ChemSusChem* **2013**, *6*, 1745–1758.
55. Wang, D.; Astruc, D. *Chem. Rev.* **2014**, *114*, 6949–6985.
56. He, D.; Horváth, I. T. *Period. Polytech. Chem. Eng.* **2021**, *65*, 1–11.

CHAPTER TWO

Biomass valorization: Catalytic approaches using benign-by-design nanomaterials

Daily Rodríguez-Padrón[a,*], Alina M. Balu[a], Antonio A. Romero[a], and Rafael Luque[a,b,*]

[a]Departamento de Química Orgánica, Universidad de Córdoba, Cordoba, Spain
[b]Peoples Friendship University of Russia (RUDN University), Moscow, Russian Federation
*Corresponding authors: e-mail address: dailydggs@gmail.com; q62alsor@uco.es

Contents

Abstract

Research on biomass valorization approaches demands endeavors from the scientific community, looking forward to more sustainable and renewable alternatives to the petro-based industry. In this regard, the group FQM-383 (NANOVAL), from University of Cordoba has dedicated many efforts, going from conventional heating approaches to continuous flow strategies, and to the use of unconventional technologies such as microwave and ultrasound irradiations. Moreover, investigations on biomimetic catalysis and photocatalysis have been also directed to the upgrading a biomass residues and derived platform molecules. Importantly, the investigations performed by our research group consider atom, cost and energy efficiency of a process, involving renewable feedstocks and catalytic steps.

Advances in Inorganic Chemistry, Volume 77
ISSN 0898-8838
https://doi.org/10.1016/bs.adioch.2020.12.003

1. Introduction

Motivated for the current need of sustainable alternatives to overcome the global warming trend and the limited fossil resources, biomass valorization strategies have greatly attracted the attention of the scientific community.[1–4] Biomass residues constitute a renewable and readily available carbon and energy source, which could be employed for the synthesis of valuable chemicals and advanced nanomaterials.[5–15] Currently, several studies have been dedicated to the production of high added-value compounds from biomass-derived platform molecules, such as ethanol, glycerol, levulinic acid, succinic acid, lactic acid, sorbitol and furans, just to name a few.[16–18]

Even if sustainable feedstocks are employed, the green character of a process also depends on other factors, including energy requirements, amount of additional reagents and toxicity of the employed solvents, among other.[16] In order to minimize energy requirements, the use of more efficient protocols, moderate reaction conditions and shorter reaction times are highly desirable. In this regard, catalytic approaches play a crucial role, looking forward to achieve the aforementioned objectives, for the development of a more sustainable chemical industry.[19–21]

Moreover, heterogeneous catalysis provides additional advantages, related to the possibility of an easy recovery and reuse of the catalysts, thereby lowering process costs.[22–25] The environmentally friendly character of the synthetic methods for the preparation of solid catalytic systems should also be considered to reach the goals on sustainability. Therefore, a sustainable catalytic system, does not only depend on the non-toxic nature and inexpensive cost of the material itself, but also on the feedstocks, solvents and protocols employed in the synthesis.[26–28] Concerning the starting materials, the use of agricultural and household waste, including biomass residues and electronic-waste, represents a highly promising approach.

Furthermore, several innovative technologies have currently attracted the attention of the scientific community, in order to design more efficient methodologies. Among them, sonochemistry,[29,30] microwave(MW)-assisted chemistry,[31,32] solar and LED (light emitting diode) technologies for photocatalysis,[33–36] and continuous flow systems should be mentioned.[37–42] It is worthwhile to highlight mechanochemistry as an emerging and green strategy for both organic synthesis and materials design. Mechanochemical methods could increase the efficiency of a process by decreasing reaction times as well as the use of additional reactants and solvents. Mechanochemical processes lead to the fracture and abrasion of the starting materials, resulting in higher surface area and hence higher probability of contact between reagents. Interestingly, the employment of mechanical forces or thermal conditions could result in different chemical behavior of the reactants. Therefore, it is expected that mechanochemistry give rise to new chemistries, not described by other known techniques.[43–46]

Through this contribution a general overview of the research topics developed by the group FQM-383 (NANOVAL), from University of Cordoba will be provided. Specially highlighted will be the research lines related to sustainable chemistry from a catalytic perspective for biomass valorization processes. Among the different methodologies that will be addressed, recent investigations on flow chemistry, microwave and ultrasound irradiation, biomimetic catalysis and photocatalysis will be reviewed and exemplified (Fig. 1).

In particular, we will focus our attention on biomass conversion toward high added-value chemicals, but also will provide examples on the possibilities of biomass-derived residues for the design of advanced nanomaterials. In this regard, the green character of the employed procedures for synthesis of catalytic materials will be also considered, paying attention to emerging and promising approaches such as mechanochemistry.

In any case, it is worth to note, that the NANOVAL group conceives its investigations, considering the 12 principles of Green Chemistry, going from the use of heterogeneous catalysis, the minimization of solvents and waste, the employment of energy-efficient protocols, and the use of inexpensive and renewable feedstocks.

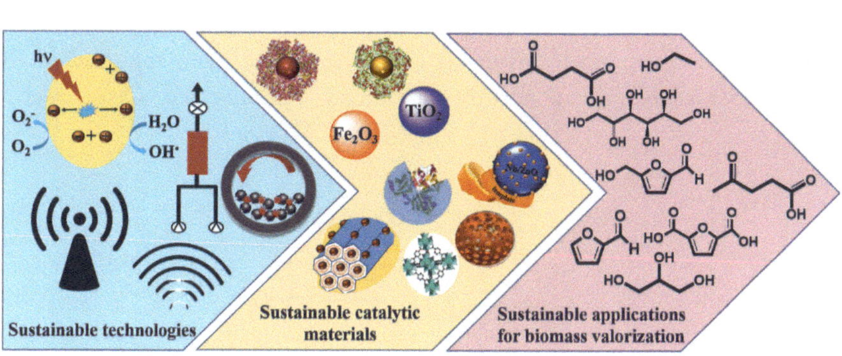

Fig. 1 Illustration of approaches for environmental catalysis, through sustainable technologies, employing sustainable materials for green applications.

2. Environmental catalysis for biomass conversion

Lignocellulosic biomass, composed of cellulose, hemicellulose and lignin, has been broadly investigated as green feedstock for the production of a wide variety of compounds with applications as flavoring molecules, pharmacophores, agrochemicals, fuels and fuel-additives.[47–51] Along this section, several examples will be described in this regard, taking into account different methodologies, going from conventional batch, to continuous flow, microwave-assisted, ultrasound-assisted, photocatalytic and biomimetic processes.

2.1 Conventional heating processes

trans-Ferulic acid is one of the biomass-derived platform molecules studied by our research group for upgrading toward vanillin (Fig. 2). Recently, Gómez-López et al. described a procedure, using conventional heating conditions at 90 °C and a short reaction time of 30 min to produce vanillin from *trans*-ferulic acid. In this process, the authors followed a catalytic approach, employing nanomaterials based on titanium and copper nanoparticles. Such catalytic systems were prepared through a one-pot hydrothermal methodology, employing different copper loadings (Cu/Ti atomic ratio) ranging from 0.01 to 1. The synthetic strategy allowed the formation of mesoporous materials with surface area values above 150 m^2/g and homogeneous particle size distribution of titanium nanoparticles. In turn, copper particle size displayed a wide distribution, being also related to the copper loading in each case. Remarkably, by using the material with 0.3 Cu/Ti ratio, a 70% of vanillin yield was achieved.[52]

Production of vanillin has been also investigated from isoeugenol, another lignin-derived compound. Several contributions have been reported in this regard under conventional heating conditions. For instance, Ostovar et al.

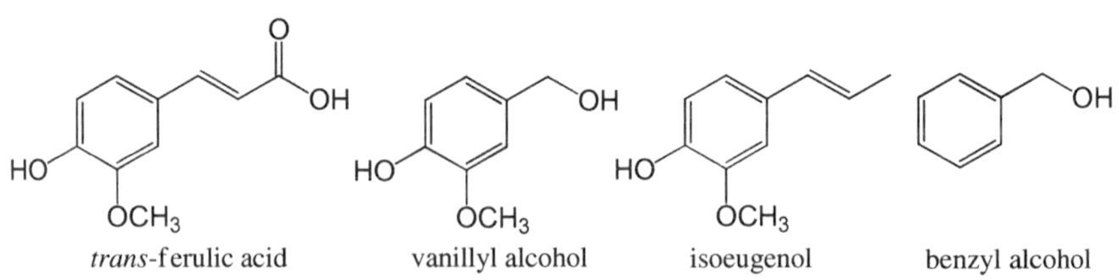

Fig. 2 Biomass-derived platform molecules employed for the preparation of flavoring molecules.

investigated the conversion of isoeugenol into vanillin, at 90 °C during 24 h, employing bifunctional catalytic systems based on surface modified SBA-15.[53] The materials were functionalized with sulfonic acid groups and iron oxide nanoparticles, via mechanochemical synthesis. In this case, the samples were additionally prepared through impregnation methods and importantly, it was observed that the materials prepared by mechanochemistry displayed a better catalytic performance. Such finding was attributed to the iron leaching and the aggregation of iron nanoparticles observed after the impregnation strategy. Moreover, the mechanochemically prepared samples exhibited a high stability over four use cycles. It is worthwhile to highlight that this study validated the suitability of mechanochemistry for the functionalization of mesoporous materials with good catalytic activity.[53] Similarly, niobium modified Al-SBA-15 mesoporous materials, synthesized via a mechanochemical approach, were used as catalytic systems for the transformation of isoeugenol into vanillin. Interestingly, a synergistic effect between either Fe or Nb metals with Al on the support was found, resulting in an enhanced activity, with conversion and selectivity values of 69% and 66%, respectively, after just 2 h of reaction.[54]

The aforementioned reaction was also investigated by Franco and coworkers employing sustainable catalytic materials derived from sewage sludge. This residue is produced as a byproduct in wastewater treatments and possesses a varied composition of organic and inorganic compounds, including proteins, carbohydrates, and lipids, among others. However, investigations on sewage sludge valorization are still incipient and require more efforts to take full advantage of its rich chemical composition. Interestingly, in this work Franco et al. employed a sewage sludge residue from the Tocantins region in Brazil, which presents a singular iron concentration. The treated waste resulted in an iron enriched biochar, which exhibited an outstanding catalytic behavior with a high conversion of isoeugenol of 80% and a selectivity toward vanillin of 65%. Moreover, such catalytic systems displayed a high stability over several reuse cycles.[55]

Isoeugenol conversion has been also studied employing magnetically separable nanomaterials based on iron oxide. Such nanocatalysts have been prepared following a mechanochemical approach, using spent coffee grounds (SCGs) as sacrificial template to control the morphology and the textural properties of the designed samples. The sacrificial template method has been broadly reported by our research group and is based on a two steps process, where firstly the metal precursor is arranged around the surface of a template, through physical or chemical interactions, and in a second step, the template is selectively eliminated. In particular, in the work reported by Rodríguez-Padrón and coworkers, a series of samples were obtained by calcining at different temperatures, from 300 to 800 °C, resulting in quasi-spherical homogeneous distributed nanoparticles (Fig. 3A and B) which combine catalytic potentialities and magnetic properties.[56] Samples calcined at 700 and 800 °C showed the best catalytic results for isoeugenol conversion, due to the high hematite content (Fig. 3C). Moreover, reusability studies displayed an unexpected increase of the catalytic activity, particularly interesting for the sample calcined at 300 °C, due to an in-situ amorphization of the material.[57]

Furthermore, vanillin synthesis has been studied from vanillyl alcohol, also derived from lignocellulosic biomass. This reaction has been investigated by Saberi et al. employing mesoporous materials based on magnetic iron oxide nanoparticles and Al-SBA-15, obtained by mechanochemical methods. In this case, the reaction was performed for 2 h, at three different

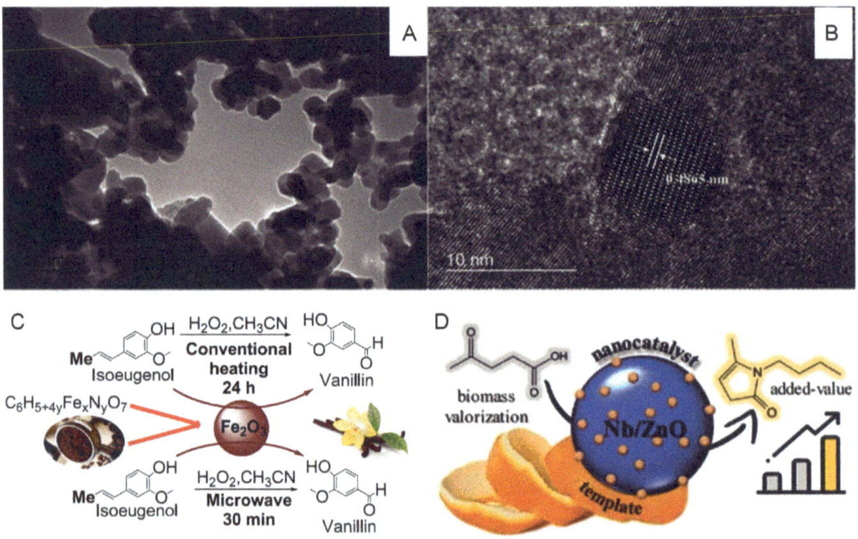

Fig. 3 (A) TEM and (B) HR-TEM images of SCGs templated magnetic nanoparticles, (C) schematic representation of catalytic applications of SCGs derived samples for vanillin production. (D) Illustration of orange peel template catalysts for levulinic acid valorization toward N-heterocycles. *Panel (C): Reprinted with permission from Rodriguez-Padron, D.; Munoz-Batista, M.J.; Li, H.; Shih, K.; Balu, A.M.; Pineda, A.; Luque, R.* ACS Sustain. Chem. Eng. ***2019,*** 7 *(20), 17030–17038. Copyright (2020) American Chemical Society. Panel (D): Reprinted with permission from Rodríguez-Padrón, D.; Zhao, D.; Carrillo-Carrion, C.; Morales-Torres, C.; Elsharif, A.M.; Balu, A.M.; Luque, R.; Len, C.* Catal. Today *2020. Copyright (2020) Elsevier.*

temperatures, namely at 90 °C, at 50 °C and at room temperature. A kinetic analysis was carried out for the three reaction temperatures, achieving a conversion higher than 99% in all cases, after 2 h of reaction, with an 88% of vanillin selectivity and 87% of vanillin yield. Remarkably, this investigation proposed an energy-efficient strategy for the synthesis of vanillin, facilitating its further scalability.[58] Unlike the previous described work, where propionic acid was employed to achieve the magnetic properties, similar samples based on mesoporous aluminosilicate materials and magnetic iron oxide have been also prepared by using a biomass waste. Indeed, mechanochemical synthesis of magnetic nanoparticles has been mainly reported in the literature employing propionic acid. The use of such chemical has several disadvantages related to its high cost and non-environmentally friendly characteristics. Therefore, the replacement of propionic acid by a biomass waste is a highly sustainable strategy. In particular, in the work developed by Marquez-Medina et al. several samples were prepared by using different iron loadings, from 30% to 50%, different calcination temperatures, of 300 and 400 °C, and different iron precursors. The authors investigated the catalytic performance of such systems in the oxidative transformation of vanillyl alcohol and isoeugenol into vanillin, displaying outstanding conversion values between 80% and 90%, with selectivity higher than 50%. Additionally, the samples showed a remarkable stability over five reuse cycles. Interestingly, studies on the structure-activity relationship demonstrated that iron superficial content and iron-aluminum interactions, play a crucial role in the acid properties and therefore in the catalytic performance.[59]

Moreover, core-shell type of materials based on silica coated iron oxide, have been reported and fully characterized by our research group, demonstrating to have a broad range of possibilities for catalytic applications. In this sense, the preparation of sulfathiazole functionalized SiO_2-coated Fe_2O_3 and its application as catalytic systems in oxidation and alkylation reactions, namely the oxidation of benzyl alcohol to benzaldehyde and the alkylation of toluene employing benzyl chloride, have been described. Through this work, the authors performed an optimization of the reaction conditions for benzyl alcohol oxidation, considering catalyst and oxidant quantities, as well as exploring the scope of the reaction for different para-substituted benzyl alcohol derivatives. These reactions were carried out at 80 °C for 2 h, reaching conversions higher than 90% and selectivity higher than 95%.[60]

In addition, silica coated iron oxide, modified with a proline analog has been tested in the production of vanillin from isoeugenol and vanillyl alcohol at 90 °C, under conventional heating. By using isoeugenol as feedstock, conversion of 65% and vanillin selectivity of 55% were reached, while employing vanillyl alcohol, conversion values higher than 99% and selectivity of 92% were achieved. It is worth to mention, that in this contribution, the synergistic effect between acidic and basic sites played an important role in the catalytic performance, as corroborated by performing the reaction employing the proton-unexchanged catalyst counterpart.[61]

Besides the aforementioned oxidation reactions, other types of processes have been investigated in our research group, employing conventional heating conditions. Among them, it could be mentioned the amination of levulinic acid toward *N*-heterocycles. In this regard, nanocatalytic systems based on niobium and zinc oxides, were tested in the reaction of levulinic acid and 1-butylamine, under solvent-free conditions (Fig. 3D). Such catalysts were prepared employing orange peel as sacrificial template by a solvent-free mechanochemical approach. Samples were prepared employing different niobium oxide contents, ranging from 2.5 to 10 wt%, revealing that the presence of niobium entities were essential for the progress of the reaction. In particular, the material modified with 10 wt% of niobium displayed the best catalytic results, achieving a conversion of 94.5% and a selectivity of 97.4% toward the formation of 1-butyl-2-(ethylideneamino)-5-methylpyrrolidin-2-ol (Fig. 4). Importantly, these kind of molecules could have several applications as pharmacophores and agrochemicals, hence finding alternative pathways for their preparation from sustainable sources, instead of from fossil feedstocks, is highly desirable.[62]

2.2 Continuous flow processes

Research on basic chemistry is highly important for the development of science. However, advances in science should not stay at laboratory scales, instead, the scientific community should take advantage of all the generated new knowledge to put it into practice, and to enhance current industrial processes and their impact on the environment and on human safety. In this regard, the development of continuous flow processes could have critical implications for the scalability of important achievements at laboratory scale.

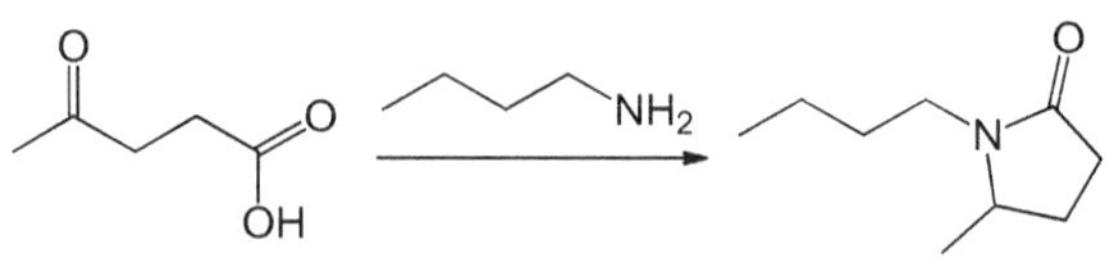

Fig. 4 Transformation of levulinic acid into 1-butyl-2-(ethylideneamino)-5-methylpyrrolidin-2-ol.

Continuous flow processes offer several advantages, including short reaction times, effective reactant mixing and efficient heat and mass transfer, which results in a better control of the reaction parameters and, therefore, in an easier scale up. Considering the aforementioned premises, several contributions from our research group have been dedicated to explore the suitability of continuous flow regimes for a wide range of catalyzed reactions.[63,64]

For instance, methyl levulinate valorization toward the formation of γ-valerolactone (GVL) has been studied by Ouyang and coworkers, through a catalytic transfer hydrogenation (CTH) pathway, under continuous flow conditions (Fig. 5). GVL possesses several applications as green solvent and fuel additive. In this contribution, the authors employed UiO-66, as catalytic system, which is a zirconium-based metal organic framework (MOF). The reaction was performed in a Phoenix reactor, a continuous flow system from ThalesNano, at a flow rate of 0.3 mL/min, at a pressure of 35 bar and at different temperatures from 180 to 240 °C. Interestingly, this contribution demonstrated the direct relationship between reaction temperature and selectivity of the process, displaying that at lower temperatures, CTH and transesterification routes were competing, with the consequent formation of transesterification byproducts. In turn, by performing the reaction at higher reaction temperatures (240 °C), selectivity to GVL achieved a maximum value of 96%. Additionally, the influence of pressure, flow rate and the long term operational stability of the employed MOF, were also investigated.[65]

The aforementioned reaction was also investigated by Lázaro and coworkers, using zirconium-based materials, including Zr-SBA-15 silicates, ZrO_2 and ZrO_2 supported on SBA-15, for a better comparison and understanding of the effect of zirconium entities on the catalytic performance. The samples were tested for the CTH of methyl levulinate under optimized conditions, as well as employing 2-propanol as hydrogen transfer solvent. The authors found out that, among the tested materials, ZrO_2 supported on SBA-15 displayed the most relevant results. This was attributed to the good

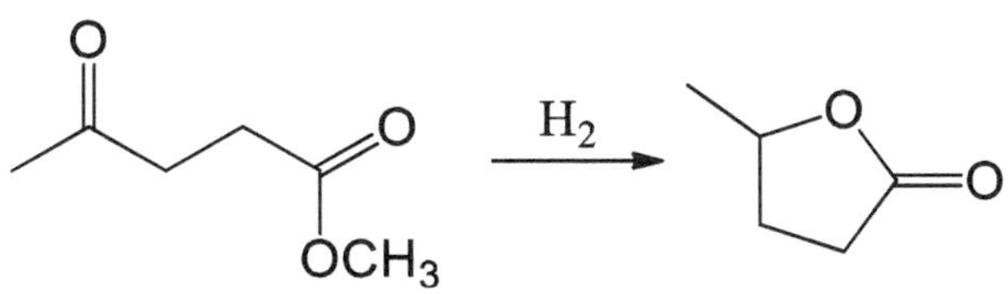

Fig. 5 Transformation of methyl levulinate into γ-valerolactone.

dispersion and lower particle size of zirconium oxide nano-entities, in comparison with bulk ZrO_2 and also to an enhanced disposition and accessibility of the active centers, as compared to the Zr-based silicate. The two previously mentioned studies revealed that zirconium-based materials could be potential candidates for their use as catalytic systems for catalytic transfer hydrogenation processes.[66]

In addition, zirconium-based materials supported on several zeolites, and prepared through mechanochemical processes, have been tested in the continuous flow synthesis of GVL from methyl levulinate. Pineda and coworkers performed an interesting study to get more insight into the role of the acidic sites in the catalytic behavior, being able to tune the acidity and therefore the catalyst response. In particular, it was found that the presence of acid sites in the supports, give rise to side reactions with a negative effect on the product selectivity. Alternatively, treating the catalysts with different bases boosted GVL selectivity by quenching undesirable side reactions.[67]

Production of GVL has also been investigated by Ouyang and coworkers, using ruthenium oxide nanoparticles supported on titanium. Different ruthenium concentrations, ranging from 1% to 5% were loaded on the titanium support, to study the influence of the Ru content on the catalytic behavior. Such analysis revealed that samples modified with 5% of ruthenium displayed an outstanding catalytic performance, with a conversion of methyl levulinate of 98% and a selectivity toward GVL of 97%, by carrying out the reaction at 200 °C and 35 bar of pressure. Moreover, this material exhibited improved results in comparison with commercial ruthenium catalysts.[68]

Methyl levulinate could be prepared from furfuryl alcohol derived from the hemicellulosic fraction of lignocellulosic biomass. In this sense, Zhao et al. have reported the use of several zeolites, such as H-ZSM-5-50, as catalytic systems for the continuous flow conversion of furfuryl alcohol into methyl levulinate at 170 °C and 50 bars (Fig. 6A). Under the aforementioned conditions, an 80% yield of methyl levulinate was achieved, with the concomitant formation of β-angelica lactone as the main byproduct. The employed zeolites displayed good stability and reusability. However, by using high furfuryl alcohol concentrations, formation of polymeric undesirable products occurs, poisoning the catalysts surface and hence requiring a further regeneration step to recover to catalytic activity.[69]

Also derived from hemicellulose, 5-hydroxymethylfurfural (HMF) could have a wide range of possibilities as sustainable feedstock. In this sense, the scientific community has proposed several strategies for its upgrading,

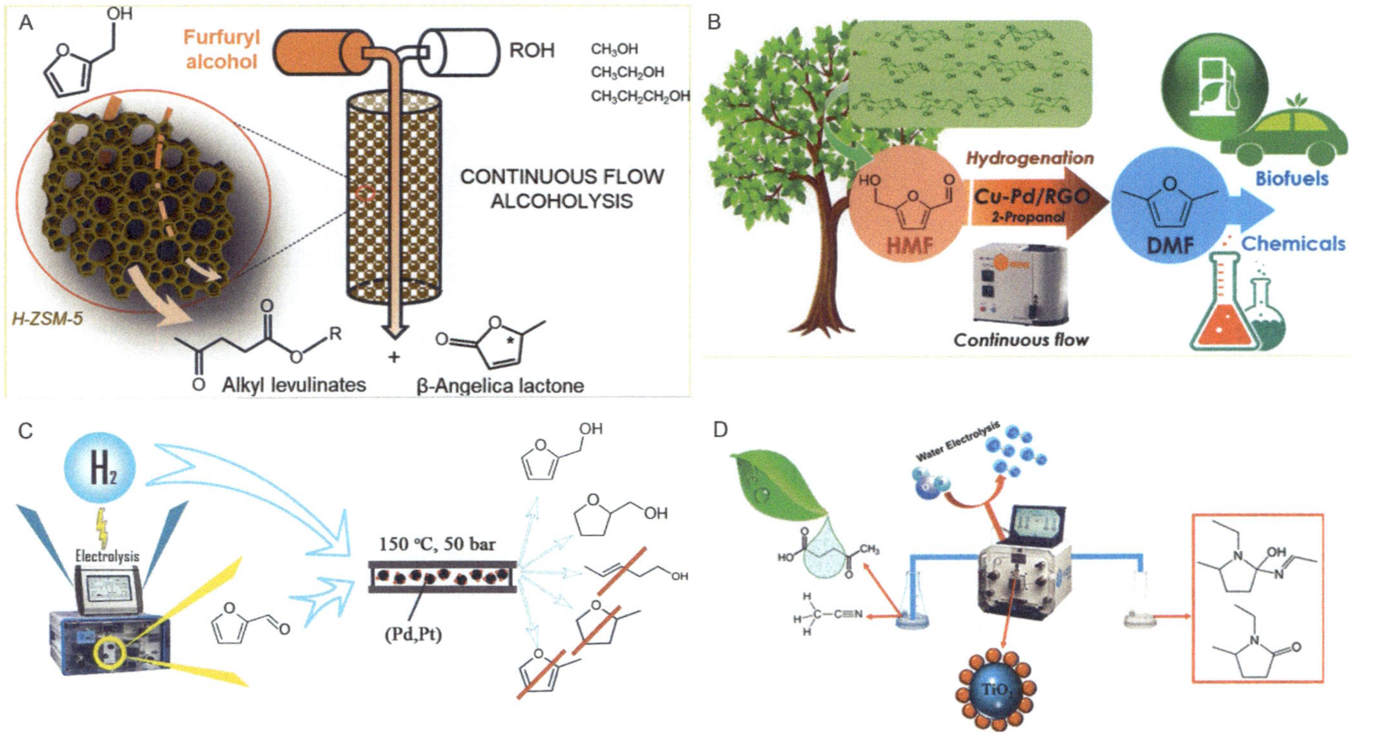

Fig. 6 (A) Continuous flow alcoholysis of furfuryl alcohol over zeolite based catalysts, (B) HMF upgrading toward DMF over Cu and Pd modified reduced graphene oxide, (C) furfural hydrogenation over palladium and platinum-based catalytic systems, and (D) levulinic acid reductive amination toward *N*-heterocycles over ruthenium supported on orange peel derived titanium. *Panel (A): Reprinted with permission from Zhao, D.; Prinsen, P.; Wang, Y.; Ouyang, W.; Delbecq, F.; Len, C.; Luque, R.* ACS Sustain. Chem. Eng. ***2018,*** 6 *(5), 6901–6909. Copyright (2020) American Chemical Society. Panel (B): Reprinted with permission from Mhadmhan, S.; Franco, A.; Pineda, A.; Reubroycharoen, P.; Luque, R.* ACS Sustain. Chem. Eng. ***2019,*** 7 *(16), 14210–14216. Copyright (2020) American Chemical Society. Panel (C): Reprinted with permission from Ouyang, W.; Yepez, A.; Romero, A.A.; Luque, R.* Catal. Today ***2018,*** 308, *32–37. Copyright (2020) Elsevier. Panel (D): Reprinted with permission from Rodriguez-Padron, D.; Puente-Santiago, A.R.; Balu, A.M.; Romero, A.A.; Muñoz-Batista, M.J.; Luque, R.* ACS Sustain. Chem. Eng. ***2018,*** 6 *(12), 16637–16644. Copyright (2020) American Chemical Society.*

for instance toward 2,5-dimethylfuran (DMF), which is a potential biofuel additive. Mhadmhan and coworkers have studied the catalytic hydrogenation of HMF to DMF under continuous flow regime, employing copper and palladium bimetallic catalysts supported on reduced graphene oxide materials, through CTH processes (Fig. 6B). The catalysts were prepared by an impregnation method employing different palladium contents. In particular, the sample modified with 1% of Pd, resulted to be the most active, displaying a conversion of 95% and a selectivity toward DMF of 96%. Further spectroscopic characterization revealed that the formation of a copper-palladium alloy favors the material stability.[70]

Palladium-based materials have been also tested in the hydrogenation of furfural to furfuryl alcohol and tetrahydrofurfuryl alcohol (THFA). In this regard, Ouyang et al. have reported the incorporation of Pd entities on different supports, including SBA-15, Al-SBA-15 and a magnetic iron oxide modified silicate, named as Pd/MAGSNC, via mechanochemical synthesis (Fig. 6C). The designed materials exhibited good catalytic performance in terms of conversion and product selectivity, and in the case of Pd/MAGSNC samples also allowed its simple recovery from the bulk solution by using a magnetic field.[71]

Continuous flow strategies have been also considered for the synthesis of a wide range of flavoring molecules. Among them, vanillin, (−)menthol and cinnamyl alcohol could be mentioned. For instance, Zuliani et al. have reported an interesting study for the sustainable preparation of (−)menthol from citronellal, being the latest compound obtained from the citral oils such as geraniol and nerol. In this contribution, the authors developed a tandem system for the continuous flow cyclisation of citronellal toward isopulegols and the subsequent hydrogenation to (−)menthol (Fig. 7). Besides the use of

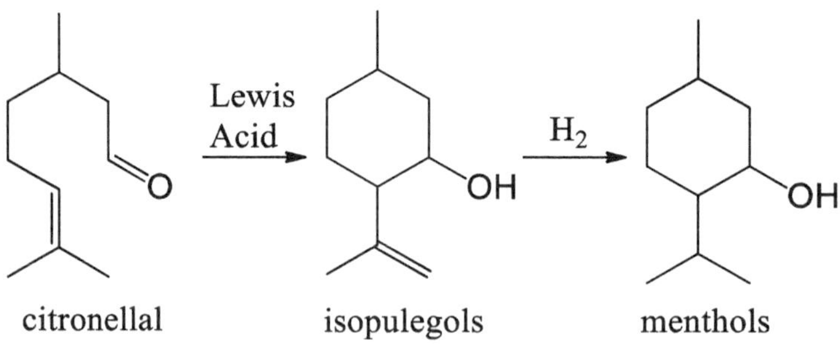

Fig. 7 Transformation of citronellal into isopulegols and the subsequent hydrogenation to (−)menthol.

an innovative tandem reaction system and a renewable feedstock, this work also proposed a strategy for the valorization of the ceramic cores of scrap catalytic converters (SCATs), taking advantage of the presence of noble metals. Such waste was used for the preparation of catalytic materials by microwave-assisted incorporation of iron oxide nanoparticles in the SCATs. This protocol allowed the formation of catalytic systems, involving Lewis acid centers due to the presence of iron entities, and hydrogenation capacity, related to the residual noble metal content of the SCATs. Under optimized conditions, a complete conversion of citronellal, with a 97% of (−)menthol yield was reached. This work is a clear example of the potentialities of flow chemistry in catalyzed processes for industrial applications.[72] A wider scope of this type of reactions, employing SCATs based catalysts under continuous flow regime, was investigated by Cova and collaborators. In particular, the reductive upgrading of isopulegol, isoeugenol, vanillin, cinnamyl alcohol and benzaldehyde, was investigated. As well as, a full parametric analysis was accomplished, considering temperature, pressure, flowrate and reactants concentration.[73]

Furthermore, Cova and coworkers investigated the catalytic activity of ruthenium supported on treated SCATs for the continuous flow transformation of cinnamaldehyde into cinnamyl alcohol. The catalytic systems were prepared following a mechanochemical step, for the incorporation of ruthenium entities, and a subsequent reduction. Several ruthenium concentrations were examined, achieving the best catalytic performance (95% of cinnamaldehyde conversion and 80% of selectivity to cinnamyl alcohol) by modification with 10 wt% of the metal. In addition, a deep analysis on the effect of the hydrogen pressure, flow rate and reaction temperature, on the reaction progress was reported.[74]

Isoeugenol oxidation toward vanillin production has also been examined under continuous flow conditions. In this direction, Filiciotto et al. have studied the suitability of humins residues for the preparation of effective catalytic systems.[75] Humins are undesirable side products from process of biomass valorization toward bio-fuels and bio-plastics.[76] In this case, the authors employed humins waste as template and carbonaceous support, for the incorporation of iron oxide nanoparticles. The catalytic activity of such samples was also compared with iron oxide supported on Al-SBA-15 and a full parametric analysis was carried out, considering temperature, pressure and flow rate. More interesting, in this contribution the authors presented insights into the reaction mechanism, proposing a radical pathway for the formation of vanillin from isoeugenol.

Filiciotto and coworkers have also investigated other strategies for the valorization of humins residues, for instance toward the synthesis of added-value chemicals. The authors have studied the decomposition of humins under continuous flow conditions, in order to get a better understanding of the structure and composition of this waste material. In this study, catalytic transfer hydrogenation processes were carried out employing nickel and palladium-based catalysts, toward the formation of 5-methoxymethylfurfural and 5-hydroxymethylfurfural. Moreover, a qualitative analysis of the decomposition products revealed the presence of furanics, sugar derived molecules, aromatics and other small organic acids and oxygenated molecules, displaying a high heterogeneity.[77] Further upgrading of humins-derived 5-methoxymethylfurfural has been described by Pfab et al. through a continuous flow hydrogenation route. Ruthenium and palladium-based commercial catalysts, supported on carbon, were tested, achieving full conversion of the starting reagent, with the consequent formation of 5-(methoxymethyl)-2-furyl]-methanol (MMFM) and 5-methyltetrahydrofurfuryl alcohol (MTHFA). Interestingly, Ru catalysts showed a higher selectivity to MMFM, while Pd catalytic systems resulted in the selective formation of MMFM.[78]

Furthermore, levulinic acid valorization strategies have been also described under continuous flow regimes. An interesting approach in this regard have been depicted by Rodríguez-Padrón and collaborators, focused on the transformation of levulinic acid into *N*-heterocycles in a H-cube reactor, which also include an electrolytic cell for the in-situ production of hydrogen through water splitting. The formation of *N*-heterocycles has been achieved employing nitrile-based compounds, in particular acetonitrile, through a reductive amination pathway. In a first step, acetonitrile is hydrogenated to give rise to the corresponding amine, which subsequently reacts with levulinic acid resulting in the production of the desired products. This reaction has been explored using ruthenium modified titanium, which was synthesized by using orange peel as sacrificial template for the mechanochemical-assisted formation of nanostructured titanium, and by subsequently incorporating ruthenium entities via chemical reduction method. A series of materials were tested, considering several ruthenium loadings, ranging from 1% to 3%, as well considering ruthenium supported on P25 titanium and commercial Ru onto carbon for comparison. Outstandingly, the material based on orange peel-derived titanium, modified with 3% of ruthenium, showed the best catalytic performance with a levulinic acid conversion of 79% and a selectivity toward

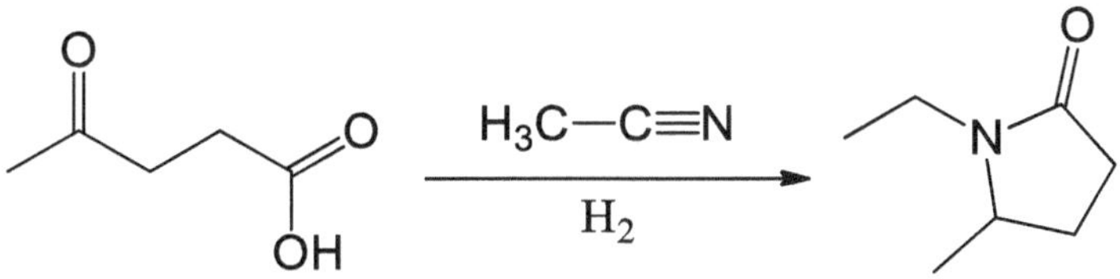

Fig. 8 Transformation of levulinic acid into 1-ethyl-2-(ethylideneamino)-5-methylpyrrolidin-2-ol.

1-ethyl-2-(ethylideneamino)-5-methylpyrrolidin-2-ol of 85% (Fig. 8). Interestingly, this material displayed a drastic change of selectivity after 60 min of reaction, which was further investigated and attributed to the deposition of organic moieties on the catalyst surface, as confirmed by X-ray diffraction (XRD) analysis and X-ray photoelectronic spectroscopy.[79] This reaction has been also investigated employing graphitic carbonitrile (g-C_3N_4) modified with 1% of platinum. A full optimization of the reaction parameters was performed using such catalytic materials, by carrying out the reaction at different pressures, temperatures and levulinic acid concentrations. A maximum catalytic activity was observed at a temperature of 90 °C, a pressure of 50 bars and using 0.3 M of levulinic acid solution. It is worth to highlight that the catalytic material displayed a high selectivity of 67.5% toward 1-ethyl-5-methylpyrrolidin-2-one product, with a remarkable stability along 3 h of reaction on-stream.[80]

A platinum-based catalyst, supported on commercial P25 titanium was also tested on levulinic acid continuous flow transformation to *N*-heterocycles, showing a notable conversion of 87% and 1-ethyl-5-methylpyrrolidin-2-one selectivity of 80%. Altuğ and coworkers have also extended these investigations to other nitriles and carbonyl containing molecules for the formation of amines. For instance, benzonitrile and ethylcyanoacetate were studied, instead of acetonitrile, demonstrating that the interactions between the phenyl ring and platinum nanoparticles, as well as the electron withdrawing characteristics of ester groups, have a direct repercussion on the reaction progress. In addition, the reaction of acetonitrile with another carbonyl containing reactant, namely benzaldehyde, was investigated, showing an outstanding conversion higher than 99% and a high selectivity toward the tertiary amine. Additionally, authors also investigated the effect of substitutions in the aromatic ring of benzaldehyde, by accomplishing the reaction with *p*-chlorobenzaldehyde and *p*-nitrobenzaldehyde, respectively. Interestingly, the platinum-supported titanium catalyst also resulted to be efficient for the reduction of –NO_2 groups.[81]

Looking forward to perform the conversion of levulinic acid toward *N*-heterocycles in a greener way, decreasing reaction costs, Defilippi and collaborators accomplished experiments using Ni-based catalytic materials. The authors employed NiN_3 and Ni nanoparticles, displaying interesting and promising results for the use of transition metals and metal nitrides for these kind of reactions. In particular, two main products, namely 1-ethyl-5-methylpyrrolidin-2-one and 1-ethyl-5-methyl-1,3-dihydro-2*H*-pyrrol-2-one were formed.[82] In summary, all the previous described examples certainly demonstrated the wide range of possibilities that continuous flow processes could offer for catalyzed organic reactions.

2.3 Microwave-assisted reactions

Exploring alternative technologies is highly necessary to achieve more cost-efficient processes. In this sense, utilization of microwave irradiation assisted approaches is a very interesting possibility, which still requires further efforts. Microwave-assisted protocols have several advantages for both catalysts preparation and organic synthesis, which include the possibilities of performing reactions in shorter times and of achieving high yields.[83] It is well known that microwave irradiation results in an improved heat transfer and in the occurrence of unique chemistries, which do not usually take place under conventional heating strategies.[84]

A recent report from our group has clearly exemplified, two different approaches for biomass upgrading, namely: (1) the transformation of biomass-derived platform molecules toward high added-value chemicals and (2) the utilization of biomass waste for materials design. Similar to the previous described work in the continuous flow section, humins were herein used as template and support for the synthesis of magnetic iron oxide nanocomposites, via mechanochemical pathway, achieving a mesoporous architecture. Such nanomaterials were successfully employed as magnetically separable catalytic materials for the conversion of isoeugenol to vanillin under microwave irradiation conditions. Conversion values higher than 87% were reached, demonstrating the suitability of humins for the preparation of active catalysts, also for microwave-assisted reactions.[85]

Moreover, Wang et al. have reported the use of a microwave-assisted strategy for upgrading furfuryl alcohol toward alkyl levulinates, remarkably using a non-metal containing activated carbon (Fig. 9A). Several parameters were optimized, including temperature, catalyst loading and furfuryl alcohol concentration. Under the optimum conditions, namely by using 0.2 M

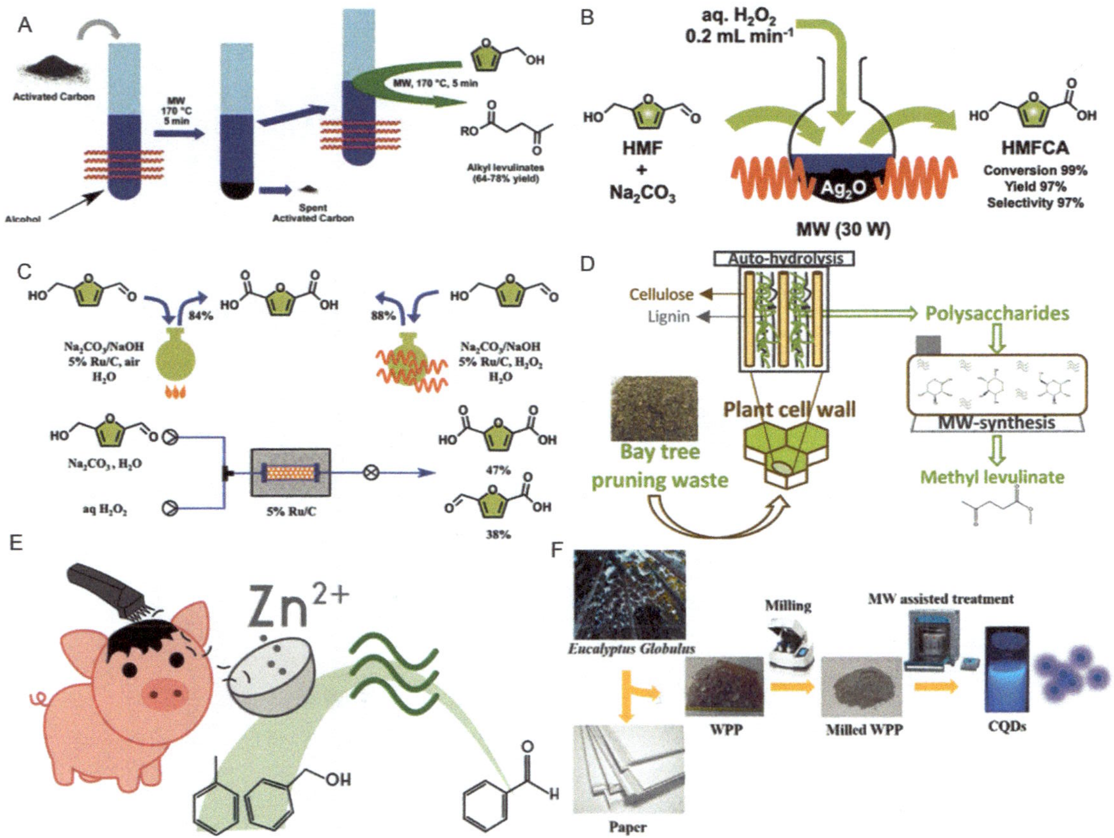

Fig. 9 See figure legend on next page.

furfuryl alcohol and 100 mg of catalyst, at 170 °C during 5 min, the scope of the reaction was investigated employing various alcohols, from methanol to *n*-butanol. Certain deactivation was observed by recycling the catalysts over three reaction cycles, which was attributed to the leaching of acid species from the carbonaceous surface.[86]

Zhao and coworkers have also explored the suitability of microwave-assisted approaches for the valorization of biomass-derived platform molecules, in particular for HMF oxidation to 5-hydroxymethyl-2-furancarboxylic acid (HMFCA) (Fig. 9B). In this work, the authors proposed the use of silver oxide as catalytic system. A comparison between conventional heating and microwave approaches was carried out, giving rise to an HMFCA yield of 98% by performing the reaction at 90 °C during 60 min, while similar HMFCA yield (97%) was achieved under MW in a reduced reaction time of 24 min. Reusability studies of silver oxide catalyst showed a partial catalyst deactivation, most likely associated with the reduction of Ag_2O and the sintering of nanoparticles after the reaction, as confirmed by XRD and TEM analysis, respectively.[87]

HMF oxidation toward 2,5-furandicarboxylic acid (FDCA), under MW irradiation has also been investigated by Zhao and coworkers, using

Fig. 9 (A) Microwave-assisted upgrading of furfuryl alcohol over non-metal containing activated carbon, (B) HMF valorization toward HMFCA over silver oxide catalyst, (C) HMF transformation under conventional heating, batch and continuous flow microwave-assisted conditions, using ruthenium-based catalysts, (D) schematic representation of bay tree fractionation and further treatment of polysaccharide fraction for methyl levulinate production, (E) oxidation of benzyl alcohol and toluene over pig bristles derived ZnS nanostructures, and (F) microwave-assisted formation of carbon dots from cellulosic residues from the paper industry. *Panel (A): Reprinted with permission from Wang, Y.; Zhao, D.; Triantafyllidis, K.S.; Ouyang, W.; Luque, R.; Len, C.* Mol. Catal. *2020,* 480, *110630. Copyright (2020) Elsevier. Panel (B): Reprinted with permission from Zhao, D.; Su, T.; Wang, Y.; Varma, R.S.; Len, C.* ACS Sustain. Chem. Eng. ***2020,*** 8 *(23), 8486–8495. Copyright (2020) American Chemical Society. Panel (C) Reprinted with permission from Zhao, D.; Rodriguez-Padron, D.; Triantafyllidis, K.S.; Wang, Y.; Luque, R.; Len, C.* ACS Sustain. Chem. Eng. ***2020,*** 8 *(8), 3091–3102. Copyright (2020) American Chemical Society. Panel (D): Reprinted with permission from Rincon, E.; Zuliani, A.; Jimenez-Quero, A.; Vilaplana, F.; Luque, R.; Serrano, L.; Balu, A.M.* ACS Sustain. Chem. Eng. ***2020,*** 8 *(29), 11016–11023. Copyright (2020) American Chemical Society. Panel (E): Reprinted with permission from Cova, C.M.; Zuliani, A.; Muñoz-Batista, M.J.; Luque, R.* ACS Sustain. Chem. Eng. ***2018,*** 7 *(1), 1300–1307. Copyright (2020) American Chemical Society. Panel (F): Reprinted with permission from Rodríguez-Padrón, D.; Algarra, M.; Tarelho, L.A.; Frade, J.; Franco, A.; de Miguel, G.; Jimenez, J.; Rodríguez-Castellon, E.; Luque, R.* ACS Sustain. Chem. Eng. ***2018,*** 6 *(6), 7200–7205. Copyright (2020) American Chemical Society.*

ruthenium-based catalysts (Fig. 9C). The reaction was tested under conventional heating in batch reactors and under a MW-continuous flow regime, for comparison. Specially, commercial ruthenium onto carbon gave rise to an 88% yield of FDCA, after 30 min of MW batch-treatment, while the conventional batch heating approach resulted in a FDCA yield of 84% after 22 h and the MW-flow strategy displayed a FDCA yield of 47%, with the concomitant formation of 5-formyl-2-furancarboxylic acid (FFCA) in a 38% yield. These studies greatly contributed to the advance of microwave technologies for furanics valorization processes.[88]

Rincón and coworkers also presented a very interesting approach for the application of microwave technologies for biomass valorization. This contribution was dedicated to the full utilization of bay tree residues, combining conventional heating and microwave-assisted strategies (Fig. 9D). First, a fractionation step was carried out by auto-hydrolysis under conventional heating, resulting in the extraction of polysaccharides, proteins and phenolics, which were further characterized in detail. In this sense, an optimization of the extraction conditions was also performed. Subsequently, the polysaccharides fraction was treated under MW irradiation, with the successful production of methyl levulinate in a 40 wt%.[89]

In another example, Cova and coworkers presented the microwave-assisted oxidation of benzyl alcohol and toluene toward benzaldehyde, at 120 °C and employing H_2O_2 as oxidant (Fig. 9E). In this contribution, a sustainable and an innovative family of materials was employed, based on ZnS. Interestingly, pig bristles were used as a sustainable sulfur source, which combined with a zinc precursor were treated during 1, 3 and 5 h to give rise to three catalytic samples, denoted as ZnS-1h, ZnS-3h and ZnS-5h. Specially, ZnS-5h catalyst exhibited the best catalytic, reaching conversion values of 63% and 29%, and selectivity of 86% and 87%, for the transformation of benzyl alcohol and toluene, respectively.[90]

Moreover, microwave approaches have been also reported for the preparation of interesting materials. It is worth to mention a relevant example developed by Franco and collaborators, for the microwave-assisted preparation of sustainable silica materials from rice husk residues. Through a MW strategy, rice husk was treated, in order to remove metal ions and to give rise to a highly pure silica. Then, the obtained silica was further modified with iron oxide by mechanochemical methods. The prepared iron oxide-silica samples were further tested as catalytic materials in MW-assisted reactions, namely in the alkylation of toluene and in the benzyl alcohol oxidation to benzaldehyde.[91] In addition, MW treatments have been also employed for the valorization of cellulosic residues from the paper industry toward the formation of carbon dots with striking fluorescent features, and a wavelength dependent emission behavior (Fig. 9F).[92]

2.4 Ultrasound-assisted reactions

Ultrasound irradiation is another unconventional energy source with great potential for a wide variety of applications, from laboratory to industrial scales. Ultrasound could give rise to improved reaction kinetics, and unexpected reaction mechanisms, in comparison with the typical pathways followed under conventional heating. Therefore, sonochemistry could modify and/or enhance products selectivity, and additionally this technology represents the less energy consuming approach with a minimum generation of byproducts.[93–97]

Our research group has also dedicated efforts in this regard. For instance, ultrasound irradiation has been employed in the valorization of valeric acid by transesterification reaction toward the formation of ethyl valerate. The reaction was carried out during 2 h, employing a valeric acid-ethanol ratio of 1:2. Remarkably, this process was catalyzed by a novel biocatalytic system based on bio-silicified lipase (from *Candida antarctica*). It is well known that the use of enzymes involves several benefits, related to their high selectivity and specificity. Nonetheless, the real time industrial application of these biomolecules still demands further efforts to simplify their recovery and to enhance their stability, under operational and storage conditions. In this regard, biosilicification strategies represent a feasible opportunity, which additionally could result in an enhancement of activity, as compared to the free enzyme. Under the previously mentioned conditions, the bio-silicified lipase showed an excellent catalytic activity, with a valeric acid conversion of 90% and a complete selectivity toward ethyl valerate product (Fig. 10A). These results were superior to those obtained for the free lipase, and more importantly, the silica-shell strategy allowed the recovery and reuse of the biocatalytic system, maintaining a relatively good activity after four reaction cycles.[98] Additionally, sonochemical strategies have also been considered for the preparation of materials, for the grafting of β-cyclodextrin (β-CD) onto silica.[99]

2.5 Biomimetic catalysis

Intrinsically related to the example described in the previous section, and according to the benefits of enzymes immobilization and encapsulation,[100–102] bio-silicified lipase has been also employed in the esterification

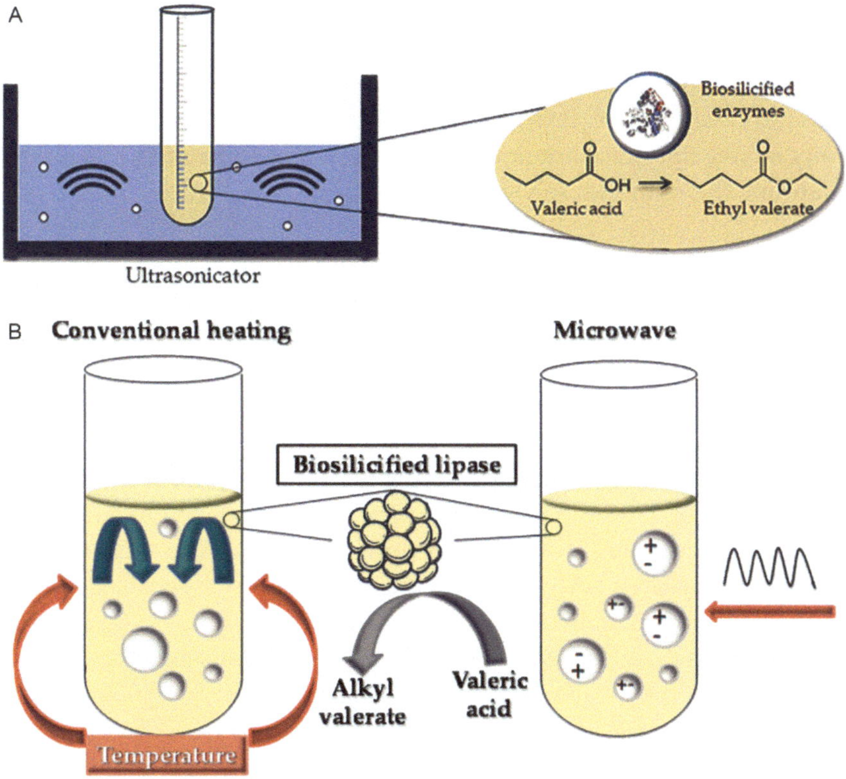

Fig. 10 (A) Ultrasound-assisted valorization of valeric acid, through a transesterification approach, using bio-silicified lipase and (B) conventional heating and microwave-assisted transesterification of valeric acid over bio-silicified lipase. *Panel (A): Reprinted with permission from Cebrián-García, S.; Balu, A.M.; Luque, R.* Front. Chem. ***2018,*** *6, 197. Copyright (2020) Frontiers. Panel (B): Reprinted with permission from Cebrián-García, S., Balu, A. M., García, A., Luque, R.* Molecules ***2018,*** 23 *(9), 2283. Copyright (2020) MDPI.*

of valeric acid under conventional heating and microwave irradiation, for comparison.[103] A full parametric analysis was carried out, taking into account catalyst loading, temperature, stirring rate and water effect, as well several alcohol media were tested to explore the scope of the reaction. Excellent ethyl valerate yields were achieved, with values above 80%, under both conventional heating and microwave irradiation (Fig. 10B). However, MW treatment also resulted in the catalyst deactivation due to the partial denaturalization of the enzyme, while the conventional heating allowed reuse of the biocatalysts.[104]

Furthermore, encapsulated laccases have been described by Pistone et al. and successfully employed in the oxidation of a wide variety of lignin-based compounds, including vanillic acid, guaiacol, syringol, syringyl acid, gallic acid and 4-vinylphenol. Remarkably, the reactions were carried out at room temperature, and it was accomplished in a series of experiments in order to determine the optimum solvent and pH.[105]

2.6 Photocatalysis

Photocatalytic reactions are another field of catalysis, which could be defined as a chemical reaction catalyzed by a solid semiconductor material under the action of light irradiation (including ultraviolet, visible and infrared irradiation). As a consequence of light irradiation, the semiconductor material gets photo-excited, generating electron and hole pairs, which can promote the initiation of a wide range of chemical reactions on the surface of the photocatalytic sample.[36,106,107] Importantly, the sustainable character of a photocatalytic process lies in several factors, including (a) the use of green protocols and starting materials for the synthesis of the photocatalyst; (b) the use of efficient light sources, and preferably inexhaustible illumination like solar irradiation; and (c) the use of photocatalysis in environmentally friendly applications, such as biomass valorization.[16,108]

The design of an optimum catalytic material needs to consider the chemical, physical and mechanical features of the resulting structure. Going beyond, the development of efficient photocatalyst also requires taking into account the electronic and optical properties of the samples, which will play a vital role on the catalytic application. Such concept was initially presented by Anderson et al. and extended to photo-electro-catalysis by our research group, as shown in Fig. 11.[16,106]

The use of household waste for the design of photocatalytic systems is a very promising option, which have been described in details by our research group in a recent contribution (Fig. 12). In that work, the main photo-active materials, including metal oxides, especially TiO_2 and ZnO, metal sulfides and carbon-based materials like graphitic carbonitrile, among others, were described, as well as the myriad possibilities for their preparation from wastes.[108]

Specially, titanium has been recognized as the most investigated semiconductor material. In this regard our research group has proposed several strategies for its greener preparation and for its modification and further applications in a variety of organic reactions.[107] For instance, Ouyang and

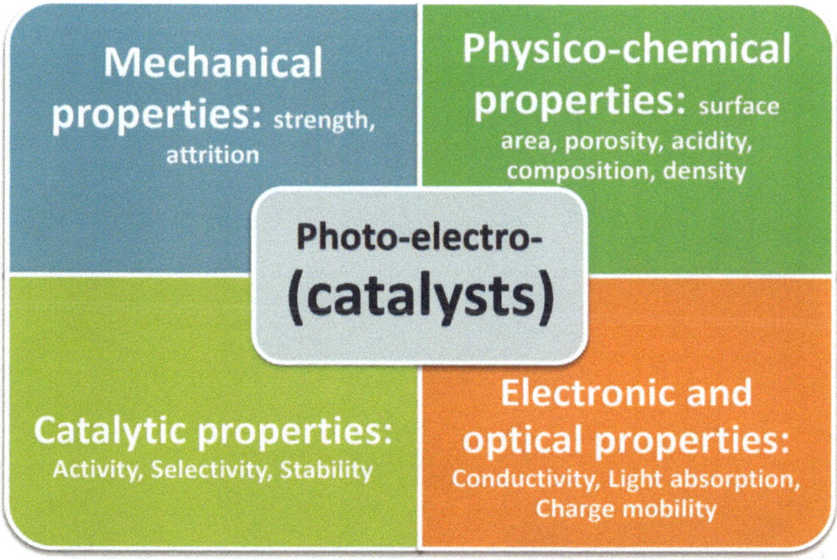

Fig. 11 Square concept including the material features that need to be considered for (photo-electro)-catalysts design. *Reprinted with permission from Rodríguez-Padrón, D.; Puente-Santiago, A.R.; Balu, A.M.; Muñoz-Batista, M.J.; Luque, R.* ChemCatChem ***2019,*** 11 *(1), 18–38. Copyright (2020) American Chemical Society.*

coworkers have studied the deposition of ruthenium nanoparticles on a pure anatase titanium and its application on the production of hydrogen from water-methanol mixtures (Fig. 13A). The authors have examined the effect of the light source, by testing ultraviolet and visible light irradiation. In this case, the best catalytic activity was observed for the sample modified with 3 wt% of ruthenium, being well correlated with the material properties and with the quantum efficiency of the sample.[109]

Titanium-based materials have been also employed for the preparation of composite materials, though an ultrasound-assisted approach, combining titanium with maghemite and silica structures. Remarkably, this procedure developed by Colmenares et al., gave rise to the formation of magnetically separable photocatalytic active samples. Moreover, the formation of hetero-junctions in the designed composite material can lead to an enhancement of the light absorption properties and avoid the recombination processes of the photo-generated electron-hole pairs. Such samples were tested in the oxidation of benzyl alcohol to benzaldehyde under UV-light irradiation (Fig. 13B), exhibiting an outstanding selectivity of 90% toward the desired product, with conversion values around 50%.[110] Similarly, materials based

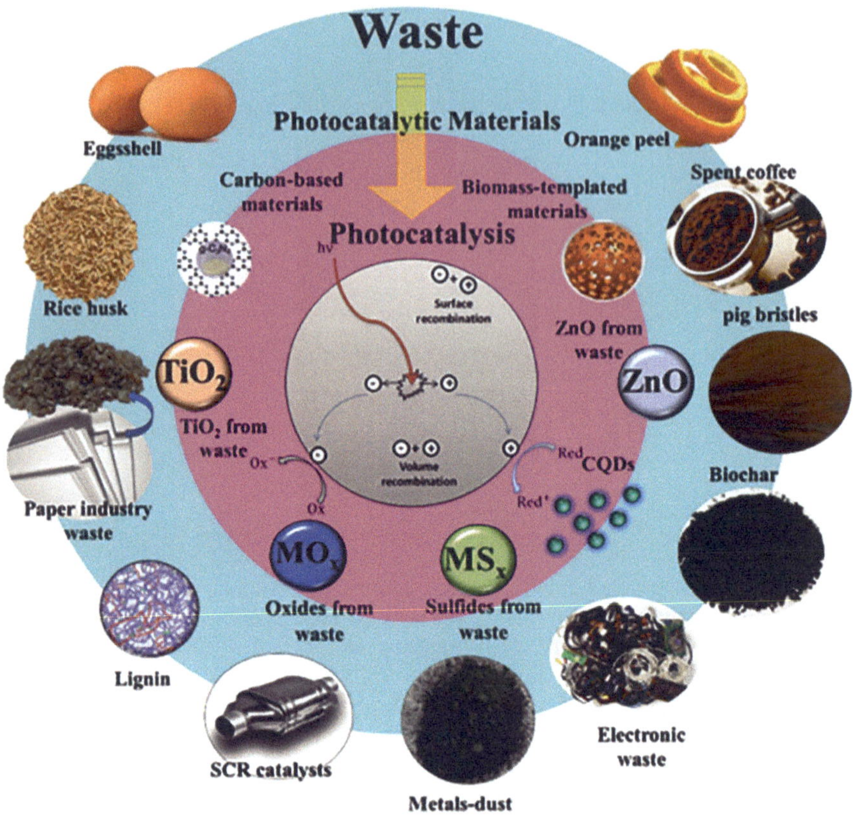

Fig. 12 Schematic illustration of waste valorization toward photocatalytic materials. *Reprinted with permission from Rodríguez-Padrón, D.; Luque, R.; Muñoz-Batista, M.J.* Top. Curr. Chem. ***2020,*** 378 *(1), 3. Copyright (2020) Springer.*

on titanium over magnetic iron oxide-silica materials have been also synthesized following a mechanochemical approach, employing different titanium loadings, from 0.5 to 5 wt%. Such catalysts displayed improved optical properties, with light absorption in the visible range. The photocatalytic activity of the samples was investigated in the oxidation of benzyl alcohol to benzaldehyde, showing conversion values of 20% and high benzaldehyde selectivity of 84%, in particular for the material modified with 5% of titanium.[111]

Agro-industrial residues have been also employed by our research group for the development of photocatalytic materials. For example, Ouyang and coworkers, have proposed the templated synthesis of nanostructured titanium employing wheat bran, which is a residue generated in enormous quantities during the milling of wheat grain to produce white wheat four.

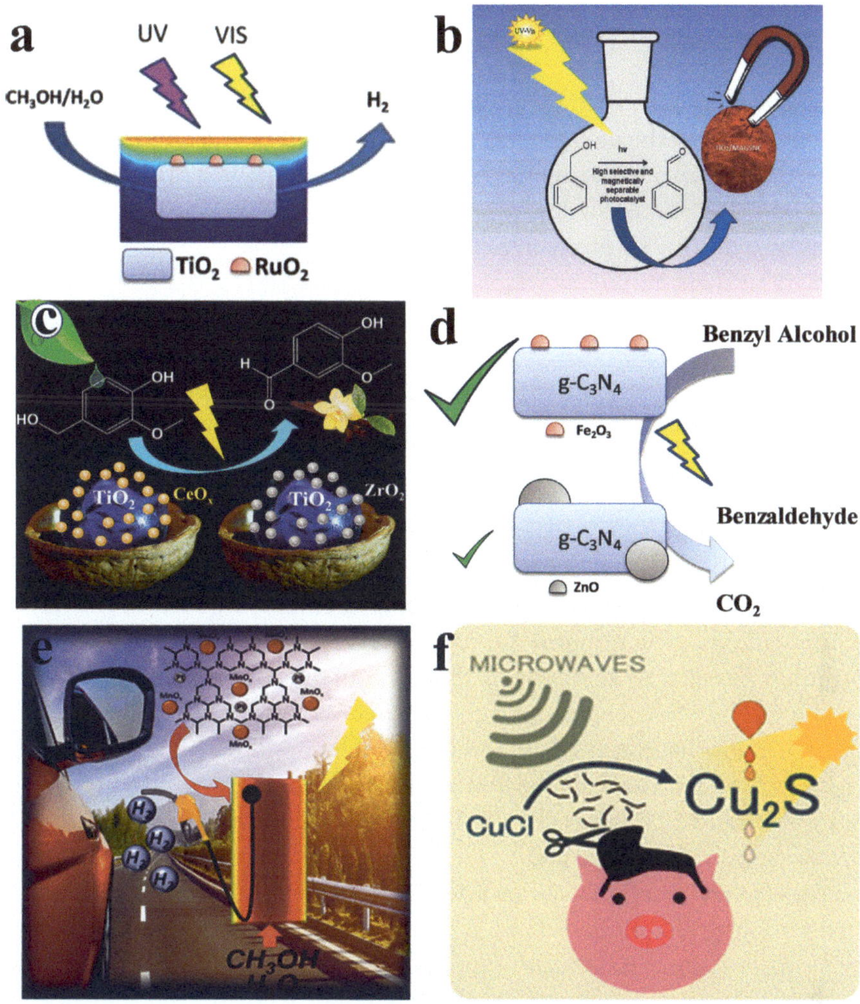

Fig. 13 (A) Photocatalytic hydrogen production over ruthenium modified titanium; (B) benzyl alcohol photo-oxidation over titanium-maghemite-silica composites; (C) walnut shell-templated titanium photocatalyst, modified with Ce and Zr, for vanillin production; (D) benzyl alcohol photo-oxidation over graphitic carbonitrile modified samples; (E) H_2 production employing platinum and manganese oxide supported on g-C_3N_4; and (F) pig bristles derived Cu_2S for dyes photo-degradation. *Panel (A): Reprinted with permission from Ouyang, W.; Munoz-Batista, M.J.; Kubacka, A.; Luque, R.; Fernández-García, M.* Appl. Catal. Environ. *2018,* 238, *434–443. Copyright (2020) Elsevier. Panel (B): Reprinted with permission from Colmenares, J.C.; Ouyang, W.; Ojeda, M.; Kuna, E.; Chernyayeva, O.; Lisovytskiy, D.; De, S.; Luque, R.; Balu, A.M.* Appl. Catal. Environ. *2016,* 183, *107–112. Copyright (2020) Elsevier. Panel (C): Reprinted with permission from Martín-Perales, A.I.; Rodríguez-Padrón, D.; García Coleto, A.; Len, C.; de Miguel, G.; Muñoz-Batista, M.J.; Luque, R.* Ind. Eng. Chem. Res. ***2020,*** 59 *(39), 17085–17093. Copyright (2020) American Chemical Society. Panel (D): Reprinted with permission from Cerdan, K.; Ouyang, W.; Colmenares, J.C.; Muñoz-Batista, M.J.; Luque, R.; Balu, A.M.* Chem. Eng. Sci. ***2019,*** 194, *78–84. Copyright (2020) Elsevier. Panel (E): Reprinted with permission from Muñoz-Batista, M.J.; Rodríguez-Padrón, D.; Puente-Santiago, A.R.; Kubacka, A.; Luque, R.; Fernández-García, M.* ChemPhotoChem ***2018,*** 2 *(10), 870–877. Copyright (2020) Wiley. Panel (F): Reprinted with permission from Zuliani, A.; Muñoz-Batista, M.J.; Luque, R.* Green Chem. ***2018,*** 20 *(13), 3001–3007. Copyright (2020) Royal Society of Chemistry.*

Such waste is mainly used for animal feed with still underexploited possibilities. In this case, the valorization of such residue toward nanomaterials was carried out by mechanochemical protocols. The obtained samples were, as well, tested in the oxidation reaction of benzyl alcohol to benzaldehyde, giving rise to 20% of product yield and 33% of conversion.[112]

Walnut shell wastes have also served as inspiration for the design of titanium nanomaterials with controlled morphology and textural properties. In this regard, Martín-Perales and collaborators have proposed a mechanochemical strategy for the preparation of titanium samples, employing walnut waste residues as sacrificial template (Fig. 13C). More interestingly, in this work the authors studied the influence of the incorporation of ceria and zirconia entities on the optical properties, and consequently in the photocatalytic behavior. The materials were tested in the photo-oxidation of vanillyl alcohol to vanillin, using a LED 12 V lamp, with a wavelength of 395 nm. Two series of samples were prepared employing cerium and zirconium loadings, ranging from 1% to 5%, respectively. In particular, for the cerium series, it was observed that the Ce content also has a direct influence on the Ce(IV)/Ce(III) ratio, which, at the same time, has an impact on the catalytic performance. It was found that samples modified with 1 wt% of Ce and 5 wt% of Zr, exhibited the best catalytic activity, with conversion values of ca. 40% and 52%, respectively. In all cases a selectivity higher that 99% toward vanillin was achieved.[113]

Moreover, graphitic carbonitrile-based materials have been recently raised as potential photocatalytic systems. These catalysts could be obtained by a simple thermal treatment of nitrogen-rich precursor, such as melamine, urea and thiourea among others. In this sense, Cerdan et al. described a one-pot solvent-less mechanochemical modification of g-C_3N_4 with ZnO and Fe_2O_3. The catalytic activity of the prepared samples was examined in the valorization of benzyl alcohol to benzaldehyde (Fig. 13D). In particular, the sample modified with iron oxide, displayed the best performance, with 20% of conversion and 70% of benzaldehyde selectivity. Importantly, the studied modification of the g-C_3N_4 surface, lead to improved results, in comparison, not just with the parent g-C_3N_4 material, but also with commercial TiO_2.[114]

The g-C_3N_4 based photo-catalysts could certainly offer a broad range of possibilities. For instance, Muñoz-Batista et al. described the synthesis of highly active samples for hydrogen production under sunlight irradiation, by simple modifications of the g-C_3N_4 surface (Fig. 13E). Remarkably, the samples were subjected to an exfoliation process, which results not only in modification of the textural, but also of the optical properties, favoring the absorption of visible light. MnO_x and metallic Pt entities were further incorporated by an impregnation method, and especially the presence of platinum proved to be crucial for the successful progress of the reaction. The ternary catalytic systems showed the best results with a high quantum efficiency, most likely attributed to the modulation of charge carriers along the reaction.[115]

Recently, our research group has also investigated the modification of g-C_3N_4 with boron nanoparticles, looking forward to favoring the absorption properties under sunlight irradiation. In this direction, Caudillo-Flores et al. described an interesting approach, combining continuous flow regime and photocatalysis to produce acetone from 2-propanol. A series of samples were prepared by mechanochemically incorporation of different boron loadings, from 1 to 10 wt%. Interestingly, it was found that up to 2.5 wt%, the samples were formed by boron supported on the g-C_3N_4, while at higher loadings, the formation of doped systems was observed. The optimum catalytic activity was reached for the samples modified with 2.5 wt% of boron.[116]

Sulfides have also been described as feasible materials for photocatalytic applications. Specially, ZnS and CdS have been widely investigated. In addition, other metal sulfides have also showed photocatalytic response, such as copper sulfides. Zuliani and coworkers have deeply investigated the potentialities of pig bristles for the preparation of sulfide-based materials, through sustainable pathways, frequently involving microwave treatment and green solvents like ethylene glycol (Fig. 13F). Such methodology has been employed for the synthesis of copper sulfide composite materials. The aforementioned samples have been used as photocatalytic systems, under green, blue, and red visible light irradiation for dyes degradation, providing new opportunities for water treatments.[117]

3. Conclusions and future perspectives

In summary, biomass valorization could be attempted from different perspectives, for the synthesis of high added-value chemicals, but also for the preparation of advanced nanomaterials with a wide range of applications. Importantly, to follow a catalytic approach could deeply influence the efficiency of a process, in terms of atom, cost and energy efficiencies. This relevant field still requires a lot of effort in order to be translated to the industrial scale and to further make possible the development of the bio-refinery as a replacement of the petro-based industry. Therefore, biomass valorization approaches employing new technologies, such as photocatalysis, continuous flow regime, microwave and ultrasound irradiations, need to be further investigated. We hope that the present contribution serves also as inspiration for the scientific community, to continue working on this topic, which is extremely crucial for the sustainable development of the world and for the environment and human safety.

Funding sources

MINECO funding project CTQ2016-78289-P, co-financed with FEDER funds'.

Acknowledgments

R.L. gratefully acknowledges MINECO for funding under project CTQ2016-78289-P, co-financed with FEDER funds and the contract for Daily Rodríguez-Padrón associated to this project. The publication has been prepared with support from RUDN University program 5-100.

References

1. Verma, N.; Kumar, V.; Bansal, M. C. *Waste Biomass Valoriz.* **2020**.
2. Kumar, R. P.; Bharathiraja, B.; Kataki, R.; Moholkar, V. S., Eds. *Biomass Valorization to Bioenergy*; Springer, 2020.
3. Sudarsanam, P.; Peeters, E.; Makshina, E. V.; Parvulescu, V. I.; Sels, B. F. *Chem. Soc. Rev.* **2019**, *48* (8), 2366–2421.
4. Margellou, A.; Triantafyllidis, K. S. *Catalysts* **2019**, *9* (1), 43.
5. Guan, W.; Tsang, C. W.; Lin, C. S. K.; Len, C.; Hu, H.; Liang, C. *Bioresour. Technol.* **2020**, *298*, 122432.
6. Gazi, S. *Appl. Catal. Environ.* **2019**, *257*, 117936.
7. Espro, C.; Mauriello, F. *Catalysts* **2019**, *9* (8), 674.
8. Zadeh, Z. E.; Abdulkhani, A.; Aboelazayem, O.; Saha, B. *Processes* **2020**, *8* (7), 799.
9. Xu, C.; Paone, E.; Rodríguez-Padrón, D.; Luque, R.; Mauriello, F. *Renew. Sustain. Energy Rev.* **2020**, *127*, 109852.
10. Luque, R.; Len, C.; Triantafyllidis, K. *Front. Chem.* **2018**, *6*, 577.
11. Costa, F. F.; de Oliveira, D. T.; Brito, Y. P.; da Rocha Filho, G. N.; Alvarado, C. G.; Balu, A. M.; Luque, R.; do Nascimento, L. A. S. *Curr. Opin. Green Sustain Chem.* **2020**, *24*, 21–25.
12. Xu, C.; Nasrollahzadeh, M.; Selva, M.; Issaabadi, Z.; Luque, R. *Chem. Soc. Rev.* **2019**, *48* (18), 4791–4822.
13. Li, H.; Riisager, A.; Saravanamurugan, S.; Pandey, A.; Sangwan, R. S.; Yang, S.; Luque, R. *ACS Catal.* **2018**, *8* (1), 148–187.
14. Yang, D. P.; Li, Z.; Liu, M.; Zhang, X.; Chen, Y.; Xue, H.; Ye, E.; Luque, R. *ACS Sustain Chem. Eng.* **2019**, *7* (5), 4564–4585.
15. Gómez-López, P.; Puente-Santiago, A.; Castro-Beltrán, A.; do Nacimiento, L. A. S.; Balu, A. M.; Luque, R.; Alvarado-Beltrán, C. G. *Curr. Opin. Green Sustain. Chem.* **2020**, *24*, 48–55.
16. Rodríguez-Padrón, D.; Puente-Santiago, A. R.; Balu, A. M.; Muñoz-Batista, M. J.; Luque, R. *ChemCatChem* **2019**, *11* (1), 18–38.

17. Xu, C.; Nasrollahzadeh, M.; Sajjadi, M.; Maham, M.; Luque, R.; Puente-Santiago, A. R. *Renew. Sustain. Energy Rev.* **2019**, *112*, 195–252.
18. Filiciotto, L.; Balu, A. M.; Van der Waal, J. C.; Luque, R. *Catal. Today* **2018**, *302*, 2–15.
19. Das, B.; Mohanty, K. *Renew. Energy* **2019**, *143*, 1791–1811.
20. Li, A. Y.; Moores, A. *ACS Sustain. Chem. Eng.* **2019**, 7 (12), 10182–10197.
21. Mahlia, T. M. I.; Syazmi, Z. A. H. S.; Mofijur, M.; Abas, A. P.; Bilad, M. R.; Ong, H. C.; Silitonga, A. S. *Renew. Sustain. Energy Rev.* **2020**, *118*, 109526.
22. Bavykina, A.; Kolobov, N.; Khan, I. S.; Bau, J. A.; Ramirez, A.; Gascon, J. *Chem. Rev.* **2020**, *120*, 8468–8535.
23. Shen, K.; Zhang, L.; Chen, X.; Liu, L.; Zhang, D.; Han, Y.; Chen, J.; Long, J.; Luque, R.; Li, Y.; Chen, B. *Science* **2018**, *359* (6372), 206–210.
24. Mark, L. O.; Cendejas, M. C.; Hermans, I. *ChemSusChem* **2020**, *13*, 5773.
25. Ndolomingo, M. J.; Bingwa, N.; Meijboom, R. *J. Mater. Sci.* **2020**, *55*, 1–47.
26. Kumar, M.; Xiong, X.; Sun, Y.; Yu, I. K.; Tsang, D. C.; Hou, D.; Gupta, J.; Bhaskar, T.; Pandey, A. *Adv. Sustain. Syst.* **2020**, *21*, 1900149.
27. Nasrollahzadeh, M.; Soheili Bidgoli, N. S.; Shafiei, N.; Soleimani, F.; Nezafat, Z.; Luque, R. *Biofuel. Bioprod. Biorefin.* **2020**, *14*, 1197–1227.
28. Lin, Y. C.; Amesho, K. T.; Chen, C. E.; Cheng, P. C.; Chou, F. C. *Sustain. Chem. Pharm.* **2020**, *17*, 100310.
29. Li, Z.; Dong, J.; Zhang, H.; Zhang, Y.; Wang, H.; Cui, X.; Wang, Z. *Nanoscale Adv.* **2020**.
30. Hiremath, L.; Nipun, S.; Sruti, O.; Kala, N. G.; Aishwarya, B. M. Sonochemistry: applications in biotechnology. In *Sonochemical Reactions*; IntechOpen, 2020.
31. Palma, V.; Barba, D.; Cortese, M.; Martino, M.; Renda, S.; Meloni, E. *Catalysts* **2020**, *10* (2), 246.
32. Mishra, S.; Kumar, P.; Samanta, S. K. *Ind. Eng. Chem. Res.* **2020**, *59* (36), 15839–15847.
33. Meng, X.; Eluagwule, B.; Wang, M.; Wang, L.; Zhang, J. *Handbook of Smart Photocatalytic Materials*; Elsevier, 2020; pp. 183–195.
34. da Costa Filho, B. M.; Araujo, A. L.; Padrão, S. P.; Boaventura, R. A.; Dias, M. M.; Lopes, J. C. B.; Vilar, V. J. *Chem. Eng. J.* **2019**, *366*, 560–568.
35. Tang, S.; Wang, Z.; Yuan, D.; Zhang, Y.; Qi, J.; Rao, Y.; Lu Li, G. B.; Wang, K.; Yin, K. *Int. J. Electrochem. Sci.* **2020**, *15* (3), 2470–2480.
36. Nair, V.; Muñoz-Batista, M. J.; Fernández-García, M.; Luque, R.; Colmenares, J. C. *ChemSusChem* **2019**, *12* (10), 2098–2116.
37. Colella, M.; Carlucci, C.; Luisi, R. *Accounts on Sustainable Flow Chemistry*; Springer: Cham, 2020; pp. 29–65.
38. Hommes, A.; Heeres, H. J.; Yue, J. *ChemCatChem* **2019**, *11* (19), 4671–4708.
39. Bennett, J. A.; Campbell, Z. S.; Abolhasani, M. *Curr. Opin. Chem. Eng.* **2019**, *26*, 9–19.
40. Gerardy, R.; Debecker, D. P.; Estager, J.; Luis, P.; Monbaliu, J. C. M. *Chem. Rev.* **2020**, *120* (15), 7219–7347.
41. Wang, Y.; Zhao, D.; Rodríguez-Padrón, D.; Len, C. *Catalysts* **2019**, *9* (10), 796.
42. Roberts, E. J.; Karadaghi, L. R.; Wang, L.; Malmstadt, N.; Brutchey, R. L. *ACS Appl. Mater. Interfaces* **2019**, *11* (31), 27479–27502.
43. Friščić, T.; Mottillo, C.; Titi, H. M. *Angew. Chem. Int. Ed.* **2020**, *59* (3), 1018–1029.
44. Tan, D.; García, F. *Chem. Soc. Rev.* **2019**, *48* (8), 2274–2292.
45. Andersen, J.; Mack, J. *Green Chem.* **2018**, *20* (7), 1435–1443.
46. Muñoz-Batista, M. J.; Rodriguez-Padron, D.; Puente-Santiago, A. R.; Luque, R. *ACS Sustain. Chem. Eng.* **2018**, *6* (8), 9530–9544.
47. Brandt, A.; Gräsvik, J.; Hallett, J. P.; Welton, T. *Green Chem.* **2013**, *15* (3), 550–583.

48. Binder, J. B.; Raines, R. T. *J. Am. Chem. Soc.* **2009**, *131* (5), 1979–1985.
49. Liu, X.; Duan, X.; Wei, W.; Wang, S.; Ni, B. J. *Green Chem.* **2019**, *21* (16), 4266–4289.
50. Wu, X.; Luo, N.; Xie, S.; Zhang, H.; Zhang, Q.; Wang, F.; Wang, Y. *Chem. Soc. Rev.* **2020**, *49* (17), 6198–6223.
51. Choe, B.; Lee, S.; Won, W. *Renew. Energy* **2020**, *162*, 242–248.
52. Gómez-López, P.; Lázaro, N.; Alvarado-Beltrán, C. G.; Pineda, A.; Balu, A. M.; Luque, R. *Molecules* **2019**, *24* (21), 3985.
53. Ostovar, S.; Franco, A.; Puente-Santiago, A. R.; Pinilla-de Dios, M.; Rodríguez-Padrón, D.; Shaterian, H. R.; Luque, R. *Front. Chem.* **2018**, *6*, 77.
54. Franco, A.; De, S.; Balu, A. M.; Romero, A. A.; Luque, R. *ChemistrySelect* **2017**, *2* (29), 9546–9551.
55. Franco, A.; Fernandes de Souza, J.; Pinheiro do Nascimiento, P. F.; Mendes Pedroza, M.; de Carvalho, L. S.; Rodriguez-Castellón, E.; Luque, R. *ACS Sustain. Chem. Eng.* **2019**, *7* (8), 7519–7526.
56. Zhang, Q.; Yang, X.; Guan, J. *ACS Appl. Nano Mater.* **2019**, *2* (8), 4681–4697.
57. Rodriguez-Padron, D.; Munoz-Batista, M. J.; Li, H.; Shih, K.; Balu, A. M.; Pineda, A.; Luque, R. *ACS Sustain. Chem. Eng.* **2019**, *7* (20), 17030–17038.
58. Saberi, F.; Rodríguez-Padrón, D.; Doustkhah, E.; Ostovar, S.; Franco, A.; Shaterian, H. R.; Luque, R. *Catal. Commun.* **2019**, *118*, 65–69.
59. Márquez-Medina, M. D.; Rodríguez-Padrón, D.; Balu, A. M.; Romero, A. A.; Muñoz-Batista, M. J.; Luque, R. *Catalysts* **2019**, *9* (3), 290.
60. Ostovar, S.; Rodríguez-Padrón, D.; Saberi, F.; Balu, A. M.; Luque, R. *Catalysts* **2019**, *9* (4), 348.
61. Saberi, F.; Rodriguez-Padrón, D.; Garcia, A.; Shaterian, H. R.; Luque, R. *Catalysts* **2018**, *8* (4), 167.
62. Rodríguez-Padrón, D.; Zhao, D.; Carrillo-Carrion, C.; Morales-Torres, C.; Elsharif, A. M.; Balu, A. M.; Luque, R.; Len, C. *Catal. Today* **2020**.
63. Baumann, M.; Moody, T. S.; Smyth, M.; Wharry, S. *Org. Process Res. Dev.* **2020**, *24*, 1802–1813.
64. Munirathinam, R.; Huskens, J.; Verboom, W. *Adv. Synth. Catal.* **2015**, *357* (6), 1093–1123.
65. Ouyang, W.; Zhao, D.; Wang, Y.; Balu, A. M.; Len, C.; Luque, R. *ACS Sustain. Chem. Eng.* **2018**, *6* (5), 6746–6752.
66. Lázaro, N.; Franco, A.; Ouyang, W.; Balu, A. M.; Romero, A. A.; Luque, R.; Pineda, A. *Catalysts* **2019**, *9* (2), 142.
67. Cabanillas, M.; Franco, A.; Lázaro, N.; Balu, A. M.; Luque, R.; Pineda, A. *Mol. Catal.* **2019**, *477*, 110522.
68. Xu, C.; Ouyang, W.; Muñoz-Batista, M. J.; Fernández-García, M.; Luque, R. *ChemSusChem* **2018**, *11* (15), 2604–2611.
69. Zhao, D.; Prinsen, P.; Wang, Y.; Ouyang, W.; Delbecq, F.; Len, C.; Luque, R. *ACS Sustain. Chem. Eng.* **2018**, *6* (5), 6901–6909.
70. Mhadmhan, S.; Franco, A.; Pineda, A.; Reubroycharoen, P.; Luque, R. *ACS Sustain. Chem. Eng.* **2019**, *7* (16), 14210–14216.
71. Ouyang, W.; Yepez, A.; Romero, A. A.; Luque, R. *Catal. Today* **2018**, *308*, 32–37.
72. Zuliani, A.; Cova, C. M.; Manno, R.; Sebastian, V.; Romero, A. A.; Luque, R. *Green Chem.* **2020**, *22* (2), 379–387.
73. Cova, C. M.; Zuliani, A.; Manno, R.; Sebastian, V.; Luque, R. *Green Chem.* **2020**, *22* (4), 1414–1423.
74. Cova, C. M.; Zuliani, A.; Muñoz-Batista, M. J.; Luque, R. *Green Chem.* **2019**, *21* (17), 4712–4722.

75. Filiciotto, L.; Márquez-Medina, M. D.; Pineda, A.; Balu, A. M.; Romero, A. A.; Angelici, C.; de Jong, E.; van der Waal, J. C.; Luque, R. *Catal. Today* **2019**.
76. Shi, N.; Liu, Q.; Cen, H.; Ju, R.; He, X.; Ma, L. *Biomass Convers. Biorefin.* **2020**, *10* (2), 277–287.
77. Filiciotto, L.; Balu, A. M.; Romero, A. A.; Angelici, C.; van der Waal, J. C.; Luque, R. *Mol. Catal.* **2019**, *479*, 110564.
78. Pfab, E.; Filiciotto, L.; Romero, A. A.; Luque, R. *Ind. Eng. Chem. Res.* **2019**, *58* (35), 16065–16070.
79. Rodriguez-Padron, D.; Puente-Santiago, A. R.; Balu, A. M.; Romero, A. A.; Muñoz-Batista, M. J.; Luque, R. *ACS Sustain. Chem. Eng.* **2018**, *6* (12), 16637–16644.
80. Rodríguez-Padrón, D.; Puente-Santiago, A. R.; Balu, A. M.; Muñoz-Batista, M. J.; Luque, R. *Front. Chem.* **2019**, 7, 103.
81. Altuğ, C.; Muñoz-Batista, M. J.; Rodríguez-Padrón, D.; Balu, A. M.; Romero, A. A.; Luque, R. *Green Chem.* **2019**, *21* (2), 300–306.
82. Defilippi, C.; Rodríguez-Padrón, D.; Luque, R.; Giordano, C. *Green Chem.* **2020**, *22* (9), 2929–2934.
83. Ji, T.; Zhu, J. *Responsive Nanomaterials for Sustainable Applications*; Springer: Cham, 2020; pp. 65–91.
84. Zuliani, A.; Balu, A. M.; Luque, R. *ACS Sustain. Chem. Eng.* **2017**, *5* (12), 11584–11587.
85. Filiciotto, L.; Balu, A. M.; Romero, A. A.; Rodríguez-Castellón, E.; van der Waal, J. C.; Luque, R. *Green Chem.* **2017**, *19* (18), 4423–4434.
86. Wang, Y.; Zhao, D.; Triantafyllidis, K. S.; Ouyang, W.; Luque, R.; Len, C. *Mol. Catal.* **2020**, *480*, 110630.
87. Zhao, D.; Su, T.; Wang, Y.; Varma, R. S.; Len, C. *ACS Sustain. Chem. Eng.* **2020**, *8* (23), 8486–8495.
88. Zhao, D.; Rodriguez-Padron, D.; Triantafyllidis, K. S.; Wang, Y.; Luque, R.; Len, C. *ACS Sustain. Chem. Eng.* **2020**, *8* (8), 3091–3102.
89. Rincon, E.; Zuliani, A.; Jimenez-Quero, A.; Vilaplana, F.; Luque, R.; Serrano, L.; Balu, A. M. *ACS Sustain. Chem. Eng.* **2020**, *8* (29), 11016–11023.
90. Cova, C. M.; Zuliani, A.; Muñoz-Batista, M. J.; Luque, R. *ACS Sustain. Chem. Eng.* **2018**, 7 (1), 1300–1307.
91. Franco, A.; De, S.; Balu, A. M.; Romero, A. A.; Luque, R. *ACS Sustain. Chem. Eng.* **2018**, *6* (9), 11555–11562.
92. Rodríguez-Padrón, D.; Algarra, M.; Tarelho, L. A.; Frade, J.; Franco, A.; de Miguel, G.; Jimenez, J.; Rodríguez-Castellon, E.; Luque, R. *ACS Sustain. Chem. Eng.* **2018**, *6* (6), 7200–7205.
93. Chatel, G. *Ultrason. Sonochem.* **2018**, *40*, 117–122.
94. Wood, R. J.; Lee, J.; Bussemaker, M. J. *Ultrason. Sonochem.* **2017**, *38*, 351–370.
95. Chatel, G. *Sonochemistry: New Opportunities for Green Chemistry*; World Scientific Publishing Company, 2017.
96. Lupacchini, M.; Mascitti, A.; Giachi, G.; Tonucci, L.; d'Alessandro, N.; Martinez, J.; Colacino, E. *Tetrahedron* **2017**, *73* (6), 609–653.
97. Chatel, G.; Colmenares, J. C. *Sonochemistry: From Basic Principles to Innovative Applications*; Springer, 2017.
98. Cebrián-García, S.; Balu, A. M.; Luque, R. *Front. Chem.* **2018**, *6*, 197.
99. Martina, K.; Calsolaro, F.; Zuliani, A.; Berlier, G.; Chávez-Rivas, F.; Moran, M. J.; Luque, R.; Cravotto, G. *Molecules* **2019**, *24* (13), 2490.
100. Bilal, M.; Zhao, Y.; Rasheed, T.; Iqbal, H. M. *Int. J. Biol. Macromol.* **2018**, *120*, 2530–2544.
101. Liang, S.; Wu, X. L.; Xiong, J.; Zong, M. H.; Lou, W. Y. *Coord. Chem. Rev.* **2020**, *406*, 213149.

102. Girelli, A. M.; Astolfi, M. L.; Scuto, F. R. *Chemosphere* **2020**, *244*, 125368.
103. Thangaraj, B.; Solomon, P. R. *ChemBioEng Rev.* **2019**, *6* (5), 157–166.
104. Cebrián-García, S.; Balu, A. M.; García, A.; Luque, R. *Molecules* **2018**, *23* (9), 2283.
105. Pistone, L.; Ottolina, G.; De, S.; Romero, A. A.; Martins, L. O.; Luque, R. *ChemSusChem* **2016**, *9* (7), 756–762.
106. Anderson, J. B. F.; Griffin, K. G.; Richards, R. E. *Chem. Tech.* **1989**, *18*, 40–44.
107. Guo, Q.; Zhou, C.; Ma, Z.; Yang, X. *Adv. Mater.* **2019**, *31* (50), 1901997.
108. Rodríguez-Padrón, D.; Luque, R.; Muñoz-Batista, M. J. *Top. Curr. Chem.* **2020**, *378* (1), 3.
109. Ouyang, W.; Munoz-Batista, M. J.; Kubacka, A.; Luque, R.; Fernández-García, M. *Appl. Catal. Environ.* **2018**, *238*, 434–443.
110. Colmenares, J. C.; Ouyang, W.; Ojeda, M.; Kuna, E.; Chernyayeva, O.; Lisovytskiy, D.; De, S.; Luque, R.; Balu, A. M. *Appl. Catal. Environ.* **2016**, *183*, 107–112.
111. Ouyang, W.; Kuna, E.; Yepez, A.; Balu, A. M.; Romero, A. A.; Colmenares, J. C.; Luque, R. *Nanomaterials* **2016**, *6* (5), 93.
112. Ouyang, W.; Reina, J. M.; Kuna, E.; Yepez, A.; Balu, A. M.; Romero, A. A.; Colmenares, J. C.; Luque, R. *J. Environ. Manage.* **2017**, *203*, 768–773.
113. Martín-Perales, A. I.; Rodríguez-Padrón, D.; García Coleto, A.; Len, C.; de Miguel, G.; Muñoz-Batista, M. J.; Luque, R. *Ind. Eng. Chem. Res.* **2020**, *59* (39), 17085–17093.
114. Cerdan, K.; Ouyang, W.; Colmenares, J. C.; Muñoz-Batista, M. J.; Luque, R.; Balu, A. M. *Chem. Eng. Sci.* **2019**, *194*, 78–84.
115. Muñoz-Batista, M. J.; Rodríguez-Padrón, D.; Puente-Santiago, A. R.; Kubacka, A.; Luque, R.; Fernández-García, M. *ChemPhotoChem* **2018**, *2* (10), 870–877.
116. Caudillo-Flores, U.; Rodríguez-Padrón, D.; Muñoz-Batista, M. J.; Kubacka, A.; Luque, R.; Fernández-García, M. *Green Chem.* **2020**, *22* (15), 4975–4984.
117. Zuliani, A.; Muñoz-Batista, M. J.; Luque, R. *Green Chem.* **2018**, *20* (13), 3001–3007.

CHAPTER THREE

The thousand faces of Cu-doped porous mixed oxides (Cu-PMO) in the conversion of renewable resources and beyond

Mario De Bruyn[a], Zhuohua Sun[b], and Katalin Barta[a,*]
[a]Institute for Chemistry, Organic and Bioorganic Chemistry, University of Graz, Graz, Austria
[b]Beijing Advanced Innovation Center for Tree Breeding by Molecular Design, Beijing Key Laboratory of Lignocellulosic Chemistry, Beijing Forestry University, Beijing, PR China
*Corresponding author: e-mail address: katalin.barta@uni-graz.at

Contents

Advances in Inorganic Chemistry, Volume 77
ISSN 0898-8838
https://doi.org/10.1016/bs.adioch.2020.12.002

Abstract

Copper doped porous metal oxides derived from synthetic hydrotalcites are an intriguing, noble metal free class of solid bases that have shown remarkable properties in the field of biomass valorization and beyond. Notably, given their activity in several reductive catalytic events, these classes of compounds are suitable catalysts for the depolymerization and further deoxygenation of lignocellulosic biomass and derived components. Notably, copper doped porous metal oxides have shown activity in methanol reforming under higher temperatures, thus the combination with supercritical methanol provides a unique catalytic environment for the very efficient processing of a wide range of renewable resources, platform chemicals but also raw lignocellulose. In this chapter, we describe our recent advances on the use of copper doped porous metal oxides of various compositions, including also other metal dopants, in various areas of biomass conversion chemistry and beyond.

Abbreviations

1256HT	(S,R)/(S,S) hexane-1,2,5,6-tetrol
1,6-HD	1,6-hexanediol
AFEX	ammonia fiber expansion
BPE	benzyl phenyl ether
COVID-19	SARS-CoV-2 (2019-nCoV) coronavirus
CPME	cyclopentyl methyl ether
Cu-PMO	Cu-doped porous mixed oxides
Cyrene™	dihydrolevoglucosenone
DA	dilute sulfuric acid treatment
DES	deep eutectic solvent
DHBF	dihydrobenzofuran
DHCA	dihydroconiferyl alcohol
DHSA	4-(3-hydroxypropyl)-2,6-dimethoxy-phenol
DMC	dimethylcarbonate
DMF	dimethylfuran
DMTHF	dimethyltetrahydrofuran
endo/exo LGOL	endo/*exo*-levoglucosanol
FA	formic acid
FDM	2,5-bis(hydroxymethyl)furan
FT-ICR	Fourier-transform ion cyclotron resonance mass spectrometry
HDO	hydrodeoxygenation
HTO	1,2,6-hexanetriol
HMF	5-hydroxymethylfurfural
LA	levulinic acid
LGO	levoglucosenone
LHW	liquid hot water
MIC	minimum inhibitory concentration
MFM	(5-methyl-2-furyl)methanol
MW	molecular weight
NADES	natural deep eutectic solvent
NMP	*N*-methylpyrrolidone
PMOs	porous mixed oxides

PPEO	2-phenoxy-1-phenylethane-1-one
SE	steam explosion
TEM	transmission electron microscope
THFDM	[5-(hydroxymethyl)tetrahydrofuran-2-yl]methanol
TRL	technology readiness level
TRPV1	transient receptor potential vanilloid 1
UCSB	University of California Santa Barbara
USY	ultrastable Y zeolite
WGSR	water-gas-shift-reaction
XRD	X-ray diffraction
XPS	X-ray photoelectron spectroscopy
ZSM-5	zeolite Socony Mobil–5

1. Introduction

1.1 General considerations regarding the valorization of lignocellulosic biomass

Present day society has been largely shaped by the abundant availability of petroleum-derived fuels, commodity chemicals and plastics.[1] With the world's petroleum and gas reserves being finite in nature, there has been an increasing interest in the use of alternative non-edible renewable resources.[2] The problem is further substantiated by the negative climate effects of fossil fuel burning, *emitting vast amounts of the greenhouse gas* CO_2, the periodic occurrence of oil crises, and the detrimental environmental impact of petroleum-derived plastics.[3–5]

Lignocellulosic biomass is by far the largest sustainable resource available on earth.[6] Its structure consists of three main components notably cellulose (30%–60%), hemicellulose (20%–40%) and lignin (15%–25%).[7,8] Many routes exist to derive value from biomass, among which the earliest ones investigated were thermal methods.[7,8] This includes the oxidative gasification (>1073 K) of biomass into CO, H_2, CO_2 and CH_4 after which the purified syngas components (CO and H_2) are converted into long chain hydrocarbons using Fischer-Tropsch technology or other products.[7,9] A final hydrocracking step then turns the latter long chain wax into a high-quality diesel.[9,10] An attractive route is also the conversion of syngas to methanol, which may serve as a liquid energy carrier in a potential future methanol economy, and of course also as a key intermediate for the production of chemicals.[7,11] Another highly attractive thermal approach is pyrolysis whereby biomass is heated in the absence of oxygen.[12] The amount of bio-oil generated depends on the reaction temperature, the reactor configuration and the residence time. This is typically followed by a second upgrading step due to the instability of pyrolysis oil itself and the still substantial oxygen content in it.[13–16]

Most biorefineries heavily focus on getting high quality cellulose mainly for the purpose of paper production[17] and more recently, second generation bioethanol production.[18] To the latter, cellulose is typically first enzymatically converted to monomeric sugars and then further fermented into bio-ethanol or alternatively transformed into selected value added platform chemicals.[19] It should be noted that the production of monomeric sugars from lignocellulose, and also cellulose itself, is not straightforward since the crystalline nature of cellulose, featuring many internal hydrogen bonds between the polysaccharide strands, makes access for enzymes challenging.[20] Due to the interwoven nature of hemicellulose, cellulose and lignin, the separation of these components is highly energy intensive and equally challenging to perform in an efficient and "clean" manner.[21] In fact, lignocellulose fractionation is the first significant bottleneck for any bio-refinery approach.

To achieve a more efficient lignocellulose fractionation, many biomass pretreatment procedures have been developed, the most common ones being steam explosion (SE),[22,23] dilute sulfuric acid treatment (DA),[22,23] organosolv,[24,25] ammonia fiber expansion (AFEX),[22,26] and liquid hot water (LHW) treatment.[26] As the production of cellulosic ethanol is expensive, it presently cannot compete with fossil fuels.[3,27] However, the possible parallel production of higher-value chemicals from the other lignocellulose constituents, such as hemicellulose or lignin, alongside the cellulosic ethanol, has the potential to offset the latter's price disadvantage.[27–30] Rosales-Calderon and Arantes recently published a comprehensive review on this topic, describing a range of compounds/processes which have reached at least a Technology Readiness Level (TRL) of 8, i.e., manufacturing level.[3] Recent efforts that target highly important bulk chemicals such as phenol or potentially high value pharmaceutically relevant compounds are interesting future directions.[31–36] It has to be noted, however, that second generation ethanol biorefineries typically focus on obtaining high quality cellulose, leaving drastically modified lignin behind. This lignin is characterized by a low content of β-O-4 linkages and more recalcitrant C—C bonds, which render such lignin streams also more difficult to valorize.[37,38]

Nonetheless, over the past two decades, and particularly inspired by the progress in the construction of second generation biorefineries, the need for more efficient lignin valorization (for the above-mentioned reasons) became very clear.[39,40] Hence, many novel lignin depolymerization approaches have been developed, the prime foci being on oxidative, reductive, carbon neutral or acid/base catalyzed methods under milder reaction conditions; and some of these have shown promise in the production of well-defined aromatic monomers.[37,38] Most of these studies are based on in-house organosolv lignins, typically isolated by different procedures and consequently yielding lignins with different β-O-4 content. Hence, no direct comparison of the efficiency of these methods in terms of monomer yields should be drawn. With mild organosolv procedures affecting the native lignin structure, the concept of the so-called "lignin-first" bio-refinery gained marked traction.[41–43] The key idea here is that lignin valorization is pursued *first*, before the one of cellulose.[44] During this process, smaller lignin fragments are liberated from the lignocellulose matrix through solvolysis or acidolysis.[45] These are in situ stabilized and further processed in the presence of a metal catalyst and hydrogen donors, typically hydrogen gas. Because in these procedures, it is largely the 'native lignin' structure which is depolymerized, and these procedures typically give very high, near theoretical yield of aromatic monomers.[36,37,41,46]

Some 10 years ago, lignin depolymerization was a discipline in its infancy compared to the one(s) of (hemi)cellulose. As such the then known biomass-derived platform molecules were near exclusively derived from the cellulosic component, and the fate of any lignin rest products was generally fuel provision by burning. After 2005, fundamentally new approaches related to lignin depolymerization appeared, and tremendous progress has been made in the past 10 years. As an example, the year 2007 saw the Ford group starting up what eventually proved to be a range of very successful investigations on lignin, cellulose and lignocellulose depolymerizations using hydrotalcite derived *copper-modified* porous metal oxide catalysts.

1.2 Synthetic hydrotalcite derived porous metal oxides

Hydrotalcite is a type of anionic clay characterized by a double-layered structure, which can be found in nature.[47] Its name relates to its resemblance to talc and equally points at the high-water content of this material.[48] Synthetic hydrotalcites are typically obtained in aqueous solutions and resemble closely the natural hydrotalcite structure.[49] Hydrotalcites have proven very successful in a range of catalytic processes and other applications.[47,50–52] Key to their on-going success is their high compositional variety, controlled surface acidity/basicity, facile exfoliation, anion-exchange

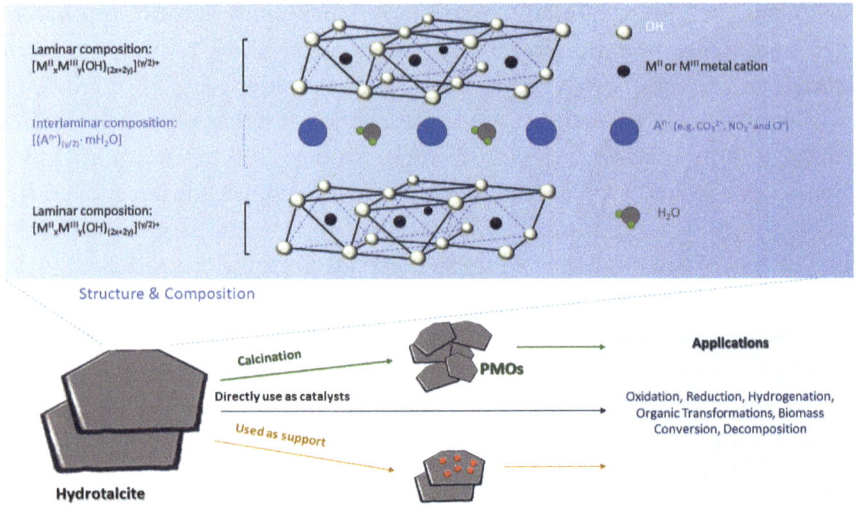

Fig. 1 General structure of hydrotalcites and their application as catalysts, catalytic supports and/or as precursors for porous metal oxides (PMOs).

capacity, etc.[50,53] More specifically, they are made up of positively charged brucite-like layers featuring $M(OH)_6$ octahedra, where the positive charge is caused by the isomorphous substitution of a portion of Mg^{2+} ions to Al^{3+} ions. The extra positive charge in the lattice is compensated by intercalated anions in the interlayers where water molecules are also present. As shown in Fig. 1, hydrotalcites are typically represented by the following general formula: $[M^{2+}{}_{1-x}\, M^{3+}{}_x\, (OH)_2]^{z+}\, A^{n-}{}_{z/n}{\cdot}mH_2O$, with M^{2+} and M^{3+} being respectively divalent and trivalent metal cations located in the host layers and A^{n-} being the exchangeable anions in the interlayer region, compensating the positive charge of the host layers.[54]

Hydrotalcites can be transformed into porous mixed oxides (PMOs) by calcination, a process which sees the characteristic layered crystalline structure of hydrotalcites collapse into an amorphous porous one. This topotactic transformation is strongly governed by heating rate, calcination temperature and the applied atmosphere. Furthermore, it has been shown that fine-tuning of the crystal facets and surface defects of PMOs, affect markedly their catalytic performance.[50] Of particular interest are transition metal-based oxides, where the presence of partially filled d-shells offers a peculiar chemisorption capacity. Exemplary are Cu-based oxides, which can display a range of Cu-oxidation states [Cu(0), Cu(I) and Cu(II)] and hence catalyze a variety of reactions.[55]

This chapter will discuss in detail our ground-breaking work on Cu-doped porous mixed oxides (Cu-PMO) applied to the valorization of biomass and its individual components. The achievements of other research groups in the area of Cu-PMO catalysis are also highlighted.

2. Cu-doped $AlMgO_x$ PMO (Cu-PMO) in the valorization of lignin

2.1 The origins of the Cu-PMO/methanol lignin depolymerization process

Lignin is a complex heterogeneous biopolymer, which is formed by random oxidative free-radical polymerization of three aromatic monolignols, notably, p-coumaryl alcohol (H), coniferyl alcohol (G) and sinapyl alcohol (S).[56] Although many variations exist, the most common linkages in lignin are aryl ethers (β-O-4 and α-O-4), β-5 (dihydrobenzofuran-like) and 4-O-5 (diaryl ether).[56] Thereby the β-O-4 makes up for ~50% of all linkages in lignin. The first strategy of the Ford group was the investigation of the reactivity of various porous metal oxides toward the scission of the simple model compound dihydrobenzofuran (DHBF).[37,57] Thereby the previewed role of the hydrotalcite-derived solid base was the sole cleavage of the β-O-4 linkage, and this without having to resort to the use of stoichiometric amounts of a homogeneous base like KOH.[57] Added motivation for the choice of DHBF was that the carbon framework remains intact, that way only yielding a limited amount of reaction products (see below), and facilitating analytics and interpretation of results. While most common reductive heterogeneous catalysts rely on the deliberate addition of hydrogen gas (H_2) or a suitable hydride donor (e.g., secondary alcohols,[58] formic acid[59]), we however decided to use the inherent ability of Cu-catalysts to decompose methanol at higher temperature into CO and $2H_2$ (Eq. 1).[60] The added benefit of this route is that any formed water in the reaction mixture will react further, through either the water-gas-shift-reaction (WGHR) (Eq. 2) and/or methanol steam reforming (Eq. 3), both reactions also supplying additional H_2 equivalents.[61,62]

$$CH_3OH \rightleftharpoons CO + 2H_2 \Delta H = 90.2\,kJ\,mol^{-1} \tag{1}$$

$$CO + H_2O \rightleftharpoons CO_2 + H_2 \Delta H = -41.2\,kJ\,mol^{-1} \tag{2}$$

$$CH_3OH + H_2O \rightleftharpoons CO_2 + 3H_2 \Delta H = 49.7\,kJ\,mol^{-1} \tag{3}$$

From the range of Cu-AlMg-PMOs, those with 10% and 20% Cu/Mg isomorphic substitution were found most active.[57] In spite of the high reaction temperature of 598 K, *a temperature at which the added MeOH is supercritical*, the DHBF reaction was sufficiently slow to consecutively identify the hydrogenolysis product *2-ethylphenol* and its fully hydrogenated equivalent *2-ethylcyclohexanol*. Crucial to this reaction sequence is the ability of the mixed oxide to promote the hydrogenolysis reaction. A notable side reaction was methylation of the aromatic ring making for the consecutive formation of the methylated equivalents of ethylphenol and ethylcyclohexanol.[57] Advancing onto organosolv lignin, and even a range of lignocellulosic biomasses, the reaction products encompassed mainly monomeric higher alcohols and ethers (C2–C6) and higher molecular weight dimers (C9–C12). Importantly, no evidence of char formation during the reaction was found. This approach, dubbed as the "UCSB Process," is depicted in Fig. 2. XPS analysis of the spent catalyst pointed at the formation of Cu(I) and Cu(0), which is consistent with its reddish color.[57,63]

2.2 Study of main reaction pathways using simple model compounds

To gain more insight in the occurring reactive pathways, we also engaged in mechanistic studies of a range of model compounds by following product formation profiles and performing kinetic modeling of the occurring reactive pathways using the free software DynaFit.[64,65] With phenol being mainly converted into methyl cyclohexanols and anisole, the main question concerned the stage at which the methylation reaction occurred. Indeed, two concurrent routes from phenol to methyl cyclohexanols could well be

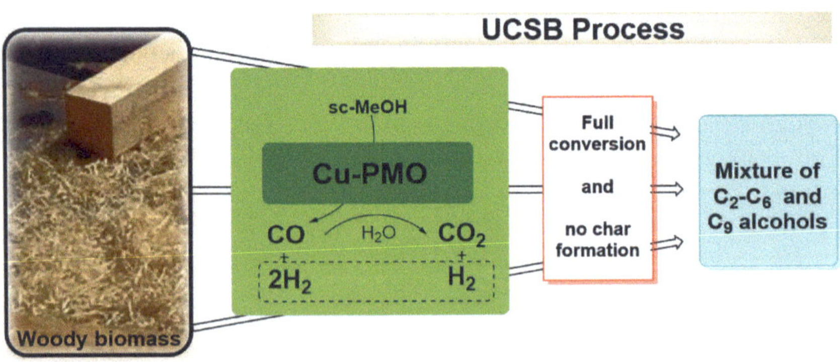

Fig. 2 Visualization of the "UCSB process."

envisaged being: (1) hydrogenation of phenol to cyclohexanol followed by its methylation and/or (2) phenol methylation to cresol followed by hydrogenation of the aromatic core. From separate studies on cresol and cyclohexanol reactivity, it could eventually be concluded that route 1 was dominant, the decisive observation being a markedly slower hydrogenation of cresol vs phenol.[64] Similar observations were made for guaiacol (2-methoxyphenol), in that the hydrogenation of veratrole (1,2-dimethoxybenzene) was found to be significantly slower compared to the hydrogenolysis pathway to phenol (and subsequent further hydrogenation to cyclohexanol). Interestingly, with alkoxy aromatics, which did not feature a separate OH group (e.g., methoxy- and ethoxy-benzene), hydrogenation of the aromatic ring was particularly slowed down, therewith making hydrodeoxygenation of the substrates to benzene the dominant pathway.[64] With oxygen-free aromatics, very little reactivity was observed; logical consequences of this being the near quantitative conversions of benzyl alcohol to toluene and styrene/acetophenone to ethylbenzene. Benzyl phenyl ether (BPE) proved to be an interesting test case with the formation of "toluene + phenol" being largely favored over the "benzene + benzyl alcohol" alternative (Scheme 1). With the more complex

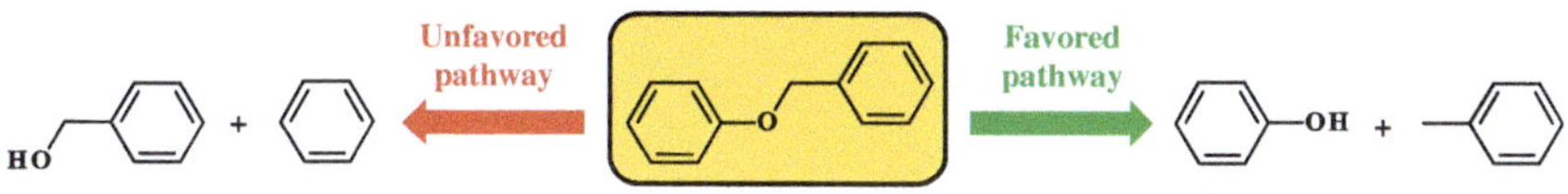

Scheme 1

substrate 2-phenoxy-1-phenylethane-1-one (PPEO), the mechanistically interesting observation was the sizeable formation of propylbenzene (aside majority ethylbenzene), which should involve methylation at a carbon α to a carbonyl group be it at the level of PPEO or the acetophenone reaction product.[64]

In advancing beyond aromatics, also the reactivity of (methyl) cyclohexanols was investigated. These substrates were invariably prone to further methylation, which is attributed to dehydrogenation to their respective cyclohexanones and consequent methylation α to the carbonyl group. Proof to this hypothesis was obtained with the clear observation of (di) methyl cyclohexanones when working directly from cyclohexanone. To the conversion of 1,2-cyclohexanediol, the occurrence of the ring contraction product cyclopentylmethanol, which is likely related to a pinacol rearrangement, is of particular note.[64]

With phenol derivatives being common outputs of reductive lignin disassembly procedures, a need existed to control their proneness to hydrogenation and (subsequent) ring methylations. In this respect, kinetic analysis provided an important key-insight, with the general observation that *O*-methylation suppressed hydrogenation of the aromatic core

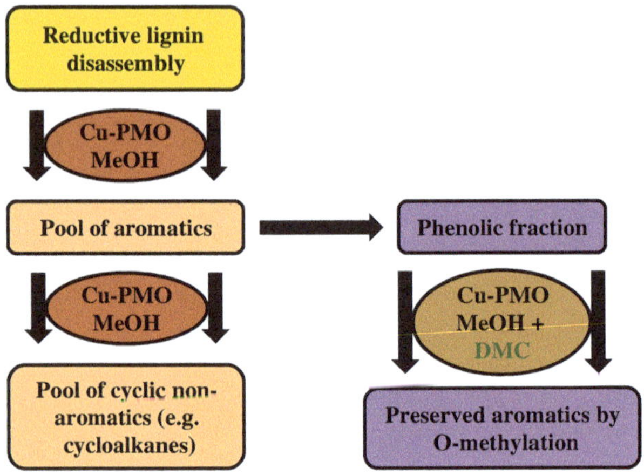

Scheme 2

(Scheme 2). Interestingly, to this issue the use of basic PMO offered a most elegant solution, in that it can catalyze *O*-methylation from dimethylcarbonate (DMC) at the higher applied reaction temperatures.[66] And indeed, in the combined presence of MeOH and DMC, significantly more anisole-(type) molecules were formed (Scheme 2).[64]

2.3 Mild depolymerization of lignin

With the original Cu-PMO/(supercritical)methanol system (hereafter "Cu-PMO/(sc)MeOH") converting lignin predominantly into substituted cyclohexanols, a need existed to tune the catalyst's reactivity and selectivity toward the formation of monomeric aromatic compounds. Indeed, while there is merit in the formation of substituted cyclohexanols from lignin, the process does forego on the formation and isolation of valuable bio-derived aromatics, the very structural component for which lignin depolymerization is envisaged in the first place. Over time a range of strategies to prevent or limit over-hydrogenation were developed, and these are discussed below.

The first attempt to control the availability of hydrogen was to modify the original Cu20PMO catalyst by additional metal dopants. A range of such catalyst compositions were prepared and tested in supercritical methanol as reaction medium. The activity/selectivity profile of the doped Cu-PMOs proved very different. In particular, on candlenut lignin, Cu20La20-PMO was found to generate appreciably more aromatic monomers than Cu20-PMO did.[67] This was attributed to the lower methanol-reforming capability of Cu20La20-PMO, which limits the amount of available hydrogen.[67]

2.3.1 Mild depolymerization of Candlenut lignin

The Candlenut tree (*Aleurites moluccanus*) is a plant which can be found in more tropical climates such as Brazil, Africa and South East Asia.[68] The Candlenut has characteristically very hard shells, a feature which links with a high lignin content (~36%).[69] The main goal here was to develop a lignin depolymerization procedure, which featured milder reaction conditions. More specifically, these concerned the subcritical regime, a zone where also methanol reforming would be largely impeded. Following a unique isolation procedure for the candlenut shell lignin, the so obtained beige colored lignin was subjected to a reaction temperature of only 413–493 K, in the presence of 40 bar of hydrogen gas.[70] Interestingly, a set of only four C9/(C10) catechols were formed (Fig. 3A–D) in a very high, *thus far unprecedented*, yield, their respective amounts depending on the applied reaction conditions.[70] Of particular interest was the major formation of catechols, which in lignin depolymerization is a rare observation. Given the mild

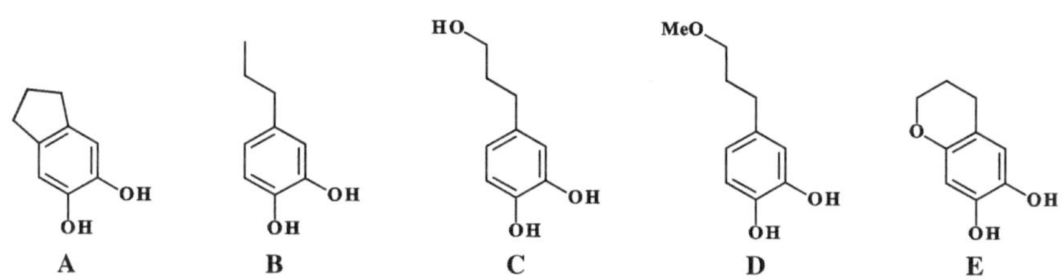

Fig. 3 Major catechol structures made from candlenut shell using Cu-PMO catalyst. (A) Indane-5,6-diol; (B) 4-propylbenzene-1,2-diol (catechylpropane); (C) 4-(3-hydroxypropyl)benzene-1,2-diol (catechylpropanol); (D) 4-(3-methoxypropyl)benzene-1,2-diol; (E) chroman-6,7-diol.

operating conditions this was unlikely the result of demethylation but rather the direct consequence of a unique lignin structure. The latter is nowadays known as C-lignin, a term referring to the presence of a homopolymer of caffeyl alcohol.[31,71,72] And indeed, to date only two other examples have been reported, notably: (a) the reductive depolymerization of lignin in the presence of hydrosilane as the reducing agent, and $B(C_6F_5)_3$ as a potent Lewis acid; a method which is capable of depolymerizing a range of lignins to either substituted catechols or pyrogallols[38,73] and (b) the Pd/C catalyzed depolymerization of C-lignin (473 K, 15 h, 40 bar H_2), as isolated from the seed coats of vanilla.[74] The latter method gave mainly rise to catechylpropanol, catechylpropane and the unique chroman-6,7-diol compound (Fig. 3B, C and E), their individual nominal yields depending on the catalyst used.[74] The Cu-PMO mediated depolymerization of Candlenut (C-)lignin yielded catechylpropanol in 43.3% yield.[70]

2.4 The flexible use of CuPMO catalyst for complete lignocellulose valorization (LignoFlex)

2.4.1 Mild depolymerization of native lignin (step 1)

Despite the gained insight on the reactivity of various biomass sources, whether in supercritical methanol or under milder depolymerization conditions, the base question still remained: *how to develop a method that would allow for a better valorization of the individual lignocellulose constituents: lignin and (hemi)cellulose?* With the (more recent) development of reductive catalytic fractionation procedures, this question has received a more substantiated answer, as this method combines lignocellulose fractionation with the catalytic depolymerization of lignin.[32,42] While many of these methods lead to high yields of aromatic monomer, the separation of the cellulose residues from the heterogeneous catalyst is challenging. We therefore explored the idea of constructing a lignocellulose conversion process, consisting of two distinct catalytic steps: (1) mild depolymerization of native lignin in methanol under relatively mild conditions (413–493 K), followed by (2) the catalytic conversion of the remaining solid residues, these mainly consisting of cellulose and unreacted lignin residues, in supercritical methanol. The key question here was whether the second step allows for the actual full conversion of the solid residues, therewith liberating the catalyst for re-use.

More specifically, the first stage concerned the mild reductive depolymerization of pine lignocellulose in the presence of the Cu-PMO catalyst. Interestingly, at 453 K this procedure uniquely led to the formation of dihydroconiferyl alcohol (DHCA—see Fig. 4A) in up to 9% yield

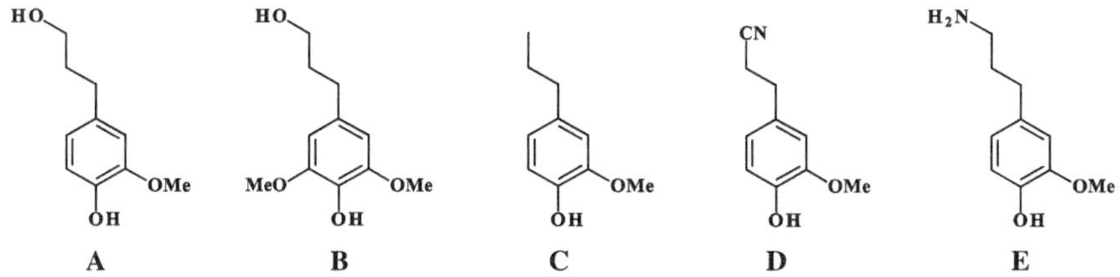

Fig. 4 Chemical structures of the different major products obtained within the LignoFlex procedure for a range of lignocellulose types or upon direct coupling of the obtained platform chemical A with ammonia; (A):4-(3-hydroxypropyl)-2-methoxy-phenol (dihydroconiferyl alcohol—DHCA); (B) 4-(3-hydroxypropyl)-2,6-dimethoxy-phenol (dihydrosinapyl alcohol—DHSA); (C) 2-methoxy-4-propyl-phenol (*n*-propylguaiacol); (D) 3-(4-hydroxy-3-methoxy-phenyl)propanenitrile; (E) 4-(3-aminopropyl)-2-methoxy-phenol.

(92% selectivity).[75] Upon a further temperature increase to 493 K the DHCA yield could be further increased to ~13% (61% selectivity), yet at the expense of additional *n*-propylguaiacol formation (Fig. 4C).[75] From a mechanistic point of view, the formation of DHCA from lignin may involve the dehydrogenation of the benzylic alcohol moiety in the β-O-4 motif, followed by C—O bond-cleavage, or alternatively, by a direct phenyl ether hydrogenolysis and subsequent hydrodeoxygenation of the benzylic ketone group—as supported by model compound studies.[75]

Furthermore, the applicability of the procedure was tested on a variety of wood lignocelluloses. Gratifyingly, high aromatic monomer yields could be obtained and then especially from poplar (36% at 42% selectivity), beech (31% at 59% selectivity) and maple (30% at 52% selectivity) derived lignocellulose. With the latter three tree species being hardwood, the main isolated compound was 4-(3-hydroxypropyl)-2,6-dimethoxy-phenol (DHSA, Fig. 4B). As expected, the softwood varieties, pine and cedar, gave DHCA as the main product, while oak surprisingly gave *n*-propylguaiacol (Fig. 4C). Overall the S/G ratio of the original lignin did reflect itself into the obtained product mixture.[75]

2.4.2 Full conversion of the solid residues in supercritical methanol (step 2)

The next phase (step 2) concerned the conversion of the unconverted (hemi) cellulose and residual lignin, a procedure which was run very similar to the previously discussed "UCSB process" (see Fig. 2). More specifically, the remaining lignocellulosic components, *surrounding the Cu-PMO catalyst*, were mixed with fresh methanol and heated to a temperature in excess of

593 K. As described before, at this temperature part of the added methanol is reformed into H_2 and CO, therewith creating ideal conditions to the complete and successful conversion of the remaining lignocellulosic components into C2–C5 aliphatic alcohols. Interestingly, it was found possible to repeat steps 1 and 2 up to 10 times with negligible reduction in catalytic activity nor any Cu, Mg or Al leaching.[75]

2.4.3 Downstream processing strategies for the obtained alcohol intermediates

In further advancements, attention was devoted to the downstream processing of both lignin and cellulose derived alcohols. As the cellulose derived small molecules concerned a majority of alcohols and ethers, a strategy was devised to convert the mixture of small aliphatic alcohols into a mixture of fuel grade alkanes. The thereto required chain extension could be achieved through a CuNi-PMO catalyzed consecutive dehydrogenation of the small aliphatic alcohols to aldehydes/ketones and their subsequent aldol condensation with cyclopentanone. Subsequent deep hydrodeoxygenation of the so obtained substituted cyclopentanones with a commercially available Ni/SiO_2-Al_2O_3 catalyst, yielded the target alkanes.[75] While the procedure worked well with neat 1-pentanol, the use of said mixture of aliphatic alcohols required removal of residual methanol through solvent-exchange with heptane. Additionally, not all alcohols coupled equally well to the cyclopentanone core making for an eventual alkane mixture of both C4–C6 and C8–C11 alkanes.[75] With cyclopentanone being directly available from hemicellulose, the here developed alkane range can be considered fully bio-derived.[76] The overall procedure is illustrated in Fig. 5.

Creating competitive bio-based processes to compounds for which already well-established petroleum-derived synthetic routes exist is not trivial. Notable requirements to success are: (1) A higher commercial value of the target compounds, that way allowing for profitability at the moderate (10–30 kton/year) to even small scale (1 kton/year); (2) Compounds with a higher degree of chemical functionality, *preferentially inherent to biomass*, therewith avoiding the many expensive and waste-producing chemical steps to their introduction and (3) Co-production of additional chemical building blocks thus improving the overall economics.[27,75,77,78]

Here we were presented with a unique opportunity in that DHCA and/or dihydrosinapyl alcohol (see Fig. 4A and B) holds chemical functionality on both the aromatic ring and the aliphatic side chain. Thereby the preserved aliphatic alcohol moiety offered a unique possibility to explore the

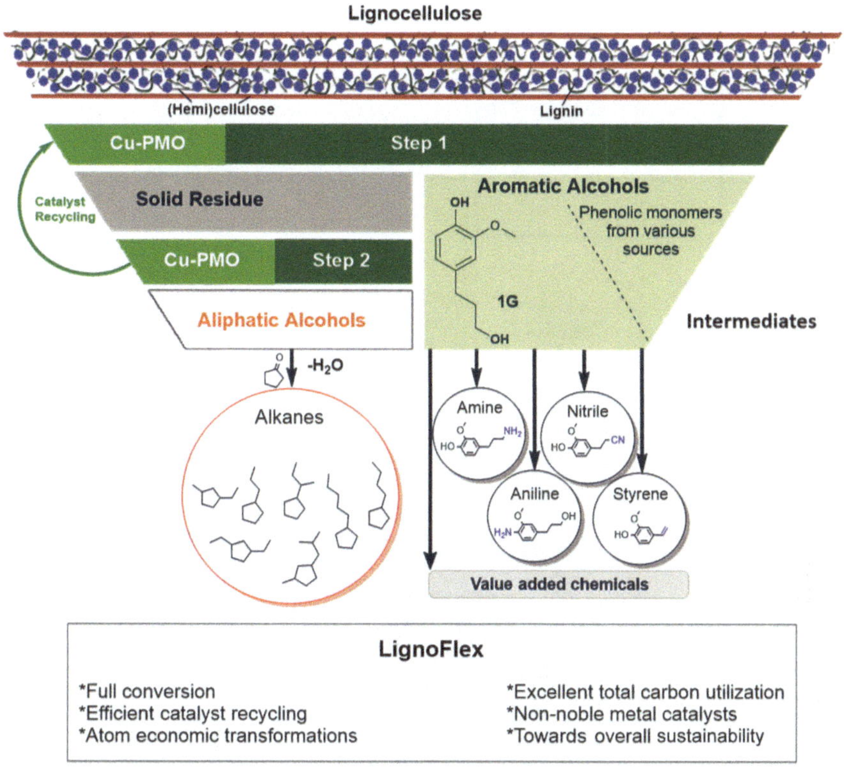

Fig. 5 An overview of the LignoFlex procedure and subsequent downstream valorization of its products.

synthesis of lignin-derived amines. The shortest route to this is the direct reaction of said alcohols with ammonia, a route which only produces water as a by-product.[75] However, the selective alkylation of ammonia to primary amines is challenging as the latter is generally more reactive than ammonia itself.[78] Consequently, to date few catalytic routes exist for this transformation; methods which are also very substrate dependent. However, using commercially available Ni/SiO_2-Al_2O_3 we were able to quantitatively transform DHCA into the corresponding nitrile (Fig. 4D) which was hydrogenated to the target amine (Fig. 4E) with high selectivity.

Given its ease to isolate and the high selectivity in which it was obtained, propanol-guaiacol (Fig. 4B) was proposed as lignin-derived platform chemical. Hence, we have developed a network of reactions, using a variety of chemo-catalytic or chemical methods that are able to convert this aromatic monomer to a range of value-added compounds. Fig. 6 provides an

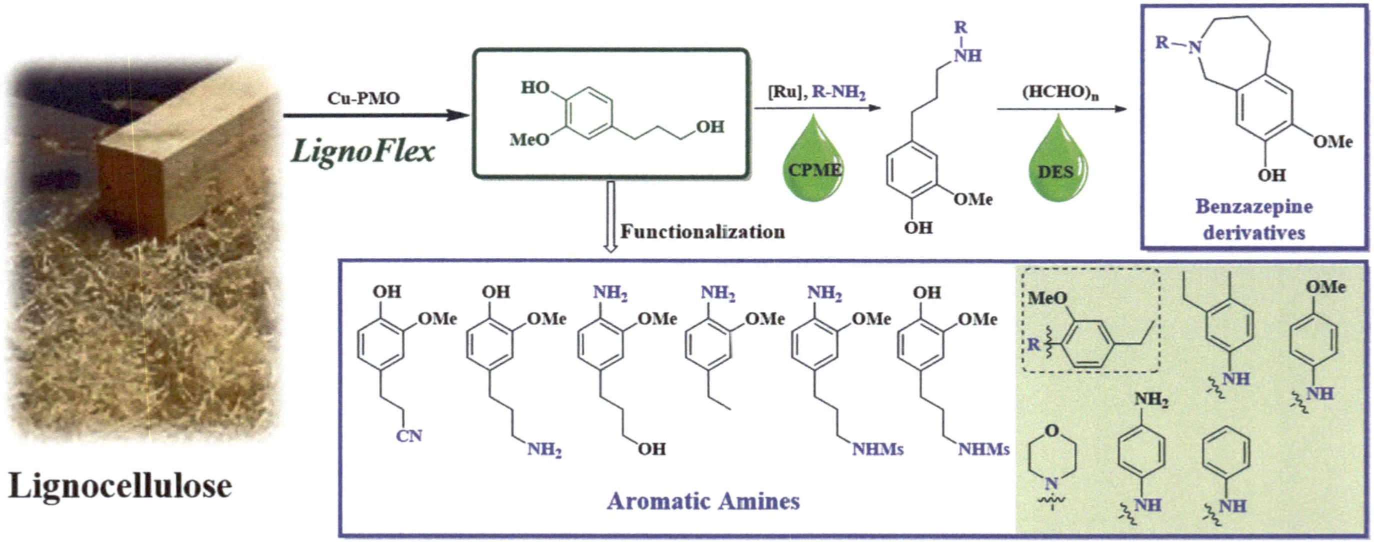

Fig. 6 An overview of the valuable aromatic monomers obtained via downstream valorization pathways of the propanol-guaiacol platform chemicals obtained from the LignoFlex procedure applied on pinewood.

overview of these different processes, while the below depicted schemes give a more detailed picture:

(a) Hydrolysis of the nitrile (Scheme 3—A) to an acid group (Scheme 3—B) followed by decarboxylation yielded 4-vinylguaiacol (Scheme 3—C)—*a valuable aromatic chemical, which also holds polymerization potential.*[79,80] Subsequent metathesis of the latter compound gave a stilbene which could serve as a starting material for bisphenol-A analogues (Scheme 3—D). Alternatively, methylation and cyclization would give 5,6-dimethoxy-2,3-dihydrobenzofuran (Scheme 3—E)—a cyclic indan, from which the high volume anti-Alzheimer drug Donepezil can be obtained.[75,78,81]

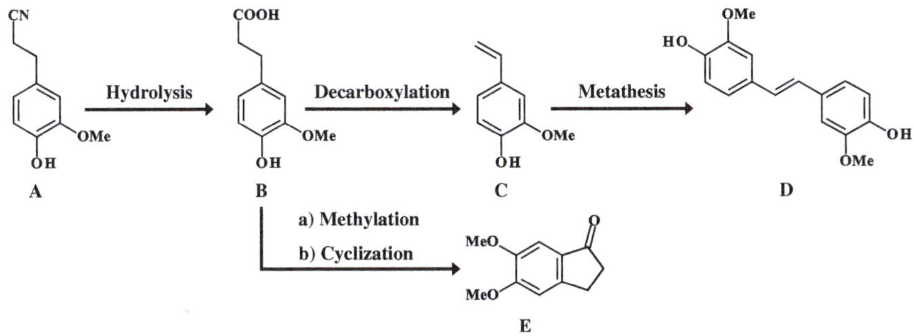

Scheme 3

(b) Using Ni/SiO_2-Al_2O_3 at higher temperature (493 K) allowed for the formation of 4-ethyl-2-methoxy-phenol (Scheme 4—A), which could be either aminated (Scheme 4—B), cross-coupled to a range of amines (Scheme 4—C), or transformed to 3-ethylanisole (Scheme 4—D).[75]

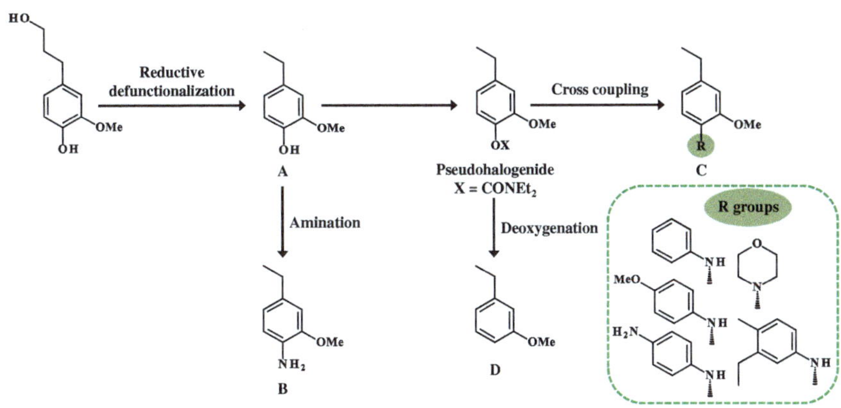

Scheme 4

(c) The acid-catalyzed cyclization of DHCA (Scheme 5—A) resulted in the formation of 7-methoxy-1-oxaspiro[4.5]deca-6,9-dien-8-one (Scheme 5—B), that allowed for the introduction of nitrogen in a stoichiometric transformation, to yield the corresponding aniline derivative (Scheme 5—C).[75]

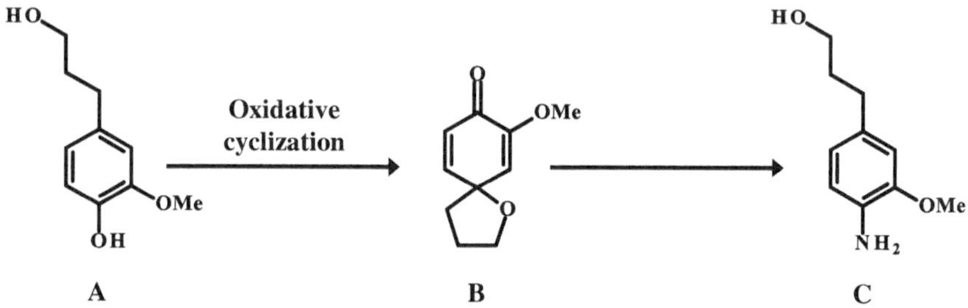

Scheme 5

(d) Amination of the aliphatic alcohol chain of DHCA (Scheme 6—A), *by a hydrogen borrowing strategy*, resulted in the corresponding N-mesyl derivative (Scheme 6—B), which was aminated through the strategy analogous to that shown in Scheme 3, in order to obtain an aliphatic-aromatic diamine (Scheme 6—C).[75]

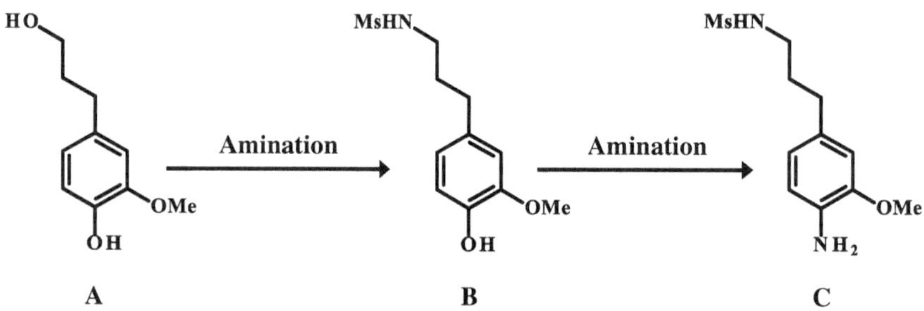

Scheme 6

Overall, these strategies lead to the formation of a range of valuable aromatic monomers, containing 0, 1 or 2 nitrogen atoms. A summary of these different products is shown in Fig. 6 below. Of particular note is the formation of N-containing heterocycles that was demonstrated later in 2019 as described below.

3. Cu-doped $AlMgO_x$ PMO (Cu-PMO) in the valorization of polysaccharides

3.1 Conversion of cellulose in supercritical methanol

Cellulose is the most abundant source of non-food-competing renewable sugar on the planet.[20] It is a linear carbohydrate polymer in which glucose units are interconnected through β-1,4 glycosidic bonds. Typically, these linear cellulosic chains are bundled into micro-fibrils, groupings of which give rise to cellulose fibers.[82] Cellulose has a very broad application range covering both (nano)materials and chemicals.[83,84] Most commonly known is its depolymerization to glucose units, which can subsequently be transformed into bioethanol.[20,27]

The application of Cu-PMO|scMeOH to cellulose was originally reported by the Ford group.[57] More specifically, the process yielded a mixture of alcohols and ethers predominantly in the range of C2–C6 carbon atoms, although higher C7+ number homologs were also detected. It was there proposed that the reaction chain consisted of rapid cellulose depolymerization, followed by retro-aldol condensations, or other alternative C—C bond scission events, dehydrations, hydrodeoxygenation, and methanol-incorporating reactions.[57,85] More recently, through a combination of isotopic labelling (D-Glucose-$^{13}C_6$), NMR analysis and state-of-the-art FT-ICR mass spectrometry, the Huber group was able to further refine the reaction scheme, largely confirming the original proposal.[86] The rapid cellulose depolymerization to C3 oxygenates (e.g., dihydroxyacetone, glycerol) was spectroscopically shown and attributed to retro-Aldol condensation. A network of follow up transformations, including reactions of the confirmed key intermediates with methanol to result in chain elongation as well as various C—C scission events were proposed. Additionally, in their hands, about 50% of the cellulose carbon was converted to C7–C25 species with in between 2–4 oxygen and 4–12 double bonds, most likely representing multiple aromatic rings.[85,86] In this respect it was also established that pre-reduced Cu-PMO yielded less of the higher MW fraction than calcined Cu-PMO.[86]

3.2 The catalytic conversion of 5-hydroxymethylfurfural in supercritical methanol and under milder reaction conditions

5-hydroxymethylfurfural (HMF) is a versatile molecule,[87] which is directly produced by the acid-catalyzed dehydration of the C6 monosaccharide fructose.[88] It can also be produced, in various yields, from a range of monomeric C6 sugars, or by consecutive hydrolysis/dehydration of polysaccharides, including cellulose.[88] With 5-HMF consisting of a furan ring, an aldehyde, and a hydroxymethyl group—*which endows it with an exceptional and multifaceted reactivity*—it has been the subject of many studies aimed at the synthesis of, among others, biofuels and renewable monomers.[89,90] 5-HMF's main drawback is its instability. Indeed, both its facile rehydration to levulinic acid (LA) and formic acid (FA), and the formation of insoluble carbonaceous by-products, named humins, are major concerns.[91] Additionally, its polar hydrophilic character makes it generally difficult to extract 5-HMF from water, its main formation medium.[92] Industrially, 5-HMF production has not yet transcended pilot scale, a matter which relates to its high manufacturing cost of ~\$2/kg. Competitiveness with fossil fuels is only expected when the plant size will exceed 100 kton/year.[88]

5-HMF hydrogenation to 2,5-bis(hydroxymethyl)furan (FDM) holds great potential as it can serve as a bio-based structural building block to the making of bio-based polyurethanes and polyesters.[93,94] Applying Cu-PMO to the hydrogenation of HMF in ethanol (50 bar H_2, 373 K), allowed for the formation of FDM in 99% yield, and as expected with this catalyst class under these conditions, the formation of [5-(hydroxymethyl) tetrahydrofuran-2-yl]methanol (THFDM), where also the furan ring is hydrogenated, was not observed.[95] At elevated temperatures though, furan ring hydrogenation also takes place. Performing the hydrogenation for 6 h at 493 K allowed for the formation of a mixture of dimethylfuran (DMF) and dimethyltetrahydrofuran (DMTHF) in 66% yield and with a DMF to DMTHF ratio of 3.6 (i.e., majority DMF). The economic rational of DMF production is fuel related as its octane number and energy density (30 kJ cm^{-3}) is higher than ethanol.[96,97] Moreover, DMF has the lowest water solubility of all mono-oxygenated C6 compounds derivable from biomass, and also displays a favorable boiling point for use as a liquid fuel.[96] Minor doping of the Cu20-PMO catalyst with Ru (Cu20Ru2-PMO—0.6 wt% Ru) improved the combined DMF-DMTHF yield to a maximum of 79% (493 K, 12 h) but did not further influence the DMF to DMTHF ratio, remaining at 3.4.[95] Thorough investigation of reaction pathways pointed to the beneficial role of small amounts of Ru dopants in accelerating hydrogenolysis of (5-methyl-2-furyl)methanol (MFM) into DMF.[95] This is in line with the high hydrogenolysis ability of Ru.[88] It was further found that the obtained DMF-DMTHF yield was solvent dependent when employing Cu20-PMO, but not so in the case of Cu20Ru2-PMO. Indeed, in the presence of Cu20-PMO a change from ethanol to 2-propanol as the reaction solvent afforded an increase in the combined DMF-DMTHF yield from 66% to 77%. While TEM of freshly prepared Cu20-PMO and Cu20Ru2-PMO revealed no observable CuO and RuO nanoparticles, which thus pointed at highly dispersed Cu and Ru into the PMO structure, distinct spherical Cu(0) particles of 10–12 nm Ø could be observed after reaction.[95] The different structures and their (potential) applications are shown in Fig. 7.

3.3 Levoglucosenone and derivatives

Levoglucosenone (LGO) is a versatile chiral molecule, which can be uniquely obtained from the pyrolysis of cellulosic feedstocks (e.g., sawdust). LGO is a highly functionalized chiral C6 molecule, comprising a cyclic acetal and an α,β-unsaturated ketone motif. While many (analytical) methods to

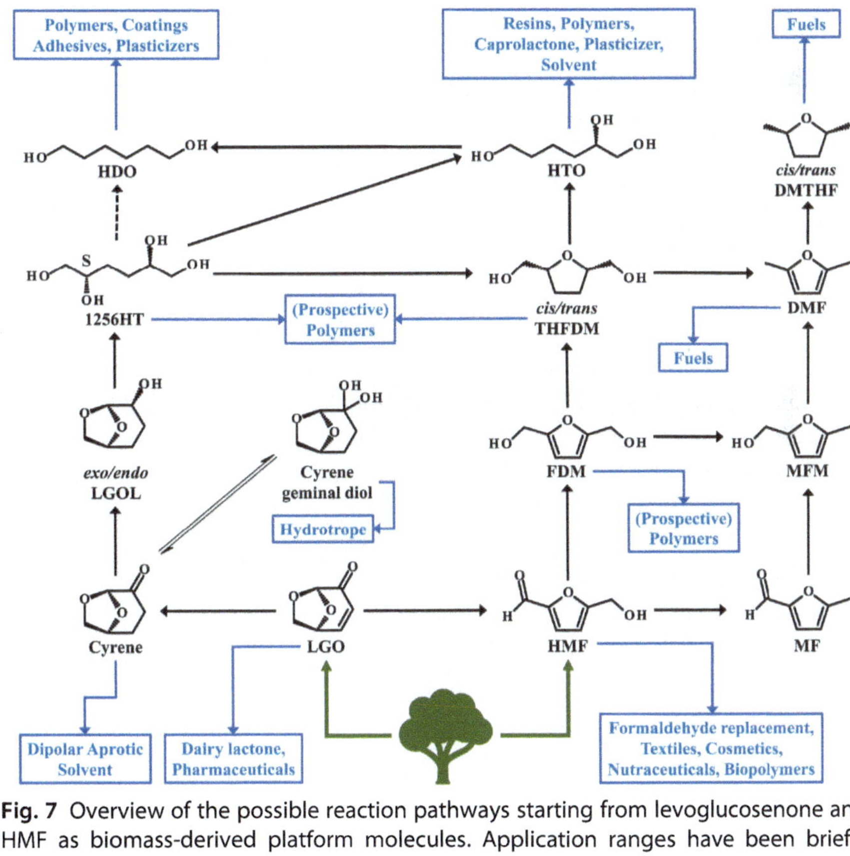

Fig. 7 Overview of the possible reaction pathways starting from levoglucosenone and HMF as biomass-derived platform molecules. Application ranges have been briefly indicated.

its synthesis have been described, only the Furacell™ technology has presently advanced to prototype scale (50 ton/year).[98,99] In this process a cellulosic feedstock is mixed with aqueous phosphoric acid and a suitable dipolar aprotic solvent (most commonly sulfolane), and subsequently pyrolyzed at 553 K in a continuous screw reactor coupled to a distillation tower.[98] Alternatively, it was recently found that LGO could be obtained at no less than 453 K from residual saccharides present in waste bio-refinery lignin.[27] In addition to lignin valorization, the higher market value of LGO could markedly improve the economics of ethanol biorefineries. A recent techno-economic analysis, based on cellulose as the starting substrate, showed that LGO could be produced at $3500/ton when produced at a scale of 30 kton/year.[100] LGO has been shown a key/hypothetical intermediate to the synthesis of dairy lactone, indinavir (anti-viral), the antibiotic 4-deazaformycin B89, the antitumoral agent (+)-varitriol90 and, *importantly*, the COVID-19 drug Remdesivir via D-ribonolactone.[77,101–103]

Partial hydrogenation of LGO leads to dihydrolevoglucosenone (Cyrene™), a liquid which displays similar properties to a range of dipolar aprotic solvents such as *N*-methylpyrrolidone (NMP), dimethylformamide, and dimethylacetamide.[104] The reduction is generally done with a heterogeneous Pd catalyst, but can also be performed using an alkene reductase, the latter way eliminating any residual Pd presence in the Cyrene product.[104,105]

Full hydrogenation of LGO yields *endo/exo*-levoglucosanol [*endo/exo* LGOL], a set of compounds from which a range of interesting bio-derived platform molecules can be synthesized such as *cis/trans* [5-(hydroxymethyl)tetrahydrofuran-2-yl]methanol (THFDM); 1,2,6-hexanetriol (HTO); 1,6-hexanediol (1,6-HD) and (S,R)/(S,S) hexane-1,2,5,6-tetrol (1256HT). The different structures and their (potential) applications are shown in Fig. 7.[106]

While Cyrene can be converted into LGOL using expensive Group VIII metals,[107] the challenge faced here was to perform the reduction with a heterogeneous base metal catalyst, in a chemoselective way (i.e., no acetal hydrolysis and subsequent reactions), in water, and without any base metal leaching. Interestingly, this conundrum could be resolved using a (mesoporous) Cu-PMO. Of particular note is the low reaction temperature (333–373 K), which is for one unusually low for copper-mediated hydrogenations, but also allowed for the use of PMO in the presence of water.[106] In batch reactions, a positive reaction order was found but, remarkably, no influence of the hydrogen pressure on the endo/exo ratio was observed. Interestingly, despite the high Cu loading of 8 wt%, TEM imaging revealed the abundant presence of small nanoparticles with only a 2–3 nm diameter.

This is in marked contrast with the 20–50 nm Cu-nanoparticles commonly observed with the Cu-PMO|scMeOH catalytic systems. Analysis of the former Cu-PMO nanoparticles using Cu 2p XPS and Auger spectroscopy revealed the clear absence of Cu(II) species, but could not establish whether the observed nanoparticles adhered to Cu(0) or rather Cu(I). Consecutive flow reactions with oxidized/reduced (ox/red) Cu-PMO revealed a mild decrease of the catalytic activity over the first three runs and stabilization over the fourth and fifth run. Importantly, the observed decrease in catalytic activity could not be tied in with any Cu or Al leaching. Very minor Mg-leaching (~9.5 ppm) was however observed, which could be explained by the acidic nature of Cyrene's geminal diol (aka Cyrene's hydrate).[106] It is however noteworthy that the observed Mg-leaching could not be corroborated to a change in the pore sizes of the used Cu-PMO catalyst. The involvement of Cyrene's acidic hydrate with the observed Mg-leaching, brings an important element forward, in that in this case batch reaction operation is likely very different from flow operation.[106,108] Indeed, in batch reaction any Mg loss is confined to the reactor, while in flow the Mg is gradually and continuously lost. Additionally, the equilibrium between Cyrene and its hydrate imposes a certain pH on the reaction mixture. While in batch this pH value increases as the Cyrene substrate gets hydrogenated into LGOL, the situation is markedly different in flow, where the catalytic bed sees a near constant stream of aqueous Cyrene at a fixed pH of ~3.75. Most surprisingly, TEM analysis of the Cu-PMO catalyst before and after reaction revealed very similarly sized Cu-nanoparticles, remaining at ~2 nm diameter. Stabilization of the catalytic activity could be obtained with only reduced Cu-PMO. This suggests for a negative influence of oxidative pretreatment of the Cu-PMO catalyst on the catalytic activity/stability. Reference reactions with 8 wt% Cu/SiO_2 and 0.4 wt% Pd/Al_2O_3 revealed respectively equal and double the catalytic activity. However, in contrast to reduced Cu-PMO, the observed initial catalytic activities with Cu/SiO_2 and 0.4 wt% Pd/Al_2O_3 could not be maintained over time.[106]

3.4 The upgrading of pyrolysis oils in supercritical methanol

An alternative way to create value from biomass is pyrolysis. Essentially this concerns the thermochemical breakdown of lignocellulosic material under an inert atmosphere, typically yielding a mixture of gas, liquid and char.[109,110] However, upon the use of fast heating rates (10^3–10^4 K s^{-1}), short residence times (<1 s) and moderate temperatures (673–873 K), biomass pyrolysis tends to favor the formation of the liquid component—termed 'bio-oil'.[111,112] The latter substance is a complex aqueous mixture of acids, alcohols, aldehydes, ketones, (oligo)saccharides, anhydrosugars, as well as phenolics and lignin-related fractions.[113] In view of bio-oil's water and oxygen content, its pH and high viscosity, but equally its instability and proneness to re-polymerization, attempts to convert bio-oil into fuels have been severely hampered.[112] To alleviate these problems, many methods for the prior stabilization of the bio-oil have been explored/described. Generally, these fall in one of the following classes: (a) bio-oil conditioning, (b) hydrodeoxygenation and/or (c) catalytic cracking. Following the successful use of Cu-PMO|scMeOH toward the valorization of HMF, we set out for the upgrading of bio-oil and fractions thereof. In a first approach, we investigated the Cu-PMO|scMeOH mediated conversion of neat levoglucosan and cellobiose, two very abundant products in cellulose pyrolysis oils.[114] While at high conversion these compounds were mainly converted into a range of C2–C7 mono-alcohols (up to 77% selectivity), the specific transformation of cellobiose allowed for the initial formation of sizeable amounts of (partly methylated) C2–C4 diols in up to 12% selectivity. In taking this further onto aqueous extracted sugar fractions from real pyrolysis oils, very similar compositions of (partly methylated) mono-alcohols and diols as those encountered with the processing of the cellobiose model compound were observed.[113] Elevating the process onto higher initial saccharide concentrations (15 wt%), therewith making the process more industrially relevant, required a change in the operating conditions. Indeed, as to control better the developing reaction pressures; a deliberate change was made from methanol to ethanol as the reaction solvent, that way however requiring the deliberate addition of 50 bar of hydrogen gas. This change had no marked influence on the overall product distribution. It was further shown that in the process residual poly/oligo-saccharides in the sugar fractions were significantly depolymerized. In conclusion, while it can be envisaged that the obtained mixtures of (partly methylated) mono-alcohols and diols can effectively be separated into their individual components (e.g., by distillation), it is also noteworthy that said mixture also offers great potential for co-feeding in zeolite-based cracking processes.[113]

3.5 The Guerbet reaction

With the wide availability of bio-ethanol from first generation biorefineries, the direct conversion of ethanol into higher alcohols, *commonly known as the*

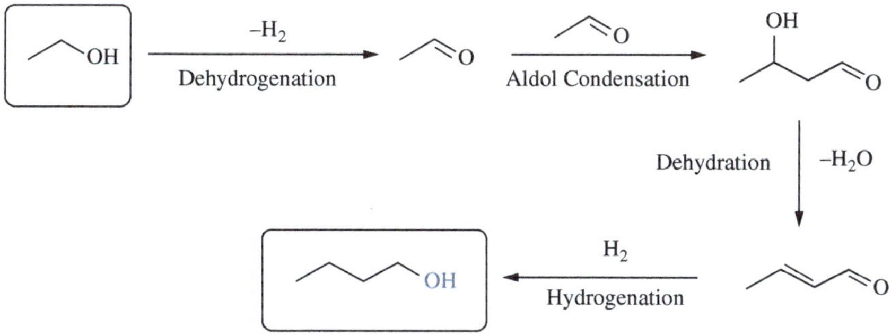

Fig. 8 Guerbet coupling of ethanol to 1-butanol.

Guerbet reaction, is a much-investigated transformation.[115–118] The simplest of the Guerbet alcohols is 1-butanol (Fig. 8), a molecule which is notably non-hygroscopic, non-corrosive, and of overall low toxicity.[119] From a fuel perspective, 1-butanol makes sense as it has an energy density of 29.3 MJ L^{-1}, which comes much closer to the one of gasoline (34.2 MJ L^{-1}) than does ethanol (24 MJ L^{-1}).[115] In this respect, it's also noteworthy that 1-butanol is fully compatible with gasoline engines as well as the existing fuel pipeline infrastructure. Butanol is fully miscible with gasoline in all ratios, and in the USA, it can be legally added to gasoline in up to 11.5%. It is also noteworthy that its lower vapor pressure and higher flash point makes it intrinsically safer to handle than gasoline.[119] Butanol also holds vast potential as a commodity and platform molecule. In this respect butyraldehyde and butyric acid are most noteworthy, while they have both mainstream applications in the solvent, polymer and specialty chemical markets. Also, butenes and butadiene hold additional opportunities as fuels and in the synthetic elastomers' markets.[119] Currently, butanol is mainly produced by the oxo process, which encompasses two steps notably (1) the cobalt/rhodium catalyzed hydroformylation of propylene to butyraldehyde and (2) the nickel-mediated hydrogenation of the latter aldehyde to form 1-butanol in 95% yield.[120,121] With cobalt being highly sought after for the production of Li-batteries, rhodium being a rare and consequently very expensive metal, and propylene being a high demand petrochemical, ample incentive exists for the development of alternative routes to 1-butanol.[119,122,123]

MgO, hydrotalcites and hydroxyapatites are among the most reported heterogeneous catalysts for the Guerbet reaction, therewith offering a clear opportunity for our Cu-PMO catalyst. Moreover, the presence of additional Cu in the hydrotalcite formulation was expected to facilitate the ethanol dehydrogenation step, which is commonly referred to as the rate-limiting step to the Guerbet reaction.[115] Setting the baseline, the sole use of AlMg-PMO at 583 K led to the formation of 5.6% 1-butanol at 30.8% ethanol conversion.[115] Using Cu-PMO led to an expected increase in conversion, yet did not change the selectivity to the desired 1-butanol. This likely relates to a markedly increased ethyl acetate formation (18.1% yield). The latter compound is formed through dehydrogenation of the hemiacetal that is produced from the reaction of ethanol and the acetaldehyde formed upon dehydrogenation. In using Ni-PMO the yield of 1-butanol increased slightly to 8.8%, but the main products remained longer chain alcohols.[115] Intriguingly, best catalytic activity was achieved with a NiCu-PMO, yielding between 18.7% and 21.1% 1-butanol at conversions levels between 47.9% and 62.4%. Best results were obtained with a Cu/Ni ratio of *one*. It was further found that both the 1-butanol yield and ethanol conversion increased with the applied temperature. However, when working in the higher T range (~593 K), the carbon balance decreased to 75% due to the formation of volatile by-products. As an additional reason to the observed 1-butanol selectivity decrease, self-condensation of formed 1-butanol can also be mentioned. Recycling experiments showed a concurrent decrease of the 1-butanol yield (21% → 14%) and the specific surface area (255.7 $m^2\,g^{-1}$ → 120.5 $m^2\,g^{-1}$).[115] Structural characterization of the CuNi-PMO with TEM showed the formation of 20–50 nm sized nanoparticles. Through XRD, the latter nanoparticles were shown to be a Cu—Ni alloy. Interestingly, the formation rate of 1-butanol increased linearly with the ratio of base to acid sites.[115]

While this work concerned the first report on the synergistic use of Cu and Ni in the Guerbet reaction, recent work by Lopez-Olmos et al. confirmed the beneficial effect of equimolar amounts of copper and nickel in the Guerbet coupling. Indeed, using a 2.5Cu2.5Ni-Mn_2O_3/graphite catalyst a selectivity of 39% to 1-butanol could be obtained at 30% ethanol conversion.[124]

Petrolini et al. investigated the continuous gas phase Guerbet reaction with Cu-PMO in the presence of hydrogen at 598 K.[125] Interestingly, the partial pressure of hydrogen had a significant effect on the reaction outcome, with atmospheric pressures of H_2 then favoring acetaldehyde (sel. 56%) over the desired C4+ alcohol and aldehyde (hereafter C4+) formation (sel. 22.1%), at an ethanol conversion level of 17.9%. Increasing the applied hydrogen pressure led to ever increasing amounts of C4+ products at the expense of acetaldehyde, reaching respective selectivity of 49.2 and 12.8% at 31 bar H_2. In contrast, the observed ethanol conversion was only found weakly dependent on the applied hydrogen pressure, reaching a maximum at 19.7%.[125] Comparing the experimentally determined equilibrium molar ratio of acetaldehyde to ethanol to tabulated thermodynamic data revealed that at low H_2 pressures the experimental ratio was far off equilibrium (0.12 vs 1.22), while at higher H_2 pressure it approached equilibrium (0.03 vs 0.04 at 31 bar). A decreasing mole fraction of acetaldehyde with increasing hydrogen pressure was thus established but given the increase in the overall pressure in the system, this also represented an increase of the partial pressure of acetaldehyde, reaching an asymptotic value of $\sim$15 kPa. Therefore, all reaction rates which depend strongly on the partial pressure of acetaldehyde (e.g., aldol condensation) are favored with increasing pressure—hence the observed increase of C4+ selectivity with the applied pressure.[125]

4. Cu-PMO catalyzed depolymerization of lignocellulosic biomass to alkanes

Petroleum-derived alkanes are at the center of our present-day society, providing around 81% of the global primary energy supply.[126] With petroleum being a finite resource, and demand for oil ever increasing, a need exists for new sources of alkanes.[127] With wind, solar and geothermal energy providing for electricity, *and not liquid fuel*, and given increasing CO_2 emission legislation, biomass is likely to play an important role as an alternative source of alkanes.[128–130] To date, many catalytic routes have been reported to convert biomass into alkanes.[131–134] As most of these routes concern the selective conversion of either cellulose or lignin, their applicability in the real world is however compromised. Indeed, the required pretreatment and fractionation processes to access cellulose or lignin will undoubtedly entail a high cost, and this both in terms of time, energy and materials.[135] With fuel alkanes holding a characteristically low profit margin, also no real margin exists to absorb potential biomass pretreatment costs.

Following our successes in the Cu-PMO-mediated transformation of polysaccharides, lignin and pyrolysis oils, we set out for a method to convert lignocellulose into alkanes. Thereby we envisaged a two-step sequence consisting of a lignocellulose depolymerization and defunctionalization step, down to (mono)alcohols, followed by deep hydrodeoxygenation (HDO), that way turning the produced alcohols into alkanes.[135] To the first step, it was found that increasing levels of Cu in the PMO led to both a higher conversion of the raw lignocellulose and higher selectivity to the target alcohols. This is in line with a more pronounced methanol reforming at the higher

Cu loadings, then providing for a higher quantity of available hydrogen gas. The use of lower copper loadings tended to accentuate the basic properties of the PMO catalyst, showing hence an increased level of aldol condensation products.[135]

To the HDO step, we followed an approach of increasing complexity, investigating consecutively the HDO of (1) cyclohexanol, (2) pre-made mixtures of secondary and primary alcohols, and (3) the true alcohol-rich reaction product from the first step. To the cyclohexanol case, the acid catalysts Nafion, ZSM-5 and USY-600 proved to be the better choices. Use of the latter acid catalysts, in combination with Pd/C as a hydrogenation catalyst, gave full conversion of the cyclohexanol substrate to cyclohexane in 4 h at 413 K. Interestingly, ZSM-5 proved the better catalyst when applying the HDO/hydrogenation catalytic system to an equimolar mixture of primary and secondary alcohols. In the case of sole primary alcohols, ZSM-5 and Nafion were able to convert the latter mixture to neat alkanes in ~85% conversion, be it at a somewhat higher reaction temperature of 433–453 K. The use of USY-600|Pd/C with primary alcohols led to the substantial formation of ethers.[135] Both ZSM-5|Pd/C and Nafion|Pd/C were found equally capable of converting the alcohol-rich product mixture obtained in step 1 to alkanes, reaching a conversion of ~83% (453 K). However, in terms of selectivity, Nafion|Pd/C showed significantly more selectivity to long chain alkanes (i.e., >C8) than ZSM-5|Pd/C, reaching respective values of 30.3% and 16.2%. This could be attributed to a difference in pore size, *with Nafion being largely mesoporous and ZSM-5 microporous*, but equally the pronounced cracking properties of ZSM-5 could have played an important role here.[135]

5. Cu-PMO mediated synthesis of *N*-heterocycles

5.1 Benzimidazoles

Next to depolymerization/decomposition of lignin and polysaccharides, the Cu-PMO|MeOH system can also be used to construct medicinally relevant scaffolds such as benzimidazoles. The latter are recurrent structural motifs in a comprehensive range of pharmaceuticals including anticancers *(Imet 3393)*, antihypertensives *(Diabazole)*, antivirals *(Maribavir)*, antifungals *(Carbendazim)*, anti-HIVs, anti-convulsants, analgesic (*Bezitramide*), anti-ulcerative *(Omeprazole)*, antihistaminic (*Astmizole),* and anti-diabetics.[136–138] Benzimidazoles are most commonly obtained by reacting *o*-phenylenediamine with carbonyl containing reagents such as acids, aldehydes,

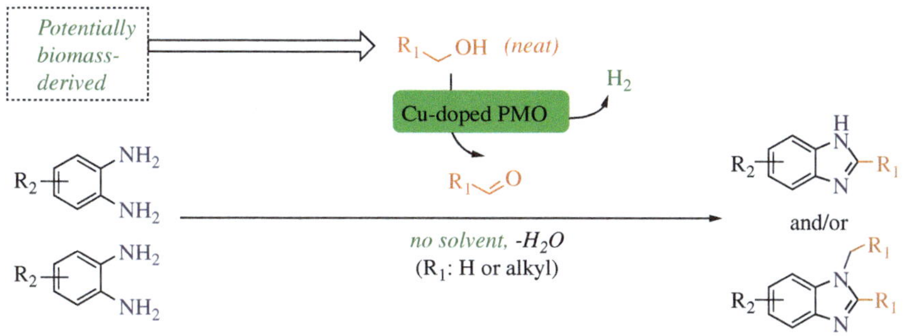

Fig. 9 Cu-PMO mediated formation of benzimidazoles from anilines and nitroanilines.

anhydrides but also other methods are known.[33,137,139] Here however an elegant opportunity presented itself in that at higher reaction temperatures Cu-PMO catalyzes the non-oxidative dehydrogenation of methanol (and other alcohols), therewith creating formaldehyde and H_2 (Fig. 9), opening the way of capturing the formed carbonyl compounds directly into the construction of the desired benzimidazole backbone.[139] And indeed, at 553 K, Cu-PMO neatly converts *o*-phenylenediamine in a mixture of benzimidazole (A) and *N*-methylbenzimidazole (B), with respective selectivity of 43% and 57% at full conversion. In applying a lower reaction temperature of 523 K, full *o*-phenylenediamine conversion showed the same selectivity to A and B, while at lower conversions (70%–80%) an A:B ratio of ~2 was found. Introducing Ni (5 wt%) into the Cu-PMO formulation allowed for a stark increase of the A:B ratio to ~4.5, while the alternative presence of Zn (5 wt%) afforded A in 50% selectivity at 92% conversion.[139]

In a further advancement we studied the formation of benzimidazoles from 2-nitroanilines (Fig. 9), the rational being that the latter compounds are more readily available than their equivalent phenylenediamines.[139] Focusing on 2-nitroaniline as the substrate, CuZn-PMO was identified as the best catalyst yielding benzimidazole in 92% selectivity at full conversion. Interestingly, neither a higher content of Zn in the catalyst formulation, nor the deliberate addition of Ru, did improve this result. It was further shown that the Cu-PMO and CuZn-PMO catalysts were active toward a range of aromatic diamines and nitroanilines.[139] It is finally important to note that the here presented catalytic system is highly advantageous as it is (1) a one pot operation, (2) can be applied in the presence of a variety of *n*-alcohols and (3) does not require the deliberate addition of additives (e.g., oxidants, bases, acids) as it essentially derives the desired carbon equivalents directly from (part of) the solvent.[57]

Mechanistically, we envisaged the occurrence of a Cu(Zn)-PMO mediated high temperature dehydrogenation of methanol to formaldehyde, followed by imine formation and subsequent cyclization. Dehydrogenation of the so formed benzimidazoline intermediate, to the aromatized benzimidazole, further drives the process. The occurrence of often sizeable amounts of methylated benzimidazoles we attribute to the potential hydrogenation of formed imines. To the case of the 2-nitroanilines, the mechanism also involves a nitro group reduction, something we found possible with Cu-PMO catalysts.[139]

5.2 Tetrahydro-2-benzazepines

N-heterocycles are omnipresent in nature ranging from alkaloids (e.g., caffeine, morphine) to DNA strands. Likewise, they are also commonly encountered in pharmaceuticals. Overall spoken, the heteroatoms comprised in them tend to engage in hydrogen bonding interactions with functional groups in proteins, but equally do they tend to affect the solubility of drugs and their metabolic profile.[140] While the main focus in medicinal chemistry has long been on the making of planar aromatic heterocycles, recent years have seen a pronounced shift toward three-dimensional heterocycles, where the heteroatom-containing ring is non-aromatic, hence non-planar.[141] The underlying reason to this is that a much wider chemical space can be reached with a saturated vs an unsaturated heterocycle.[141]

From a different angle, the making of pharmaceuticals has long been plagued by appreciable waste formation. Indeed, the many steps required to create a given medicine from a suitable petroleum-derived molecule, has imparted particularly low E-factors (25–100) to the overall chemical processes.[142] Hence a significant drive exists to create shorter and more efficient routes to pharmaceuticals and especially heterocyclic compounds. Ideally, this should involve biologically derived compounds, while they tend to hold already a high degree of complexity.

Following the LignoFlex invention (see above, Fig. 6),[75,78] we set out on a strategy to create value from DHCA and DHSA by transforming these into medicinally relevant heterocycles (Fig. 10). Firstly, this required the transformation of DHCA/DHSA's aliphatic alcohol into a suitable secondary amine. We have previously found that Shvo's catalyst is highly active in the *N*-alkylation of unprotected amino acids with alcohols.[143] Hence, anticipating potential catalyst compatibility issues with the phenolic moiety present in the lignin-derived DHCA/DHSA, we started our investigations

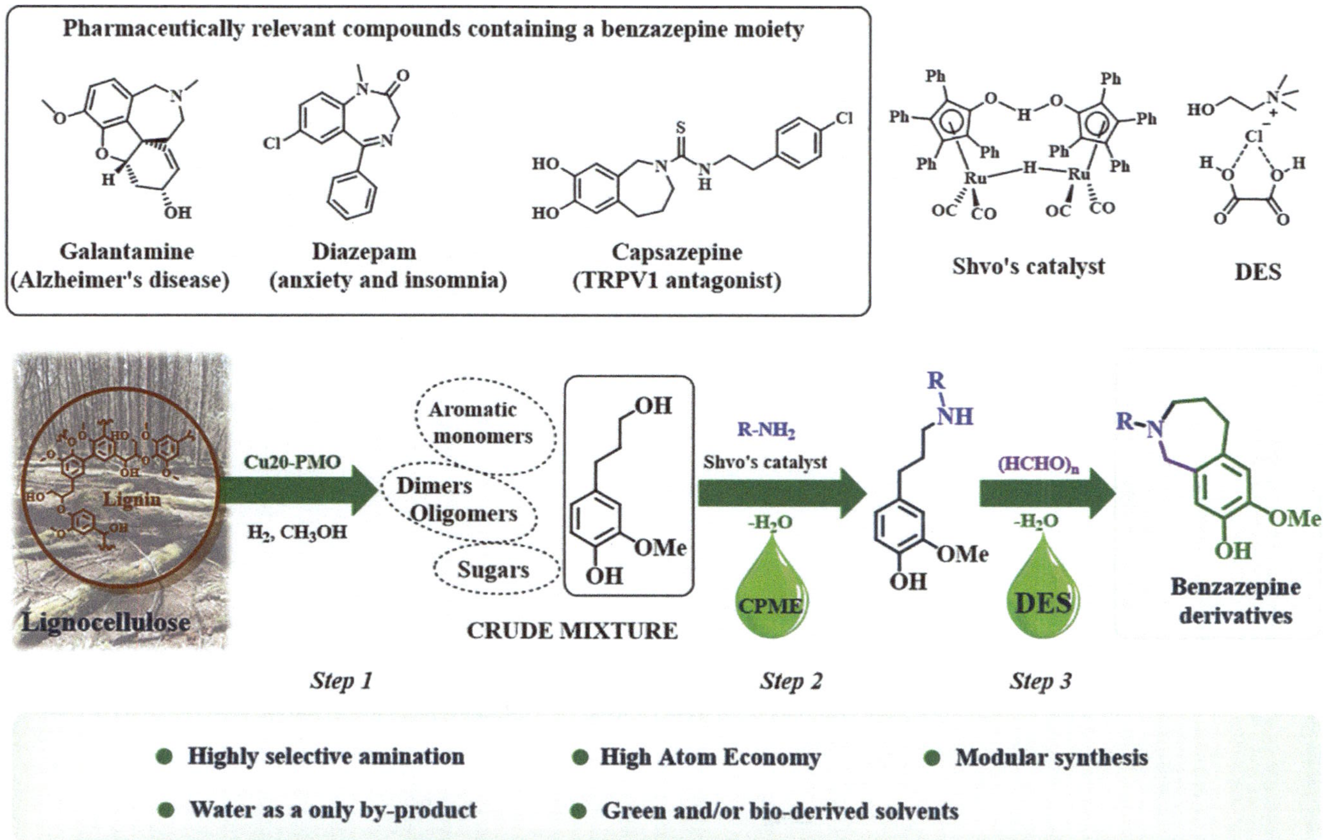

Fig. 10 Obtaining tetrahydro-2-benzazepines directly from lignocellulose in three steps using highly atom economic transformations and non-toxic alternative solvents. *Adapted from ACS Cent. Sci. 2019, 5, 10, 1707–1716 (DOI: 10.1021/acscentsci.9b00781).*

using this particular catalyst for our direct amination attempts via the borrowing hydrogen methodology. Indeed, in utilizing Shvo's catalyst in the presence of cyclopentyl methyl ether (CPME), *as a non-toxic solvent*, we were able to transform the aliphatic alcohol in DHCA into a secondary amine with almost full selectivity, without any over-alkylation when using aniline as coupling partner.[144] Then, depending on the type of amine used, isolated yields of the target/desired secondary amines ranged from 43% to 97%. It is noteworthy that the process displayed outstanding functional group tolerance, allowing for the use of anilines with a range of reducible functional groups. Also, the synthesis of tertiary amines, which could subsequently be quaternated, belonged to the possibilities. To the specific synthesis of 3-aminopropyl guaiacol, Shvo's catalyst proved ineffective, and only Raney Nickel turned out to be an effective catalyst.[144]

Importantly, it could also be shown that the above described procedure was applicable to crude mixtures of DHCA/DHSA—as directly derived from Cu-PMO/MeOH processed pine and poplar lignocellulose. Due to the presence of sugars and sugar-derived small molecules, and lignin oligomers, this required the use of a larger amount of the Shvo catalyst (10 wt%) and excess of the primary amine (4 eq.). Additionally, since the methanol used in the LignoFlex process interfered with the Shvo catalyst, it was necessary to exchange the latter solvent for CPME.

The next stage concerned the ring cyclization, which could in principle be obtained using a Pictet-Spengler reaction.[145] However, in starting from paraformaldehyde, and in screening a wide range of Bronsted/Lewis acid catalysts and organic solvents, the reaction yielded not only the desired seven-membered N-heterocycle, but equally an unusual nine-membered N-heterocycle. The latter compound features a methylene bridge between the two involved aromatic rings. Another typical side reaction is the methylation of the starting secondary amine, which relates to the competing Eschweiler–Clarke reaction.[144,146,147] The occurrence of the Eschweiler-Clarke reaction is not always undesired as sometimes it's specifically sought out to methylate tetrahydro-2-benzazepines and other *N*-heterocycles.[148]

Following the successful use of deep eutectic solvents (DESs) in the Pictet-Spengler reaction of tryptamine and p-anisaldehyde,[149] we decided to explore its applicability to our case. To our delight, the use of choline chloride/oxalic acid led uniquely to the formation of the target tetrahydro-2-benzazepines with isolated yields ranging from 41% to 95%, at full conversion, and this at temperatures as low as 323 K. The rationale behind this success lays in the fact that DES not only functions as a powerful solvent, but equally as an acid catalyst and a highly ionic reaction medium, which very likely stabilizes charged reaction intermediates, thereby facilitating the cyclization step. Furthermore, presumably the hydrogen donor abilities of the used acid component also play an important role. Indeed, DES are typically comprised of two or more components, *these being a hydrogen bond donor and a hydrogen bond acceptor*, the association of which creating an eutectic mixture. Furthermore, depending on the composition of the DES it can display Lewis acidity (= metal-involved, e.g., Zn) or Bronsted acidity (e.g., COOH groups). In case all the components of a DES are naturally derived they are termed natural deep eutectic solvents or NADES.[150] Noteworthy advantages of (NA)DESs are their low cost, their straightforward preparation and their biodegradability, low volatility and non-toxicity.[151,152]

To summarize, our approach to obtain tetrahydro-2-benzazapine derivatives aims to create sustainable pathways to this pharmaceutically important class of compounds. Firstly, the method recognizes the inherent functionality present in lignin, which inspires the two subsequent waste free steps: alcohol to amine transformation via the borrowing hydrogen method where the only side-product is water and the subsequent cyclization to obtain the challenging seven-membered N-heterocycle in an alternative, naturally derived and non-toxic reaction medium.

The relevance of tetrahydro-2-benzazepines lays in their use as dopamine receptor agonists and antagonists, but equally in their effectiveness in combating neurological disorders such as Parkinson's and Alzheimer's disease.[153] The most commonly known tetrahydro-2-benzazepine is (-)-Galantamine (Fig. 10) which is the active ingredient in the anti-Alzheimer drug Reminyl®.[154] The here presented lignin-derived tetrahydro-2-benzazepines were evaluated toward their anti-infective activities. Comparison with erythromycin, *a known antibiotic toward gram-positive bacteria*, gave a 10-fold higher "minimum inhibitory concentration" (MIC). In the specific case of *E. coli* TolC, the MIC was only 5.5 times higher than obtained with the reference antibiotic, chloramphenicol. Regarding their anti-cancer activities, a very similar picture emerged with recorded activities being ~44 times lower than the reference compound, doxorubicin.[144] These are however very encouraging and promising findings and the more while none of our compounds has been optimized.

6. Conclusions and future perspectives

In conclusion, in the above we have illustrated the wide applicability of the Cu-PMO catalyst to the production of fuels, commodity chemicals and pharmaceuticals. While Cu-PMO|scMeOH processing of cellulose or pyrolysis-derived bio-oils typically leads to the formation of primary alcohols, its application on lignin tends to lead to a mixture of mainly cyclic secondary alcohols/diols. Of significant interest thereby was also the incorporation of methanol, be it as methoxy units or through straight methylations. The function of the latter did however not limit itself to the introduction of valuable branching in the molecular chain but proved also a way to control reaction chemo-selectivity.

Aromatic monomers from organosolv lignin can be obtained with controlling the amount of added H_2 gas, and under milder conditions, temperatures at which the methanol solvent is no longer supercritical. Applied to raw lignocellulose, an efficient way forward was for a two-stage process, called *LignoFlex*, consisting of a low T reductive depolymerization of native lignin, and the subsequent depolymerization of process residues consisting mainly of (hemi)-cellulose mixed with the catalyst, under supercritical methanol conditions whereby reductive equivalents originate from the solvent itself—this latter step also serves as unique way of catalyst regeneration and gives access to aliphatic small molecules from the sugar platform which can be channeled to jet-fuel range alkanes.

Uniquely, the LignoFlex procedure led to the clean formation of dihydroconiferyl alcohol (DHCA) from the lignin. Among the many successful research accounts on lignin depolymerization, the formation of DHCA is *rarer yet full of synthetic opportunity*, as demonstrated with obtaining a range of aromatic amines and other important building blocks that can be further explored as precursors for the formation of pharmaceutical compounds, polymers, materials. An illustrative example was the formation of tetrahydro-2-benzazepines by first turning DHCA's primary alcohol into a secondary amine, *using Shvo's catalyst*, and cyclizing the so obtained secondary amine into the desired seven-membered N-heterocycle using a Pictet-Spengler reaction in DES. The Cu(Zn)-PMO mediated conversion of phenylenediamine (2-nitroaniline) into neat and *N*-methyl substituted benzimidazoles, illustrates the general applicability of the Cu-PMO catalytic system toward the synthesis of heterocycles.

Also, at the lower temperature range Cu-PMO proved to be a very useful catalyst and most notably toward the chemo-selective conversion of 5-HMF into 2,5-bis(hydroxymethyl)furan or dimethylfuran, and the aqueous hydrogenation of dihydrolevoglucosenone (Cyrene™) into levoglucosanol. Additionally, the important Guerbet reaction, *in which ethanol is transformed into higher alcohols*, was shown feasible with Cu-PMO (gas phase) and CuNi-PMO (batch—scMeOH).

Closing, the Cu-PMO|scMeOH catalytic system is unique in that it combines full conversion of lignocellulosic biomass to products, in situ hydrogen production by methanol reforming, water elimination through the water gas shift reaction, and the use of copper as a non-expensive base metal—in a one pot reaction. Its future looks bright as methanol can be derived from both fossil and biomass sources, *by turning these into syn gas*, but equally also from the hydrogenation of the greenhouse gas CO_2.[155,156]

Acknowledgments

K.B. acknowledges the European Research Council's for the ERC Starting Grant CatASus (2015) under grant agreement No. 638076 and for the ERC Proof of Concept grant PURE (2019). K.B. also thanks the Netherlands Organization for Scientific Research (NWO) for part funding of the Vidi research program Talent scheme (No. 723.015.005).

References

1. Service, R. F. *Science* **2019**. https://www.sciencemag.org/news/2019/09/can-world-make-chemicals-it-needs-without-oil.
2. Tuck, C. O.; Perez, E.; Horvath, I. T.; Sheldon, R. A.; Poliakoff, M. *Science* **2012**, *337* (6095), 695–699.
3. Rosales-Calderon, O.; Arantes, V. *Biotechnol. Biofuels* **2019**, *12* (1), 240.
4. F&D Finance & Development, 2019. https://www.imf.org/external/pubs/ft/fandd/2019/12/pdf/fd1219.pdf.
5. *World Development Report: Development and Climate Change; The World Bank*; 2010. https://openknowledge.worldbank.org/handle/10986/4387.
6. Dahmen, N.; Lewandowski, I.; Zibek, S.; Weidtmann, A. *Glob. Change Biol. Bioenergy* **2019**, *11* (1), 107–117.
7. Huber, G. W.; Iborra, S.; Corma, A. *Chem. Rev.* **2006**, *106* (9), 4044–4098.
8. Corma, A.; Iborra, S.; Velty, A. *Chem. Rev.* **2007**, *107* (6), 2411–2502.
9. van Steen, E.; Claeys, M. *Chem. Eng. Technol.* **2008**, *31* (5), 655–666.
10. Leckel, D.; Liwanga-Ehumbu, M. *Energy Fuel* **2006**, *20* (6), 2330–2336.
11. Dapsens, P. Y.; Mondelli, C.; Perez-Ramirez, J. *ACS Catal.* **2012**, *2* (7), 1487–1499.
12. Goyal, H. B.; Seal, D.; Saxena, R. C. *Renew. Sustain. Energy Rev.* **2008**, *12* (2), 504–517.
13. Bridgwater, A. V. *Biomass Bioenergy* **2012**, *38*, 68–94.
14. Bridgwater, A. V.; Meier, D.; Radlein, D. *Org. Geochem.* **1999**, *30* (12), 1479–1493.
15. Huber, G. W.; Chheda, J. N.; Barrett, C. J.; Dumesic, J. A. *Science* **2005**, *308* (5727), 1446–1450.
16. Venderbosch, R. H.; Ardiyanti, A. R.; Wildschut, J.; Oasmaa, A.; Heeresb, H. J. *J. Chem. Technol. Biotechnol.* **2010**, *85* (5), 674–686.
17. Gomes, F. J. B.; Santos, F. A.; Colodette, J. L.; Demuner, I. F.; Batalha, L. A. R. *Nat. Resour. J.* **2014**, *5* (9), 419–432.
18. Menon, V.; Rao, M. *Prog. Energy Combust. Sci.* **2012**, *38* (4), 522–550.
19. Takkellapati, S.; Li, T.; Gonzalez, M. A. *Clean Technol. Envir.* **2018**, *20* (7), 1615–1630.
20. Fan, J. J.; De Bruyn, M.; Budarin, V. L.; Gronnow, M. J.; Shuttleworth, P. S.; Breeden, S.; Macquarrie, D. J.; Clark, J. H. *J. Am. Chem. Soc.* **2013**, *135* (32), 11728–11731.
21. Lee, H. R.; Kazlauskas, R. J.; Park, T. H. *Sci. Rep.* **2017**, 7, 12216.
22. Mosier, N.; Wyman, C.; Dale, B.; Elander, R.; Lee, Y. Y.; Holtzapple, M.; Ladisch, M. *Bioresour. Technol.* **2005**, *96* (6), 673–686.

23. Hendriks, A.; Zeeman, G. *Bioresour. Technol.* **2009**, *100* (1), 10–18.
24. Zhao, X. B.; Cheng, K. K.; Liu, D. H. *Appl. Microbiol. Biotechnol.* **2009**, *82* (5), 815–827.
25. Zhang, K.; Pei, Z. J.; Wang, D. H. *Bioresour. Technol.* **2016**, *199*, 21–33.
26. Kumar, B.; Bhardwaj, N.; Agrawal, K.; Chaturvedi, V.; Verma, P. *Fuel Process. Technol.* **2020**, *199*, 106244.
27. De bruyn, M.; Fan, J.; Budarin, V. L.; Macquarrie, D. J.; Gomez, L. D.; Simister, R.; Farmer, T. J.; Raverty, W. D.; McQueen-Mason, S. J.; Clark, J. H. *Energ. Environ. Sci.* **2016**, *9* (8), 2571–2574.
28. Liu, Z. H.; Olson, M. L.; Shinde, S.; Wang, X.; Hao, N. J.; Yoo, C. G.; Bhagia, S.; Dunlap, J. R.; Pu, Y. Q.; Kao, K. C.; Ragauskas, A. J.; Jin, M. J.; Yuan, J. S. *Green Chem.* **2017**, *19* (20), 4939–4955.
29. Ji, H. R.; Dong, C. H.; Yang, G. H.; Pang, Z. Q. *ACS Sustainable Chem. Eng.* **2018**, *6* (11), 15306–15315.
30. Ferreira, J. A.; Taherzadeh, M. J. *Bioresour. Technol.* **2020**, *299*, 13. 122695.
31. Chen, F.; Tobimatsu, Y.; Havkin-Frenkel, D.; Dixon, R. A.; Ralph, J. *Proc. Natl. Acad. Sci. U. S. A.* **2012**, *109* (5), 1772–1777.
32. Qiu, S.; Wang, M.; Fang, Y.; Tana, T. *Sustain. Energy Fuels* **2020**, *4*, 5588–5594.
33. Brasche, G.; Buchwald, S. L. *Angew. Chem. Int. Ed.* **2008**, *47* (10), 1932–1934.
34. Liao, Y. H.; Koelewijn, S. F.; Van den Bossche, G.; Van Aelst, J.; Van den Bosch, S.; Renders, T.; Navare, K.; Nicolai, T.; Van Aelst, K.; Maesen, M.; Matsushima, H.; Thevelein, J. M.; Van Acker, K.; Lagrain, B.; Verboekend, D.; Sels, B. F. *Science* **2020**, *367* (6484), 1385–1390.
35. Ouyang, X. H.; Huang, X. M.; Boot, M. D.; Hensen, E. J. M. *ChemSusChem* **2020**, *13* (7), 1705–1709.
36. Questell-Santiago, Y. M.; Galkin, M. V.; Barta, K.; Luterbacher, J. S. *Nat. Rev. Chem.* **2020**, *4* (6), 311–330.
37. Sun, Z.; Fridrich, B.; de Santi, A.; Elangovan, S.; Barta, K. *Chem. Rev.* **2018**, *118* (2), 614–678.
38. Schutyser, W.; Renders, T.; Van den Bosch, S.; Koelewijn, S. F.; Beckham, G. T.; Sels, B. F. *Chem. Soc. Rev.* **2018**, *47* (3), 852–908.
39. Clark, J. H. *J. Chem. Technol. Biotechnol.* **2007**, *82* (7), 603–609.
40. Ragauskas, A. J.; Beckham, G. T.; Biddy, M. J.; Chandra, R.; Chen, F.; Davis, M. F.; Davison, B. H.; Dixon, R. A.; Gilna, P.; Keller, M.; Langan, P.; Naskar, A. K.; Saddler, J. N.; Tschaplinski, T. J.; Tuskan, G. A.; Wyman, C. E. *Science* **2014**, *344* (6185), 709. 1246843.
41. Renders, T.; Van den Bosch, S.; Koelewijn, S. F.; Schutyser, W.; Sels, B. F. *Energ. Environ. Sci.* **2017**, *10* (7), 1551–1557.
42. Renders, T.; Van den Bossche, G.; Vangeel, T.; Van Aelst, K.; Sels, B. *Curr. Opin. Biotechnol.* **2019**, *56*, 193–201.
43. Sun, Z.; Cheng, J. L.; Wang, D. S.; Yuan, T. Q.; Song, G. Y.; Barta, K. *ChemSusChem* **2020**, *13* (19), 5199–5212.
44. Van den Bosch, S.; Schutyser, W.; Vanholme, R.; Driessen, T.; Koelewijn, S. F.; Renders, T.; De Meester, B.; Huijgen, W. J. J.; Dehaen, W.; Courtin, C. M.; Lagrain, B.; Boerjan, W.; Sels, B. F. *Energ. Environ. Sci.* **2015**, *8* (6), 1748–1763.
45. Ouyang, X. H.; Huang, X. M.; Zhu, J. D.; Boot, M. D.; Hensen, E. J. M. *ACS Sustainable Chem. Eng.* **2019**, 7 (16), 13764–13773.
46. Deuss, P. J.; Lancefield, C. S.; Narani, A.; de Vries, J. G.; Westwood, N. J.; Barta, K. *Green Chem.* **2017**, *19* (12), 2774–2782.
47. Cavani, F.; Trifiro, F.; Vaccari, A. *Catal. Today* **1991**, *11* (2), 173–301.
48. Debek, R.; Motak, M.; Grzybek, T.; Galvez, M. E.; Da Costa, P. *Catalysts* **2017**, 7 (1), 32–57.

49. Tichit, D.; Layrac, G.; Gerardin, C. *Chem. Eng. J.* **2019**, *369*, 302–332.
50. Xu, M.; Wei, M. *Adv. Funct. Mater.* **2018**, *28* (47), 1802943.
51. Nishimura, S.; Takagaki, A.; Ebitani, K. *Green Chem.* **2013**, *15* (8), 2026–2042.
52. Kaneda, K.; Mizugaki, T. *Green Chem.* **2019**, *21* (6), 1361–1389.
53. Wang, Q.; O'Hare, D. *Chem. Rev.* **2012**, *112* (7), 4124–4155.
54. Mishra, G.; Dash, B.; Pandey, S. *Appl. Clay Sci.* **2018**, *153*, 172–186.
55. Konsolakis, M.; Lykaki, M. *Catalysts* **2020**, *10* (2), 160–213.
56. Barrett, J. A.; Gao, Y.; Bernt, C. M.; Chui, M.; Tran, A. T.; Foston, M. B.; Ford, P. C. *ACS Sustainable Chem. Eng.* **2016**, *4* (12), 6877–6886.
57. Barta, K.; Ford, P. C. *Acc. Chem. Res.* **2014**, *47* (5), 1503–1512.
58. De Bruyn, M.; De Vos, D. E.; Jacobs, P. A. *Adv. Synth. Catal.* **2002**, *344* (10), 1120–1125.
59. Toledano, A.; Serrano, L.; Labidi, J.; Pineda, A.; Balu, A. M.; Luque, R. *ChemCatChem* **2013**, *5* (4), 977–985.
60. Rozovskii, A. Y.; Lin, G. I. *Top. Catal.* **2003**, *22* (3–4), 137–150.
61. Gokhale, A. A.; Dumesic, J. A.; Mavrikakis, M. *J. Am. Chem. Soc.* **2008**, *130* (4), 1402–1414.
62. Lindstrom, B.; Pettersson, L. J. *Int. J. Hydrogen Energy* **2001**, *26* (9), 923–933.
63. Xu, F. G.; Deng, M.; Li, G. Y.; Chen, S. H.; Wang, L. *Electrochim. Acta* **2013**, *88*, 59–65.
64. Bernt, C. M.; Bottari, G.; Barrett, J. A.; Scott, S. L.; Barta, K.; Ford, P. C. *Cat. Sci. Technol.* **2016**, *6* (9), 2984–2994.
65. P. Kuzmic, *Anal. Biochem.* **237**, 1996, 260–273.
66. Tundo, P.; Memoli, S.; Herault, D.; Hill, K. *Green Chem.* **2004**, *6* (12), 609–612.
67. Warner, G.; Hansen, T. S.; Riisager, A.; Beach, E. S.; Barta, K.; Anastas, P. T. *Bioresour. Technol.* **2014**, *161*, 78–83.
68. CABI. *Invasive Species Compendium.* https://www.cabi.org/isc/datasheet/4124 (Accessed 22 October 2020).
69. Darmawan, S.; Wistara, N. J.; Pari, G.; Maddu, A.; Syafii, W. *BioResources* **2016**, *11* (2), 3561–3574.
70. Barta, K.; Warner, G. R.; Beach, E. S.; Anastas, P. T. *Green Chem.* **2014**, *16* (1), 191–196.
71. Tobimatsu, Y.; Chen, F.; Nakashima, J.; Escamilla-Trevino, L. L.; Jackson, L.; Dixon, R. A.; Ralph, J. *Plant Cell* **2013**, *25* (7), 2587–2600.
72. Chen, F.; Tobimatsu, Y.; Jackson, L.; Nakashima, J.; Ralph, J.; Dixon, R. A. *Plant J.* **2013**, *73* (2), 201–211.
73. Feghali, E.; Carrot, G.; Thuery, P.; Genre, C.; Cantat, T. *Energ. Environ. Sci.* **2015**, *8* (9), 2734–2743.
74. Li, Y. D.; Li, S.; Kim, H.; Motagamwala, A. H.; Mobley, J. K.; Yue, F.; Tobimatsu, Y.; Havkin-Frenkel, D.; Chen, F.; Dixon, R. A.; Luterbacher, J. S.; Dumesic, J. A.; Ralph, J. *Sci. Adv.* **2018**, *4* (9), eaau2968.
75. Sun, Z.; Bottari, G.; Afanasenko, A.; Stuart, M. C. A.; Deuss, P. J.; Fridrich, B.; Barta, K. *Nat. Catal.* **2018**, *1* (1), 82–92.
76. Wang, W.; Li, N.; Li, G. Y.; Li, S. S.; Wang, W. T.; Wang, A. Q.; Cong, Y.; Wang, X. D.; Zhang, T. *Chem. Eng.* **2017**, *5* (2), 1812–1817.
77. Krishna, S. H.; Huang, K. F.; Barnett, K. J.; He, J. Y.; Maravelias, C. T.; Dumesic, J. A.; Huber, G. W.; De Bruyn, M.; Weckhuysen, B. M. *AIChE J.* **2018**, *64* (6), 1910–1922.
78. Sun, Z.; Barta, K. *Chem. Commun.* **2018**, *54* (56), 7725–7745.
79. Takeshima, H.; Satoh, K.; Kamigaito, M. *Macromolecules* **2017**, *50* (11), 4206–4216.
80. Williamson, J. J.; Bahrin, N.; Hardiman, E. M.; Bugg, T. D. H. *Biotechnol. J.* **2020**, *15* (7), 1900571.
81. Dubey, S. K.; Kharbanda, M.; Dubey, S. K.; Mathela, C. S. *Chem. Pharm. Bull.* **2010**, *58* (9), 1157–1160.
82. Trache, D.; Hussin, M. H.; Chuin, C. T. H.; Sabar, S.; Fazita, M. R. N.; Taiwo, O. F. A.; Hassan, T. M.; Haafiz, M. K. M. *Int. J. Biol. Macromol.* **2016**, *93*, 789–804.

83. Foster, E. J.; Moon, R. J.; Agarwal, U. P.; Bortner, M. J.; Bras, J.; Camarero-Espinosa, S.; Chan, K. J.; Clift, M. J. D.; Cranston, E. D.; Eichhorn, S. J.; Fox, D. M.; Hamad, W. Y.; Heux, L.; Jean, B.; Korey, M.; Nieh, W.; Ong, K. J.; Reid, M. S.; Renneckar, S.; Roberts, R.; Shatkin, J. A.; Simonsen, J.; Stinson-Bagby, K.; Wanasekara, N.; Youngblood, J. *Chem. Soc. Rev.* **2018**, *47* (8), 2609–2679.
84. Shrotri, A.; Kobayashi, H.; Fukuoka, A. *Acc. Chem. Res.* **2018**, *51* (3), 761–768.
85. Galebach, P. H.; Thompson, S.; Wittrig, A. M.; Buchanan, J. S.; Huber, G. W. *ChemSusChem* **2018**, *11* (23), 4007–4017.
86. Galebach, P. H.; McClelland, D. J.; Eagan, N. M.; Wittrig, A. M.; Buchanan, J. S.; Dumesic, J. A.; Huber, G. W. *ACS Sustainable Chem. Eng.* **2018**, *6* (3), 4330–4344.
87. van Putten, R. J.; van der Waal, J. C.; de Jong, E.; Rasrendra, C. B.; Heeres, H. J.; de Vries, J. G. *Chem. Rev.* **2013**, *113* (3), 1499–1597.
88. Chen, S.; Wojcieszak, R.; Dumeignil, F.; Marceau, E.; Royer, S. *Chem. Rev.* **2018**, *118* (22), 11023–11117.
89. Mascal, M. *ACS Sustainable Chem. Eng.* **2019**, 7 (6), 5588–5601.
90. Hu, L.; Lin, L.; Wu, Z.; Zhou, S. Y.; Liu, S. J. *Renew. Sustain. Energy Rev.* **2017**, *74*, 230–257.
91. van Zandvoort, I.; Wang, Y. H.; Rasrendra, C. B.; van Eck, E. R. H.; Bruijnincx, P. C. A.; Heeres, H. J.; Weckhuysen, B. M. *ChemSusChem* **2013**, *6* (9), 1745–1758.
92. Mascal, M. *ChemSusChem* **2015**, *8* (20), 3391–3395.
93. Zhang, L.; Michel, F. C.; Co, A. C. *J. Polym. Sci. A Polym. Chem.* **2019**, *57* (14), 1495–1499.
94. Jiang, Y.; Woortman, A. J. J.; van Ekenstein, G.; Petrovic, D. M.; Loos, K. *Biomacromolecules* **2014**, *15* (7), 2482–2493.
95. Kumalaputri, A. J.; Bottari, G.; Erne, P. M.; Heeres, H. J.; Barta, K. *ChemSusChem* **2014**, 7 (8), 2266–2275.
96. Roman-Leshkov, Y.; Barrett, C. J.; Liu, Z. Y.; Dumesic, J. A. *Nature* **2007**, *447* (7147), 982–985.
97. Bohre, A.; Dutta, S.; Saha, B.; Abu-Omar, M. M. *ACS Sustainable Chem. Eng.* **2015**, *3* (7), 1263–1277.
98. Lawrence, C. H.; Raverty, W. D.; Duncan, A. J., 2016, US20120111714A1.
99. Richardson, D. E.; Raverty, W. D. *Appita J.* **2016**, *69* (4), 344–351.
100. He, J. Y.; Liu, M. J.; Huang, K. F.; Walker, T. W.; Maravelias, C. T.; Dumesic, J. A.; Huber, G. W. *Green Chem.* **2017**, *19* (15), 3642–3653.
101. Bonneau, G.; Peru, A. A. M.; Flourat, A. L.; Allais, F. *Green Chem.* **2018**, *20* (11), 2455–2458.
102. Flourat, A. L.; Haudrechy, A.; Allais, F.; Renault, J. H. *Org. Process Res. Dev.* **2020**, *24* (5), 615–636.
103. Szymkuć, S.; Gajewska, E. P.; Molga, K.; Wołos, A.; Roszak, R.; Beker, W.; Moskal, M.; Dittwalda, P.; Grzybowski, B. A. *Chem. Sci.* **2020**, *11*, 6736–6744.
104. Sherwood, J.; Constantinou, A.; Moity, L.; McElroy, C. R.; Farmer, T. J.; Duncan, T.; Raverty, W.; Hunt, A. J.; Clark, J. H. *Chem. Commun.* **2014**, *50* (68), 9650–9652.
105. Mouterde, L. M. M.; Allais, F.; Stewart, J. D. *Green Chem.* **2018**, *20* (24), 5528–5532.
106. De Bruyn, M.; Sener, C.; Petrolini, D. D.; McClelland, D. J.; He, J. Y.; Ball, M. R.; Liu, Y. F.; Martins, L.; Dumesic, J. A.; Huber, G. W.; Weckhuysen, B. M. *Green Chem.* **2019**, *21* (18), 5000–5007.
107. Krishna, S. H.; McClelland, D. J.; Rashke, Q. A.; Dumesic, J. A.; Huber, G. W. *Green Chem.* **2017**, *19* (5), 1278–1285.
108. De Bruyn, M.; Budarin, V. L.; Misefari, A.; Shimizu, S.; Fish, H.; Cockett, M.; Hunt, A. J.; Hofstetter, H.; Weckhuysen, B. M.; Clark, J. H.; Macquarrie, D. J. *ACS Sustainable Chem. Eng.* **2019**, 7 (8), 7878–7883.
109. Mohan, D.; Pittman, C. U.; Steele, P. H. *Energy Fuel* **2006**, *20* (3), 848–889.

110. Kan, T.; Strezov, V.; Evans, T. J. *Renew. Sustain. Energy Rev.* **2016**, *57*, 1126–1140.
111. Zhang, X. S.; Yang, G. X.; Jiang, H.; Liu, W. J.; Ding, H. S. *Sci. Rep.* **2013**, *3*, 1120.
112. Valle, B.; Remiro, A.; Garcia-Gomez, N.; Gayubo, A. G.; Bilbao, J. *J. Chem. Technol. Biotechnol.* **2019**, *94* (3), 670–689.
113. Yin, W.; Venderbosch, R. H.; Bottari, G.; Krawzcyk, K. K.; Barta, K.; Heeres, H. J. *Appl Catal B* **2015**, *166*, 56–65.
114. Yu, Y.; Chua, Y. W.; Wu, H. W. *Energy Fuel* **2016**, *30* (5), 4145–4149.
115. Sun, Z.; Vasconcelos, A. C.; Bottari, G.; Stuart, M. C. A.; Bonura, G.; Cannilla, C.; Frusteri, F.; Barta, K. *ACS Sustainable Chem. Eng.* **2017**, *5* (2), 1738–1746.
116. Gabriels, D.; Hernandez, W. Y.; Sels, B.; Van Der Voort, P.; Verberckmoes, A. *Cat. Sci. Technol.* **2015**, *5* (8), 3876–3902.
117. Kozlowski, J. T.; Davis, R. J. *ACS Catal.* **2013**, *3* (7), 1588–1600.
118. Wu, X. Y.; Fang, G. Q.; Tong, Y. Q.; Jiang, D. H.; Liang, Z.; Leng, W. H.; Liu, L.; Tu, P. X.; Wang, H. J.; Ni, J.; Li, X. N. *ChemSusChem* **2018**, *11* (1), 71–85.
119. Mascal, M. *Biofuels Bioprod. Biorefin.* **2012**, *6* (4), 483–493.
120. Angelici, C.; Weckhuysen, B. M.; Bruijnincx, P. C. A. *ChemSusChem* **2013**, *6* (9), 1595–1614.
121. Uyttebroek, M.; Van Hecke, W.; Vanbroekhoven, K. *Catal. Today* **2015**, *239*, 7–10.
122. Sener, C.; Petrolini, D. D.; McClelland, D. J.; He, J.; Ball, M. R.; Liu, Y.; Martins, L.; Dumesic, J. A.; Huber, G. W.; Weckhuysen, B. M. *Green Chem.* **2019**, *21*, 5000–5007.
123. Rhodium, n.d., https://www.rsc.org/periodic-table/element/45/rhodium.
124. Lopez-Olmos, C.; Guerrero-Ruiz, A.; Rodríguez-Ramos, I. *Catal. Today* **2019**, *357*, 132–142.
125. Petrolini, D. D.; Eagan, N.; Ball, M. R.; Burt, S. P.; Hermans, I.; Huber, G. W.; Dumesic, J. A.; Martins, L. *Cat. Sci. Technol.* **2019**, *9* (8), 2032–2042.
126. Hook, M.; Davidsson, S.; Johansson, S.; Tang, X. *Phil. Trans. R. Soc. A* **2014**, *372*, 20120448.
127. Gowdy, J.; Julia, R. *Energy* **2007**, *32* (8), 1448–1454.
128. Fantazzini, D.; Hook, M.; Angelantoni, A. *Energy Policy* **2011**, *39* (12), 7865–7873.
129. *EU Climate Policy Explained*; Routledge, 2015.
130. Bond, J. Q.; Upadhye, A. A.; Olcay, H.; Tompsett, G. A.; Jae, J.; Xing, R.; Alonso, D. M.; Wang, D.; Zhang, T. Y.; Kumar, R.; Foster, A.; Sen, S. M.; Maravelias, C. T.; Malina, R.; Barrett, S. R. H.; Lobo, R.; Wyman, C. E.; Dumesic, J. A.; Huber, G. W. *Energ. Environ. Sci.* **2014**, 7 (4), 1500–1523.
131. Han, X. W.; Guo, Y.; Liu, X. H.; Xia, Q. N.; Wang, Y. Q. *Catal. Today* **2019**, *319*, 2–13.
132. Alonso, D. M.; Bond, J. Q.; Dumesic, J. A. *Green Chem.* **2010**, *12* (9), 1493–1513.
133. De, S.; Saha, B.; Luque, R. *Bioresour. Technol.* **2015**, *178*, 108–118.
134. Muldoon, J. A.; Harvey, B. G. *ChemSusChem* **2020**, *13* (22), 5777–5807.
135. Sun, Z.; Buwalda, D.; Barta, K. *RSC Adv.* **2019**, *9* (41), 23727–23734.
136. Rajasekhar, S.; Maiti, B.; Balamurali, M. M.; Chanda, K. *Curr. Org. Synth.* **2017**, *14* (1), 40–60.
137. Alaqeel, S. I. *J. Saudi Chem. Soc.* **2017**, *21* (2), 229–237.
138. Kanwal, A.; Ahmad, M.; Aslam, S.; Naqvi, S. A. R.; Saif, M. J. *Pharm. Chem. J.* **2019**, *53* (3), 179–187.
139. Sun, Z.; Bottari, G.; Barta, K. *Green Chem.* **2015**, *17* (12), 5172–5181.
140. O'Brien, P. *Open Access Government*; 2018.
141. Lovering, F.; Bikker, J.; Humblet, C. *J. Med. Chem.* **2009**, *52* (21), 6752–6756.
142. Sheldon, R. A. *Green Chem.* **2017**, *19* (1), 18–43.
143. Yan, T.; Feringa, B. L.; Barta, K. *Sci. Adv.* **2017**, *3* (12), eaao6494.
144. Elangovan, S.; Afanasenko, A.; Haupenthal, J.; Sun, Z.; Liu, Y. Z.; Hirsch, A. K. H.; Barta, K. *ACS Cent. Sci.* **2019**, *5* (10), 1707–1716.

145. Stockigt, J.; Antonchick, A. P.; Wu, F. R.; Waldmann, H. *Angew. Chem. Int. Ed.* **2011**, *50* (37), 8538–8564.
146. Baltzly, R. *J. Am. Chem. Soc.* **1953**, *75* (23), 6038–6039.
147. Kumar, R. S.; Almansour, A. I.; Arumugam, N.; Kumar, R. R. *Tetrahedron Lett.* **2020**, *61* (11), 151606.
148. Berney, D.; Jauner, T. *Helv. Chim. Acta* **1976**, *59* (2), 623–626.
149. Handy, S.; Wright, M. *Tetrahedron Lett.* **2014**, *55* (23), 3440–3442.
150. Paiva, A.; Craveiro, R.; Aroso, I.; Martins, M.; Reis, R. L.; Duarte, A. R. C. *ACS Sustainable Chem. Eng.* **2014**, *2* (5), 1063–1071.
151. Unlu, A. E.; Arikaya, A.; Takac, S. *Green Processes Synth.* **2019**, *8* (1), 355–372.
152. Zhang, Q. H.; Vigier, K. D.; Royer, S.; Jerome, F. *Chem. Soc. Rev.* **2012**, *41* (21), 7108–7146.
153. Garcia, D.; Foubelo, F.; Yus, M. *Eur. J. Org. Chem.* **2010**, *2010* (15), 2893–2903.
154. Hasebein, P.; Aulinger, K.; Schepmann, D.; Wunsch, B. *Tetrahedron* **2013**, *69* (23), 4552–4562.
155. Olah, G. A. *Angew. Chem. Int. Ed.* **2005**, *44* (18), 2636–2639.
156. Liu, W. C.; Baek, J.; Somorjai, G. A. *Top. Catal.* **2018**, *61* (7–8), 530–541.

CHAPTER FOUR

Sequential oxidation-depolymerization strategies for lignin conversion to low molecular weight aromatic chemicals

Yanbin Cui[a,†], Shannon L. Goes[a,†], and Shannon S. Stahl[a,b,*]
[a]Department of Chemistry, University of Wisconsin-Madison, Madison, WI, United States
[b]Wisconsin Energy Institute, University of Wisconsin–Madison, Madison, WI, United States
*Corresponding author: e-mail address: stahl@chem.wisc.edu

Contents

Abstract

Lignin is the largest source of bio-derived aromatic chemicals, and oxidative conversion of this polymeric material can generate valuable oxygenated aromatic compounds. Oxidative depolymerization of lignosulfonate feedstocks under alkaline conditions is used commercially for vanillin production. Recent studies have led to other oxidation methods that access different aromatic products from lignin in good yields. A particularly effective strategy that accesses some of the highest yields of aromatic monomers to date features a two-step process in which the oxidation of specific alcohol groups in lignin is followed by an oxidative, reductive, or redox-neutral step that cleaves

† These authors contributed equally.

Advances in Inorganic Chemistry, Volume 77
ISSN 0898-8838
https://doi.org/10.1016/bs.adioch.2021.02.003

the polymer into aromatic monomers and oligomers. Studies of model compounds have provided crucial mechanistic insights and contributed to the development of effective lignin depolymerization methods. This review provides a general overview of lignin depolymerization methods, followed by a survey of oxidation-depolymerization methods that access oxygenated aromatic monomers in good yields.

Abbreviations

ACT	4-acetamido-TEMPO
bpy	2,2′-bipyridine
BDE	bond dissociation energy
CEL	cellulolytic enzyme lignin
Cp*	1,2,3,4,5-pentamethylcyclopentadiene
Cu-AHP	copper-catalyzed alkaline hydrogen peroxide
DAIB	(diacetoxyiodo)benzene
DDQ	2,3-dichloro-5,6-dicyano-1,4-benzoquinone
$DDQH_2$	2,3-dichloro-5,6-dicyanohydroquinone
DIPEA	N,N-diisopropylethylamine
DMSO	dimethyl sulfoxide
DPA-BP	4,4′-bis(diphenylamino)benzophenone
dtbbpy	4,4′-di-tert-butyl-2,2′-bipyridyl
EA	extractive ammonia
G	guaiacyl
GPC	gel permeation chromatography
GVL	γ-valerolactone
H	*para*-hydroxyphenyl
HAT	hydrogen atom transfer
HSQC	heteronuclear single quantum coherence
LDH	layered double hydroxide
MCF	mesoporous cellular silica foam
MeCN	acetonitrile
NHC	N-heterocyclic carbene
NHPI	N-hydroxyphthalimide
OCF	oxidative catalytic fractionation
phen	1,10-phenanthroline
PINO	phthalimide-N-oxyl
POM	polyoxometalates
ppy	2-phenylpyridine
RCF	reductive catalytic fractionation
S	syringyl
SET	single-electron-transfer
***t*BuOOH**	*tert*-butyl hydroperoxide
***t*BuONO**	*tert*-butyl nitrite
TEMPO	2,2,6,6-tetramethylpiperidine-1-oxyl
TTPFeCl	meso-tetraphenylporphyrin iron(III) chloride

1. Introduction

Increasing global demand for sustainable alternatives to organic chemicals derived from coal and petroleum is driving efforts to leverage biomass-derived feedstocks. Lignin is a significant component of lignocellulosic biomass and represents an abundant source of functionalized, renewable aromatics that could be used directly as valuable products or transformed into fuels or chemicals. Many tons of lignin are produced as a byproduct of the pulp and paper and biofuel industries that prioritize acquisition of cellulose;[1] however, the majority of this material is burned as low value fuel to provide energy.[2] Strategies to convert lignin into value-added chemicals could play an important role in supporting the economic viability of future biorefineries.

Lignin is a non-repeating, mostly linear biopolymer created by radical coupling of monolignols, consisting of *p*-hydroxy-3-phenylpropanol-derived building blocks.[3] Lignin is composed of different quantities of *para*-hydroxyphenyl (H), guaiacyl (G) and syringyl (S) aromatic subunits linked via C—C and C—O bonds comprising β-O-4, α-O-4, β-1, β-5, β-β, and 5-5 units (Fig. 1). Many plant sources contain other aromatic units, such as *p*-hydroxybenzoate, appended to the alcohols in these linker units. The relative abundance of each type of subunit and linkage varies between plant species, and the amount of lignin can vary from 15% to 35% of the overall biomass weight.[4] Multiple methods have been developed to extract lignin from biomass, including Kraft, enzymatic, organosolv, and mild acidolysis methods,[5] and each can lead to modifications of the lignin structure that influence subsequent depolymerization. Some methods of lignin depolymerization, such as pyrolysis, use harsh conditions to break many inter-unit linkages and generate a wide array of small-molecule products.[6] Other depolymerization methods are tailored to cleave specific linkages and lead to a smaller distribution of products. The β-O-4 ether unit is the most common linkage in lignin (40–70 wt%) and is among the most readily cleaved. It contains comparatively weak C—C/C—O bonds with bond dissociation energies (BDE) as low as 67.0 kcal/mol.[7] The prevalence and reactivity of β-O-4 lignin units have made them a target for many depolymerization strategies and provide a basis for generation of S- and G-derived aromatic products in up to 45–55% (hardwoods) or 20–30%

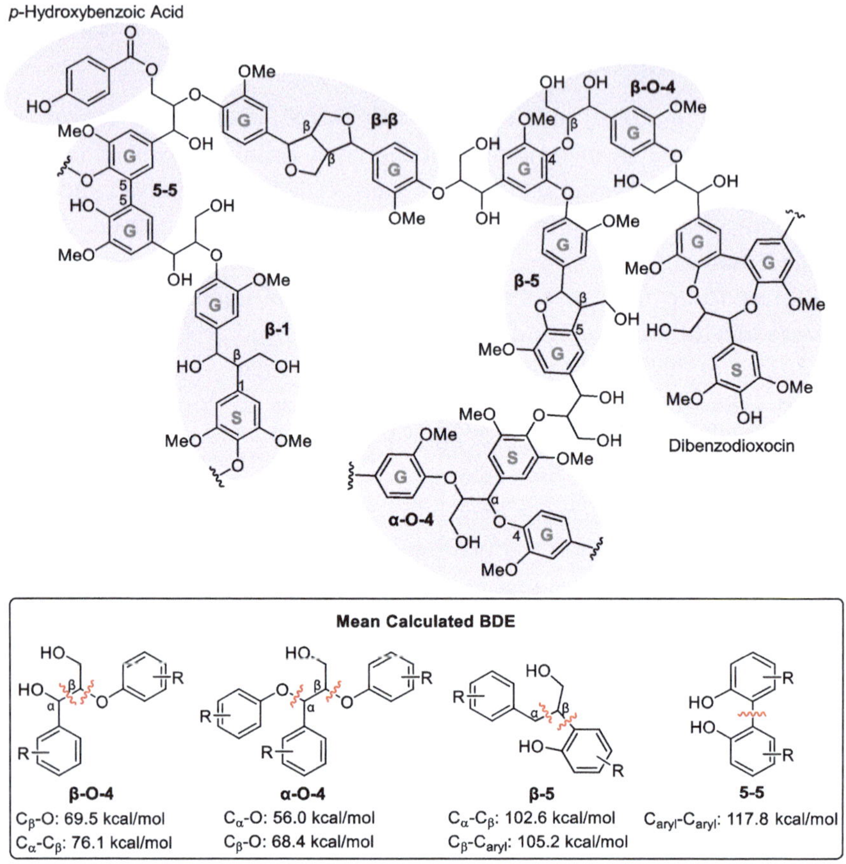

Fig. 1 Lignin structure (top) and bond strengths (bottom) with typical lignin linkages.

(softwood) monomer yields,[8,9] often supplemented by *p*-hydroxybenzoate and other aromatic groups appended to the naturally occurring polymer.

Reduction methods have been the focus of extensive attention for lignin depolymerization. These have been reviewed elsewhere,[8,10,11] but reductive catalytic fractionation (RCF) methods are particularly noteworthy because they are capable of producing near theoretical yields of aromatics in one step.[12–18] These methods combine heterogeneous supported metal catalysts with solubilized lignin or raw biomass under elevated temperature (180–250 °C) and H_2 pressure (5–15 MPa). RCF conditions lead to hydrogenation and hydrogenolysis pathways that stabilize reactive lignin intermediates and prevent lignin degradation and condensation in the process of generating propylphenol-derived aromatics.

Oxidative lignin depolymerization strategies are appealing because they provide access to a different set of value-added aromatics relative to those obtained from reduction methods. A prominent example is vanillin, an aromatic aldehyde derived from the G subunit that is widely used in the flavor, fragrance, and cosmetic industries.[19] It has been produced commercially through oxidative lignin depolymerization using copper-based catalysts under strong alkaline conditions.[20–22] Recent research has focused on identifying new oxidative depolymerization strategies to generate other lignin-derived aromatics.[23,24] Oxidized aromatics derived from lignin have strategic benefits for microbial funneling of product mixtures into a single value-added chemical, owing to the higher solubility of these precursors in the aqueous growth media.[25,26]

Research into the development of oxidative lignin depolymerization processes has featured two general strategies: top-down and bottom-up (Fig. 2). *Top-down* strategies directly use raw lignocellulosic biomass or extracted lignin as the feedstock in efforts to develop new depolymerization methods. In contrast, *bottom-up* strategies start with well-defined model compounds in an effort to identify new catalytic methods and characterize mechanisms of reactions capable of cleaving strategic linkages in lignin, such as β-O-4 units. Successful examples are then tested on authentic lignin samples to assess their efficacy with complex feedstocks. This review provides a brief overview of methods that have been developed for oxidative conversion of lignin and related models into aromatic monomers, followed by an extensive presentation of two-step alcohol oxidation-bond cleavage methods that have proven to be particularly effective for lignin conversion into oxygenated aromatic monomers. These processes correspond to bottom-up strategies originating from systematic studies of reactions with model compounds that were successfully translated to depolymerization of lignin in high yield.[27–29]

2. Top-down vs. bottom-up strategies for lignin depolymerization

2.1 Top-down strategies for efficient aromatic monomer production

Top-down strategies for lignin depolymerization (Fig. 2) empirically optimize methods to obtain monomers from biomass or extracted lignin, frontloading challenges associated with using a complex polymer substrate, such as solubility, source variation, and product analysis. Intermediates generated during reactions with the complex feedstock are difficult to detect or characterize, and thus limit mechanistic insights that could be used to

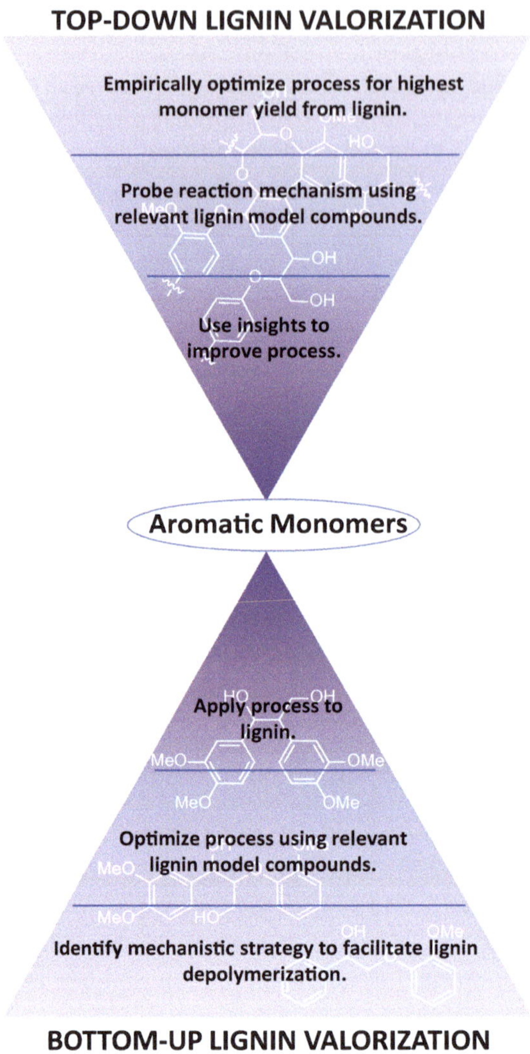

Fig. 2 Top-down and bottom-up strategies for lignin depolymerization.

guide process optimization. Nonetheless, top-down approaches provide the most direct test of the utility of different methods for lignin conversion into aromatic monomers.

Top-down strategies have led to several effective oxidative methods to obtain vanillin from lignin. In the 1940s, stoichiometric oxidants under basic conditions were found to convert lignin materials into vanillin (Fig. 3A–i/ii).[30,31] Mechanistic studies implicate a single-electron-transfer (SET) pathway that generates radicals in alkaline solution, which result in

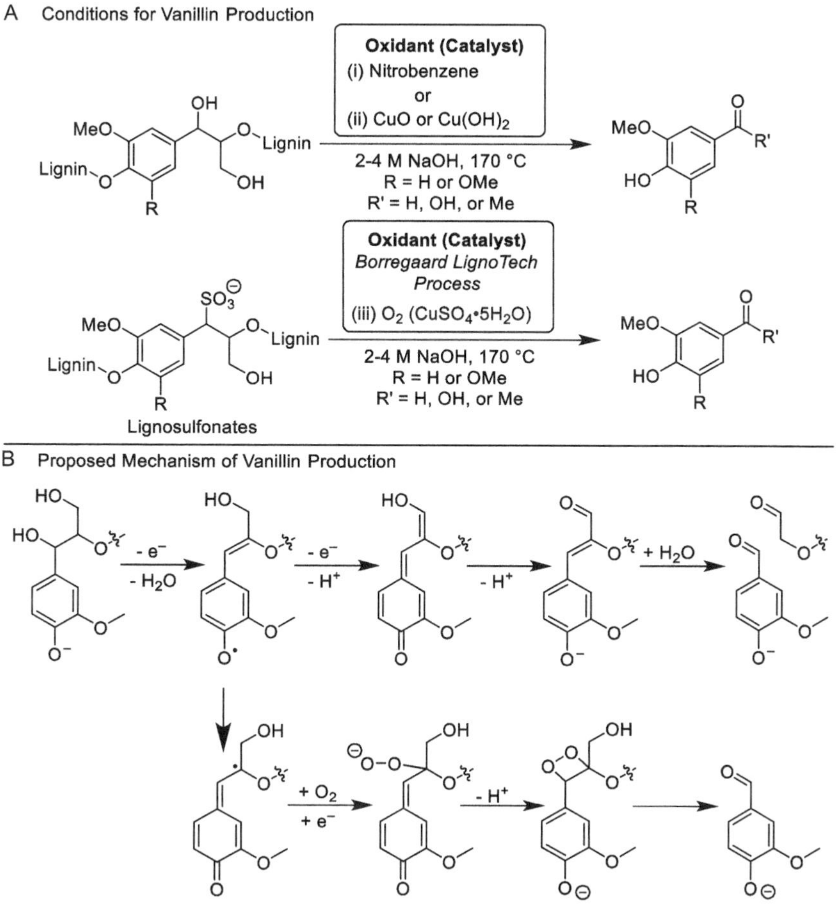

Fig. 3 (A) Typical reaction conditions to depolymerize lignin or lignosulfonates through (i) nitrobenzene, (ii) copper-based catalyst, or (iii) commercial vanillin production used by Borregaard; (B) proposed mechanism of vanillin formation under alkaline oxidative conditions.

homolytic cleavage of the C_α—C_β bond of β-O-4 linkages to form the aldehyde product (Fig. 3B).[5,32,33] As a representative protocol, nitrobenzene (2–4 M NaOH, 160–180 °C, 2–3 h) yields up to 16 wt% vanillin from softwood sulfite lignin as the major product.[20,34] The cost and toxic by-products of nitrobenzene relegate its use to laboratory scale applications, for example, to quantify the β-O-4 linkages and/or the relative amounts of uncondensed H, G, and S units of various lignin sources.[35] CuO was also identified as an effective oxidant, although it was used at rather high loading (up to 600 wt%) and the vanillin is slowly converted to vanillic acid under the oxidizing alkaline conditions.[22,36]

Molecular oxygen (O_2) is an ideal oxidant, and high yields of vanillin are possible by controlling the O_2 partial pressure, reaction temperature, and reaction pH in the presence of $CuSO_4$ as a catalyst (Fig. 3A-iii).[37] A variation of this method was developed by Borregaard for conversion of lignosulfonates, which provides the basis for commercial production of bio-based vanillin.[20,22,38] Other variations of this method have been used for direct treatment of raw biomass, enabling direct conversion to low molecular weight chemicals.[39–41]

2.2 Bottom-up strategy for mechanistic insight into lignin depolymerization

Bottom-up lignin depolymerization (Fig. 2) refers to studies that start with low molecular weight model substrates containing representative fragments of native lignin (e.g., β-O-4 units). These lignin-like models allow for straightforward testing and analysis of chemical reactivity and facilitate mechanistic studies. Numerous homogeneous transition-metal catalysts have been used for selective cleavage of C—C and C—O bonds in β-O-4 model compounds via oxidative and redox neutral pathways (Fig. 4). As summarized in Fig. 5 and elaborated below, a number of these methods generate aromatic monomers in good yield.

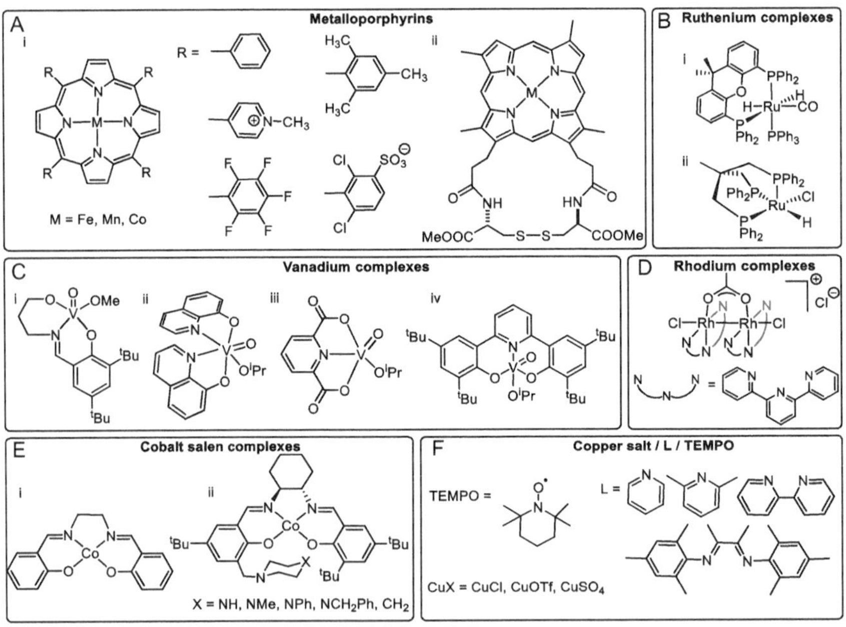

Fig. 4 Metal complex catalysts used for selective bond cleavage of lignin model compounds.

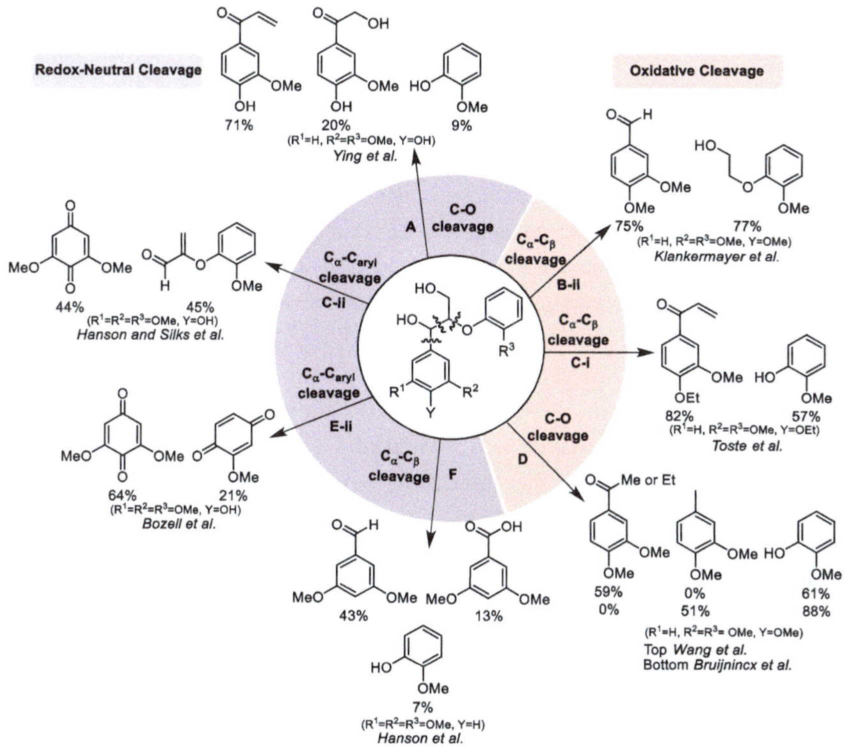

Fig. 5 Yields of aromatic monomer products through selective bond cleavage of β-O-4 model compounds via different metal complex catalysts listed in Fig. 4.

Naturally occurring enzymes, such as Fe and Mn-based peroxidases, contribute to fungal degradation of lignin,[42,43] and biomimetic Fe, Mn, and Co porphyrins have been widely tested as synthetic catalysts for lignin and lignin-model degradation (Fig. 4A–i). Various oxidants such as H_2O_2, *tert*-butyl hydroperoxide (tBuOOH), and oxone have been used to generate high-valent metal-oxo reactive species.[23,44] The metal-oxo species are proposed to promote SET from electron-rich phenol and non-phenolic compounds to generate a radical cation that initiates further reaction and degradation of the lignin material. Early studies showed that sterically-hindered porphyrins mimic enzyme activity and limit oxidative degradation of the porphyrin scaffold.[45,46] More recently, Ying and coworkers demonstrated that cobalt-containing deuteroporphyrin derivatives with tethered disulfide groups (Fig. 4A-ii) depolymerize phenolic lignin models, accessing >90% yield of monomers from $C_β$-O cleavage monomers with oxone as the oxidant (Fig. 5A).[47] A lower yield of guaiacol monomer reflects overoxidation of the phenol, resulting in ring cleavage. Subsequent application of this catalyst to various lignin samples generated significant quantities (>30 wt% in several cases) of a lignin oil containing H/G/S-derived aromatics and aliphatic material.

Bergman and Ellman showed that a Ru-based complex, $Ru(H)_2(CO)(PPh_3)$(xantphos) (Fig. 4B–I) effects C—O bond cleavage in simplified β-O-4 model compounds that lack the $-C_\gamma H_2OH$ functional group.[48] The catalyst mediates tandem catalytic dehydrogenation/C—O bond cleavage in a redox-neutral process. Incorporation of the $C_\gamma H_2OH$ fragment that is present in lignin, however, blocks dehydrogenation and prevents subsequent C—O bond cleavage.[49,50] Modified methods from other groups, using KOH addition or acetylation of the γ-hydroxyl group, were able to overcome this problem.[51,52] A Ru-triphos catalyst system (Fig. 4B-ii) was developed by Klankermayer and coworkers to cleave the C_α—C_β bond in β-O-4 model compounds containing a $-C_\gamma H_2OH$ moiety. The process involves primary alcohol dehydrogenation followed by retro-aldol reaction to achieve C—C bond cleavage (Fig. 5B-ii).[53]

Oxovanadium catalysts have been used for cleavage of lignin model compounds in both redox-neutral and oxidative methods.[54] Toste and coworkers used Schiff-base vanadium catalysts (Fig. 4C-i) to generate monomeric enone and guaiacol products from dimeric and trimeric β-O-4 models (Fig. 5C-i). The authors proposed that the reaction is initiated by hydrogen atom transfer (HAT) from the β-O-4 secondary alcohol, followed by subsequent homolysis of the C_β—O bond.[55] When the catalytic oxovanadium system was applied to organosolv grass lignin, a decrease in the lignin polymer molecular weight was observed and selective β-O-4 degradation was demonstrated by 2D NMR spectroscopy, but less than 3 wt% yield of identified monomers was recovered.[56] Under similar reaction conditions using V-quinolinate catalysts (Fig. 4C-ii), Hanson, Silks, and coworkers propose an oxidative mechanism in which the catalyst promotes SET oxidation of phenolic model compounds to generate a phenoxy radical, followed by subsequent C_{aryl}—C_α bond cleavage (Fig. 5C-ii).[57] Baker et al. showed that V-quinolinate complexes catalyze oxidation of mixed-hardwood organosolv lignin, decreasing the average molecular weight by 77%, although monomer yields were not reported.[58] Other types of vanadium-based catalysts, (dipic)V^V(O)OiPr (Fig. 4C-iii) and V-bisphenolate (Fig. 4C-iv) are less selective and generate a mixture of C_α—C_β and redox-neutral C_β—O bond cleavage products depending on substrate structure and reaction conditions.[59,60]

A binuclear Rh complex, reported by Li and Wang (Fig. 4D), was used to cleave a series of lignin model compounds and authentic lignin via hydrogen-transfer methods. Rh-catalyzed dehydrogenation of the benzylic alcohol produces a ketone intermediate that undergoes subsequent C_{β}—O bond cleavage (110 °C, Ar, NaOH) to afford 61–64% guaiacol and 59–61% methyl- and ethylarene products (Fig. 5D).[61] Almost complete deconstruction of raw basswood powder sample resulted in 26.6 wt% aromatic oil, although monomer yields of only 2.3 wt% (based on the starting raw basswood powder) were generated. Bruijnincx and coworkers used $Sc(OTf)_3$ as a Lewis acid catalyst to promote cleavage of ether bonds. When paired with Rh-catalyzed decarbonylation of aliphatic aldehydes, a Rh/$Sc(OTf)_3$ catalyst system provided stable aromatic product mixtures. This chemistry led to good yields of methylarenes (51%) and guaiacols (88%) from model compounds and also accessed aromatic monomers from pine (6.8 wt%) and poplar (9.8%) sawdust.[62]

Cobalt-Schiff base complexes have been studied extensively for phenol oxidation with O_2.[63,64] Bozell and coworkers have investigated Co(salen) complexes (Fig. 4E) that react with O_2 to form a superoxide adduct that can abstract a hydrogen atom from phenols, generating a phenoxyl radical that can induce cleavage of C_{α}—C_{aryl} bonds in the lignin models.[65,66] Catalysts from a family of unsymmetrical Co-Schiff base complexes, each bearing a bulky heterocyclic nitrogen base on the ligand (Fig. 4E-ii), convert phenolic lignin models to benzoquinones in yields up to 64% (Fig. 5E-ii).[66] When applied to organosolv poplar lignin, however, only small quantities of monomers were obtained due to the low phenolic content of organosolv lignin.

Copper/nitroxyl catalyst systems for aerobic alcohol oxidation[67] have been investigated with lignin model compounds. Initial studies by Baker, Wu, Hanson and coworkers used 2,2,6,6-tetramethylpiperidine-1-oxyl (TEMPO) as the nitroxyl catalyst in combination with CuCl, pyridine, and O_2 at 100 °C. Limited activity for aerobic oxidative cleavage of simplified lignin models was attributed to instability of the active copper catalyst (Fig. 4F).[54,59] Use of stoichiometric CuCl/TEMPO with an S/G lignin β-O-4 dimer led to 89% conversion to cleavage products, albeit with only 7–43% yield of monomeric aromatic products (Fig. 5F). A modified Cu/bpy/TEMPO catalyst system (bpy = 2,2′-bipyridine),[68] investigated by Stahl and coworkers,[27] is somewhat more effective and will be revisited below.

This brief survey of catalytic methods is not comprehensive, but it illustrates the types of bottom-up strategies that have been pursued using well-defined model compounds. The attempts to achieve direct conversion of dimeric models into aromatic monomers in a single step has led to mixed success, and oxidative methods derived from these studies have not (yet) led to lignin depolymerization strategies that are competitive with reductive methods or previously known oxidative methods (cf. Section 2.1 and Fig. 3).

3. Sequential oxidation-cleavage methods for lignin depolymerization

3.1 Concepts and initial discovery of the alcohol oxidation-cleavage strategy

Bottom-up approaches to lignin depolymerization necessarily incorporate uncertainty because the model compounds lack the complexity intrinsic to natural biomass or extracted lignin samples. But, studies of relevant model compounds provide a means to access valuable insights that could be applied to biomass conversion. This principle is illustrated by a different bottom-up strategy initiated within the Stahl laboratory in 2011. The central hypothesis in this approach was that selective oxidation of either alcohol within the β-O-4 unit of lignin could facilitate a subsequent cleavage step to generate monomeric aromatic products (Fig. 6A). For example, oxidation of the primary (1°) alcohol would form an aldehyde susceptible to retro-aldol cleavage, while the oxidation of the secondary (2°) benzylic alcohol could be cleaved by various redox and non-redox reaction pathways. Computational studies of lignin suggest that alcohol oxidation weakens the β-O-4 C—O bond by

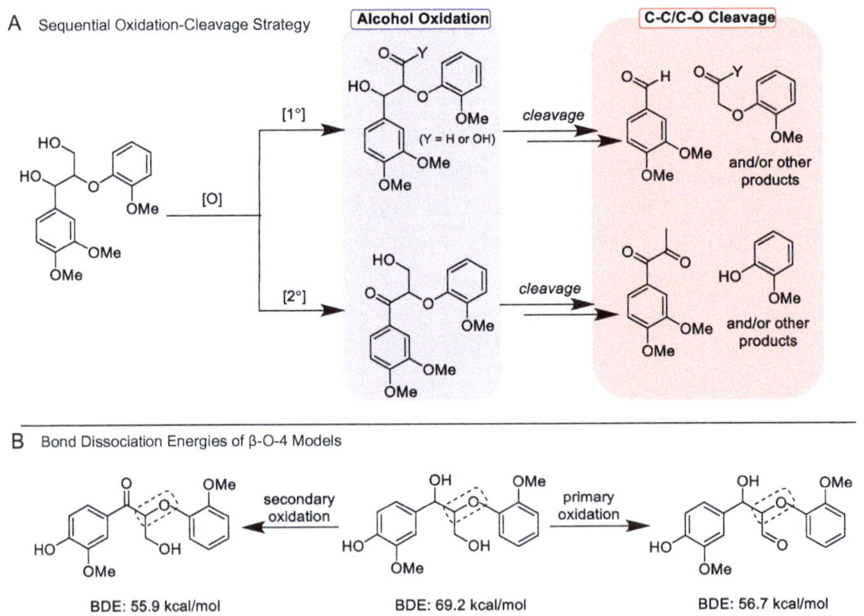

Fig. 6 Sequential oxidation-cleavage strategies (A) selectively oxidize β-O-4 alcohol groups to weaken the C—O ether bond (B) to facilitate subsequent bond cleavage in a separate step.

12–13 kcal/mol, thereby facilitating homolytic cleavage pathways for lignin depolymerization (Fig. 6B).[7,69–71]

Stahl and coworkers tested the reaction of numerous common stoichiometric oxidants and catalytic alcohol oxidation methods with a β-O-4 dimeric model compound, and methods employing nitroxyl catalysts proved to be the most effective (Fig. 7A).[27] Nitroxyl catalysts, such as TEMPO and 4-acetamido-TEMPO (ACT), undergo reversible 1 e^- electrochemical

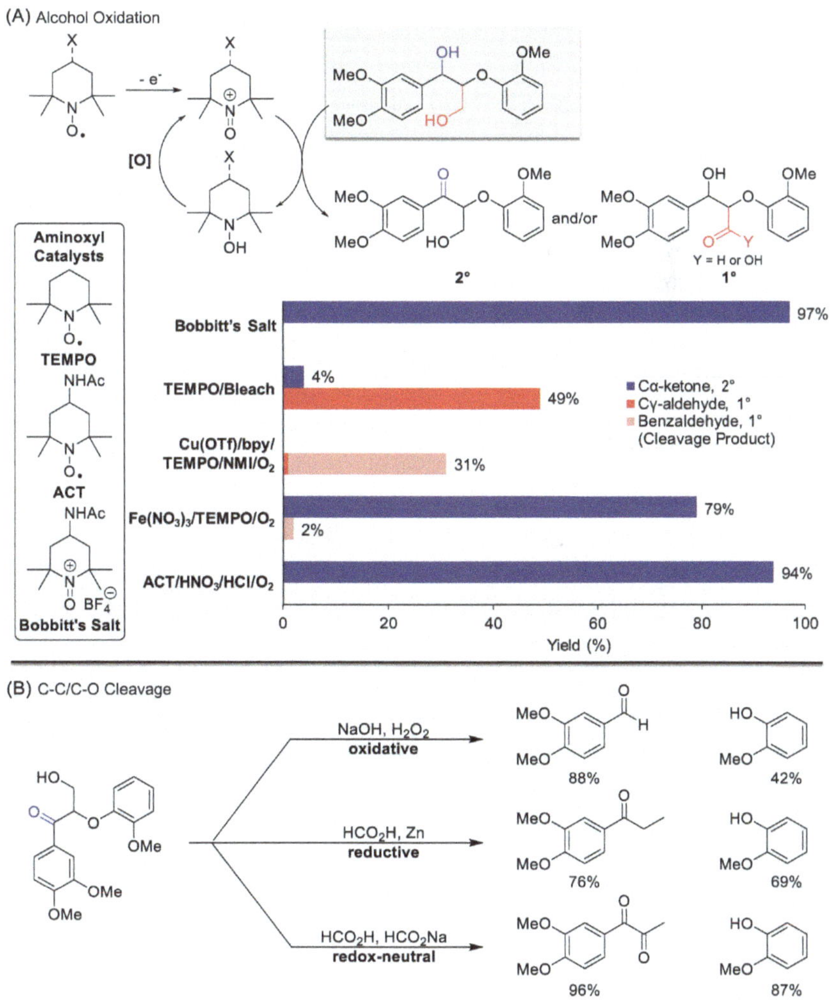

Fig. 7 (A) Selective primary or secondary alcohol oxidation through various reaction systems; (B) different bond cleavage methods on secondary alcohol oxidized lignin models.

oxidation to the corresponding oxoammonium and a quasi-reversible 1 $e^-/1H^+$ reduction to hydroxylamines.[72] The TEMPO radical is generally unreactive with organic species, but various chemical and electrochemical oxidation methods may be used to generate the reactive oxoammonium species from the nitroxyl or hydroxylamine form of the catalyst.[73] The oxoammonium, generated in situ or isolated as a salt, mediates hydride-transfer oxidation of alcohols. This reactivity is evident in the reaction with stoichiometric Bobbitt's salt, which oxidizes the 2° benzylic alcohol to the corresponding ketone in 97% yield. In contrast, use of TEMPO/bleach (NaOCl) under mildly basic conditions leads to preferential oxidation of the more sterically accessible 1° alcohol, affording the aldehyde product in 49% yield, with only 4% yield of the benzylic ketone. An aerobic Cu^I/TEMPO catalyst system also shows selectivity for 1° alcohol oxidation, and, while proceeding in only moderate yield, induces retro-aldol reactivity to generate aromatic monomers as the major products. Two other aerobic oxidation methods, employing $(FeNO_3)_3$/TEMPO and ACT/HNO_3/HCl, led to good yields and selectivity for formation of the ketone, with the latter method generating a 94% yield of ketone. The ACT/HNO_3/HCl oxidation condition was then tested on poplar lignin extracted using cellulolytic enzyme lignin (CEL), which retains much of the native lignin structure. Analysis of the oxidized lignin by 2D heteronuclear single quantum coherence (HSQC) NMR spectroscopy suggests that nearly all of the benzylic alcohols adjacent to G and S aromatic units are oxidized in the final lignin material.

An oxidized β-O-4 model compound bearing a benzylic ketone was shown to undergo cleavage into aromatic monomers under oxidizing, reducing, and redox-neutral conditions (Fig. 7B).[27,28] Alkaline hydrogen peroxide led to partial degradation of the phenol monomer,[27] while a redox neutral transformation in the presence of aqueous formic acid/formate led to the best results (Fig. 7B).[28] Excellent yields of both products were obtained from S-, G-, and H-derived β-O-4 dimer models (89–92% isolated yield of diketone monomer).

ACT/HNO_3/HCl oxidation of poplar CEL, followed by treatment with formic acid/formate, induced lignin depolymerization to afford >60 wt% yield of low molecular weight aromatic compounds, including a 52 wt% yield of characterized aldehyde, carboxylic, and diketone aromatics (Fig. 8).[28] These results were later extended to lignin samples obtained from other biomass sources (poplar, maize, and maple) and extracted by different methods, including mild acidolysis,[74] γ-valerolactone/dilute sulfuric acid (GVL),[75]

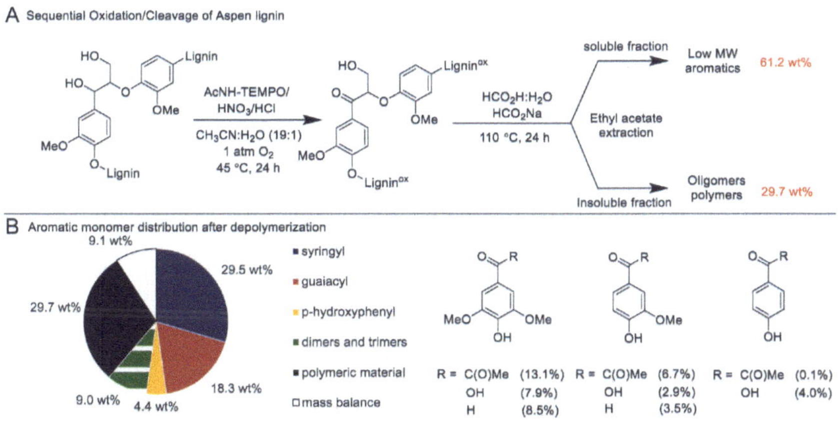

Fig. 8 (A) Depolymerization of aspen lignin with formic acid following oxidation; (B) monomer distribution from depolymerization of oxidized aspen lignin.

extractive ammonia (EA),[76,77] and copper-catalyzed alkaline hydrogen peroxide (Cu-AHP).[78] Aromatic yields of 3–42% were obtained when this two-step oxidation-depolymerization sequence was applied to these different lignin samples, with the best yield of aromatics obtained from a poplar sample extracted via mild acidolysis (42%) or Cu-AHP conditions (31%).[29]

This sequential oxidation-cleavage method demonstrated the best yields of aromatics to date for a lignin depolymerization method originating from a bottom-up strategy. These results provided a foundation for numerous other studies employing a two-step sequence to access aromatic monomers from lignin. These studies include the development of different alcohol oxidation methods, including those with selectivity for the 2° and 1° positions in β-O-4 units (Sections 3.2 and 3.3), and different methods for bond cleavage/depolymerization from the oxidized lignin and lignin model compounds (Section 3.4). A summary of these various methods is provided in Fig. 9.

3.2 Secondary alcohol oxidation

Oxidation methods that exhibit selective oxidation of 2° alcohols in lignin include those employing TEMPO, N-hydroxyphthalimide (NHPI), 2,3-dichloro-5,6-dicyano-1,4-benzoquinone (DDQ), and transition-metal catalyst systems. Some of these oxidants are used in stoichiometric quantities, while others are components of catalyst systems.

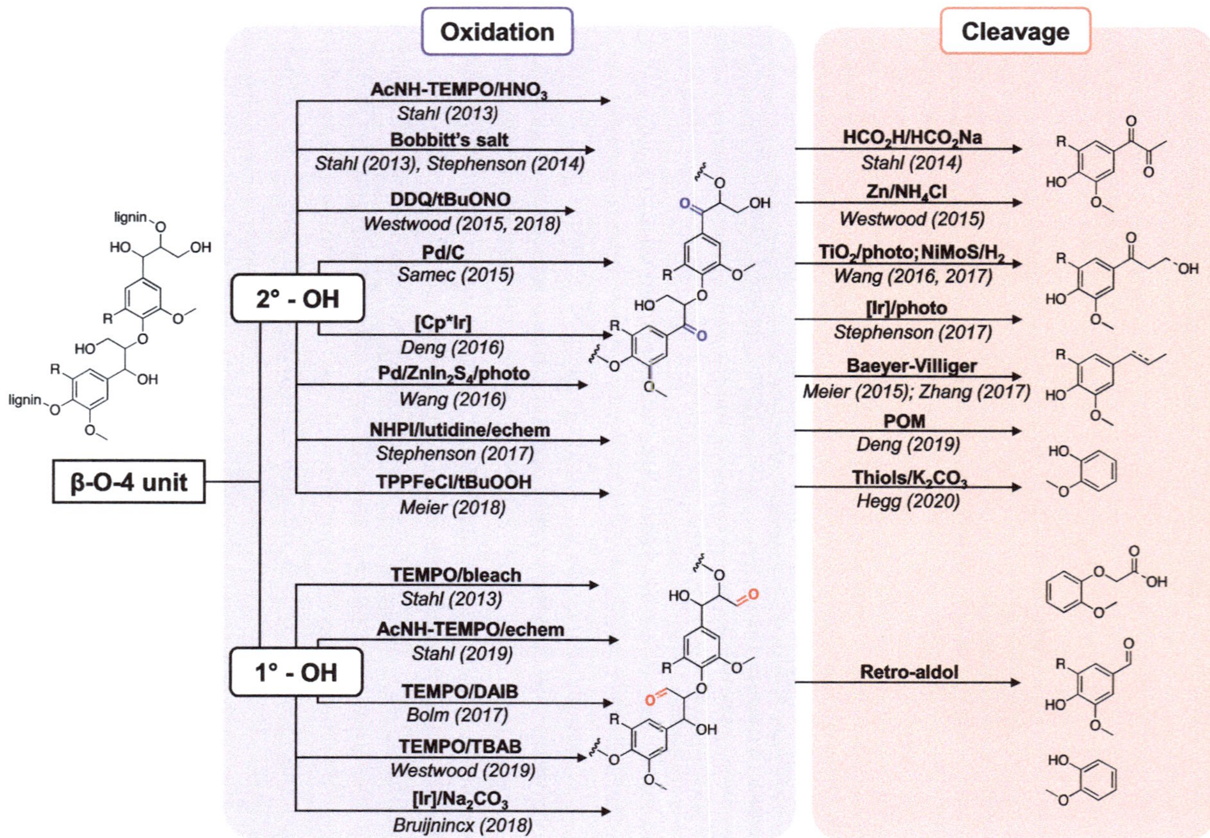

Fig. 9 Overview of the sequential oxidation-cleavage approaches applied in real lignin.

3.2.1 TEMPO and TEMPO derivatives

TEMPO derivatives are among the most effective reagents and catalyst systems available for 2° alcohol oxidation in lignin.[27] As noted above, Stahl and coworkers showed that stoichiometric Bobbitt's salt and ACT/HNO_3/HCl (O_2, MeCN/H_2O, 45 °C, 20 h) produced up to 97% yields of oxidized lignin model compounds (Fig. 10A and B).[27] Meier et al. obtained similar results, albeit with somewhat lower yields, for similar model compounds using modified catalytic conditions (TEMPO/$NaNO_2$/HCl/NaCl, O_2, dichloromethane, 19 h).[79]

Wang and coworkers developed an aerobic system using $VOSO_4$ as a co-catalyst to generate the oxoammonium species (Fig. 10C), followed by aerobic C—C bond cleavage over a Cu/1,10 phenanthroline (phen) catalyst for production of aromatic acids and phenols. Oxidation of non-phenolic β-O-4 lignin models (20% $VOSO_4$/TEMPO, 0.4 MPa O_2, MeCN, 90 °C, 6 h) generated the corresponding β-O-4 ketone product in 53% yield, together with a 20% yield of the retro-aldol cleavage product 2-phenoxyacetophenone. When the $VOSO_4$/TEMPO system was tested on organosolv Alcell lignin, 2D HSQC NMR analysis revealed partial oxidation of the β-O-4 linkages in lignin. Details on the cleavage step are elaborated below (see Section 3.4.2).[80]

Bolm et al. achieved benzylic oxidation (>94% yield) of non-phenolic β-O-4 lignin model compounds through mechanochemical methods using catalytic amounts of HO-TEMPO/KBr in the presence of oxone as terminal oxidant (20% HO-TEMPO/KBr, 25 Hz, 90 min) (Fig. 10D).[81] Dimeric models containing phenols oxidatively cleaved under these conditions, likely reflecting oxone-promoted generation of phenoxy radicals that led to formation of benzoquinone (82–91%) and guaiacol (76–82%) products.

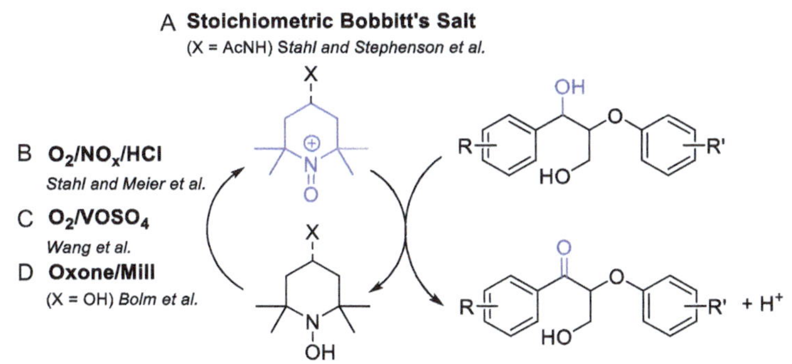

Fig. 10 TEMPO and its derivatives mediated benzylic alcohol oxidation in lignin and its models.

Optimized reaction conditions were applied to organosolv beechwood lignin. Gel permeation chromatography (GPC) analysis and HSQC NMR and IR spectroscopic data supported a high degree of oxidation and some cleavage of β-O-4 linkages of lignin.

Stephenson and coworkers developed a sequential alcohol oxidation-photochemical reductive cleavage method using stoichiometric Bobbitt's salt for the alcohol oxidation step (Fig. 10A).[82] Initial efforts to use catalytic ACT/HNO_3/HCl oxidation method[27] were not compatible with the subsequent photochemical cleavage step, elaborated below in Section 3.4.3. Non-phenolic β-O-4 dimers were oxidized, the reaction was filtered, solvent was removed, and the mixture was subjected to depolymerization conditions without further purification. Hegg and coworkers explored a sequential oxidation/depolymerization method with poplar lignin extracted using the Cu-AHP method.[83] Although some alcohols are oxidized during the oxidative Cu-AHP extraction process ($CuSO_4$/bpy, H_2O_2, NaOH, 30 °C), treatment of the extracted lignin with Bobbitt's salt led to an approximately five-fold increase in oxidation at the benzylic position, as revealed by HSQC-NMR spectroscopic analysis. The latter material proved much more susceptible to cleavage by a biomimetic thiol-mediated process than the lignin obtained directly from the Cu-AHP process (see Section 3.4.3 for discussion of the cleavage process).

3.2.2 NHPI

NHPI is the stable precursor to another type of nitroxyl mediator, phthalimide-N-oxyl (PINO), used in organic oxidation reactions (Fig. 11). Unlike TEMPO, the imidoxyl radical PINO is generated from NHPI in situ. Although it has limited stability in solution, PINO mediates efficient HAT from weak C—H bonds, such as those in benzylic alcohols.[72,84] A number of different NHPI oxidation methods have been employed to achieve NHPI/PINO-catalyzed oxidation of lignin and lignin model compounds (Fig. 11).

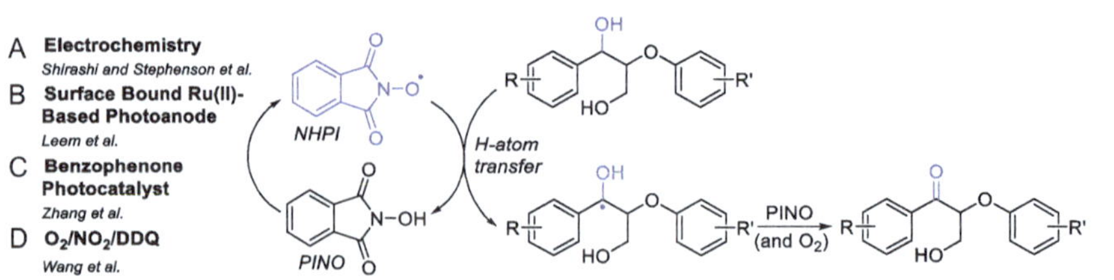

Fig. 11 NHPI mediated benzylic alcohol oxidation in lignin and its models. DDQ = 2,3-dichloro-5,6-dicyano-1,4-benzoquinone.

Shiraishi and coworkers reported an NHPI-mediated electrochemical oxidation of a non-phenolic β-O-4 dimer into the C_α-ketone derivative in excellent yield (93%; conditions: $LiClO_4$, 2,6-lutidine, MeCN, 0.7 V vs. Ag/Ag^+) (Fig. 11A).[85] The lutidine base facilitates electrochemical reoxidation of NHPI to PINO. Stephenson et al. used a similar NHPI/2,6-lutidine electrocatalytic oxidation method,[86] and showed that inclusion of O_2 in the reaction mixture led to increased yield of the ketone for a series of model compounds. O_2 reacts rapidly with benzylic radicals generated by PINO-mediated HAT, contributing to oxidation at this site.[87–89] Analysis of simplified model compounds suggested excellent reactivity and selectivity for benzylic alcohol oxidation; no primary alcohol oxidation was detected. Pine dioxosolv-lignin was subjected to electrochemical NHPI oxidation using a modified acetone:DMSO (98:2) solvent system to solubilize the lignin, and characteristic signals for oxidation of β-O-4 to the C_α-ketone were observed by HSQC NMR spectroscopy. This reactivity was paired with a photochemical reduction process to achieve partial depolymerization of the lignin (see Section 3.4.3 for the cleavage step). A photo-electrochemical variant of NHPI-catalyzed alcohol oxidation in lignin and lignin model compounds was recently developed by Leem et al. using electrode-immobilized Ru^{II}-polypyridyl complexes (Fig. 11B).[90]

Aromatic ketones, such as benzophenone and its derivatives, have been widely used for photochemically initiated HAT, including reactions with weak O—H bonds.[91,92] Zhang and coworkers utilized donor-substituted aromatic ketones 4,4′-bis(diphenylamino)benzophenone (DPA-BP) to generate PINO through chemical photocatalysis (Fig. 11C).[93] The DPA-BP/NHPI/O_2 photocatalytic system exhibits good reactivity and selectivity for the aerobic oxidation of phenolic and non-phenolic β-O-4 lignin models. The benzylic ketone products were obtained in excellent isolated yields (81–99%).

Finally, the combination of NHPI, DDQ, and O_2 is a known catalyst system for oxidation of benzylic C—H bonds.[94] Wang and coworkers supplemented this co-catalyst mixture with a NO_x source to achieve oxidation of benzylic ketones from β-O-4 model compounds in 35–45% yields (Fig. 11D).[95]

3.2.3 DDQ

DDQ is a common reagent for dehydrogenation of benzylic alcohols.[96] Lignin oxidation with stoichiometric or catalytic DDQ has been proposed to be initiated by SET, HAT, or hydride transfer,[97,98] although the reaction

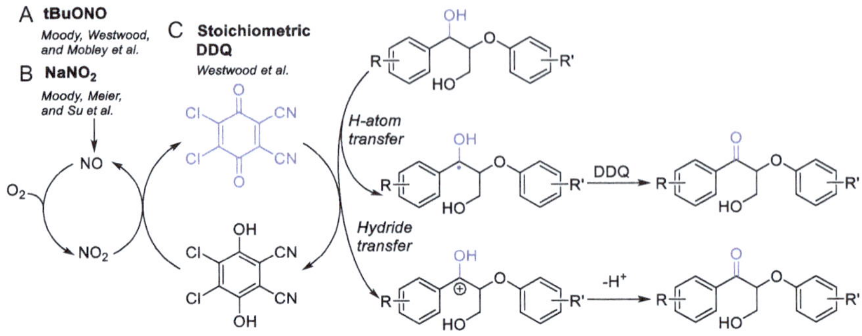

Fig. 12 DDQ-mediated benzylic alcohol oxidation in lignin and its models.

mechanisms have seldom been the focus of specific investigation. Aerobic oxidation of the benzylic alcohol in lignin β-O-4 model compounds was developed by employing NO_x-based co-catalysts (e.g., $NaNO_2$, tert-butyl nitrite (*t*BuONO), HNO_3) to promote re-oxidation of the reduced quinone (2,3-dichloro-5,6-dicyanohydroquinone, $DDQH_2$).[99–101] Moody et al. reported oxidation of lignin model compounds at ambient temperature using catalytic amounts of DDQ and $NaNO_2$ or *t*BuONO under visible light irradiation, affording excellent yield (>92%) of the benzylic ketone in β-O-4 model compounds (Fig. 12A and B).[102] Photoexcitation of DDQ to a long-lived triplet state was proposed to initiate electron transfer from the substrate.[103]

Westwood et al. applied the DDQ/*t*BuONO catalytic system for the oxidation of simple β-O-4 model compounds and polymeric models at elevated temperature (80 °C) rather than promoting the reaction with light. The reactions produce ketones with high yield and complete selectivity.[104] The optimized catalytic conditions were applied to the oxidation of organosolv birch lignin, and the desired β-O-4 ketone was identified by 2D HSQC NMR spectroscopy. This study was the first to use DDQ in the sequential oxidation/depolymerization of authentic lignin. 6 wt% aromatic monomers were isolated following a Zn-mediated reductive cleavage method, elaborated in Section 3.4.3 below (Fig. 12A). The same group oxidized organsolv beech lignin with varying amounts of stoichiometric DDQ (i.e., no re-oxidation method was employed), and used 2D HSQC NMR analysis to track linkage reactivity with increasing amounts of oxidant (Fig. 12C).[105] This stoichiometric DDQ-based method generated oxidized lignin with six hardwood dioxosolv lignins, including those from beech, oak, maple, hickory, cherry, and birch, and the protocol was validated at 20 g scale.

DDQ is also capable of oxidizing benzylic ethers.[98,106] Luterbacher and coworkers have developed an aldehyde-protecting group strategy that enables much higher yields of lignin to be extracted via acidolysis from various biomass sources.[107] They subsequently showed that the acetal-protected lignin is amenable to stoichiometric or catalytic DDQ-promoted oxidation of the benzylic position to afford the corresponding oxidized lignin.[108] The oxidation method was demonstrated on a series of β-O-4 model compounds and then applied to the oxidation of acetal-protected birch and genetically-modified, poplar lignin enriched in syringyl units.

A number of subsequent studies have employed stoichiometric DDQ and catalytic DDQ/*t*BuONO or $NaNO_2/O_2$ catalyst systems to carry out lignin oxidation.[79,109] Bugg et al. generated oxidized lignin from multiple plant sources using stoichiometric DDQ oxidation and evaluated the monomer yield from a Zn-based depolymerization method.[110] The best results were obtained with poplar organosolv lignin, which had the highest β-O-4 content of the investigated lignins. Other examples will be discussed in Section 3.4.2 in connection with depolymerization studies of these materials.

3.2.4 Additional oxidation methods

Several other metal-catalyzed methods have been evaluated for lignin (model) oxidation. Meier et al. investigated the use of meso-tetraphenylporphyrin iron(III) chloride (TPPFeCl) and *t*BuOOH for catalytic oxidation of benzylic alcohols in lignin model compounds.[79,111] Ball-milling treatment in KOH and toluene resulted in an increase in the carbonyl content in Kraft lignin, and the mechanochemical processing with TPPFeCl-catalyzed oxidation has a synergistic, positive effect on the depolymerization of lignin. When the TPPFeCl/*t*BuOOH oxidation method was applied to Kraft lignin, 2D HSQC NMR spectra suggested that the β-O-4 benzylic alcohol groups were oxidized, and a Baeyer-Villiger oxidation method was used in a subsequent depolymerization method to generate aromatic monomers (cf. Section 3.4.2).

Deng, Fu, and coworkers employed Cp*Ir catalysts bearing a cooperative bipyridone ligand to promote dehydrogenation of β-O-4 lignin model compounds (Fig. 13).[112] These reactions promote thermal dehydrogenation of the benzylic alcohols, generating H_2. Good conversions and yields were observed with a series of model compounds.

Samec and coworkers demonstrated the selective oxidation of benzylic alcohols in lignin model compounds using carbon supported Pd nanoparticles under mild conditions (Fig. 14A).[113,114] Use of oxidizing

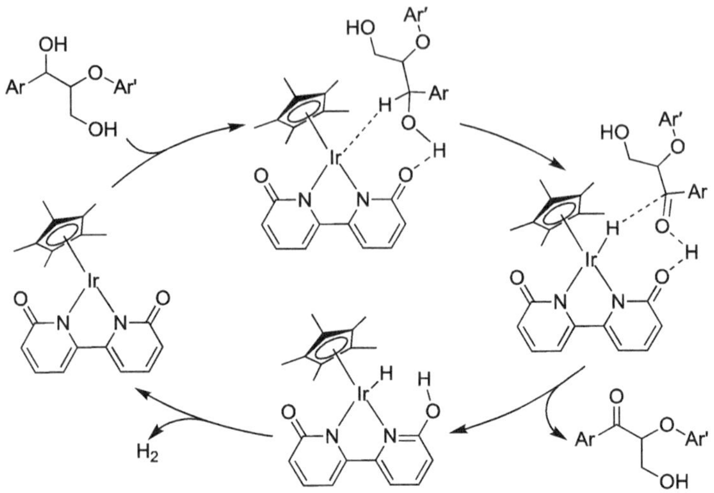

Fig. 13 Transition metal mediated secondary alcohol oxidation in lignin models using [Ir(Cp*)L], where L = 2,2′-bipyridine-6,6′-diol.

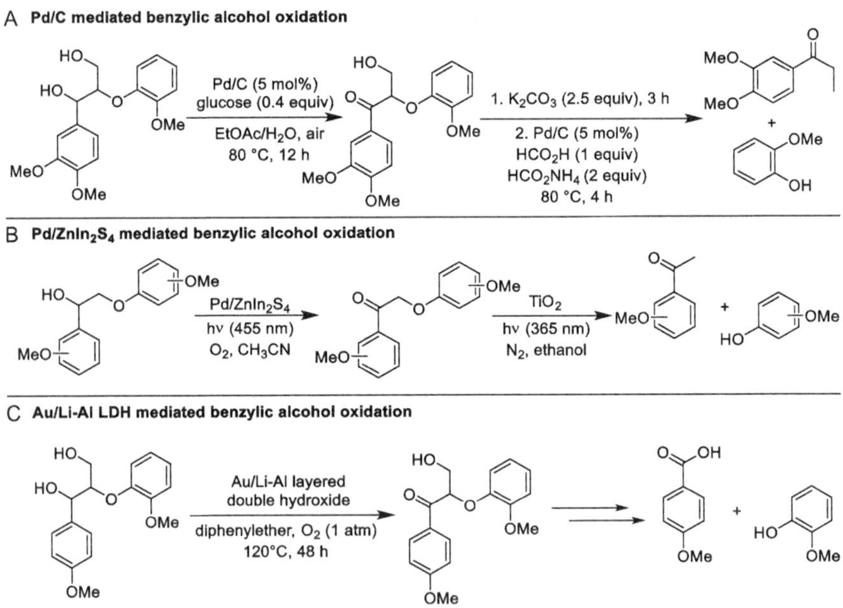

Fig. 14 Heterogeneous catalyst-mediated secondary alcohol oxidation in lignin model compounds using (A) Pd/C, (B) Pd/ZnIn$_2$S$_4$, and (C) Au/Li-Al LDH.

conditions led to benzylic ketone formation without cleavage of C—O or C—C bonds or dehydrogenation of the γ 1° alcohol. The highest yields of ketone were obtained in the presence of glucose, which was proposed to stabilize the Pd catalyst to prevent aggregation. Zhang, Wang and coworkers studied cerium oxide-supported Pd catalyst (Pd/CeO_2) for the aerobic oxidation of various model compounds, in addition to a source of organosolv lignin. Low yields of monomers were obtained from the latter process: vanillin (5.2%), guaiacol (0.87%) and 4-hydroxybenzaldehyde (2.4%).[115]

$ZnIn_2S_4$ is a ternary chalcogenide semiconducting material with a band gap that supports photochemistry with visible light irradiation. Wang et al. found that $Pd/ZnIn_2S_4$ promoted oxidation of benzylic alcohols with lignin models upon illumination with 455 nm light under ambient conditions (Fig. 14B). This step was paired with TiO_2-mediated photochemical cleavage (365 nm illumination) of the models into aromatic monomers.[80,116]

Crocker et al. employed the heterogeneous catalyst Au/Li–Al layered double hydroxide (LDH) to catalyze aerobic oxidation of dimeric lignin models with 1 atm O_2 and observed excellent activity for benzylic alcohol oxidation in β-O-4 models (Fig. 14C).[117] The high surface area and strong basicity of the Li—Al LDH support facilitated substrate adsorption and reaction on the Au surface. Lignin, one sample sourced from the γ-valerolactone extraction method[75] and another from the Kraft process, were subjected to aerobic oxidation conditions with this catalyst. Alkaline hydrolysis of the resulting oxidized lignin materials led to lignin depolymerization, with monomer yields as high as 40% for γ-valerolactone extracted lignin and 10% for Kraft lignin.

3.3 Primary alcohol oxidation

Under basic conditions, oxoammonium generated from TEMPO or ACT mediates alcohol oxidation by an inner sphere mechanism that favors reaction with sterically unhindered primary alcohols.[73] This selectivity was demonstrated by using bleach/TEMPO conditions at pH 9 (cf. Fig. 7A).[27] The Cu/TEMPO-catalyzed aerobic oxidation method, which features a different mechanism,[118] also shows selectivity for primary alcohol oxidation.[27]

Bolm et al. reported that the hypervalent iodine reagent (diacetoxyiodo) benzene (DAIB), in combination with TEMPO, leads to selective oxidation of non-phenolic β-O-4 primary hydroxyl groups to aldehydes in 43% yield (Fig. 15A).[119] Stoichiometric DIAB and a TEMPO catalyst were used to oxidize organosolv beech lignin. A reduction of C_γ-H_γ signals was revealed

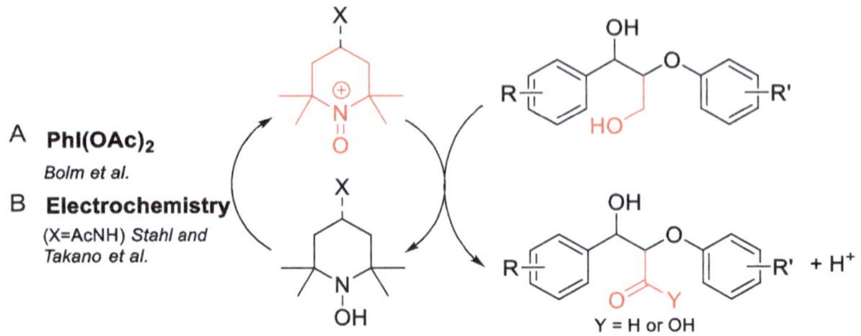

Fig. 15 Mechanism of TEMPO mediated primary alcohol oxidation of a lignin model compound.

by 2D HSQC NMR analysis, corresponding to an increase in carbonyl chemical shifts typical for aldehydes. Depolymerization of this material is presented in Section 3.4.1 below.

The groups of Takano and Stahl have used electrochemical methods to support TEMPO-catalyzed oxidation of lignin and/or lignin model compounds. Stahl et al. used ACT as the mediator for the oxidation of β-O-4 model substrates lacking free phenols under basic conditions (5% ACT, pH 10, 70:30 H_2O:MeCN, 0.8 V vs. Ag/AgCl). Higher turnover frequencies were observed for models containing primary alcohols with β-ether fragments, similar to those in the β-O-4 unit in lignin.[120] A systematic comparison of various 1° and 2° alcohol substrates shows that 1° alcohols are much more reactive under these conditions, and β-O-4 models with both types of alcohols undergo selective oxidation of the primary alcohol to the carboxylic acid in >90% yield (Fig. 15B). Takano and coworkers observed comparable yields and selectivity with similar β-O-4 model compounds under modified conditions 90:10 dioxane:phosphate buffer (pH 7, 20 mol% ACT) conditions.[121] Stahl and coworkers also tested the reactivity of a mild-acidolysis poplar lignin material under their optimized reaction conditions. 2D HSQC NMR spectroscopy showed a decrease in characteristic β-O-4 primary alcohol β- and γ- signals, with no change to the secondary alcohol signals. An acid-base titration of oxidized lignin revealed an approximately 12-fold increase in acidic functionality, attributed to the introduction of carboxylic acid functional groups in the polymer. Some lignin cleavage is observed by GPC during the oxidation step, but the oxidized polymer was also amenable to depolymerization under acidic conditions, as described below (see Section 3.4.1).

Lancefield, Bruijnincx and coworkers reported an efficient catalytic dehydrogenation system using a Cp*Ir-bipyridonate complex (same as that in Fig. 13) in a 1,4-dioxane/water solution at pH ~ 11. This approach exhibited selectivity for primary over secondary alcohol dehydrogenation and led to cleavage of the C_α—C_β bond. This catalytic system was applied to enzyme-extracted lignin and dioxosolv softwood lignins and led to significant reduction in the lignin molecular weight and low yields of mono-aromatic compounds.[122]

3.4 Bond cleavage methods

The two-step oxidation/depolymerization strategy allows for systematic optimization of the two steps, whereby selective oxidation of the primary or secondary alcohols in the backbone may be subjected to different conditions that enable efficient cleavage of the C_β—O aryl, C_α—C_β, or other bonds in the linkers between aromatic units. A variety of methods, including oxidative, reductive, and redox-neutral reactions, have been developed for this second-step in the process (Fig. 16). Secondary alcohol oxidation has been paired with both redox and non-redox depolymerization methods, while primary alcohol oxidation has been typically paired with redox-neutral processes.

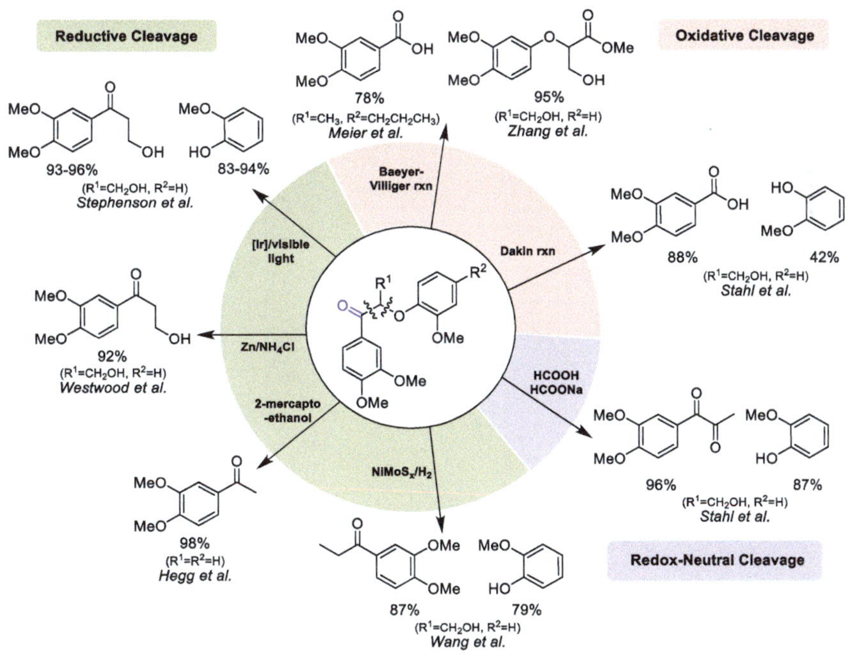

Fig. 16 Different bond cleavage patterns in lignin model with C_α ketone structure.

3.4.1 Redox-neutral cleavage

A retro-aldol reaction following selective primary alcohol oxidation enables cleavage of the C_{α}—C_{β} bond, resulting in lignin depolymerization.[53] In some cases, this reactivity proceeds directly under the oxidative reaction conditions (cf. Fig. 7A), while other cases (emphasized here) feature treatment of the oxidized lignin under modified conditions to promote cleavage. Lignin model compounds with terminal C_{γ}-aldehydes undergo facile retro-aldol under basic conditions to generate benzylaldehyde and an aryl ether fragment (Fig. 17A). Bolm and coworkers use DL-proline as an

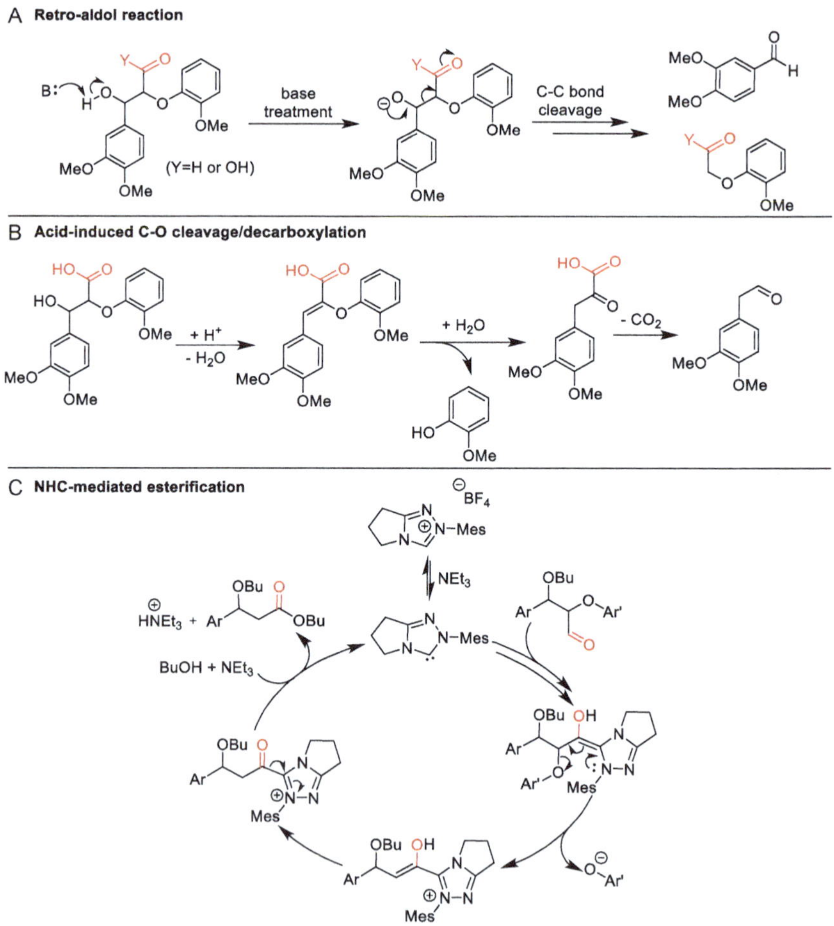

Fig. 17 Mechanisms of redox-neutral bond cleavage method of primary alcohol oxidized lignin.

organocatalyst to promote this reactivity at ambient temperature, generating veratraldehyde and guaiacol in 70% and 26% yields, respectively.[119] Treatment of oxidized organosolv beech lignin (cf. Section 3.3) led to reduction in the molecular weight of the polymer, consistent with bond cleavage in the backbone.

Stahl and coworkers explored acidic conditions to promote depolymerization of poplar lignin containing C_γ-carboxylic acids. This approach generated up to 30 wt% yields of monomeric aromatic compounds from an oxidized poplar lignin sample.[120] Various aqueous acids proved effective (HCl, H_2SO_4, and HCO_2H) for cleavage of the oxidized model compounds, but the highest monomer yields from lignin were obtained with a 9:1 mixture of HCO_2H and H_2O. These products can arise from acid-induced elimination of the benzylic hydroxyl group to afford a vinyl ether that is susceptible to hydrolysis, resulting in generation of Hibbert's ketone-like α-keto-acid and phenol aromatic monomers (Fig. 17B).[123] Subsequent decarboxylation of the keto-acid generates the observed aromatic monomer containing an aliphatic aldehyde. The aldehyde is susceptible to subsequent aldol condensation, resulting in the formation of by-products. The latter condensation reactions present major challenges for lignin depolymerization methods that generate aliphatic aldehydes, although strategies to protect the aldehydes and minimize the byproduct formation have been developed.[124]

Westwood and coworkers developed an N-heterocyclic carbene (NHC) -mediated internal redox esterification method for depolymerizing oxidized lignin containing C_γ-aldehydes (generated from a modified method of Einhorn et al.)[125] (Fig. 17C).[126] Incorporation of butylethers at the C_α position of lignin during the "butanosolv" extraction method supports this process as it prevents the competing retro-aldol reaction. The proposed mechanism involves initial formation of a Breslow-type intermediate from the reaction of the NHC with the C_γ-aldehyde, followed by elimination of the phenolate and formation of the enol intermediate. Tautomerization leads to an acylazolium that reacts with butanol.[127] The process takes advantage of the inherent structure of the β-aryl ether units present in the butanosolv lignin and delivers novel functionalized aromatic monomers in modest yield (6 wt%).

Redox-neutral processes also support cleavage of lignin oxidized at the 2° benzylic alcohol position (see Section 3.1 for oxidation conditions). Stahl and coworkers showed that aqueous formic acid/formate supports bond cleavage of such structures. In an application with poplar CEL, ACT/

HNO_3/HCl oxidation of the lignin followed by HCO_2H/HCO_2Na treatment led to a 52 wt% yield of aromatic monomers. Diketones are the major product of this reaction (cf. Fig. 8), and mechanistic studies supported the mechanism shown in Fig. 18. Acylation of the 1° C_γ-OH group generates a formic ester intermediate that undergoes formate-induced elimination to generate the aryl vinyl ether species. Hydrolysis of the vinyl ether releases the phenol and diketone products (Fig. 18).[28] This redox neutral process is noteworthy because it avoids degradation of phenols that can occur under oxidative cleavage conditions. As noted in Section 3.1, this method was also used successfully in the oxidation/depolymerization of lignin samples from various biomass sources and pretreatment protocols.[29]

The formic acid/formate depolymerization method developed by Stahl et al. has been employed in other two-step oxidation/depolymerization methods to afford aromatic monomers from oxidized lignin with benzylic ketones. Luterbacher and coworkers employed a HCO_2H/HCO_2Na depolymerization step following DDQ-mediated oxidation of their extracted acetal-protected lignin (see Section 3.2.3).[108] A yield of up to 36% aromatic monomers (diketones, carboxylic acids, and aldehydes) was obtained from propylidene acetal protected birch lignin by using a two-step DDQ oxidation/HCO_2H-HCO_2Na depolymerization sequence. A 52% yield of monomers was obtained with F5H poplar lignin (a genetically modified poplar with high syringyl content). The syringyl propanedione monomer was the major

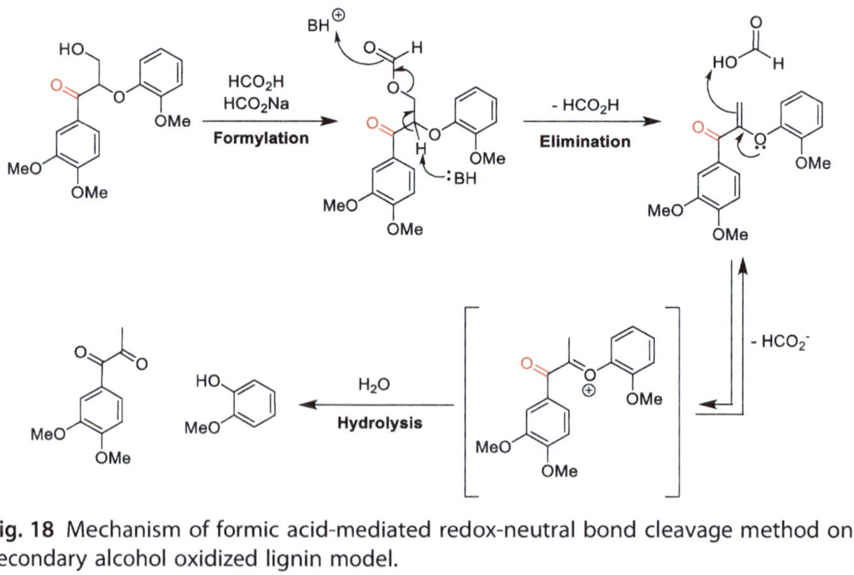

Fig. 18 Mechanism of formic acid-mediated redox-neutral bond cleavage method on secondary alcohol oxidized lignin model.

product in both cases, consistent with the mechanism shown in Fig. 18. Mobley and Crocker found that the HCO_2H/HCO_2Na depolymerization process was most effective for conversion of DDQ-oxidized Kraft pine lignin.[109]

3.4.2 Oxidative cleavage

Oxidative conditions also support cleavage of oxidized lignin; however, these methods are susceptible to further oxidation and/or oxidative degradation of the monomeric products. This reactivity was noted above in the use of H_2O_2 under basic conditions to cleave benzylic ketone model compounds, resulting in an 88% yield of veratric acid but only 42% yield of guaiacol (Figs. 7B).[27,65] Baeyer-Villiger oxidation methods provide an alternative strategy for cleavage of the oxidized lignin (models) (Fig. 19). Meier et al. employed in situ generated performic acid to convert benzylic ketones into esters,[79] and ester hydrolysis resulted in 78% of aromatic acids from models resembling oxidized β-O-4 structures. The corresponding aldehyde and phenol cleavage products were not recovered, likely reflecting their oxidative degradation under the reaction conditions. Oxidized Kraft lignin was subjected to these cleavage conditions (8 eq. H_2O_2, 4 eq. HCOOH, DCE, 50 °C, 24 h), and the resulting aromatic carboxylic acids were esterified in a work-up step to give a 10 wt% yield of methyl ester products.[111] Major products included methyl vanillate 7.8 wt% and methyl 5-carbomethoxyvanillate 2.2 wt%. Zhang et al. used modified Baeyer-Villiger conditions, and paired this method with alcoholysis (K_2CO_3, 45 °C, MeOH) to achieve high yields (>90%) in the cleavage of various model compounds to phenol and methyl benzoate derivatives (Fig. 19B).[128]

A variety of aerobic cleavage methods have also been explored using organometallic catalysts. Lignin models oxidized to the ketone by Wang et al. were subjected to oxidative cleavage by a variety of Cu salts bearing nitrogen donor ligands.[80] The $Cu(OAc)_2$/1,10-phenanthroline complex led to selective $C_\alpha—C_\beta$ bond cleavage to form aromatic acids in 92% yield

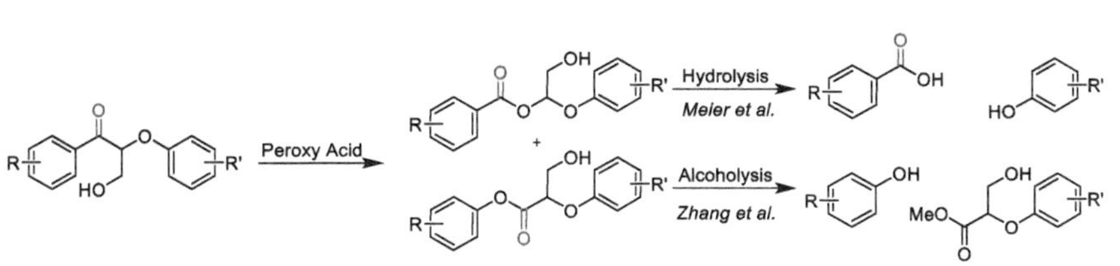

Fig. 19 Baeyer-Villiger bond cleavage for β-O-4 linkages.

under mild conditions (0.4 MPa O_2, 150 °C, MeOH, 6 h). Deng et al. employed polyoxometalates (POM) as a regenerable catalyst to depolymerize pre-oxidized dioxosolv lignin. The optimized oxidative conditions (150 °C, 10 bar O_2, MeOH) used a phosphomolybdic acid ($H_3PMo_{12}O_{40}$) catalyst to generate monomer yields up to 33 wt%. Aromatic monomers account for 19.4 wt% of these products, with other products derived from aromatic ring-opening. Overoxidation also converts low molecular weight depolymerization products to CO, CO_2, and H_2O.[129]

Su and coworkers reported a mechanochemical approach for selective cleavage of C_β—O bonds and C_α—C_β bonds of lignin β-O-4 ketones via NaOH-promoted oxidation under milling conditions (5 equiv. NaOH, 25 Hz milling, 2 h).[130] In contrast to other oxidative cleavage methods, the yield of phenol (87–92%) is always 13–22% *higher* than the yield of the corresponding aromatic acid. Dioxosolv birch lignin milled with DDQ/$NaNO_2$ (see Section 3.2.3 for oxidation conditions) led to a 9.1 wt% yield of aromatic monomers. When the lignin was independently oxidized with the DDQ/tBuONO conditions of Westwood (see Section 3.2.3 for oxidation conditions),[104] mechanochemical cleavage generated a 15.5 wt% yield of aromatic monomers, with syringate as the major product (7.5 wt%).

3.4.3 Reductive cleavage

Oxidized β-O-4 fragments with the benzylic ketone correspond to O-aryl acyloin derivatives that are susceptible to reductive cleavage. Zinc is an effective reductant in these reactions,[131–133] and could initiate single-electron transfer or direct two-electron reactivity (Fig. 20A). This reactivity was among those demonstrated by Stahl and coworkers using β-O-4 model compounds in their initial report on two-step oxidation/depolymerization of lignin.[28] Westwood and co-workers used Zn/NH_4Cl at 80 °C after DDQ-mediated oxidation of benzylic alcohol in both lignin and model compounds.[104] The approach generates β-hydroxy phenyl ketone products in >80% yield for different types of β-O-4 model compounds, and modest yields of phenolic monomers were also obtained from the oxidation/reductive depolymerization of birch lignin. Deng and Fu et al. used this method for lignin depolymerization following Cp*Ir-catalyzed dehydrogenation of benzyl alcohols in lignin (cf. Fig. 13).[112]

Reductive cleavage of oxidized lignin and lignin model compounds also may be achieved via photoredox-initiated SET. Stephenson and coworkers reported that an $[Ir(ppy)_2(dtbbpy)]PF_6$ photocatalyst (ppy = 2-phenylpyridine,

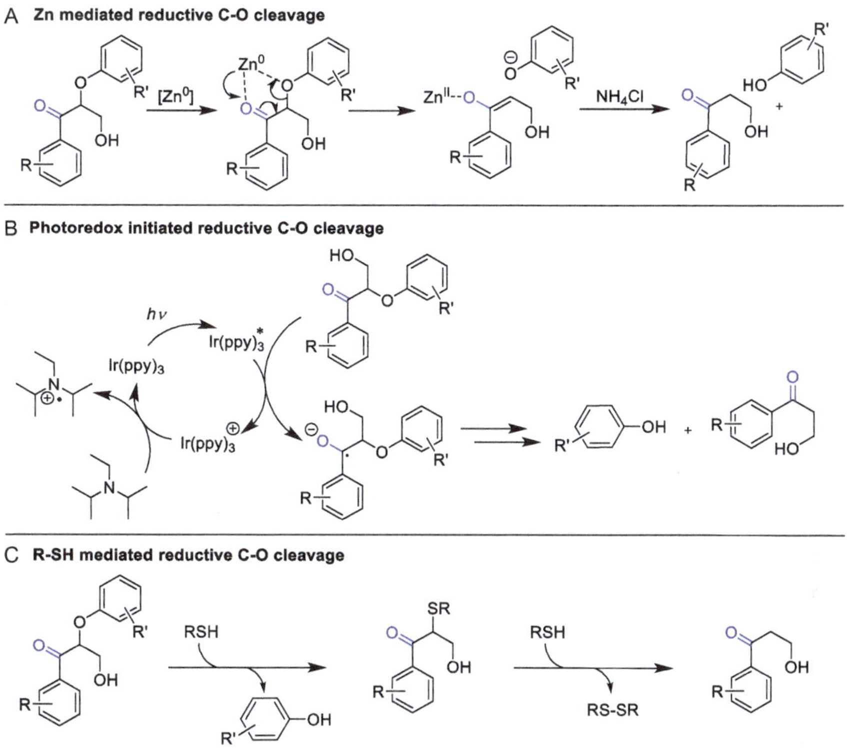

Fig. 20 Mechanisms of reductive bond cleavage methods.

dtbbpy=4,4′-di-tert-butyl-2,2′-bipyridyl) under blue LED irradiation could promote C_{α}—O bond cleavage, generating phenols (yield >70%) and β-hydroxyaryl ketones (yield >81%) from oxidized lignin model compounds (Fig. 20B).[82,86] Bond cleavage is proposed to proceed via fragmentation of the ketone radical anion generated by SET from the photocatalyst. N,N-diisopropylethylamine (DIPEA) and formic acid provide a source of reducing equivalents for the reaction. Modest yields of aromatic monomers were also obtained from a two-step oxidation/photoreductive cleavage sequence when applied to purified dioxosolv pine lignin (2.4 wt% guaiacol and phenyl ketone monomers were recovered).[86] (Fig. 20A). The latter process was implemented as a batch-to-flow sequence involving electrochemical NHPI-mediated lignin oxidation in batch, followed by photochemically initiated reductive cleavage in flow.[86] Zhang and coworkers expanded on this work by developing a thiol-functionalized mesoporous cellular silica foam (MCF) with immobilized $Ir(ppy)_2(bpy)$ photocatalyst.[134] The $Ir(ppy)_2(bpy)$-MCF material was effective for the reductive cleavage of oxidized lignin β-O-4 model compounds under mild reaction conditions (yield >90%). The photocatalyst could be reused at least six times without a significant drop in performance.

Wang and coworkers showed that oxidized lignin models are amenable to hydrogenolytic cleavage with a $NiMoS_x$ catalyst. The C_β—OPh bond is cleaved without hydrogenation of the aromatic ring or the C_α ketone to afford phenols and benzylic ketones. The authors propose direct cleavage of the C_β—OPh bond on the $NiMoS_x$ catalyst surface. This process was combined with an initial oxidation step using a $NaNO_2$/DDQ/NHPI oxidation system to afford a 32% yield of aromatic monomers from birch powder.[95]

Wood-digesting bacteria feature enzymes that use thiol-containing tripeptide glutathione as a redox active cofactor to promote aryl ether cleavage in lignin. Examples include the Lig enzymes E, P, F (etherases) and G (lyase) *Sphingobium* sp. strain SYK 6.[135,136] In this context, Hegg and coworkers demonstrated complementary chemical reactivity, exploiting the nucleophilic and reductive reactivity of small organic thiols to cleave aryl alkyl ether linkages in oxidized lignin model compounds (Fig. 20C). Organic thiols, such as β-mercaptoethanol and dithiothreitol can promote cleavage in nearly quantitative yields with certain β-O-4 model compounds in acetonitrile containing K_2CO_3 as a basic additive.[137] They further employed small thiols (such as propanedithiol, 2-mercaptoethanol, and dithiothreitol) to support cleavage of model polymers and oxidized lignin materials bearing the C_α-ketone functionality. Thiol-dependent depolymerization is proposed to proceed via S_N2 attack adjacent to the ketone to release a phenol. The thiol is regenerated by nucleophilic attack of another thiol on the S-atom of the thioether to form a disulfide and ketone. Poplar lignin derived from a Cu-AHP extraction method[78] was oxidized using Bobbitt's salt (cf. Fig. 10) and then incubated with β-mercaptoethanol. This oxidation/reductive cleavage sequence resulted in approximately 65% reduction of the lignin molecular weight.[83]

4. Conclusion and future outlook

Lignin valorization remains a persistent challenge that is complicated by the complex structure of lignin and the difficulty in obtaining high-quality native-like lignin materials from biomass fractionation methods. Nonetheless, new methods for conversion of lignin into low molecular weight aromatic chemicals continue to emerge. The content outlined herein highlights the merits of bottom-up strategies to address this long-standing challenge. Consideration of the fundamental reactivity of individual functional groups prevalent in lignin—1° aliphatic and 2° benzylic alcohols—has enabled the development of sequential oxidation/depolymerization methods that successfully convert lignin into oxygenated aromatic chemicals. These efforts greatly benefitted from fundamental studies of well-defined model compounds that incorporate structural features directly present in lignin, especially those present in the common β-O-4 linkage. The development of selective, high-yielding methods for oxidation of the 1° or 2° alcohol in these compounds sets the stage for systematic development of various bond cleavage pathways that are enabled by the initial oxidation step. Insights gained from the well-defined model compounds were then translated into successful lignin depolymerization methods employing similar two-step methods. This approach has been used with a diverse array of lignin materials derived from different biomass plant sources, albeit favoring hardwoods, and different pretreatment/lignin-extraction methods. An appealing feature of this sequential depolymerization strategy is the ability to optimize each step independently, allowing the overall process to be optimized more effectively for individual feedstocks to maximize the yield of aromatic monomers.

Future work will benefit from studies focused on cleaving a broader array of linkages, as the majority of studies to date have focused on the alcohol-containing β-O-4 unit. Although the β-O-4 fragment is prevalent in lignin, the development of methods capable of cleaving other linkages could greatly improve the monomer yield.

Another future priority will be the development of improved methods for direct oxidative catalytic fractionation (OCF) of biomass,[40,138,139] resembling existing methods for RCF.[12–18] Essentially all processes for lignin extraction damage the native lignin, generating recalcitrant subunits that limit the yields of aromatics that may be obtained from lignin. For example, under acidic conditions commonly used for lignin extraction, protonation of the benzylic alcohol in the β-O-4 unit can lead to the loss of water and formation of a benzyl cation that can cross-link the lignin polymer (i.e., via Friedel-Crafts alkylation). This reactivity converts a β-O-4 fragment, which is amenable to cleavage, into a new fragment that blocks depolymerization. OCF methods that oxidize the benzylic alcohol while lignin is still integrated within the biomass could provide a means to "protect" the lignin, while generating a material that is more amenable to depolymerization once the lignin is separated from the sugar. This concept and related approaches take inspiration from the two-step oxidation/depolymerization studies outlined herein and could provide the basis for cost-effective valorization of whole biomass.

Acknowledgments

The lignin-based research in our laboratory and preparation of this article was supported in part by the Great Lakes Bioenergy Research Center, U.S. Department of Energy, Office of Science, Office of Biological and Environmental Research under Award Numbers DE-SC0018409 and DE-FC02-07ER64494 and in part by the U.S. Department of Energy, Office of Energy Efficiency & Renewable Energy under Contract No. DE-EE0008148.

References

1. Ragauskas, A. J.; Beckham, G. T.; Biddy, M. J.; Chandra, R.; Chen, F.; Davis, M. F.; Davison, B. H.; Dixon, R. A.; Gilna, P.; Keller, M.; Langan, P.; Naskar, A. K.; Saddler, J. N.; Tschaplinski, T. J.; Tuskan, G. A.; Wyman C. E. *Science* **2014**, *344*. 1246843.
2. Rinaldi, R.; Jastrzebski, R.; Clough, M. T.; Ralph, J.; Kennema, M.; Bruijnincx, P. C. A.; Weckhuysen, B. M. *Angew. Chem. Int. Ed.* **2016**, *55*, 8164–8215.
3. Boerjan, W.; Ralph, J.; Baucher, M. *Annu. Rev. Plant Biol.* **2003**, *54*, 519–546.
4. Vanholme, R.; Morreel, K.; Ralph, J.; Boerjan, W. *Curr. Opin. Plant Biol.* **2008**, *11*, 278–285.
5. Schutyser, W.; Renders, T.; Van den Bosch, S.; Koelewijn, S.-F.; Beckham, G. T.; Sels, B. F. *Chem. Soc. Rev.* **2018**, *47*, 852–908.
6. Mu, W.; Ben, H.; Ragauskas, A.; Deng, Y. *Bioenergy Res.* **2013**, *6*, 1183–1204.
7. Parthasarathi, R.; Romero, R. A.; Redondo, A.; Gnanakaran, S. *J. Phys. Chem. Lett.* **2011**, *2*, 2660–2666.
8. Galkin, M. V.; Samec, J. S. M. *ChemSusChem* **2016**, *9*, 1544–1558.
9. Phongpreecha, T.; Hool, N. C.; Stoklosa, R. J.; Klett, A. S.; Foster, C. E.; Bhalla, A.; Holmes, D.; Thies, M. C.; Hodge, D. B. *Green Chem.* **2017**, *19*, 5131–5143.
10. Renders, T.; Van den Bossche, G.; Vangeel, T.; Van Aelst, K.; Sels, B. *Curr. Opin. Biotechnol.* **2019**, *56*, 193–201.
11. Cheng, C.; Shen, D.; Gu, S.; Luo, K. H. *Catal. Sci. Technol.* **2018**, *8*, 6275–6296.
12. Yan, N.; Zhao, C.; Dyson, P. J.; Wang, C.; Liu, L.-T.; Kou, Y. *ChemSusChem* **2008**, *1*, 626–629.
13. Ferrini, P.; Rinaldi, R. *Angew. Chem. Int. Ed.* **2014**, *126*, 8778–8783.
14. Parsell, T.; Yohe, S.; Degenstein, J.; Jarrell, T.; Klein, I.; Gencer, E.; Hewetson, B.; Hurt, M.; Im Kim, J.; Choudhari, H.; Saha, B.; Meilan, R.; Mosier, N.; Ribeiro, F.; Deglass, W. N.; Chapple, C.; Kenttämaa, H. I.; Agrawal, R.; Abu-Omar, M. M. *Green Chem.* **2015**, *17*, 1492–1499.
15. Abu-Omar, M. M.; Barta, K.; Beckham, G. T.; Luterbacher, J. S.; Ralph, J.; Rinaldi, R.; Román-Leshkov, Y.; Samec, J. S. M.; Sels, B. F.; Wang, F. *Energy Environ. Sci.* **2021**, *14*, 262–292.
16. Schutyser, W.; Van den Bosch, S.; Dijkmans, J.; Turner, S.; Meledina, M.; Van Tendeloo, G.; Debecker, D. P.; Sels, B. F. *ChemSusChem* **2015**, *8*, 1805–1818.
17. Anderson, E. M.; Katahira, R.; Reed, M.; Resch, M. G.; Karp, E. M.; Beckham, G. T.; Román-Leshkov, Y. *ACS Sustainable Chem. Eng.* **2016**, *4*, 6940–6950.
18. Zhai, Y.; Li, C.; Xu, G.; Ma, Y.; Liu, X.; Zhang, Y. *Green Chem.* **2017**, *19*, 1895–1903.
19. Walton, N. J.; Mayer, M. J.; Narbad, A. *Phytochemistry* **2003**, *63*, 505–515.
20. Evju, H. *US Patent 4151207A*; 1979.

21. Pinto, P. C. R.; Da Silva, E. A. B.; Rodrigues, A. E. Lignin as a Source of Fine Chemicals: Vanillin and Syringaldehyde. In *Biomass Conversion*; Springer: Berlin, 2012; pp. 381–420.
22. Fache, M.; Boutevin, B.; Caillol, S. *ACS Sustainable Chem. Eng.* **2016**, *4*, 35–46.
23. Ma, R.; Xu, Y.; Zhang, X. *ChemSusChem* **2015**, *8*, 24–51.
24. Yu, X.; Wei, Z.; Lu, Z.; Pei, H.; Wang, H. *Bioresour. Technol.* **2019**, *291*, 121885.
25. Beckham, G. T.; Johnson, C. W.; Karp, E. M.; Salvachúa, D.; Vardon, D. R. *Curr. Opin. Biotechnol.* **2016**, *42*, 40–53.
26. Perez, J. M.; Kontur, W. S.; Alherech, M.; Coplien, J.; Karlen, S. D.; Stahl, S. S.; Donohue, T. J.; Noguera, D. R. *Green Chem.* **2019**, *21*, 1340–1350.
27. Rahimi, A.; Azarpira, A.; Kim, H.; Ralph, J.; Stahl, S. S. *J. Am. Chem. Soc.* **2013**, *135*, 6415–6418.
28. Rahimi, A.; Ulbrich, A.; Coon, J. J.; Stahl, S. S. *Nature* **2014**, *515*, 249–252.
29. Das, A.; Rahimi, A.; Ulbrich, A.; Alherech, M.; Motagamwala, A. H.; Bhalla, A.; da Costa Sousa, L.; Balan, V.; Dumesic, J. A.; Hegg, E. L.; Dale, B. E.; Ralph, J.; Coon, J. J.; Stahl, S. S. *ACS Sustainable Chem. Eng.* **2018**, *6*, 3367–3374.
30. Creighton, R. H. J.; McCarthy, J. L.; Hibbert, H. *J. Am. Chem. Soc.* **1941**, *63*, 3049–3052.
31. Pearl, I. A. *J. Am. Chem. Soc.* **1942**, *64*, 1429–1431.
32. Gierer, J. *Wood Sci. Technol.* **1985**, *19*, 289–312.
33. Tarabanko, V. E.; Petukhov, D. V.; Selyutin, G. E. *Kinet. Catal.* **2004**, *45*, 569–577.
34. Tarabanko, V. E.; Tarabanko, N. *Int. J. Mol. Sci.* **2017**, *18*, 2421.
35. Chen, C.-L. Nitrobenzene and Cupric Oxide Oxidations. In Methods in Lignin Chemistry; Springer: Berlin, 1992; pp. 301–321.
36. Villar, J. C.; Caperos, A.; García-Ochoa, F. *J. Wood Chem. Technol.* **1997**, *17*, 259–285.
37. Tarabanko, V. E.; Fomova, N. A.; Kuznetsov, B. N.; Ivanchenko, N. M.; Kudryashev, A. V. *React. Kinet. Catal. Lett.* **1995**, *55*, 161–170.
38. Bjørsvik, H.-R.; Minisci, F. *Org. Process Res. Dev.* **1999**, *3*, 330–340.
39. Klinke, H. B.; Ahring, B. K.; Schmidt, A. S.; Thomsen, A. B. *Bioresour. Technol.* **2002**, *82*, 15–26.
40. Schutyser, W.; Kruger, J. S.; Robinson, A. M.; Katahira, R.; Brandner, D. G.; Cleveland, N. S.; Mittal, A.; Peterson, D. J.; Meilan, R.; Román-Leshkov, Y.; Beckham, G. T. *Green Chem.* **2018**, *20*, 3828–3844.
41. Kuznetsov, B. N.; Sudakova, I. G.; Garyntseva, N. V.; Tarabanko, V. E.; Yatsenkova, O. V.; Djakovitch, L.; Rataboul, F. *Catal. Today* **2020**. https://doi.org/10.1016/j.cattod.2020.05.044.
42. Hatakka, A. *FEMS Microbiol. Rev.* **1994**, *13*, 125–135.
43. Eggert, C.; Temp, U.; Dean, J. F. D.; Eriksson, K.-E. L. *FEBS Lett.* **1996**, *391*, 144–148.
44. Lange, H.; Decina, S.; Crestini, C. *Eur. Polym. J.* **2013**, *49*, 1151–1173.
45. Artaud, I.; Ben-Aziza, K.; Mansuy, D. *J. Org. Chem.* **1993**, *58*, 3373–3380.
46. Dolphin, D.; Traylor, T. G.; Xie, L. Y. *Acc. Chem. Res.* **1997**, *30*, 251–259.
47. Zhu, C.; Ding, W.; Shen, T.; Tang, C.; Sun, C.; Xu, S.; Chen, Y.; Wu, J.; Ying, H. *ChemSusChem* **2015**, *8*, 1768–1778.
48. Nichols, J. M.; Bishop, L. M.; Bergman, R. G.; Ellman, J. A. *J. Am. Chem. Soc.* **2010**, *132*, 12554–12555.
49. Wu, A.; Patrick, B. O.; Chung, E.; James, B. R. *Dalton Trans.* **2012**, *41*, 11093–11106.
50. Chmely, S. C.; Kim, S.; Ciesielski, P. N.; Jiménez-Osés, G.; Paton, R. S.; Beckham, G. T. *ACS Catal.* **2013**, *3*, 963–974.
51. Huo, W.; Li, W.; Zhang, M.; Fan, W.; Chang, H.-M.; Jameel, H. *Catal. Lett.* **2014**, *144*, 1159–1163.
52. Wu, A.; Lauzon, J. M.; James, B. R. *Catal. Lett.* **2015**, *145*, 511–518.

53. vom Stein, T.; den Hartog, T.; Buendia, J.; Stoychev, S.; Mottweiler, J.; Bolm, C.; Klankermayer, J.; Leitner, W. *Angew. Chem. Int. Ed.* **2015**, *54*, 5859–5863.
54. Hanson, S. K.; Baker, R. T. *Acc. Chem. Res.* **2015**, *48*, 2037–2048.
55. Son, S.; Toste, F. D. *Angew. Chem. Int. Ed.* **2010**, *49*, 3791–3794.
56. Chan, J. M. W.; Bauer, S.; Sorek, H.; Sreekumar, S.; Wang, K.; Toste, F. D. *ACS Catal.* **2013**, *3*, 1369–1377.
57. Hanson, S. K.; Wu, R.; Silks, L. A. P. *Angew. Chem. Int. Ed.* **2012**, *51*, 3410–3413.
58. Díaz-Urrutia, C.; Chen, W.-C.; Crites, C.-O.; Daccache, J.; Korobkov, I.; Baker, R. T. *RSC Adv.* **2015**, *5*, 70502–70511.
59. Sedai, B.; Diaz-Urrutia, C.; Baker, R. T.; Wu, R.; Silks, L. A. P.; Hanson, S. K. *ACS Catal.* **2011**, *1*, 794–804.
60. Zhang, G.; Scott, B. L.; Wu, R.; Silks, L. A. P.; Hanson, S. K. *Inorg. Chem.* **2012**, *51*, 7354–7361.
61. Liu, Y.; Li, C.; Miao, W.; Tang, W.; Xue, D.; Li, C.; Zhang, B.; Xiao, J.; Wang, A.; Zhang, T.; Wang, C. *ACS Catal.* **2019**, *9*, 4441–4447.
62. Jastrzebski, R.; Constant, S.; Lancefield, C. S.; Westwood, N. J.; Weckhuysen, B. M.; Bruijnincx, P. C. A. *ChemSusChem* **2016**, *9*, 2074–2079.
63. Basolo, F.; Hoffman, B. M.; Ibers, J. A. *Acc. Chem. Res.* **1975**, *8*, 384–392.
64. Zombeck, A.; Drago, R. S.; Corden, B. B.; Gaul, J. H. *J. Am. Chem. Soc.* **1981**, *103*, 7580–7585.
65. Bozell, J. J.; Hames, B. R.; Dimmel, D. R. *J. Org. Chem.* **1995**, *60*, 2398–2404.
66. Biannic, B.; Bozell, J. J. *Org. Lett.* **2013**, *15*, 2730–2733.
67. Ryland, B. L.; Stahl, S. S. *Angew. Chem. Int. Ed.* **2014**, *53*, 8824–8838.
68. Hoover, J. M.; Stahl, S. S. *J. Am. Chem. Soc.* **2011**, *133*, 16901–16910.
69. Cho, D. W.; Parthasarathi, R.; Pimentel, A. S.; Maestas, G. D.; Park, H. J.; Yoon, U. C.; Dunaway-Mariano, D.; Gnanakaran, S.; Langan, P.; Mariano, P. S. *J. Org. Chem.* **2010**, *75*, 6549–6562.
70. Elder, T. *Holzforschung* **2010**, *64*, 435–440.
71. Kim, S.; Chmely, S. C.; Nimlos, M. R.; Bomble, Y. J.; Foust, T. D.; Paton, R. S.; Beckham, G. T. *J. Phys. Chem. Lett.* **2011**, *2*, 2846–2852.
72. Nutting, J. E.; Rafiee, M.; Stahl, S. S. *Chem. Rev.* **2018**, *118*, 4834–4885.
73. Bobbitt, J. M.; Brückner, C.; Merbouh, N. *Org. React.* **2009**, *74*, 103–424.
74. Hibino, T.; Shibata, D.; Ito, T.; Tsuchiya, D.; Higuchi, T.; Pollet, B.; Lapierre, C. *Phytochemistry* **1994**, *37*, 445–448.
75. Luterbacher, J. S.; Rand, J. M.; Alonso, D. M.; Han, J.; Youngquist, J. T.; Maravelias, C. T.; Pfleger, B. F.; Dumesic, J. A. *Science* **2014**, *343*, 277–280.
76. da Costa Sousa, L.; Foston, M.; Bokade, V.; Azarpira, A.; Lu, F.; Ragauskas, A. J.; Ralph, J.; Dale, B.; Balan, V. *Green Chem.* **2016**, *18*, 4205–4215.
77. da Costa Sousa, L.; Jin, M.; Chundawat, S. P. S.; Bokade, V.; Tang, X.; Azarpira, A.; Lu, F.; Avci, U.; Humpula, J.; Uppugundla, N.; Gunawan, C.; Pattathil, S.; Cheh, A. M.; Kothari, N.; Kumar, R.; Ralph, J.; Hahn, M. G.; Wyman, C. E.; Singh, S.; Simmons, B. A.; Dale, B. E.; Balan, V. *Energy Environ. Sci.* **2016**, *9*, 1215–1223.
78. Bhalla, A.; Bansal, N.; Stoklosa, R. J.; Fountain, M.; Ralph, J.; Hodge, D. B.; Hegg, E. L. *Biotechnol. Biofuels* **2016**, *9*.
79. Patil, N. D.; Yao, S. G.; Meier, M. S.; Mobley, J. K.; Crocker, M. *Org. Biomol. Chem.* **2015**, *13*, 3243–3254.
80. Wang, M.; Lu, J.; Zhang, X.; Li, L.; Li, H.; Luo, N.; Wang, F. *ACS Catal.* **2016**, *6*, 6086–6090.
81. Dabral, S.; Wotruba, H.; Hernández, J. G.; Bolm, C. *ACS Sustainable Chem. Eng.* **2018**, *6*, 3242–3254.

82. Nguyen, J. D.; Matsuura, B. S.; Stephenson, C. R. J. *J. Am. Chem. Soc.* **2014**, *136*, 1218–1221.
83. Klinger, G. E.; Zhou, Y.; Foote, J. A.; Wester, A. M.; Cui, Y.; Alherech, M.; Stahl, S. S.; Jackson, J. E.; Hegg, E. L. *ChemSusChem* **2020**, *13*, 4394–4399.
84. Sheldon, R. A.; Arends, I. W. C. E. *Adv. Synth. Catal.* **2004**, *346*, 1051–1071.
85. Shiraishi, T.; Takano, T.; Kamitakahara, H.; Nakatsubo, F. *Holzforschung* **2012**, *66*, 311–315.
86. Bosque, I.; Magallanes, G.; Rigoulet, M.; Kärkäs, M. D.; Stephenson, C. R. J. *ACS Cent. Sci.* **2017**, *3*, 621–628.
87. Masui, M.; Hosomi, K.; Tsuchida, K.; Ozaki, S. *Chem. Pharm. Bull.* **1985**, *33*, 4798–4802.
88. Masui, M.; Hara, S.; Ozaki, S. *Chem. Pharm. Bull.* **1986**, *34*, 975–979.
89. Hruszkewycz, D. P.; Miles, K. C.; Thiel, O. R.; Stahl, S. S. *Chem. Sci.* **2017**, *8*, 1282–1287.
90. Li, S.; Li, Z.-J.; Yu, H.; Sytu, M. R.; Wang, Y.; Beeri, D.; Zheng, W.; Sherman, B. D.; Yoo, C. G.; Leem, G. *ACS Energy Lett.* **2020**, *5*, 777–784.
91. Leigh, W. J.; Lathioor, E. C.; St. Pierre, M. J. *J. Am. Chem. Soc.* **1996**, *118*, 12339–12348.
92. Galian, R. E.; Litwinienko, G.; Pérez-Prieto, J.; Ingold, K. U. *J. Am. Chem. Soc.* **2007**, *129*, 9280–9281.
93. Luo, J.; Zhang, J. *J. Org. Chem.* **2016**, *81*, 9131–9137.
94. Recupero, F.; Punta, C. *Chem. Rev.* **2007**, *107*, 3800–3842.
95. Zhang, C.; Li, H.; Lu, J.; Zhang, X.; MacArthur, K. E.; Heggen, M.; Wang, F. *ACS Catal.* **2017**, 7, 3419–3429.
96. Walker, D.; Hiebert, J. D. *Chem. Rev.* **1967**, *67*, 153–195.
97. Wurche, F.; Sicking, W.; Sustmann, R.; Klärner, F.-G.; Rüchardt, C. *Chem. Eur. J.* **2004**, *10*, 2707–2721.
98. Wendlandt, A. E.; Stahl, S. S. *Angew. Chem. Int. Ed.* **2015**, *54*, 14638–14658.
99. Shen, Z.; Dai, J.; Xiong, J.; He, X.; Mo, W.; Hu, B.; Sun, N.; Hu, X. *Adv. Synth. Catal.* **2011**, *353*, 3031–3038.
100. Wang, L.; Li, J.; Yang, H.; Lv, Y.; Gao, S. *J. Org. Chem.* **2012**, 77, 790–794.
101. Katsina, T.; Clavier, L.; Giffard, J.-F.; Macedo Portela da Silva, N.; Fournier, J.; Tamion, R.; Copin, C.; Arseniyadis, S.; Jean, A. *Org. Process Res. Dev.* **2020**, *24*, 856–860.
102. Walsh, K.; Sneddon, H. F.; Moody, C. J. *Org. Lett.* **2014**, *16*, 5224–5227.
103. Ohkubo, K.; Fujimoto, A.; Fukuzumi, S. *J. Am. Chem. Soc.* **2013**, *135*, 5368–5371.
104. Lancefield, C. S.; Ojo, O. S.; Tran, F.; Westwood, N. J. *Angew. Chem. Int. Ed.* **2015**, *54*, 258–262.
105. Guo, H.; Miles-Barrett, D. M.; Neal, A. R.; Zhang, T.; Li, C.; Westwood, N. J. *Chem. Sci.* **2018**, *9*, 702–711.
106. Lee-Ruff, E.; Ablenas, F. J. *Can. J. Chem.* **1989**, *67*, 699–702.
107. Shuai, L.; Amiri, M. T.; Questell-Santiago, Y. M.; Héroguel, F.; Li, Y.; Kim, H.; Meilan, R.; Chapple, C.; Ralph, J.; Luterbacher, J. S. *Science* **2016**, *354*, 329–333.
108. Lan, W.; de Bueren, J. B.; Luterbacher, J. S. *Angew. Chem. Int. Ed.* **2019**, *58*, 2649–2654.
109. Song, Y.; Motagamwala, A. H.; Karlen, S. D.; Dumesic, J. A.; Ralph, J.; Mobley, J. K.; Crocker, M. *Green Chem.* **2019**, *21*, 3940–3947.
110. Lancefield, C. S.; Rashid, G. M. M.; Bouxin, F.; Wasak, A.; Tu, W.-C.; Hallett, J.; Zein, S.; Rodríguez, J.; Jackson, S. D.; Westwood, N. J.; Bugg, T. D. H. *ACS Sustainable Chem. Eng.* **2016**, *4*, 6921–6930.
111. Yao, S. G.; Mobley, J. K.; Ralph, J.; Crocker, M.; Parkin, S.; Selegue, J. P.; Meier, M. S. *ACS Sustainable Chem. Eng.* **2018**, *6*, 5990–5998.

112. Zhu, R.; Wang, B.; Cui, M.; Deng, J.; Li, X.; Ma, Y.; Fu, Y. *Green Chem.* **2016**, *18*, 2029–2036.
113. Dawange, M.; Galkin, M. V.; Samec, J. S. M. *ChemCatChem* **2015**, 7, 401–404.
114. Galkin, M. V.; Sawadjoon, S.; Rohde, V.; Dawange, M.; Samec, J. S. M. *ChemCatChem* **2014**, *6*, 179–184.
115. Deng, W.; Zhang, H.; Wu, X.; Li, R.; Zhang, Q.; Wang, Y. *Green Chem.* **2015**, *17*, 5009–5018.
116. Chen, Z.; Xu, J.; Ren, Z.; He, Y.; Xiao, G. *J. Solid State Chem.* **2013**, *205*, 134–141.
117. Song, Y.; Mobley, J. K.; Motagamwala, A. H.; Isaacs, M.; Dumesic, J. A.; Ralph, J.; Lee, A. F.; Wilson, K.; Crocker, M. *Chem. Sci.* **2018**, *9*, 8127–8133.
118. Ryland, B. L.; McCann, S. D.; Brunold, T. C.; Stahl, S. S. *J. Am. Chem. Soc.* **2014**, *136*, 12166–12173.
119. Dabral, S.; Hernández, J. G.; Kamer, P. C. J.; Bolm, C. *ChemSusChem* **2017**, *10*, 2707–2713.
120. Rafiee, M.; Alherech, M.; Karlen, S. D.; Stahl, S. S. *J. Am. Chem. Soc.* **2019**, *141*, 15266–15276.
121. Sannami, Y.; Kamitakahara, H.; Takano, T. *Holzforschung* **2017**, *71*, 109–117.
122. Lancefield, C. S.; Teunissen, L. W.; Weckhuysen, B. M.; Bruijnincx, P. C. A. *Green Chem.* **2018**, *20*, 3214–3221.
123. Miles-Barrett, D. M.; Neal, A. R.; Hand, C.; Montgomery, J. R. D.; Panovic, I.; Ojo, O. S.; Lancefield, C. S.; Cordes, D. B.; Slawin, A. M. Z.; Lebl, T.; Westwood, N. J. *Org. Biomol. Chem.* **2016**, *14*, 10023–10030.
124. Deuss, P. J.; Scott, M.; Tran, F.; Westwood, N. J.; de Vries, J. G.; Barta, K. *J. Am. Chem. Soc.* **2015**, *137*, 7456–7467.
125. Einhorn, J.; Einhorn, C.; Ratajczak, F.; Pierre, J.-L. *J. Org. Chem.* **1996**, *61*, 7452–7454.
126. Xiao, G.; Lancefield, C. S.; Westwood, N. J. *ChemCatChem* **2019**, *11*, 3182–3186.
127. Ling, K. B.; Smith, A. D. *Chem. Commun.* **2011**, *47*, 373–375.
128. Wang, Y.; Wang, Q.; He, J.; Zhang, Y. *Green Chem.* **2017**, *19*, 3135–3141.
129. Yang, W.; Du, X.; Liu, W.; Tricker, A. W.; Dai, H.; Deng, Y. *Energy Fuels* **2019**, *33*, 6483–6490.
130. Sun, C.; Zheng, L.; Xu, W.; Dushkin, A. V.; Su, W. *Green Chem.* **2020**, *22*, 3489–3494.
131. Yanagita, M.; Yamakawa, K. *J. Org. Chem.* **1959**, *24*, 903–909.
132. Ibuka, T.; Hayashi, K.; Minakata, H.; Inubushi, Y. *Tetrahedron Lett.* **1979**, *2*, 159–160.
133. Yao, Z.; Ye, D.; Liu, H.; Chen, K.; Jiang, H. *Synth. Commun.* **2007**, *37*, 149–156.
134. Hao, Z.; Li, S.; Sun, J.; Li, S.; Zhang, F. *Appl. Catal., B* **2018**, *237*, 366–372.
135. Pereira, J. H.; Heins, R. A.; Gall, D. L.; McAndrew, R. P.; Deng, K.; Holland, K. C.; Donohue, T. J.; Noguera, D. R.; Simmons, B. A.; Sale, K. L.; Ralph, J.; Adams, P. D. *J. Biol. Chem.* **2016**, *291*, 10228–10238.
136. Kamimura, N.; Takahashi, K.; Mori, K.; Araki, T.; Fujita, M.; Higuchi, Y.; Masai, E. *Environ. Microbiol. Rep.* **2017**, *9*, 679–705.
137. Klinger, G. E.; Zhou, Y.; Hao, P.; Robbins, J.; Aquilina, J. M.; Jackson, J. E.; Hegg, E. L. *ChemSusChem* **2019**, *12*, 4775–4779.
138. Zhu, Y.; Liao, Y.; Lv, W.; Liu, J.; Song, X.; Chen, L.; Wang, C.; Sels, B. F.; Ma, L. *ACS Sustainable Chem. Eng.* **2020**, *8*, 2361–2374.
139. Vangeel, T.; Schutyser, W.; Renders, T.; Sels, B. F. Perspective on Lignin Oxidation: Advances, Challenges, and Future Direction. In *Lignin Chemistry*; Springer: Cham, 2020; pp. 53–68.

CHAPTER FIVE

Lignin extraction and valorization using heterogeneous transition metal catalysts

Baoyuan Liu[a] and Mahdi M. Abu-Omar[a,b,*]
[a]Department of Chemistry and Biochemistry, University of California, Santa Barbara, CA, United States
[b]Department of Chemical Engineering, University of California, Santa Barbara, CA, United States
*Corresponding author: e-mail address: mabuomar@ucsb.edu

Contents

Abstract

Lignocellulosic biomass is the most abundant renewable carbon source on our planet. It offers an alternative as well as a complementary source to petrochemical refining for energy and chemical production. The market of renewable energy and chemicals has been rapidly growing over the past two decades. However, utilization of biomass is still underdeveloped. Energy production from biomass has seen only moderate increases.

Today's second generation biorefinery only uses the carbohydrate components (cellulose and hemicellulose) from the biomass and lignin is generated as a waste byproduct to be used for its low value heat. In contrast, the concept of lignin valorization can improve the economics of biorefining by producing value-added products from

Advances in Inorganic Chemistry, Volume 77
ISSN 0898-8838
https://doi.org/10.1016/bs.adioch.2021.02.001

lignin. This can be accomplished by changing the pretreatment of biomass to provide fractionation and upgrading of lignin first into valuable products, "lignin-first" biorefinery concept. Alternatively, if the pretreatment can provide a protected technical lignin byproduct which can be valorized to chemicals and/or hydrocarbon biofuels. Monomeric phenols are the major products of lignin valorization through heterogeneous catalysis. The complex structure of lignin, impurities from its preparation, and catalyst selection are among the key factors restricting yield of products. This chapter presents and contrasts preparation techniques of technical lignin, reviews the use of inorganic transition metal heterogeneous catalysts for lignin valorization into chemicals and fuels, and lastly demonstrated examples of subsequent applications of lignin derived monomers.

Abbreviations

BDE	bond dissociation energy
BTX	benzene toluene xylene
CDL	catalytic depolymerization of lignin
^{13}C NMR	carbon-13 nuclear magnetic resonance
DFT	density function theory
DHE	dihydrogeneugenol
DHEO	4-propylcatechol
2D HSQC NMR	two-dimensional heteronuclear single quantum coherence nuclear magnetic resonance
DMPP	2,6-dimethoxy-4-propylphenol
EXAFS	extended X-ray absorption fine structure analysis
FA/AA	formic acid/acetic acid
GC-MS	gas chromatography-mass spectrometry
GPa	gigapascal, 1GPa = one billion pascals
GPC	gel permeation chromatography
G unit	guaiacyl lignin unit
HDO	hydrodeoxygenation
High-S poplar	poplar wood with higher content of syringyl lignin unit
5-HMF	5-hydroxymethylfurfural
H unit	*p*-hydroxyphenyl lignin unit
INS	inelastic neutron scattering
Low-S poplar	poplar wood with lower content of syringyl lignin unit
MPL	organosolv lignin prepared from methanol/sulfuric acid treatment
OPL	organosolv lignin prepared from methanol/hydrochloric acid treatment
PG-OH	dihydroconiferyl alcohol
PG-diol	1-(4-hydroxy-3-methoxyphenyl)propane-1,3-diol
PS-OH	dihydrosinapyl alcohol
RCF	reductive catalytic fractionation
SEM	scanning electron microscopy
S unit	syringyl lignin unit
TPs	triphenylmethane-type polyphenols

1. Introduction

Lignocellulose is sustainably produced by photosynthesis which converts solar energy into stored chemical energy on Earth. Lignocellulosic biomass is the most abundant renewable organic feedstock for energy consumption, yielding 50–85 EJ of energy per year.[1] An annual production of ca. 1.5 billion tons of lignocellulosic biomass can be achievable by 2030.[2] Meanwhile the value of global biomass power market will be doubled from the current value of 52 billon U.S. dollars (USD) to more than 100 billion USD by 2027. However, only 11% of the total energy consumption in the U.S. is renewable energy in 2019, 43% of which comes from biomass energy (Fig. 1).[3] During the past decade (2009–19), energy consumption in the U.S. from biomass increased at a moderate rate of less than 3% per year.[3,4] The major energy sources remain non-renewable fossil fuels: petroleum, natural gas, and coal. The continued reliance on fossil fuels has many negative ramifications including greenhouse emission, toxic gas release, and water pollution. Thus, a more rapid implementation of renewable resources including lignocellulosic biomass is necessary to meet a sustainable future.

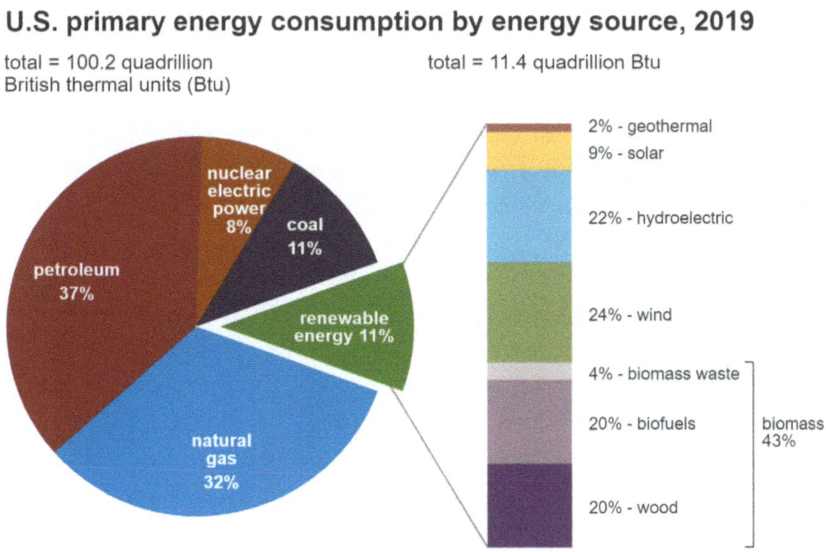

Fig. 1 Chart of the U.S. energy consumption by energy source, 2019. *Adapted from U.S. Energy Information Administration (EIA).* Monthly Energy Review, *Table 1.3 and 10.1, April 2020 (Accessed on Oct 2020 at: https://www.eia.gov/energyexplained/us-energy-facts/).*

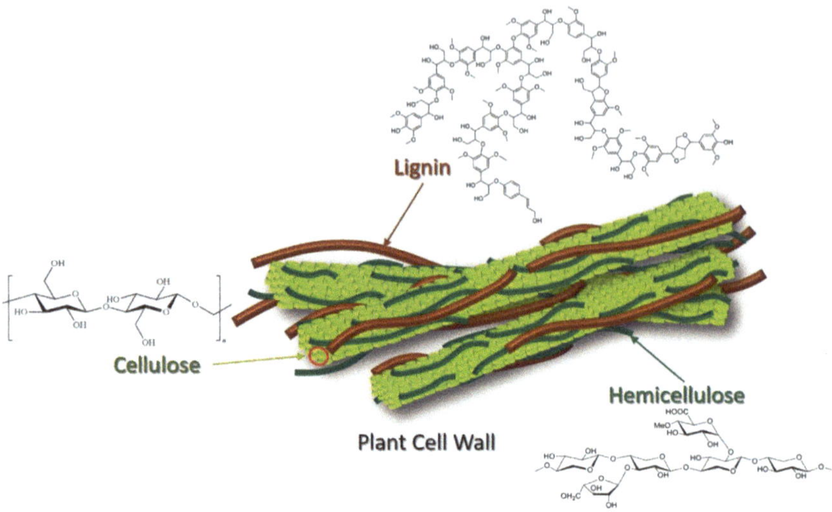

Fig. 2 Structure of lignocellulosic biomass. *Adapted from Jensen, C. U.; Guerrero, J. K. R.; Karatzos, S.; Olofsson, G.; Iversen, S. B.* Biomass Convers. Bioref. ***2017***, 7 *(4), 495–509.*

Lignocellulosic biomass is composed of three major biological polymers: cellulose (50%), hemicellulose (25%), and lignin (25%).[5] They are the major building blocks of the secondary cell walls (Fig. 2).[6] The crystalline cellulose forms the fibrous trestle of plant cell wall and is bundled with hemicellulose and lignin, which provides rigidity, ability of plants to grow against gravity, and a mechanism for water transport.[7] While cellulose and hemicellulose are polysaccharides, lignin is the only natural polymer made of aromatic monomers.[8] Interestingly, the most common industrial method of biomass utilization is direct combustion, which does not make use of the chemical differences between polysaccharides and lignin. Recent advances in fermentation of lignocellulose (second generation biorefining) to bio-ethanol has attracted commercial attention for the utilization of carbohydrates from cellulose and hemicellulose on a large scale. However, in the second generation biorefineries lignin is a waste by-product and used only for its heat value.[9–12] Similarly, in the paper and pulp industries, 50–70 million tons of lignin is produced every year, of which 98% is discarded into direct combustion.[5,13] Burning lignin for heat energy is an inefficient way of biomass utilization. It impedes lignin's potential valorization toward high energy density fuels and chemicals. Lignin because of its composition has high carbon content and the potential to provide platform aromatic chemicals and fuels.[14–18] Hence, development of efficient lignin valorization to chemicals can be a game changer for the future of biorefining.

The idea of efficient lignin valorization is to depolymerize the complex three-dimensional lignin polymer into smaller molecules that can be selectively upgraded to chemicals, fuels, and materials for further applications.[19] This approach is distinctly different from the conventional biorefinery methods where lignin is treated as the "left over" waste that can be used for heat value only. Focusing on successful lignin utilization/valorization has been referred to as "lignin first," a concept which targets chemistry that affords selective depolymerization of lignin into low molecular weight phenolic compounds.[20–22] Since native lignin is composed mostly of C—O ether bonds and chemistry for cleaving C—O ether bonds is well known, "lignin first" targets lignin valorization directly from lignocellulosic biomass.[23] In contrast, lignin liquor obtained from harsh pre-treatment in conventional biorefining (acid, base, steam explosion, and ammonia treatment) is composed of re-condensed lignin products with non-native C—C bonds, which are recalcitrant and cannot be upgraded selectively to valuable chemicals/molecules. However, "lignin first" must contend with impurities in the biomass, limitations on process intensification because of the use of solids and low density of biomass, tolerate carbohydrates (cellulose and hemicellulose), and challenges in separations.

A third approach is extraction and separation of lignin in its native form, a pretreatment that protects C—O bonds and prevents re-condensation of lignin into recalcitrant C—C bonds. In this case, lignocellulosic biomass is pre-treated under conditions and/or in the presence of specific reagents to allow for lignin isolation in which the "native" chemical connectivity is maintained as much as possible. This isolated lignin is referred to as "technical lignin." Depending on the pre-treatment, technical lignin is divided into several categories. Kraft and soda lignin refer to aqueous alkali (NaOH and Na_2SO_4) process.[24] Klason lignin is obtained after the cellulose and hemicellulose are dissolved in 72% sulfuric acid.[25] Organosolv lignin results from using various organic solvents such as formic acid, acetic acid, methanol, and acetone.[5,26,27] Lignin is first dissolved and extracted in an organic solvent, and recovered by precipitation. Compared to alkali (Kraft and soda) and acid (Klason) pre-treatment, organosolv lignin retains chemical connectivity (C—O bonds) present in "native lignin."[5] As a result, organosolv lignin can be valorized to chemicals with reasonable yields. This contrasting approaches for handling lignin are summarized in Fig. 3.

Recent studies have investigated many lignin valorization methods including pyrolysis,[28–30] gasification,[31,32] enzymatic catalysis,[33,34] and inorganic catalysis,[35–38] into a broad range of high-value aromatic chemicals, such as

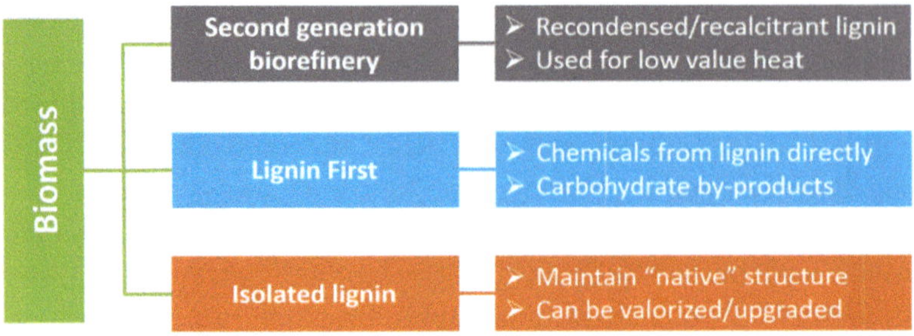

Fig. 3 Comparison of biomass treatment with the respect to lignin separation and use.

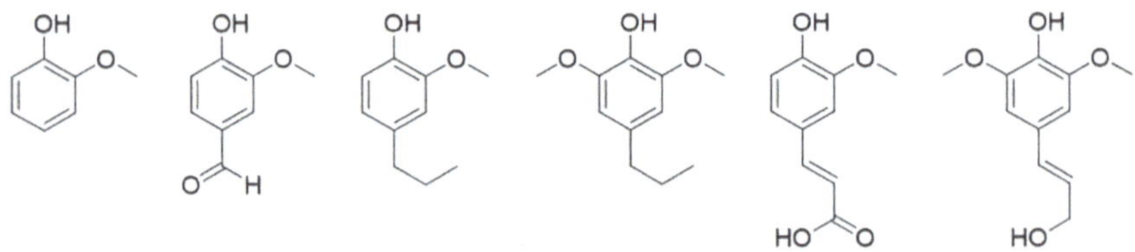

Fig. 4 Example of lignin monomers. (A) Guaiacol. (B) Vanillin. (C) Dihydroeugenol (DHE). (D) 2,6-Dimethoxy-4-propylphenol (DMPP). (E) Ferulic acid. (F) Sinapyl alcohol.

guaiacol,[5,35,36,39–41] vanillin,[42–45] dihydroeugenol (DHE),[5,35–37,46,47] 2,6-dimethoxy-4-propylphenol (DMPP),[5,35,37,48,49] ferulic acid,[39,50,51] and sinapyl alcohol,[39,52,53] (Fig. 4). Among those lignin valorization methods, the use of heterogeneous metal catalysts is a promising technology for making value-added chemicals from lignin.[35,54,55]

Among its attractive features are, high lignin conversion, relatively mild reaction conditions, ease of product separation, production of cellulose and hemicellulose as unscathed byproducts that can be used further in making biofuels and chemicals, and catalyst stability. The application of various metals has been demonstrated. Pd, Pt, Rh, and Ru are the most common used noble metals that have been studied to selectively cleave C—O bonds (in some instances C—C bonds) of the lignin polymer.[37,47,56–58] On one hand, these noble metals are usually supported on activated carbon to increase the surface area for improved catalyst activity. On the other hand, the integration of noble metals with an acidic zeolite or metal oxide support, creates a bifunctional catalyst system which can improve selectivity to deoxygenated aromatic hydrocarbons.[59] To satisfy scientific curiosity, sustainability and economics, inexpensive Earth-abundant metals have been favored in recent investigations. Cu,[60,61] Ni,[35,36,62,63] and Co[64,65] catalysts have been shown to give comparable results to noble metals. More interestingly, lignin depolymerization can occur under supercritical conditions in ethanol/isopropanol medium without the need for a transition metal catalyst.[66]

In this chapter, we review "lignin first" valorization and compare it to isolation and upgrading of technical lignin with a focus on organosolv lignin. An atomic level characterization of lignin will be discussed to understand and contrast the properties of technical lignin compared to recalcitrant and not upgradable lignin. A discussion of mechanistic aspects of lignin conversion will be provided. We will end the chapter by showing how platform molecules obtained from lignin can serve as versatile monomers to make renewable plastics and as feedstock for making specialty hydrocarbon fuels, mapping a complete supply chain from lignocellulosic biomass to value-added bioproducts.

2. Technical lignin

Lignin is an aromatic polymer made of three major monolignols: p-coumaryl alcohol, coniferyl alcohol, and sinapyl alcohol (Fig. 5A). Through biological polymerization, the monolignols are linked by C—O and C—C bonds forming three key lignin units: *p*-hydroxyphenyl (H unit), guaiacyl (G unit), and syringyl (S unit) (Fig. 5B).[5] In general, softwood lignin is composed of 90–95% G unit with a few percentage of H unit, while hardwood lignin is 50–75% S unit and 25–50% G unit.[22,67] Even though the content of key units differ from species to species, the lignin interunit linkages are similar among all types of lignocellulose. The most common linkage is the *β*-O-4 ether bond, which accounts for 50–60% of the total linkages (Fig. 6).[68,69] Other linkages commonly found include α-O-4 and 4-O-5; connected by C—C covalent bond are 5-5, *β*-5, β-1, and β-β. The

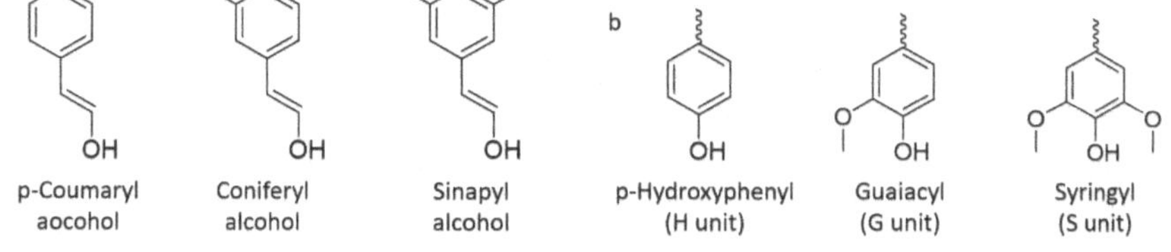

Fig. 5 (A): the three monolignols that make up lignin units; (B): the key units in lignin polymer.

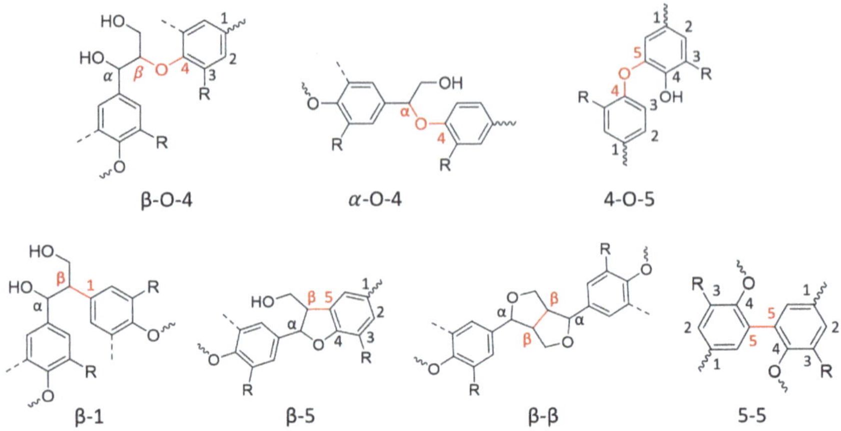

Fig. 6 Common lignin interunit linkages (labeled in red color) in lignin polymer. *Reproduced from: (1) Patil, N. D.; Tanguy, N. R.; Yan, N. In* Lignin in Polymer Composites, *Faruk, O.; Sain, M., Eds. William Andrew Publishing, 2016; pp. 27–47. (2) Luo, H.; Abu-Omar, M. M. In* Encyclopedia of Sustainable Technologies, *Abraham, M. A., Ed. Elsevier: Oxford, 2017; pp. 573–585.*

Table 1 Proportions of linkages in softwood and hardwood lignin.

Linkage type	Softwood (%)	Hardwood (%)
β-O-4	45–50	50–65
α-O-4	6–8	4–8
4-O-5	4–8	6–7
β-1	7–10	5–7
β-5	9–12	4–6
β-β	3	3–7
5-5	18–25	4–10

Values were obtained from: (1) Chakar, F. S.; Ragauskas, A. J. *Ind. Crop. Prod.* **2004,** *20,* 2, 131–141. (2) Zakzeski, J.; Bruijnincx, P. C. A.; Jongerius, A. L.; Weckhuysen, B. M. *Chem. Rev.* **2010,** *110,* 6, 3552–3599. (3) Mei, Q.; Shen, X.; Liu, H.; Han, B. *Chin. Chem. Lett.* **2019,** *30,* 1, 15–24.

approximate content of each lignin linkage in softwood and hardwood is summarized in Table 1.[24,69–71]

Making technical lignin is driven by the distinct solubility of biomass components. One approach is dissolution of carbohydrates and leaving the lignin as a solid residue, such as Klason lignin. Another strategy is to extract and dissolve the lignin in aqueous alkali or organic solvent and leave other components of the biomass (cellulose) as a solid residue. Examples of the second method include Kraft, soda, and organosolv lignin. Regardless of what method is used, production of technical lignin generally requires elevated temperature and the use of acid/base to cleave the intermolecular linkages between lignin and carbohydrates, and separation of the different components. Because acid/base also catalyze lignin inter-units cleavage, technical lignin is always a modified structure from the native form and the extent of modification varies from method to method. The differentiating characteristics between different technical lignins are type of linkages preserved and introduced, molecular weight, and elemental composition. For instance, the β-O-4 ether linkage is selectively cleaved during Kraft pulping by sulfonation and result in increased content of non-native C—C bonds. Compared to organosolv, Kraft lignin exhibits higher molecular weight and higher sulfur content.[24,72,73] Different pulping methods and the resulting lignins will be discussed and compared in this section.

2.1 Kraft lignin

The most abundant production of technical lignin is via Kraft processing, which produces 55 million tons of lignin every year in the U.S.[13] The Kraft pulping process was initially developed in the paper industry by the German chemist C.F. Dahl in 1879. Wood chips are cooked under alkaline condition with sodium hydroxide (NaOH) and sodium sulfide (Na_2S) at 170 °C for 2 h to dissolve lignin and hemicellulose in the aqueous solution. The residual cellulose from Kraft pulping is a brown solid. After bleaching with hydrogen peroxide (H_2O_2) and chlorine dioxide (ClO_2) the cellulose is used for paper production. Lignin dissolved in the aqueous alkaline fraction is recovered by decreasing the pH.[74]

Studies have proven that the solubilization of lignin into the aqueous phase is accompanied with the cleavage of both α and β ether bonds (α-O-4 and β-O-4 linkages).[75] Content of α-O-4 linkage is relatively low in most lignin structures (6–8%). Therefore, one characteristic of Kraft lignin is the low content of β-O-4 linkage. Advanced 2D (^{1}H and ^{13}C) Heteronuclear Single Quantum Coherence NMR (2D HSQC NMR) spectroscopy has been widely used to analyze linkages and functional groups on the lignin framework.[76] Quantitative HSQC NMR indicates up to 86% of β-O-4 linkages are cleaved during Kraft pulping.[77] Scheme 1 illustrates the mechanism of cleaving β-O-4 linkage in the presence of sulfide and hydroxide ions. The cleavage of C—O ether bonds is initiated by

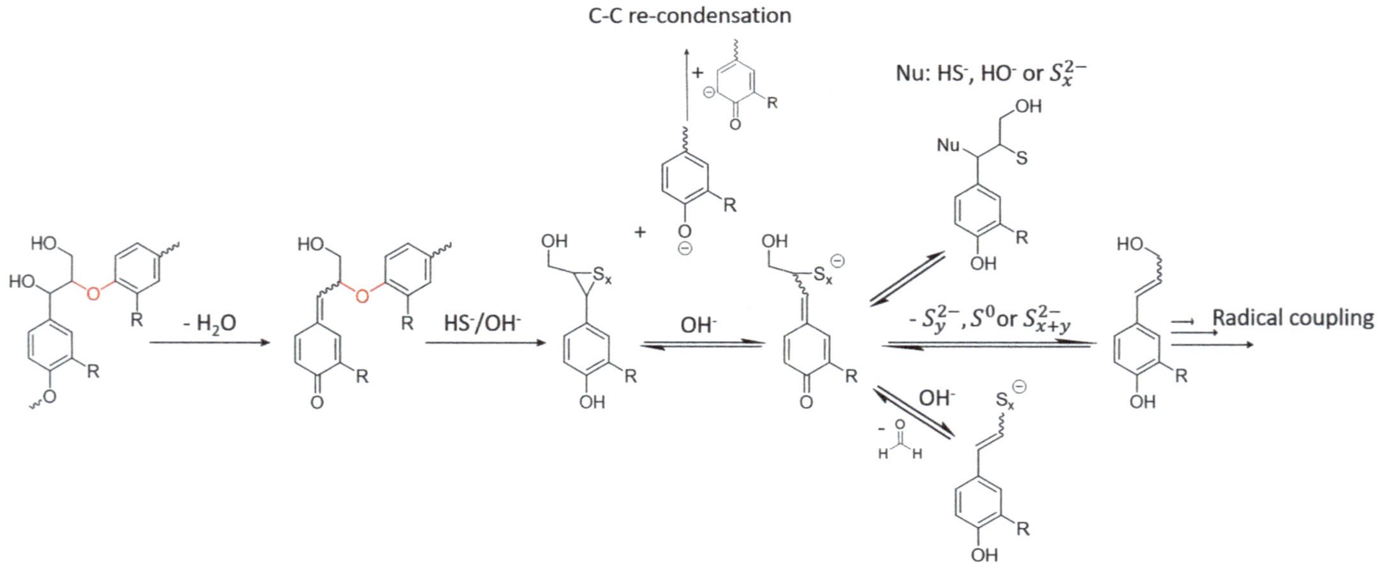

Scheme 1 Cleavage of phenolic β ether bond in Kraft pulping process. *Reproduced from: Rinaldi, R.; Jastrzebski, R.; Clough, M. T.; Ralph, J.; Kennema, M.; Bruijnincx, P. C. A.; Weckhuysen, B. M.* Angew. Chem. Int. Edit. ***2016***, 55, *29, 8164–8215.*

dehydration and ring rearrangement to form quinone methides and thiirane intermediates resulting in a coniferyl alcohol type fragment.[77,78] The sulfonation and sulfur extrusion promote an increase in sulfur content as well as recondensation in Kraft lignin. This mechanism explains the high sulfur content and increased amount of recalcitrant C—C bonds in Kraft lignin. Giummarella et al. recently reported their findings of C—C direct coupling between S and G units in Eucalyptus Kraft lignin.[79] Analysis by 1D ^{13}C NMR indicated non-native C—C bond formation is common through retro-aldol and subsequent radical condensation between aromatic carbons. Oxidized quinone-like byproduct from the condensation reaction gives Kraft lignin its amber color. Besides the direct coupling of aromatic carbons, Lancefield et al. demonstrated another possibility of repolymerization by a different type of C—C linkage between homovanillin and formaldehyde to a lactone structure.[78] This complex condensation reaction can result in over 100 new C—C linkages in Kraft lignin. Rinaldi et al. have summarized the bond dissociation energy (BDE) of both β-O-4 linkages in native lignin and C—C linkages in Kraft lignin. Accordingly, a typical BDE of β-O-4 is 54–72 kcal/mol while the BDE of C—C linkage falls into a higher value range 86–118 kcal/mol.[77] As a result, the valorization of Kraft lignin is hindered by its recalcitrant C—C bonds formed during the extraction process.

Another characteristic of Kraft lignin is its high sulfur content. Sulfur exists in both organic and inorganic structures during the growth of native biomass.[80,81] In general, raw biomass contains less than 0.1% of sulfur.[82] However, the Kraft process increases the sulfur content up to 7% in lignin.[13,68,83,84] The sulfonation within Kraft lignin is caused by the residual sulfur from sodium sulfide. It forms a variety of sulfur-containing structures in the Kraft lignin including: thiol (-SH), sulfide (-S-), and disulfide bonds (-S-S-) (Fig. 7).[85] Although several desulfurization methods have been

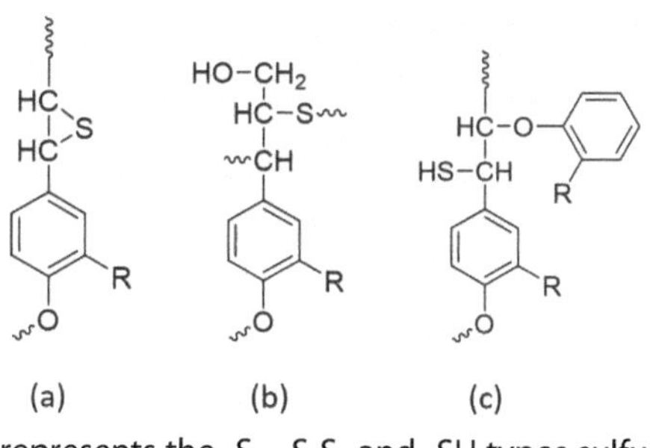

Fig. 7 (A), (B) and (C) represents the -S-, -S-S- and -SH types sulfur structure in Kraft lignin. *Reproduced from: Evdokimov, A. N.; Kurzin, A. V.; Fedorova, O. V.; Lukanin, P. V.; Kazakov, V. G.; Trifonova, A. D.* Wood. Sci. Technol. ***2018***, 52, *4, 1165–1174.*

investigated, such as sulfur extraction by organic solvent, sulfur removal by O_2 oxidation, reduction by Raney nickel, and sodium sulfite treatment etc., the typical sulfur content remains between 1% and 3% in Kraft lignin. Inwood et al. studied the elemental analysis and proposed the chemical formula of Kraft lignin to be $C_9H_{10.09}O_{3.23}S_{0.53}$, $C_9H_{7.98}O_{5.67}S_{0.59}$, and $C_9H_{9.96}O_{3.76}S_{0.19}$.[84] The increased sulfur content can also prevent Kraft lignin from further valorizations. Osada et al. reported sulfur-free lignin could be completely converted to syngas (mixture of methane, carbon monoxide, and hydrogen) over Ru/TiO_2 catalyst.[86] However, the yield of syngas is significantly lowered when sulfur is present because of catalyst poisoning. As a result, only catalysts resistant to sulfur poisoning can be used with Kraft lignin.[87] Instead, fast pyrolysis at high temperature (up to 850 °C) is a common way to utilize Kraft lignin. The thermal decomposition at high temperature overcomes sulfur poisoning; however, sulfur byproducts such as SO_2, H_2S, CH_3SH, CH_3SCH_3, and CH_3SSCH_3 must be scrubbed and managed to avoid pollution.[88]

2.2 Lignosulfonates

Lignin made with sulfite ions (SO_3^{2-} or HSO_3^-) is referred to as lignosulfonate. It is a byproduct from the paper industry. The process of making lignosulfonates, sulfite pulping, has been used as a delignification technique since the 1930s. It is the second lignin source in the market after Kraft lignin. One million tons of lignosulfonate solids is produced annually.[89] Lignosulfonate pulping also generates sulfur-containing lignin. The sulfur content in dry lignosulfonate is around 3–8%.[68,90] While sulfur in Kraft lignin is mainly covalent sulfur, in lignosulfonates it is anionic as salt of typical ions such as sodium (Na^+), potassium (K^+), and calcium (Ca^{2+}).[68,91] Fig. 8 displays

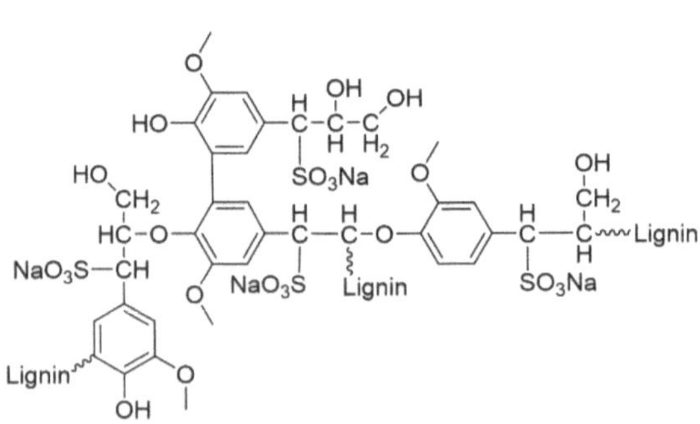

Fig. 8 Illustration of main building blocks in sodium lignosulfonate molecule. *Reproduced from: Flatt, R.; Schober, I. In* Understanding the Rheology of Concrete, *Roussel, N., Ed. Woodhead Publishing, 2012; pp. 144–208.*

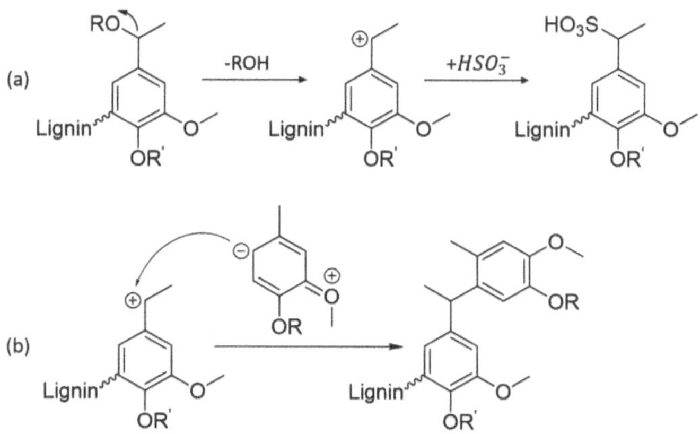

Scheme 2 Illustration of (A) sulfonation and (B) condensation pathways in acidic sulfite pulping. *Reproduced from: Aro, T.; Fatehi, P.* Chemsuschem ***2017***, 10, *9, 1861–1877.*

an example of sodium lignosulfonate.[92] Because of its highly charged structure, lignosulfonate is water-soluble.

Sulfite pulping takes up to 14 h.[68] After that, the lignosulfonate is extracted into an aqueous phase. In most cases, acidic (pH 1–5) and neutral (pH 5–7) conditions are applied to the sulfite pulping process. Reactions described in Scheme 2 represent the major mechanism of sulfonation and condensation pathways during acidic sulfite pulping.[90,93] The α ether linkages are first cleaved through hydrolysis catalyzed by acid (H^+) at 130–160 ° C followed by sulfonation on the resulting benzylic cations. Competitive condensation of the benzylic cation with the aromatic ring of another lignin unit forms C—C bonds. In contrast to acidic conditions, the β ether bond is selectively cleaved under neutral conditions.[93] Therefore, the properties of lignosulfonate are highly dependent on pH and the sulfite reagent used in pulping. The large variety makes lignosulfonate a broad range of molecular weight (10–50 kg/mol) with distinct properties. For example, lignosulfonate made with ammonium-based sulfite is more condensed and results in higher molecular weight product. On the other hand, sodium lignosulfonates exhibit the lowest viscosity because of the strong electro-kinetic repulsive force of sodium ions.[90]

Compared to Kraft lignin, lignosulfonates are generally of low purity. Carbohydrates, ash, and inorganic salts can take up to 30% by mass of lignosulfonate. The high sulfur content, highly condensed structure, and low lignin purity make lignosulfonate a less attractive feedstock for production of chemicals. Instead, utilizing lignosulfonate polymer directly is more prevalent. For example, the sodium lignosulfonate is a well-known dye dispersant.[94,95] Besides, a recent study by Huang et al. demonstrated electrostatic repulsion and steric hindrance of the anionic surface makes lignosulfonate a good water repellent additive improving the quality of concrete.[96]

2.3 Soda lignin

Soda lignin is made from the soda pulping process. Compared to Kraft lignin and lignosulfonates, soda lignin is sulfur-free and has higher purity.[89,97,98] The soda pulping method is mainly used for processing annual crops such as straws, bagasse, and hardwood. In fact, the alkaline process of delignification in soda pulping is quite comparable to the Kraft method with the exception of sulfur content. Sodium hydroxide (NaOH) is used to generate an aqueous alkaline medium. Delignification reaction occurs within the hot alkaline solution and can modify several lignin inter-units such as β-O-4 and β-5 linkages (Scheme 3).[77,99] The chemical treatment in soda lignin generates many vinyl ether and p-hydroxyl units.[68,97] Condensation within in soda lignin often occurs on the vinyl ether units. Nevertheless, most soda lignin still has low molecular weight of 0.3–3 kg/mol.[68] The low end corresponds to a lignol dimer.

After delignification, the dissolved soda lignin can be recovered from the alkaline solution by lowering the pH. Mousavioun et al. reported an interesting two-stage acid precipitation in which the soda lignin collected at pH 5.5 gives higher purity and larger particle size than that produced at pH 3.[98] Because of its low sulfur content and high purity, soda lignin is more suited

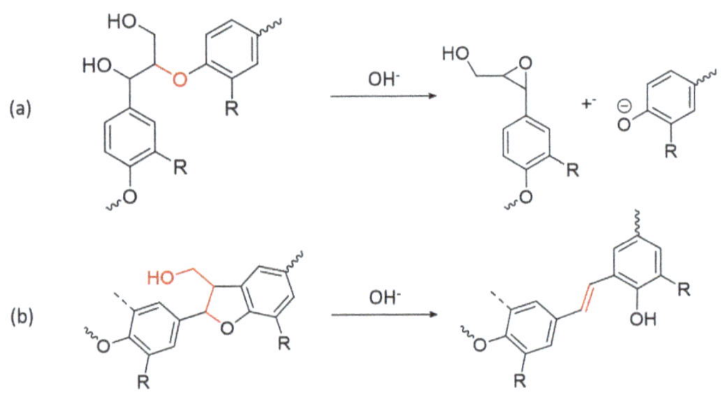

Scheme 3 Illustration of lignin linkages modified in soda pulping. (A) Describes the reaction of β-O-4 linkage in alkaline condition. (B) Represents the change of β-5 linkage in alkaline condition. *Reproduced from: (1) Schutyser, W.; Renders, T.; Van den Bossche, G.; Van den Bosch, S.; Koelewijn, S.-F.; Ennaert, T.; Sels, B., Catalysis in Lignocellulosic Biorefineries. 2017; pp. 537–584. (2) Rinaldi, R.; Jastrzebski, R.; Clough, M. T.; Ralph, J.; Kennema, M.; Bruijnincx, P. C. A.; Weckhuysen, B. M.* Angew. Chem. Int. Edit. ***2016***, 55, *29, 8164–8215.*

for chemical upgrading to functional biopolymers,[100] phenols and hydrocarbons[101] than Kraft lignin and lignosulfonates. Besides, soda lignin can also serve as a natural feed additive for monogastric animals.[102]

2.4 Organosolv lignin

The organosolv process is another important method for making sulfur-free lignin from biomass. To date, there is no commercial organosolv lignin on the market. Most efforts are still on the laboratory scale. The principle of organosolv method is to extract lignin by solubilization in organic solvent or solvent mixtures. Methanol,[5,103] ethanol,[104] acetone,[5,105] ethylene glycol,[106] and 1,4-dioxane have been among the most commonly used organic solvent for lignin extraction and isolation.[107] Organosolv lignin is highly soluble in organic solvents but insoluble in water.[108] Due to this hydrophobic characteristic, organosolv lignin can be easily precipitated and recovered by adding water to the solution.

Besides the simple recovery, organosolv lignin is attractive for the following reasons: (1) organosolv methodology affords complete fractionation of biomass into the three major components: cellulose, hemicellulose, and lignin into three separate streams. Cellulose is collected as the leftover solid, lignin is recovered from the organic solvent phase, and hemicellulose (and its derivatives) are washed out in the aqueous phase. This complete fractionation of biomass establishes the possibility of utilizing every component of the biomass efficiently. (2) Organosolv lignin has excellent chemical characteristics. It is sulfur-free, high in lignin purity, and low in ash content (1.75 wt%).[13] Thus, organosolv lignin is an ideal precursor for production of chemicals and biopolymers.[5,109–111] (3) The preparation of organosolv lignin requires moderate conditions. (4) Compared to Kraft, sulfite, and soda lignin, organosolv lignin shows the least chemical modifications and resembles "native" lignin the most.[76,112,113] (5) The production of organosolv lignin is environmentally friendly.[114,115] Organosolv lignin is generally made with low energy input, reduced use of strong acid/base and metal catalyst. The organic solvent can be recycled and reused. As a result, organosolv treatment exhibits low cost, no water pollution, and no sulfur dioxide (SO_2) emission, while providing high quality lignin.

Organosolv lignin from different solvent systems may share similar properties such as high lignin purity and low molecular weight.[68,89] However, studies have shown the quality of organosolv lignin is highly dependent on the solvent system. A recent study from our group[5] compared formic acid/acetic acid (FA/AA), acetone/formaldehyde, and methanol with dilute sulfuric acid (0.045N) for the preparation of poplar wood organosolv lignin. A narrow range of molecular weight was observed (1.8–2.5 kg/mol) for the isolated organosolv lignins. Despite this similarity in molecular weight average, the aggregation and shape of the resulting organosolv lignin varied significantly. Scanning electron microscopy (SEM) images showed organosolv lignin from methanol to be highly regular and spherical in structure while those from FA/AA and acetone to irregular (Fig. 9).

Two-dimensional (^{13}C, ^{1}H) HSQC-NMR analysis revealed the presence of alcohol (methoxy group) on the α position of the β-O-4 linkages (Scheme 4A). Another unique signal at around δ_C/δ_H 94/5 ppm was only

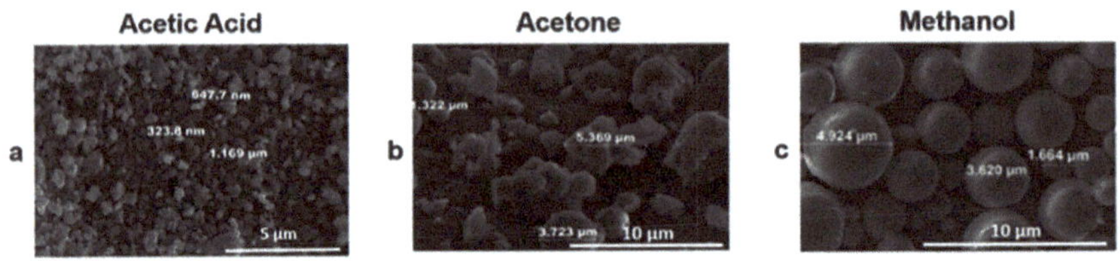

Fig. 9 Scanning electron microscopy (SEM) images of the isolated organosolv lignin from wild-type poplar using different solvent extraction methods. (A) Lignin sample from FA/AA treatment. (B) Lignin sample from acetone/formaldehyde treatment. (C) Lignin sample from methanol treatment. *Image was acquired from: Luo, H.; Abu-Omar, M. M.* Green Chem. ***2018***, 20, *3*, *745–753.*

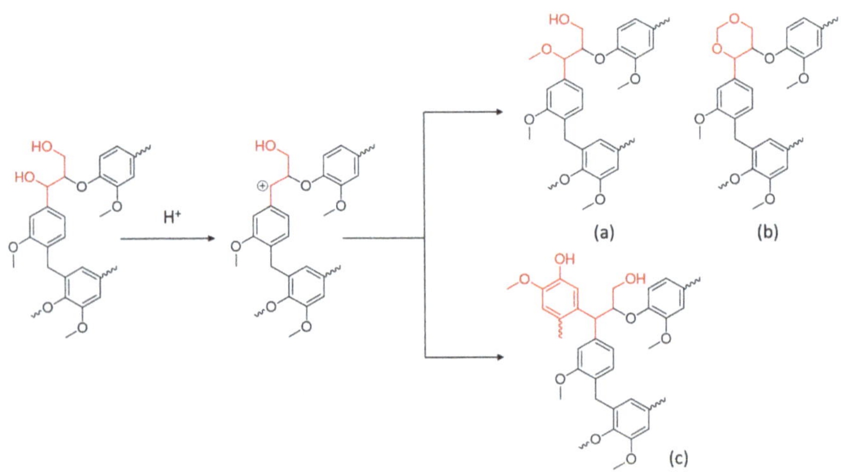

Scheme 4 Illustration of acid catalyzed modification of lignin linkages within different organic solvents. (A) Nucleophile protected α-aryl ether bond in methanol solution. (B) Dioxane structure by formaldehyde addition. (C) Condensation occurs on the α carbon through recalcitrant C—C bond. *Adapted from: Luo, H.; Abu-Omar, M. M.* Green Chem. ***2018***, 20, *3*, *745–753.*

detected in the lignin made from acetone/formaldehyde mixture. This signal was due to the dioxane structure formed by addition of formaldehyde to the lignin linkage (Scheme 4B). Therefore, methanol and formaldehyde serve as protection groups to trap the active α carbon cation intermediate and prevent formation of recalcitrant C—C bond (Scheme 4C).

Overall, the organosolv extraction of lignocellulosic biomass is an attractive alternative to Kraft, sulfite, and soda methods. It has potential for making high quality lignin and cellulose fractions for further upgrading. Many studies have shown promising yields of phenolic products from lignin by various heterogeneous catalysts. Compared to direct catalytic conversion of lignin from lignocellulosic biomass where lignin is only a 20–30 wt% component, catalysis of organosolv lignin has the advantage of process intensification because organosolv lignin is pure lignin and it is soluble in most common organic solvents. Studies of heterogeneous catalysis for lignin valorization to phenolic monomers will be discussed next in terms of feedstock, reaction mechanism, and catalyst selection.

3. Catalytic depolymerization of lignin

"Lignin first" refers to fractionation of lignocellulosic biomass and in the same step depolymerizing lignin and upgrading it to bioproduct-chemicals and/or fuels.[23] The lignin first approach begins with the raw biomass directly. Lignin valorization can also start from technical lignin and simplify products separation.

Catalytic depolymerization of lignin (CDL) is one of the most common strategies within the "lignin first" valorization to produce value-added lignin monomers.[5] As described in Fig. 4, lignin monomers are aromatic phenols with aldehyde or alcohol attached to the end of the alkyl sidechain. Transition metal heterogeneous catalysts have been well-studied for the conversion of lignin into monomeric phenol derivatives. The approach of reductive lignin depolymerization dates back to the 1940s when it was used to characterize the lignin structure in woody biomass.[22] In the presence of a transition metal catalyst and under reductive conditions, lignin selectively undergoes CDL at temperatures <250 °C and the cellulosic fibers are left intact.[35,116,117] The catalytic conversion requires elevated temperature to overcome the activation barriers of the abundant C—O and C—C lignin linkages.[35,118–120] The high temperature and pressure may limit the scalability of reductive lignin valorization.[121] Therefore, the use of hydrogen transfer reagents instead of hydrogen gas promotes milder conditions. Alcohols, such as methanol, ethanol, and 2-propanol can serve as a hydrogen source in lignin valorization.[23,47,122] These alcohols can provide hydrogen through reforming over a metal catalyst. Thus reforming catalysts such as Pt, Ru and Raney-Ni can be ideal for CDL.[47,123] Due to the easy recovery, low cost, and non-toxic characteristics, methanol and ethanol make good solvents for CDL processes. They have also been found to reduce lignin re-condensation and avoid char formation.[122,124,125] Formic acid as an additive to ethanol has been shown as a good source of in situ hydrogen.[126,127]

Regardless of what H-source is used, catalytic fractionation of carbohydrate along with lignin depolymerization is generally referred to as reductive catalytic fractionation (RCF), Fig. 10.[23] RCF is a convenient way of biomass valorization in a one-pot reaction. However, the low-density of biomass (150–205 kg/m^3)[128] and complexity of products separation may limit the future scalability of RCF. In contrast, a concentrated lignin feedstock can be obtained from the organosolv process, and its subsequent conversion via CDL can be intensified and as a result more amenable to industrial scaling. Most catalysts that are effective for RCF were found to work well in CDL. Accordingly, upgrading lignin is protected by catalyst from re-condensation. In this section, we review metal-catalyzed lignin valorizations that apply to both RCF and CDL.

3.1 Palladium

Palladium is well known for its ability to catalyze hydrogenation and hydrogenolysis reactions. Previous work by Pandarus et al. showed that supported metallic palladium was active for direct C—O hydrogenolysis to cleave benzylic ether bonds.[129] They investigated several model compounds and achieved 100% C—O cleavage with palladium catalyst in

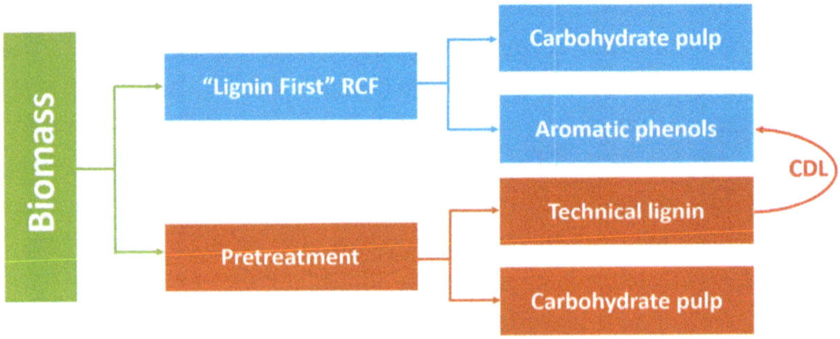

Fig. 10 Illustration of reductive catalytic fractionation (RCF) and technical lignin valorization via catalytic depolymerization of lignin (CDL).

methanol under hydrogen pressure. Up to 65% of lignin linkages are composed to C—O ether bonds (Table 1). Pd/C has been shown as an effective catalyst for lignin valorization to cleave β-O-4, 4-O-5, α-O-4, and even β-β linkages.[22,130]

We employed model compounds and poplar woody biomass to understand the mechanism of β-O-4 cleavage over Pd/C catalyst.[37] The catalytic conversion was performed in methanol under hydrogen pressure at 225 °C. The primary result indicated the phenolic products were mainly PG-OH and PS-OH when the monometallic Pd/C catalyst was used; product yields were 59%. The two products were consistent from both poplar wood and model compounds. The molecules defined as PG-OH and PS-OH are the DHE and DMPP with a hydroxyl group (OH) attached to the γ carbon on propyl sidechain (Fig. 11).[37] This result confirmed the Pd/C catalyst is active in breaking the C—O ether bond in the lignin polymer. Our findings also indicated the OH at the α position of β-O-4 linkage is cleaved by Pd/C as well. The experimental results agreed with computational prediction from density function theory (DFT) for C—O ether model compound and Pd (111) species by Lu et al.[131]

However, Pd/C did not catalyze the hydrogenolysis of the hydroxyls at the γ carbon of β-O-4 linkage. This hypothesis was confirmed by the reaction of PG-diol and Pd/C which yielded PG-OH. A modified Pd/C catalyst was also studied to understand the effect of introducing Lewis acids to the reaction mixture. $ZnCl_2$ and $Zn(OAc)_2{\cdot}2H_2O$ were added as a co-catalysts. Hydrogenolysis of the γ carbon hydroxyl of β-O-4 linkages was observed in the presence of Zn^{2+}. DHE and DMPP were the major products. The addition of a Lewis acid facilitated the hydrogenolysis of γ C-OH. However, applying this Pd-Zn^{2+} co-catalyst system to PG-diol, the major product was still PG-OH (63% yield). DHE was only observed as minor product. Therefore, PG-OH was not an intermediate and the hydroxyl group at

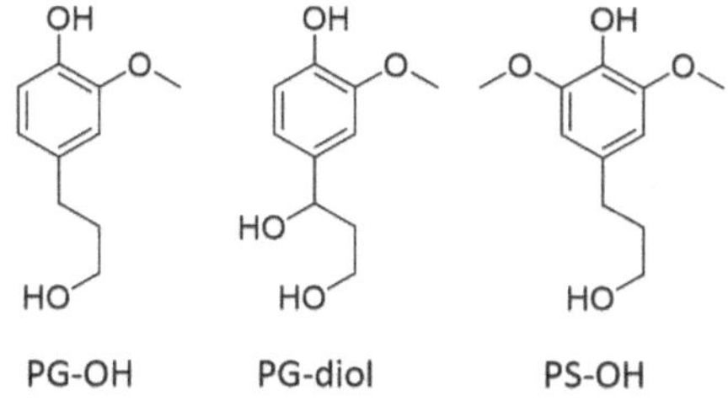

Fig. 11 Molecular structure of PG-OH, PG-diol, and PS-OH. *Reproduced from: Klein, I.; Marcum, C.; Kenttamaaa, H.; Abu-Omar, M. M.* Green Chem. ***2016***, *18, 8, 2399–2405.*

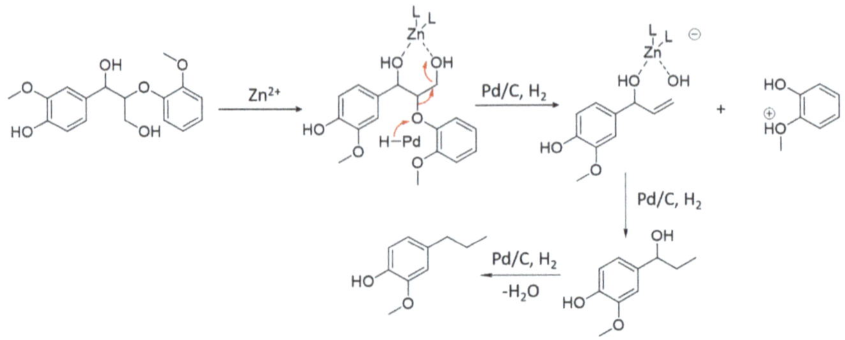

Scheme 5 Proposed mechanism for cleavage and HDO of β-O-4 ether linkage using Pd/C and Zn^{2+}. *Reproduced from: Klein, I.; Marcum, C.; Kenttamaaa, H.; Abu-Omar, M. M.* Green Chem. ***2016***, 18, *8, 2399–2405.*

the γ position must be activated and cleaved by Zn^{2+} prior to hydroxyl removal from the α position by Pd/C. Also, PG-OH was unreactive with the Pd-Zn^{2+} co-catalyst.

Further investigations and spectroscopic evidence for coordination between the α/γ OH and Zn^{2+} led to a proposed mechanism for β-O-4 cleavage by the Pd-Zn^{2+} co-catalyst system (Scheme 5). Hydride transfer from the Pd surface to the C—O ether bond initiates the reaction. Coordination between Zn^{2+} and the substrate made the γ OH a better leaving group. Lastly, Pd/C catalyzed hydrogenolysis cleaved the benzylic OH on the α carbon and hydrogenation of the terminal C=C bond completed the reaction. Other Lewis acids such as $FeCl_3$, $NiCl_2$, and $AlCl_3$ were also investigated and shown to be affective for producing DHE and DMPP from lignin.

Another study investigated a prepared Pd—Zn bimetallic catalyst over carbon support for lignin valorization instead of a physical mixture of Pd/C and a zinc salt.[132] The molar ratio between Pd and Zn in the synthesized catalyst was 1:1. The Pd—Pd coordination was detected by the Pd K-edge extended X-ray absorption fine structure (EXAFS) analysis while Pd—Zn coordination was absent in the bimetallic catalyst. Thus, no Pd—Zn alloy nor direct interaction was created during the in situ catalyst preparation. The zinc was hypothesized to have adsorbed onto the activated carbon via coordination to —OH groups on the carbon's surface. When the catalyst was heated to 225 °C with lignin substrate and methanol, the Zn^{2+} could be activated and released from the carbon surface to coordinate with hydroxyl groups of the lignin substrate. Therefore, the cleavage of lignin catalyzed by this Zn/Pd/C bimetallic catalyst would undergo the same reaction pathway as the physical mixture co-catalyst system. Compared to the physical mixture catalyst, the Zn/Pd/C catalyst showed similar activity and selectivity for poplar wood. 54% Total yield of DHE and DMPP was observed. Meanwhile the solid carbohydrate residue containing 85% cellulose was recovered.[46]

3.2 Ruthenium

Ruthenium (Ru) has been widely used for biomass conversion. Similar to Pd, most of the active Ru catalysts are metallic ruthenium on porous support. Ru/C is one of the most used catalysts for lignin valorization. It exhibits the best selectivity for 4-ethylpehnol from lignin.[133]

The Sels group contrasted Ru/C and Pd/C for lignin hydrogenolysis using birch wood as substrate.[134] The reactions were performed in methanol under hydrogen pressure. Ru/C gave comparable results to Pd/C, lignin conversion, total yield of monomers, and retention of carbohydrates fraction (Table 2). The striking difference was the selectivity of γ-OH. Pd/C gave primarily PG-OH and PS-OH with 91% total selectivity. In contrast, 75% of the phenolic monomers by Ru/C were DHE and DMPP. More —OH was removed by Ru/C indicating that it is a more efficient in catalyzing hydrogenolysis. Interestingly, the hydrogenolysis of lignin by Ru/C was found to be insignificantly affected by hydrogen pressure. As reported, when the Ru/C reaction was carried out under 1 bar N_2 in methanol, 40% yield of lignin monomers were still obtained. This result is attributed to the activity of Ru/C in catalyzing methanol reforming to produce in situ hydrogen. In

Table 2 Comparison of Ru/C and Pd/C catalyzed RCF on birch wood.

	Ru/C	Pd/C
Delignification (wt%)	85	90
Monomer yield (C%)	48	49
Dimer yield (C%)	13	15
DHE + DMPP selectivity (%)	75	4
PGOH + PSOH selectivity (%)	19	91
C5 carbohydrate retention (C%)	69	81
C6 carbohydrate retention (C%)	93	94

From Van den Bosch, S.; Schutyser, W.; Koelewijn, S. F.; Renders, T.; Courtin, C. M.; Sels, B. F. *Chem. Commun.* **2015**, *51*, 67, 13158–13161.

contrast, without added hydrogen, 4-ethylguaiacol and 4-ethylsyringol became the major products in the Pd/C reaction. This finding was attributed to C—C hydrogenolysis by Pd/C through consecutive dehydrogenation/decarbonylation in the absence of hydrogen gas.[135] Overall, Ru/C is a good catalyst for reductive fractionation of lignin from biomass to make phenolic monomers (DHE and DMPP) without requiring added hydrogen. This could lower the cost and improve the safety and sustainability of lignin valorization.

Additional investigations from the Sels lab revealed further products arising from lignin including dimers and oligomers.[136] GPC and GC-MS analysis showed the phenolic monomers represent 50% yield of lignin carbon while another 18% fall into dimeric products. 2D HSQC NMR showed cleavage of all C—O ether bonds within the lignin network by Ru/C. However, Ru/C was ineffective in converting dimers and oligomers connected by C—C bond, Scheme 6 for example. Moreover, the free-ortho position on the G unit is active to form the 5-5 bond between two phenyl groups. The 5-5 bond is the most common and has the strongest BDE among all C—C interunit linkages in the lignin polymer.[77,137] As a result, Ru/C does not achieve complete depolymerization of lignin because of native and non-native C—C bonds.

A modified ruthenium catalyst reported in 2019 may provide ways to overcome the conventional limit of incomplete C—C bond cleavage. Dong et al. introduced a mesoporous Ru/$NbOPO_4$ catalyst which gave excellent activity toward cleaving recalcitrant C—C interunit linkages in Kraft lignin.[137] Instead of making phenolic monomers, the monocyclic hydrocarbons including benzene, toluene, ethylbenzene, propylbenzene, and their corresponding cyclohexanes were obtained as the major products. This one-pot conversion was done in dodecane solvent with 5 bar H_2 at 310 °C over 40 h. Under this condition, 68% selectivity to monocyclic

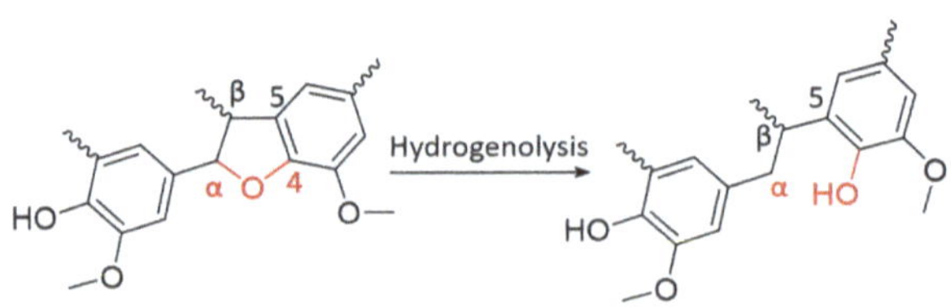

Scheme 6 Illustration of breaking α-O-4 unit on β-5 linkage. *Reproduced from: Van den Bosch, S.; Schutyser, W.; Vanholme, R.; Driessen, T.; Koelewijn, S. F.; Renders, T.; De Meester, B.; Huijgen, W. J. J.; Dehaen, W.; Courtin, C. M.; Lagrain, B.; Boerjan, W.; Sels, B. F.* Energ. Environ. Sci. ***2015***, *8, 6, 1748–1763.*

arenes was achieved. In this Ru/$NbOPO_4$ catalyst, the abundant Brønsted acid sites on the phosphate-based support are possibly responsible for the superior C—C bond cleaving ability. Inelastic neutron scattering (INS) and DFT analysis indicated the acidic $NbOPO_4$ support shows stronger binding of phenyl structures than conventional Nb_2O_5 and other zeolitic materials. In cleaving the 5-5 bond, the biphenyl is first adsorbed on the $NbOPO_4$ support and undergoes partial hydrogenation. The BDE of the 5-5 bond is reduced by converting the stable sp^2-sp^2 to an active sp^2-sp^3 bond. After this step, the partially reduced 5-5 bond is broken through direct hydrogenolysis catalyzed by Ru particles. The design of this multifunctional ruthenium catalyst suggested a pathway for improved phenyl binding onto a catalyst with abundant acid sites to promote C—C interunit cleavage. Compared to the monometallic Ru/C catalyst, the products from this Ru/$NbOPO_4$ system are fully deoxygenated hydrocarbons. Although these are less valuable than phenolic monomers, C6-C9 hydrocarbons are potential drop-in fuels. The arene byproducts can be used as fuel additives or BTX replacements.[138] More studies on the catalyst design are necessary to tune the selectivity between hydrocarbons and monomeric phenols.

3.3 Earth-abundant Ni catalyst

Nickle (Ni) is an Earth-abundant element and has shown promising performance in valorizing lignin into phenolic products. Compared with precious noble metal catalysts, the low-cost Ni catalyst is ecofriendly for large scale applications. Supported monometallic Ni/C catalyst has been studied extensively by our group with various biomass/lignin feedstock. An early study by Klein et al. used amorphous Ni/C to catalyze lignin valorization in methanol at 200 °C starting from various woods, poplar, birch, and eucalyptus without added hydrogen.[36] Overall, birch performed best and gave DHE and DMPP as major products. It was also noted that Ni/C is active in reforming methanol to produce hydrogen in situ.[139] Miscanthus, a grass biomass, was also investigated.[35] The one-pot RCF was carried out in methanol under hydrogen pressure at 225 °C. Notably, besides DHE and DMPP, two molecules containing the methyl ferulate ester structure were observed (Fig. 12). These two molecules originated from the unique ferulate/diferulate linkages that are present in grasses but absent from wood biomass. The total yield of the four lignin monomers was 69%. This higher yield indicates that lignin in grass species is potentially more accessible than in woody biomass.

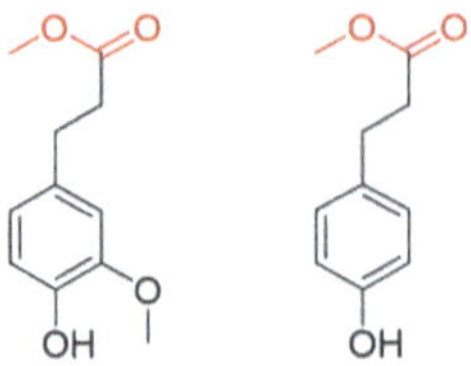

Fig. 12 The unique methyl ferulate ester (labeled in red color) products only obtained from grassy lignin. *Reproduced from: Luo, H.; Klein, I. M.; Jiang, Y.; Zhu, H. Y.; Liu, B. Y.; Kenttamaa, H. I.; Abu-Omar, M. M.* ACS Sustain. Chem. Eng. ***2016***, 4, *4, 2316–2322.*

Besides, 61 wt% of the starting miscanthus substrate was recovered as a solid residue. This solid residue was found to be carbohydrate rich, 56% cellulose and 21% xylan (monomer of hemicellulose). Moreover, this recovered polysaccharide pulp could be further converted to furfural and levulinic acid by a $FeCl_3$ catalyst. Because Ni/C exhibits the abilities to fractionate the native biomass and to depolymerize/upgrade lignin, it is another excellent catalyst for lignin-first processes.

In order to understand the ability of Ni/C to valorize technical lignin, Luo et al. studied several organosolv lignins from poplar wood.[5] The Ni/C was an effective catalyst for valorizing organosolv lignin to phenolic monomers with yields comparable to those observed from "lignin first" RCF processing. Genetically engineered poplar substrates were also used in this study. The gene modification tuned the content of S unit in the lignin network. Compared to wild-type poplar, one mutant contained higher content of S unit (high-S poplar) and another mutant had less S unit (low-S poplar). Interestingly, compared to the wild-type lignin the monomer yield obtained from high-S lignin was slightly improved. This was because the S unit, having two occupied ortho positions does not form 5-5 carbon linkages.

Taking the series of studies on Ni/C together, it has clearly been shown that Ni/C is a versatile catalyst for lignin valorization from biomass directly as well as from protected organosolv lignin. It is also an attractive catalyst because it can catalyze methanol reforming, thereby removing the requirement of added hydrogen.

3.4 Without an added transition metal catalyst

Cheng et al. studied hydrogenolysis of organosolv lignin in ethanol/isopropanol medium under supercritical condition without the addition of a transition metal catalyst.[140] So far, no other studies have mentioned using supercritical alcohol mixtures for lignin depolymerization in the absence of an added catalyst. Organosolv lignins made from poplar biomass prepared by two different solvent treatments: methanol/sulfuric acid (MPL) and methanol/HCl (OPL) were employed as the feedstocks. Lignin depolymerization was performed under supercritical condition at 270 °C with 10 bar N_2 for 4 h. In this study, isoeugenol and 4-propenyl syringol were obtained as the major products. Isopropanol acted as hydrogen-donor as evidenced by the formation of acetone in the product mixture. HSQC NMR of pre and post reaction mixtures showed significant disappearance of *β*-O-4 signal suggesting sufficient cleavage of C—O ether bonds while retaining *β*-5 and *β*-*β* signals. Therefore, C—C cross-links were not cleaved under supercritical conditions in EtOH/^{i}PrOH. The optimal ratio between the co-solvents was found to be 1:1 EtOH/^{i}PrOH, which gave 48% total yield of phenolic monomers. The yield of products and the selectivity for bond cleavages under these supercritical conditions were comparable to those observed with transition metal catalysts.[5,134,141] Notably, the mechanism of lignin depolymerization under supercritical conditions remains unclear. A study of the lignin *β*-O-4 linkage model compound showed no conversion under the same supercritical reaction conditions, suggesting that something in the lignin itself is essential for the reaction. Conversion of the model compound was only observed when a small amount of organosolv lignin or sodium chloride was added to the reaction mixture. Thus, the organosolv lignin itself appeared to serve as a catalyst to break *β*-O-4 bonds. It was suggested that minor ions or acids introduced along with the lignin from the organosolv treatment must be the actual catalyst(s) in this unique depolymerization/hydrogenolysis system.

In summary, the crucial activity of metal catalysts for lignin valorization is to catalyze hydrogenolysis of C—O and C—C bonds. Many carbon-supported monometallic catalysts are active toward cleavage of C—O ether linkages but leave C—C linkages intact. The recalcitrance of C—C bonds is attributed to their high BDEs. Thus, the C—C bonds in lignin limit the yields of phenolic monomers. Although the content of natural C—C interunit linkages in native lignin is relatively small, more C—C crosslinks form during the pulping of technical lignin and occur as side reactions along reductive catalytic upgrading. Additionally, the use of alcohol solvents is advantageous for lignin valorization because it prevents lignin re-condensation and can serve as a hydrogen donor eliminating the need for added hydrogen. Considering the future development of lignin valorization on a large industrial scale, technical lignin is a more desirable feedstock because it is more amenable to process intensification than raw biomass.

4. Upgrading of lignin derived phenols

Lignin monomers are mainly methoxylated, hydroxylated, and alkylated benzenes. The highly functionalized structure makes them attractive developing applications. Taking advantage of these existing functional groups, lignin monomers can be further utilized as precursors for drug and biopolymer synthesis. For instance, Blondiaux et al. developed a 5-step synthesis from lignin-derived DHE to 3,4-dialkoxyanilines and alkyl propionates.[142] Aniline is a key molecule in the pharmaceutical industry. 3,4-Dialkoxyanilines can be used as drop-in chemicals in the synthesis of anticancer drugs such as Gefitinib and Erlotinib.[143,144] Jiang et al. synthesized polyphenol-furan thermoset polymers by using 4-methylcatechol, furfural, and 5-hydroxymethylfurfural (5-HMF) as the building blocks.[145] 4-methylcatechol can be obtained from reductive lignin valorization. 5-HMF and furfural are well-known platform chemicals from cellulose and hemicellulose, respectively.[146,147] Thus, polyphenol-furan based thermosets is 100% renewable from lignocellulosic biomass. Besides organic synthesis, heterogeneous catalysis is another effective approach to increase the intrinsic value of lignin derived phenols. Hydrodeoxygenation (HDO) reactions catalyzed by transition metals can efficiently remove oxygen and retain the carbon structure to produce hydrocarbon fuels from lignin. The transformation of phenols to hydrocarbons increases the energy density and widens applications to larger fuel markets.

Zhang et al. recently published such a strategy using Ru on an acid support, $Ru(SO_4^{2-})/ZrO_2\text{-}CeO_2$, to catalyze HDO of lignin monomers into arenes, cyclic alkanes, and linear alkenes.[148] The lignin-based C6 to C9 hydrocarbons are within the range of carbon numbers in gasoline fuel. Utilization of lignin monomers for renewable biopolymer and biofuels has potential to reduce the demand on fossil energy and realize a more sustainable future. In this section we provide examples from our own work of lignin-derived biopolymers and biofuels.

4.1 Renewable thermoset plastics

Thermoset is a big family of crosslinked polymers such as phenolic and urea formaldehyde resins, unsaturated polyesters, and polyepoxides.[149] 20% of all commercial polymers are thermoset, of which, 70% is polyepoxide.[150]Generally, thermosetting plastics are synthesized in a liquid solution that irreversibly leads to solid material during the curing step.

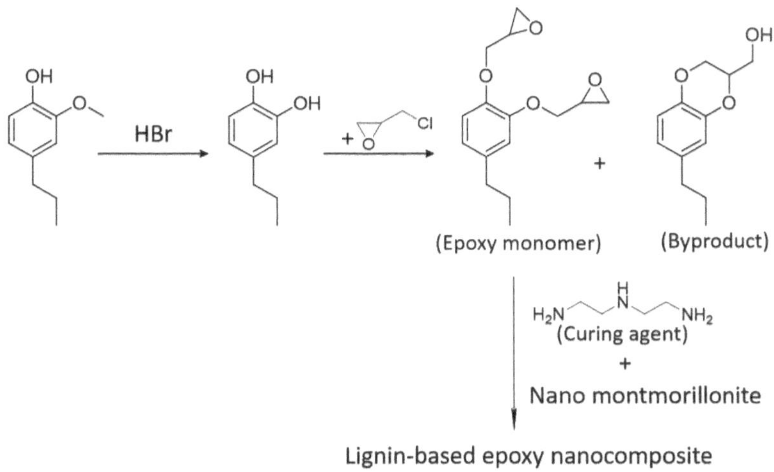

Scheme 7 Synthesis route of lignin-based epoxy nanocomposite. *Reproduced from: Zhao, S.; Abu-Omar, M. M.* Biomacromolecules ***2015****,* 16*, 7, 2025–2031.*

Heating and UV irradiation are commonly applied to promote cross-linking during thermoset synthesis. Because of the cross-linked structure, thermoset polymers exhibit outstanding mechanical properties, thermal stability, and solvent resistance compared to thermoplastics.[151]

Zhao et al. introduced several pathways for making epoxy-based thermosetting plastics from lignin monomers (Scheme 7).[152] Monomers for making thermoset usually contain two to three hydroxyl groups, whereas the methoxy group limits the reactivity of lignin-based phenols such as DHE. To increase the content of hydroxyls, DHE was modified through ortho-demethylation to yield 4-propylcatechol (DHEO), which can be glycidylated with epichlorohydrin to generate a suitable epoxy monomer. With amine curing agents, lignin-based epoxy nanocomposite are made. More importantly, since lignin-derived monophenols have similar structures (Fig. 3), this strategy can be applied to synthesize various types of lignin-based thermosets.

Zhao et al. demonstrated in further studies the general applicability of this synthetic route.[151,153] Scheme 8 illustrates an optimized strategy for making epoxy thermosets from two lignin monomers.[153] In this work, ortho-demethylation of DHE could activate its para and ortho sites. The activated catechol molecule can undergo condensation reactions with aldehydes to form triphenylmethane-type polyphenols (TPs). Vanillin, another lignin monomer was employed as the aldehyde precursor to produce a fully lignin-based TPs. According to the highly functioned TPs structure, five

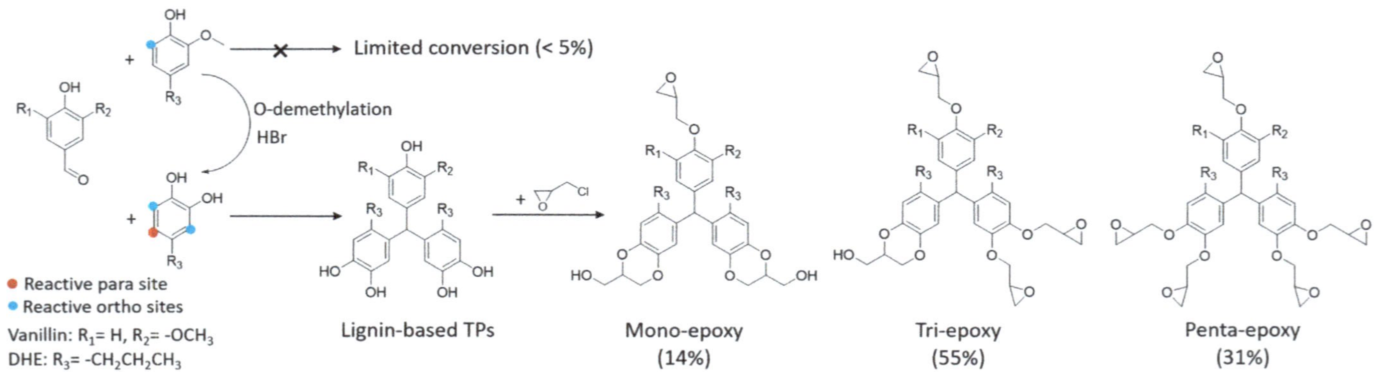

Scheme 8 Synthesis route of renewable TPs-epoxy from lignin derived aromatic aldehyde and phenols. *Reproduced from: Zhao, S.; Abu-Omar, M. M.* Macromolecules ***2017***, *50, 9, 3573–3581*.

epoxides could be attached to its framework. The newly made TP-epoxy thermoset exhibited excellent storage modulus (12.3 GPa), glass transition temperature (167 °C), and thermal stability among other biobased epoxy thermosets.[154–156] The improved mechanical properties of the TP-epoxy thermoset are attributed to its rigid framework and high cross-link density.[157]

4.2 Bio-hydrocarbon fuels

Liquid hydrocarbons are the main energy sources consumed around the world. In 2019, Americans used 142 billion gallons of motor gasoline.[158] Gasoline on today's market is made from non-renewable fossil carbons. Production of gasoline from non-fossil energy sources will expand energy production in the future. Lignocellulose is by far the most abundant renewable carbon source. Biomass pyrolysis is a thermochemical process that has been developed to produce syngas and heavy oil for energy consumption.[159] The syngas (carbon monoxide and hydrogen gas) can be transformed by Fischer-Tropsch chemistry to produce liquid hydrocarbons.[160] However, typical pyrolysis requires high energy input and the synthesis of liquid hydrocarbons from syngas have low efficiency and the resulting bio-oils have undesirable chemical properties. Due to this dilemma, preparation of hydrocarbon fuels directly from biomass through low-energy cost, high efficiency, and simple processing is an attractive area of research.

Lignin derived C6 to C9 phenolic monomers are good precursors to generate biogasoline through HDO reactions. Lu and the co-workers introduced guaiacol conversion over Pd/TiO_2 catalyst to cyclohexane.[161] The HDO reaction was performed between 200 °C and 280 °C in n-dodecane solvent under 20 bar of hydrogen gas. In this study, several palladium catalysts were compared. Pd/C, for example, did not affect HDO and instead yielded 2-methoxycyclohexanol via ring hydrogenation. The TiO_2 support was necessary for C—O scission to induce HDO. It was also noted that the Pd/TiO_2 catalyst effectiveness in HDO reactions depended on the catalyst preparation temperature. For instance, the Pd/TiO_2 synthesized at 200 °C showed better conversion and higher selectivity for fully deoxygenated products than when it was prepared at 50 °C. This was mainly because more Ti^{4+} was reduced to a lower oxidation state at higher temperatures. The partially reduced titanium exhibited stronger binding with oxygenated substrates. Although Pd/TiO_2 prepared at 500 °C still gave high selectivity toward the cyclohexane, conversion of guaiacol was reduced. This hindrance of catalyst activity was rationalized by suggesting that reduced TiO_x species migrated to the Pd surface and suppressed H_2 adsorption. Therefore, Pd/TiO_2 prepared at 200 °C performed the best. However, some improvements were still necessary including the need for added hydrogen and use of alkane solvents.[162,163]

A recent advance toward making C9 hydrocarbons as drop-in biofuel in water (a green solvent) was described by our group.[164] In this system, full deoxygenation was achieved by adding a small amount of methanol to serve as an in situ source of hydrogen. The HDO reaction was catalyzed by a physical mixture of Ru/C and Nb_2O_5. The preparation temperature of the Nb_2O_5 was key to preserving its acidity.[165] The mechanism of the co-catalyst system was elucidated through analysis of intermediates and isotope labeling experiments. Scheme 9 describes the proposed reaction steps for the conversion of DHE to propylcyclohexane (**1**), the green highlighted pathway. In the absence of Nb_2O_5 conversion to propylphenol and propylcyclohexanol was more prominent.[166] The stability of this co-catalyst system was demonstrated through multiple recycling tests.

Besides the gasoline range hydrocarbons, the heavier fuel molecules such as jet fuel and biodiesel are also achievable from lignin. Wang et al. have studied a catalyst mixture of Ru/Al_2O_3 and acidic H^+-Y zeolite.[167]One advantage of their work was the use of lignin directly. This one-pot strategy combined lignin depolymerization and HDO for biofuel production. As well, water was used as solvent. The large-porous H^+-Y zeolite with abundant acid sites showed promising activity in catalyzing lignin depolymerization. Coupling the Ru/Al_2O_5 with the H^+-Y together did not only enhance the HDO reaction, but it also promoted the alkylation and dimerization reactions between monomers to yield higher carbon numbers. More than 80% conversion of lignin was achieved within a relatively short time (4 h) at 250 °C with 85% selectivity for C12—C18 hydrocarbons (jet fuel range). More importantly, the product distribution among the hydrocarbons was tunable by adjusting the catalyst amount and reaction temperature. For example, C6—C11 cyclic hydrocarbons were the main product when only Ru/Al_2O_3 was used; and arenes were dominant when only H^+-Y acidic zeolite was used. Higher temperature promoted formation of non-cyclic alkanes with more than 18 carbons on average. Overall, the carbon efficiency of lignin conversion to hydrocarbons in this system was 38%, which compares well with other catalytic systems limited by 42–48 wt% conversion based on lignin.[168]

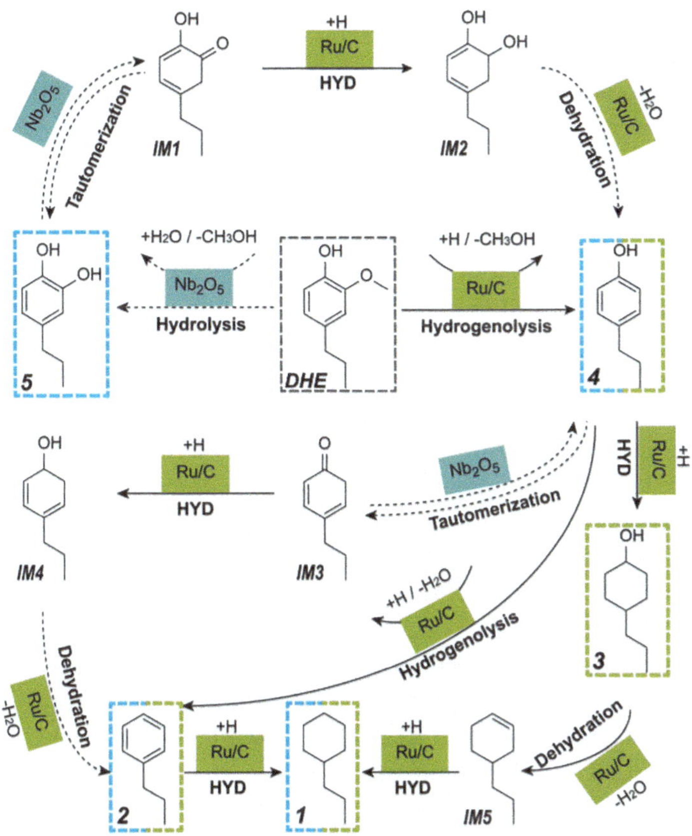

Scheme 9 Proposed sequence of mechanistic steps for conversion of DHE by the catalyst mixture Ru/C and Nb_2O_5. Steps catalyzed by Ru/C is labeled in green color. The steps labeled in blue color indicate function of Nb_2O_5. *Reaction scheme was obtained from: Li, S.; Liu, B.; Truong, J.; Luo, Z.; Ford, P. C.; Abu-Omar, M. M.* Green Chem. ***2020****, 22, 21, 7406–7416.*

5. Conclusions and further perspectives

Lignin valorization can be achieved via catalytic reductive pathways either directly from the biomass itself, "lignin first" approach, or through catalytic depolymerization and upgrading of protected lignin such as organosolv. Both noble metal catalysts as well as Earth-abundant metals such as nickel have been successful in these endeavors. Preservation of the lignin's phenol monomer group can have advantages as these molecules can be used in making polymers, pharmaceutical molecules, or still be deoxygenated to make hydrocarbon biofuels. Biomass consumes CO_2 from the atmosphere. Thus, valorization of lignin enhances the carbon efficiency of biomass utilization. The intrinsic value of lignin is being recognized and realized. With more efficient methods and chemistries for lignin valorization, biomass use for renewable energy and chemicals will expand.

Despite impressive recent advances, lignin valorization to monomeric phenols and chemicals has been limited owing to recalcitrant C—C bonds found in native lignin or formed via re-condensation during the lignin extraction process. Either reduction in C—C bond formation or new catalysts that can target specific C—C bonds without total pyrolysis are necessary. The use of hydrogen limits commercial utility for cost and safety reasons. Laboratory scale reactions have shown the use of alcohol solvents coupled with reforming catalysts can circumvent this limitation. However, the implication on cost as well as industrial practice of such a strategy remains to be vetted. Nevertheless, phenolic monomers that can be obtained from lignin have been used in making renewable plastics, chemicals used in drug syntheses, and hydrocarbon biofuels. In many instances, these chemistries have been demonstrated with commercially available pure molecules from non-renewable carbon sources rather than with the bio-based molecules coming from biomass processing. This presents a major challenge for separations and utilization of products derived directly from biomass or lignin samples.

Acknowledgments

Our research in the area of biomass valorization has been supported by the Office of Science, Basic Energy Sciences, US Department of Energy, through award no. DE-SC0019161 and DE-SC0000997 (Center for Catalytic Conversion of Biomass to Biofuels), and the Mellichamp Sustainability Initiative at UCSB.

References

1. Lange, J. P. *Biofuels Bioprod. Biorefin.* **2007**, *1*, 39–48.
2. Langholtz, M. H.; Stokes, B. J.; Eaton, L. M.; US Department of Energy. *Economic Availability*, Vol. 1; Oak Ridge National Laboratory, 2016.
3. U. S. Energy Information Administration (EIA). https://www.eia.gov/energyexplained/us-energy-facts/:.
4. U. S. Energy Information Administration (EIA). https://www.eia.gov/totalenergy/data/annual/archive/038409.pdf:.
5. Luo, H.; Abu-Omar, M. M. *Green Chem.* **2018**, *20*, 745–753.
6. Jensen, C. U.; Guerrero, J. K. R.; Karatzos, S.; Olofsson, G.; Iversen, S. B. *Biomass Conv. Bioref.* **2017**, *7*, 495–509.
7. Liu, Q. Q.; Luo, L.; Zheng, L. Q. *Int. J. Mol. Sci.* **2018**, *19*, 335.

8. Bertella, S.; Luterbacher, J. S. *Trends Chem.* **2020**, *2*, 440–453.
9. Zheng, Y.; Zhang, R. H. In *Bioethanol: Production, Benifits and Economics*; Erbaum, J. B., Ed.; Nova Science: New York, 2009; pp. 1–48.
10. Waghmare, P. R.; Watharkar, A. D.; Jeon, B. H.; Govindwar, S. P. *3 Biotech* **2018**, *8*, 158.
11. Gnansounou, E.; Dauriat, A. *J. Sci. Ind. Res. India* **2005**, *64*, 809–821.
12. Busic, A.; Mardetko, N.; Kundas, S.; Morzak, G.; Belskaya, H.; Santek, M. I.; Komes, D.; Novak, S.; Santek, B. *Food Technol. Biotechnol.* **2018**, *56*, 289–311.
13. Bajwa, D. S.; Pourhashem, G.; Ullah, A. H.; Bajwa, S. G. *Ind. Crop Prod.* **2019**, *139*, 111526.
14. Singh, R.; Prakash, A.; Dhiman, S. K.; Balagurumurthy, B.; Arora, A. K.; Puri, S. K.; Bhaskar, T. *Bioresour. Technol.* **2014**, *165*, 319–322.
15. Singh, S. K.; Ekhe, J. D. *RSC Adv.* **2014**, *4*, 27971–27978.
16. Zhao, C.; Xie, S. X.; Pu, Y. Q.; Zhang, R.; Huang, F.; Ragauskas, A. J.; Yuan, J. S. *Green Chem.* **2016**, *18*, 1306–1312.
17. Luo, Z. C.; Qin, S. F.; Chen, S.; Hui, Y.; Zhao, C. S. *Green Chem.* **2020**, *22*, 1842–1850.
18. Oregui-Bengoechea, M.; Miletic, N.; Hao, W. M.; Bjornerback, F.; Rosnes, M. H.; Garitaonandia, J. S.; Hedin, N.; Arias, P. L.; Barth, T. *ACS Sustain. Chem. Eng.* **2017**, *5*, 11226–11237.
19. Ren, T. Y.; Qi, W.; Su, R. X.; He, Z. M. *ChemCatChem* **2019**, *11*, 639–654.
20. Huang, Y.; Duan, Y. J.; Qiu, S.; Wang, M.; Ju, C.; Cao, H.; Fang, Y. M.; Tan, T. W. *Sustainable Energy Fuels* **2018**, *2*, 637–647.
21. Yang, H. B.; Zhang, X. M.; Luo, H.; Liu, B. Y.; Shiga, T. M.; Li, X.; Kim, J. I.; Rubinelli, P.; Overton, J. C.; Subramanyam, V.; Cooper, B. R.; Mo, H. P.; Abu-Omar, M. M.; Chapple, C.; Donohoe, B. S.; Makowski, L.; Mosier, N. S.; McCann, M. C.; Carpita, N. C.; Meilan, R. *Biotechnol. Biofuels* **2019**, *12*, 171.
22. Koranyi, T. I.; Fridrich, B.; Pineda, A.; Barta, K. *Molecules* **2020**, *25*, 2815.
23. Abu-Omar, M. M.; Barta, K.; Beckham, G. T.; Luterbacher, J.; Ralph, J.; Rinaldi, R.; Roman-Leshkov, Y.; Samec, J.; Sels, B.; Wang, F. *Energ. Environ. Sci.* **2021**, *14*, 262–292.
24. Chakar, F. S.; Ragauskas, A. J. *Ind. Crop Prod.* **2004**, *20*, 131–141.
25. Abdel Rahman, N.; Galiwango, E. *J. Food Process Eng.* **2018**, *57*, 46–58.
26. Başakçılardan Kabakcı, S.; Tanış, M. H. *Biomass Conv. Bioref.* **2020**. https://doi.org/10.1007/s13399-020-00677-2.
27. Nitsos, C.; Rova, U.; Christakopoulos, P. *Energies* **2018**, *11*, 50.
28. Saraeian, A.; Aui, A.; Gao, Y.; Wright, M. M.; Foston, M.; Shanks, B. H. *Green Chem.* **2020**, *22*, 2513–2525.
29. Chu, J. Y.; Jiang, W. K.; Wu, S. B. *BioResources* **2017**, *12*, 7241–7254.
30. Lazaridis, P. A.; Fotopoulos, A. P.; Karakoulia, S. A.; Triantafyllidis, K. S. *Front. Chem.* **2018**, *6*, 295.
31. Zhang, Z. L.; Zijlstra, D. S.; Lahive, C. W.; Deuss, P. J. *Green Chem.* **2020**, *22*, 3791–3801.
32. Sricharoenchaikul, V. *Bioresour. Technol.* **2009**, *100*, 638–643.
33. Gall, D. L.; Kontur, W. S.; Lan, W.; Kim, H.; Li, Y. D.; Ralph, J.; Donohue, T. J.; Noguera, D. R. *Appl. Environ. Microbiol.* **2018**, *84* (3), e02076-17. https://doi.org/10.1128/AEM.02076-17.
34. Lee, S.; Kang, M.; Bae, J. N.; Sohn, J. N.; Sung, B. H. *Front. Bioeng. Biotechnol.* **2019**, *7*, 209.
35. Luo, H.; Klein, I. M.; Jiang, Y.; Zhu, H. Y.; Liu, B. Y.; Kenttamaa, H. I.; Abu-Omar, M. M. *ACS Sustain. Chem. Eng.* **2016**, *4*, 2316–2322.

36. Klein, I.; Saha, B.; Abu-Omar, M. M. *Cat. Sci. Technol.* **2015**, *5*, 3242–3245.
37. Klein, I.; Marcum, C.; Kenttamaaa, H.; Abu-Omar, M. M. *Green Chem.* **2016**, *18*, 2399–2405.
38. Ma, R. S.; Guo, M.; Zhang, X. *ChemSusChem* **2014**, 7, 412–415.
39. Cherubini, F.; Stromman, A. H. In *Biofuels: Alternative Feedstocks and Conversion Processes*; Pandey, A., Larroche, C., Ricke, S. C., Dussap, C. G., Gnansounou, E., Eds.; 2011; pp. 3–24.
40. Azadfar, M.; Gao, A. H.; Bule, M. V.; Chen, S. L. *Int. J. Biol. Macromol.* **2015**, *75*, 58–66.
41. Shen, X. J.; Meng, Q. L.; Mei, Q. Q.; Liu, H. Z.; Yan, J.; Song, J. L.; Tan, D. X.; Chen, B. F.; Zhang, Z. R.; Yang, G. Y.; Han, B. X. *Chem. Sci.* **2020**, *11*, 1347–1352.
42. Sainsbury, P. D.; Hardiman, E. M.; Ahmad, M.; Otani, H.; Seghezzi, N.; Eltis, L. D.; Bugg, T. D. H. *ACS Chem. Biol.* **2013**, *8*, 2151–2156.
43. Cheng, C. B.; Wang, J. Z.; Shen, D. K.; Xue, J. T.; Guan, S. P.; Gu, S.; Luo, K. H. *Polymers (Basel, Switz.)* **2017**, *9*, 240.
44. Araujo, J. D. P.; Grande, C. A.; Rodrigues, A. E. *Catal. Today* **2009**, *147*, S330–S335.
45. Fache, M.; Boutevin, B.; Caillol, S. *ACS Sustain. Chem. Eng.* **2016**, *4*, 35–46.
46. Parsell, T.; Yohe, S.; Degenstein, J.; Jarrell, T.; Klein, I.; Gencer, E.; Hewetson, B.; Hurt, M.; Kim, J. I.; Choudhari, H.; Saha, B.; Meilan, R.; Mosier, N.; Ribeiro, F.; Delgass, W. N.; Chapple, C.; Kenttamaa, H. I.; Agrawal, R.; Abu-Omar, M. M. *Green Chem.* **2015**, *17*, 1492–1499.
47. Ouyang, X. H.; Huang, X. M.; Zhu, J. D.; Boot, M. D.; Hensen, E. J. M. *ACS Sustain. Chem. Eng.* **2019**, 7, 13764–13773.
48. McVeigh, A.; Bouxin, F. P.; Jarvis, M. C.; Jackson, S. D. *Cat. Sci. Technol.* **2016**, *6*, 4142–4150.
49. Huang, X. M.; Ludenhoff, J. M.; Dirks, M.; Ouyang, X. H.; Boot, M. D.; Hensen, E. J. M. *ACS Catal.* **2018**, *8*, 11184–11190.
50. Xu, Z. X.; Lei, P.; Zhai, R.; Wen, Z. Q.; Jin, M. *J. Biotechnol. Biofuels* **2019**, *12*, 32.
51. Linger, J. G.; Vardon, D. R.; Guarnieri, M. T.; Karp, E. M.; Hunsinger, G. B.; Franden, M. A.; Johnson, C. W.; Chupka, G.; Strathmann, T. J.; Pienkos, P. T.; Beckham, G. T. *Proc. Natl. Acad. Sci. U. S. A.* **2014**, *111*, 12013–12018.
52. Davis, K. M.; Rover, M.; Brown, R. C.; Bai, X. L.; Wen, Z. Y.; Jarboe, L. R. *Energies* **2016**, *9*, 808.
53. Renders, T.; Cooreman, E.; Van den Bosch, S.; Schutyser, W.; Koelewijn, S. F.; Vangeel, T.; Deneyer, A.; Van den Bossche, G.; Courtin, C. M.; Sels, B. F. *Green Chem.* **2018**, *20*, 4607–4619.
54. Karkas, M. D.; Matsuura, B. S.; Monos, T. M.; Magallanes, G.; Stephenson, C. R. J. *Org. Biomol. Chem.* **2016**, *14*, 1853–1914.
55. Wang, H.; Tucker, M.; Ji, Y. *J. Appl. Chem.* **2013**, *2013*, 838645.
56. Li, H. L.; Song, G. *ACS Catal.* **2019**, *9*, 4054–4064.
57. Galkin, M. V.; Smit, A. T.; Subbotina, E.; Artemenko, K. A.; Bergquist, J.; Huijgen, W. J. J.; Samec, J. S. M. *ChemSusChem* **2016**, *9*, 3280–3287.
58. McCallum, C. S.; Strachan, N.; Bennett, S. C.; Forsythe, W. G.; Garrett, M. D.; Hardacre, C.; Morgan, K.; Sheldrake, G. N. *Green Chem.* **2018**, *20*, 2702–2705.
59. Laskar, D. D.; Tucker, M. P.; Chen, X. W.; Helms, G. L.; Yang, B. *Green Chem.* **2014**, *16*, 897–910.
60. Barrett, J. A.; Gao, Y.; Bernt, C. M.; Chui, M.; Tran, A. T.; Foston, M. B.; Ford, P. C. *ACS Sustain. Chem. Eng.* **2016**, *4*, 6877–6886.
61. Chui, M.; Metzker, G.; Bernt, C. M.; Tran, A. T.; Burtoloso, A. C. B.; Ford, P. C. *ACS Sustain. Chem. Eng.* **2017**, *5*, 3158–3169.
62. Zhai, Y. X.; Li, C.; Xu, G. Y.; Ma, Y. F.; Liu, X. H.; Zhang, Y. *Green Chem.* **2017**, *19*, 1895–1903.

63. Sturgeon, M. R.; O'Brien, M. H.; Ciesielski, P. N.; Katahira, R.; Kruger, J. S.; Chmely, S. C.; Hamlin, J.; Lawrence, K.; Hunsinger, G. B.; Foust, T. D.; Baldwin, R. M.; Biddy, M. J.; Beckham, G. T. *Green Chem.* **2014**, *16*, 824–835.
64. Patil, P. T.; Armbruster, U.; Richter, M.; Martin, A. *Energy Fuel* **2011**, *25*, 4713–4722.
65. Rautiainen, S.; Di Francesco, D.; Katea, S. N.; Westin, G.; Tungasmita, D. N.; Samec, J. S. M. *ChemSusChem* **2019**, *12*, 404–408.
66. Cheng, C. B.; Truong, J.; Barrett, J. A.; Shen, D. K.; Abu-Omar, M. M.; Ford, P. C. *ACS Sustain. Chem. Eng.* **2020**, *8*, 1023–1030.
67. Edmunds, C. W.; Peralta, P.; Kelley, S. S.; Chiang, V. L.; Sharma-Shivappa, R. R.; Davis, M. F.; Harman-Ware, A. E.; Sykes, R. W.; Gjersing, E.; Cunningham, M. W.; Rottmann, W.; Miller, Z. D.; Peszlen, I. *Cellul.* **2017**, *24*, 1901–1914.
68. Abu-Omar, M. M. In *Encyclopedia of Sustainable Technologies*; Luo, H., Abraham, M. A., Eds.; Elsevier: Oxford, 2017; pp. 573–585.
69. Patil, N. D.; Tanguy, N. R.; Yan, N.; Faruk, O.; Sain, M., Eds. *Lignin in Polymer Composites*; William Andrew Publishing, 2016; pp. 27–47.
70. Zakzeski, J.; Bruijnincx, P. C. A.; Jongerius, A. L.; Weckhuysen, B. M. *Chem. Rev.* **2010**, *110*, 3552–3599.
71. Mei, Q.; Shen, X.; Liu, H.; Han, B. *Chin. Chem. Lett.* **2019**, *30*, 15–24.
72. Uraki, Y.; Koda, K.; Yamada, T.; Oikawa, C.; Aso, T. *ACS Symp. Ser.* **2012**, *1107*, 243–254.
73. Galkin, M. V.; Samec, J. S. M. *ChemSusChem* **2016**, *9*, 1544–1558.
74. Zhu, W. Z.; Theliander, H. *16th International Symposium on Wood, Fiber and Pulping Chemistry, Proceedings*, Vols I & Ii; 2011; pp. 195–199.
75. Vázquez, G.; Antorrena, G.; González, J.; Freire, S. *J. Wood Chem. Technol.* **1997**, *17*, 147–162.
76. Constant, S.; Wienk, H. L. J.; Frissen, A. E.; de Peinder, P.; Boelens, R.; van Es, D. S.; Grisel, R. J. H.; Weckhuysen, B. M.; Huijgen, W. J. J.; Gosselink, R. J. A.; Bruijnincx, P. C. A. *Green Chem.* **2016**, *18*, 2651–2665.
77. Rinaldi, R.; Jastrzebski, R.; Clough, M. T.; Ralph, J.; Kennema, M.; Bruijnincx, P. C. A.; Weckhuysen, B. M. *Angew. Chem. Int. Ed.* **2016**, *55*, 8164–8215.
78. Lancefield, C. S.; Wienk, H. L. J.; Boelens, R.; Weckhuysen, B. M.; Bruijnincx, P. C. A. *Chem. Sci.* **2018**, *9*, 6348–6360.
79. Giummarella, N.; Pylypchuk, I. V.; Sevastyanova, O.; Lawoko, M. *ACS Sustain. Chem. Eng.* **2020**, *8*, 10983–10994.
80. Knudsen, J. N.; Jensen, P. A.; Lin, W. G.; Frandsen, F. J.; Dam-Johansen, K. *Energy Fuel* **2004**, *18*, 810–819.
81. Rennenberg, H.; Huber, B.; Schroder, P.; Stahl, K.; Haunold, W.; Georgii, H. W.; Slovik, S.; Pfanz, H. *Plant Physiol.* **1990**, *92*, 560–564.
82. Saleh, S. B.; Flensborg, J. P.; Shoulaifar, T. K.; Sarossy, Z.; Hansen, B. B.; Egsgaard, H.; DeMartini, N.; Jensen, P. A.; Glarborg, P.; Dam-Johansen, K. *Energy Fuel* **2014**, *28*, 3738–3746.
83. Kai, D.; Tan, M. J.; Chee, P. L.; Chua, Y. K.; Yap, Y. L.; Loh, X. J. *Green Chem.* **2016**, *18*, 1175–1200.
84. Inwood, J. P. W.; Pakzad, L.; Fatehi, P. *BioResources* **2018**, *13*, 53–70.
85. Evdokimov, A. N.; Kurzin, A. V.; Fedorova, O. V.; Lukanin, P. V.; Kazakov, V. G.; Trifonova, A. D. *Wood Sci. Technol.* **2018**, *52*, 1165–1174.
86. Osada, M.; Hiyoshi, N.; Sato, O.; Arai, K.; Shirai, M. *Energy Fuel* **2007**, *21*, 1854–1858.
87. Agarwal, S.; Chowdari, R. K.; Hita, I.; Heeres, H. J. *ACS Sustain. Chem. Eng.* **2017**, *5*, 2668–2678.
88. Han, T.; Sophonrat, N.; Evangelopoulos, P.; Persson, H.; Yang, W. H.; Jonsson, P. *J. Anal. Appl. Pyrolysis* **2018**, *133*, 162–168.

89. Vishtal, A.; Kraslawski, A. *BioResources* **2011**, *6*, 3547–3568.
90. Aro, T.; Fatehi, P. *ChemSusChem* **2017**, *10*, 1861–1877.
91. Zulfikar, M. A.; Wahyuningrum, D.; Lestari, S. *Sep. Sci. Technol.* **2013**, *48*, 1391–1401.
92. Flatt, R.; Schober, I. In *Understanding the Rheology of Concrete*; Roussel, N., Ed.; Woodhead Publishing, 2012; pp. 144–208.
93. Matsushita, Y. *J. Wood Sci.* **2015**, *61*, 230–250.
94. Yu, L. X.; Yu, J.; Mo, W. J.; Qin, Y. L.; Yang, D. J.; Qiu, X. Q. *RSC Adv.* **2016**, *6*, 70863–70869.
95. Yang, D. J.; Li, H. J.; Qin, Y. L.; Zhong, R. S.; Bai, M. X.; Qiu, X. Q. *J. Dispers. Sci. Technol.* **2015**, *36*, 532–539.
96. Huang, C. X.; Ma, J. M.; Zhang, W. Y.; Huang, G.; Yong, Q. *Polymers (Basel, Switz.)* **2018**, *10*, 841.
97. Kumar, A.; Anushree; Kumar, J.; Bhaskar, T. *J. Energy Inst.* **2020**, *93*, 235–271.
98. Mousavioun, P.; Doherty, W. O. S. *Ind. Crop Prod.* **2010**, *31*, 52–58.
99. Schutyser, W.; Renders, T.; Van den Bossche, G.; Van den Bosch, S.; Koelewijn, S.-F.; Ennaert, T.; Sels, B. In *Nanotechnology in Catalysis: Applications in the hemical Industry, Energy Development, and Environment Protection*; Van de Voorde, M., Sels, B., Eds.; Wiley-VCH Verlag GmbH & Co. KGaA, 2017; pp. 537–584. https://doi.org/10.1002/9783527699827.
100. Tanase-Opedal, M.; Espinosa, E.; Rodriguez, A.; Chinga-Carrasco, G. *Materials* **2019**, *12*, 3006.
101. Huang, X. M.; Atay, C.; Koranyi, T. I.; Boot, M. D.; Hensen, E. J. M. *ACS Catal.* **2015**, *5*, 7359–7370.
102. Baurhoo, B.; Ruiz-Feria, C. A.; Zhao, X. *Anim. Feed Sci. Technol.* **2008**, *144*, 175–184.
103. Barta, K.; Matson, T. D.; Fettig, M. L.; Scott, S. L.; Iretskii, A. V.; Ford, P. C. *Green Chem.* **2010**, *12*, 1640–1647.
104. Hochegger, M.; Cottyn-Boitte, B.; Cezard, L.; Schober, S.; Mittelbach, M. *Int. J. Chem. Eng.* **2019**, *2019*, 1734507.
105. Sadeghifar, H.; Wells, T.; Le, R. K.; Sadeghifar, F.; Yuan, J. S.; Ragauskas, A. J. *ACS Sustain. Chem. Eng.* **2017**, *5*, 580–587.
106. Yu, O.; Yoo, C. G.; Kim, C. S.; Kim, K. H. *ACS Omega* **2019**, *4*, 16103–16110.
107. Saha, M.; Saynik, P. B.; Borah, A.; Malani, R. S.; Arya, P.; Shivangi; Moholkar, V. S. *Bioresource technology reports*, Vol. 5; 2019; pp. 206–211.
108. Hodasova, L.; Jablonsky, M.; Skulcova, A.; Haz, A. *Wood Res* **2015**, *60*, 973–986.
109. Wanmolee, W.; Laosiripojana, N.; Daorattanachai, P.; Moghaddam, L.; Rencoret, J.; del Rio, J. C.; Doherty, W. O. S. *ACS Sustain. Chem. Eng.* **2018**, *6*, 3010–3018.
110. Parto, S. G.; Jorgensen, E. K.; Christensen, J. M.; Pedersen, L. S.; Larsen, D. B.; Duus, J. O.; Jensen, A. D. *Sustainable Energy Fuels* **2020**, *4*, 1844–1854.
111. Wang, S. X.; Yang, L. P.; Stubbs, L. P.; Li, X.; He, C. B. *ACS Appl. Mater. Interfaces* **2013**, *5*, 12275–12282.
112. Rumpf, J.; Do, X. T.; Burger, R.; Monakhova, Y. B.; Schulze, M. *Biomacromolecules* **2020**, *21*, 1929–1942.
113. Trubetskaya, A.; Lange, H.; Wittgens, B.; Brunsvik, A.; Crestini, C.; Rova, U.; Christakopoulos, P.; Leahy, J. J.; Matsakas, L. *Processes* **2020**, *8*, 860.
114. de la Torre, M. J.; Moral, A.; Hernandez, M. D.; Cabeza, E.; Tijero, A. *Ind. Crop Prod.* **2013**, *45*, 58–63.
115. Akgul, M.; Kirci, H. *J. Environ. Biol.* **2009**, *30*, 735–740.
116. Liu, F.; Liu, Q. Y.; Wang, A. Q.; Zhang, T. *ACS Sustain. Chem. Eng.* **2016**, *4*, 3850–3856.
117. Hita, I.; Deuss, P. J.; Bonura, G.; Frusteri, F.; Heeres, H. J. *Fuel Process. Technol.* **2018**, *179*, 143–153.

118. Huang, J.-B.; Wu, S.-B.; Cheng, H.; Lei, M.; Liang, J.-J.; Tong, H. *J. Fuel Chem. Technol.* **2015**, *43*, 429–436.
119. Ma, Q. Z.; Liu, Q. Y.; Dou, X. M.; Li, W. Z.; Fan, W.; Liu, M. H. *BioResources* **2018**, *13*, 8024–8040.
120. Struven, J. O.; Meier, D. *ACS Sustain. Chem. Eng.* **2016**, *4*, 3712–3721.
121. Zhao, B.; Hu, Y.; Gao, J.; Zhao, G.; Ray, M. B.; Xu, C. C. *Ind. Eng. Chem. Res.* **2020**, *59*, 16987–17007.
122. Huang, X. M.; Koranyi, T. I.; Boot, M. D.; Hensen, E. J. M. *Green Chem.* **2015**, *17*, 4941–4950.
123. Feng, J. F.; Yang, Z. Z.; Hse, C. Y.; Su, Q. L.; Wang, K.; Jiang, J. C.; Xu, J. M. *Renew. Energy* **2017**, *105*, 140–148.
124. Tekin, K.; Hao, N. J.; Karagoz, S.; Ragauskas, A. J. *ChemSusChem* **2018**, *11*, 3559–3575.
125. Kim, K. H.; Kim, C. S. *Front. Energy Res.* **2018**, *6*, 92.
126. Huang, S. H.; Mahmood, N.; Tymchyshyn, M.; Yuan, Z. S.; Xu, C. B. *Bioresour. Technol.* **2014**, *171*, 95–102.
127. Ouyang, X. P.; Huang, X. Z.; Zhu, Y.; Qiu, X. Q. *Energy Fuel* **2015**, *29*, 5835–5840.
128. Arakadiusz, G.; Monika, A.; Kinga, C. *Annals of Warsaw University of Life Sciences—SGGW, Agriculture*, Vol. 67; Warsaw University of Life Sciences Press, 2016; pp. 101–111.
129. Pandarus, V.; Beland, F.; Ciriminna, R.; Pagliaro, M. *ChemCatChem* **2011**, *3*, 1146–1150.
130. Kim, J. K.; Lee, J. K.; Kang, K. H.; Song, J. C.; Song, I. K. *Appl. Catal. Gen.* **2015**, *498*, 142–149.
131. Lu, J. M.; Wang, M.; Zhang, X. C.; Heyden, A.; Wang, F. *ACS Catal.* **2016**, *6*, 5589–5598.
132. Parsell, T. H.; Owen, B. C.; Klein, I.; Jarrell, T. M.; Marcum, C. L.; Haupert, L. J.; Amundson, L. M.; Kenttamaa, H. I.; Ribeiro, F.; Miller, J. T.; Abu-Omar, M. M. *Chem. Sci.* **2013**, *4*, 806–813.
133. Ye, Y. Y.; Zhang, Y.; Fan, J.; Chang, J. *Bioresour. Technol.* **2012**, *118*, 648–651.
134. Van den Bosch, S.; Schutyser, W.; Koelewijn, S. F.; Renders, T.; Courtin, C. M.; Sels, B. F. *Chem. Commun.* **2015**, *51*, 13158–13161.
135. Davis, J. L.; Barteau, M. A. *Surf. Sci.* **1987**, *187*, 387–406.
136. Van den Bosch, S.; Schutyser, W.; Vanholme, R.; Driessen, T.; Koelewijn, S. F.; Renders, T.; De Meester, B.; Huijgen, W. J. J.; Dehaen, W.; Courtin, C. M.; Lagrain, B.; Boerjan, W.; Sels, B. F. *Energ. Environ. Sci.* **2015**, *8*, 1748–1763.
137. Dong, L.; Lin, L.; Han, X.; Si, X.; Liu, X.; Guo, Y.; Lu, F.; Rudić, S.; Parker, S. F.; Yang, S.; Wang, Y. *Chem* **2019**, *5*, 1521–1536.
138. Hocking, M. B. In *Handbook of Chemical Technology and Pollution Control*; Hocking, M. B., Ed.; 3rd ed.; Academic Press: San Diego, 2005; pp. 593–636.
139. Song, Q.; Wang, F.; Cai, J.; Wang, Y.; Zhang, J.; Yu, W.; Xu, J. *Energ. Environ. Sci.* **2013**, *6*, 994–1007.
140. Cheng, C. B.; Truong, J.; Barrett, J.; Shen, D. K.; Abu-Omar, M.; Ford, P. *ACS Sustainable Chem. Eng.* **2020**, *8*, 1023–1030.
141. Perras, F. A.; Luo, H.; Zhang, X. M.; Mosier, N. S.; Pruski, M.; Abu-Omar, M. M. *J. Phys. Chem. A* **2017**, *121*, 623–630.
142. Blondiaux, E.; Bomon, J.; Smolen, M.; Kaval, N.; Lemiere, F.; Sergeyev, S.; Diels, L.; Sels, B.; Maes, B. U. W. *ACS Sustain. Chem. Eng.* **2019**, 7, 6906–6916.
143. Augustin, A.; Lamerz, J.; Meistermann, H.; Golling, S.; Scheiblich, S.; Hermann, J. C.; Duchateau-Nguyen, G.; Tzouros, M.; Avila, D. W.; Langen, H.; Essioux, L.; Klughammer, B. *Mol. Cancer Ther.* **2013**, *12*, 520–529.

144. Zhang, Y. M.; Tortorella, M. D.; Liao, J. X.; Qin, X. C.; Chen, T. T.; Luo, J. F.; Guan, J. T.; Talley, J. J.; Tu, Z. C. *ACS Med. Chem. Lett.* **2015**, *6*, 1086–1090.
145. Jiang, Y.; Ding, D. C.; Zhao, S.; Zhu, H. Y.; Kenttamaa, H. I.; Abu-Omar, M. M. *Green Chem.* **2018**, *20*, 1131–1138.
146. Zhang, X. M.; Murria, P.; Jiang, Y.; Xiao, W. H.; Kenttamaa, H. I.; Abu-Omar, M. M.; Mosier, N. S. *Green Chem.* **2016**, *18*, 5219–5229.
147. Jiang, Y.; Yang, L. N.; Bohn, C. M.; Li, G. N.; Han, D.; Mosier, N. S.; Miller, J. T.; Kenttamaa, H. I.; Abu-Omar, M. M. *Org. Chem. Front.* **2015**, *2*, 1388–1396.
148. Zhang, X.; Yan, H.; Zhu, L. J.; Li, T.; Wang, S. R. *Adv. Sustain. Syst.* **2020**, *4*, 1900136.
149. Ng, F. F.; Couture, G.; Philippe, C.; Boutevin, B.; Caillol, S. *Molecules* **2017**, *22*, 149.
150. Auvergne, R.; Caillol, S.; David, G.; Boutevin, B.; Pascault, J. P. *Chem. Rev.* **2014**, *114*, 1082–1115.
151. Zhao, S.; Abu-Omar, M. M. *Macromolecules* **2019**, *52*, 3646–3654.
152. Zhao, S.; Abu-Omar, M. M. *Biomacromolecules* **2015**, *16*, 2025–2031.
153. Zhao, S.; Abu-Omar, M. M. *Macromolecules* **2017**, *50*, 3573–3581.
154. Fache, M.; Boutevin, B.; Caillol, S. *Green Chem.* **2016**, *18*, 712–725.
155. Shimasaki, T.; Yoshihara, S.; Shibata, M. *Polym. Compos.* **2012**, *33*, 1840–1847.
156. Hernandez, E. D.; Bassett, A. W.; Sadler, J. M.; La Scala, J. J.; Stanzione, J. F. *ACS Sustain. Chem. Eng.* **2016**, *4*, 4328–4339.
157. Zhao, S.; Abu-Omar, M. M. *ACS Sustain. Chem. Eng.* **2016**, *4*, 6082–6089.
158. U. S. Energy Information Administration (EIA)., **2020**. https://www.eia.gov/energyexplained/gasoline/use-of-gasoline.php.
159. Fahmy, T. Y. A.; Fahmy, Y.; Mobarak, F.; El-Sakhawy, M.; Abou-Zeid, R. E. *Environ. Dev. Sustain.* **2020**, *22*, 17–32.
160. Revankar, S. T. In *Storage and Hybridization of Nuclear Energy*; Bindra, H., Revankar, S., Eds.; Academic Press, 2019; pp. 177–227.
161. Lu, M. H.; Du, H.; Wei, B.; Zhu, J.; Li, M. S.; Shan, Y. H.; Song, C. S. *Energy Fuel* **2017**, *31*, 10858–10865.
162. Wang, W. C.; Bai, C. J.; Thapaliya, N. *Int. J. Energy Res.* **2015**, *39*, 1083–1093.
163. Jin, W.; Pastor-Perez, L.; Villora-Pico, J. J.; Pastor-Blas, M. M.; Sepulveda-Escribano, A.; Gu, S.; Charisiou, N. D.; Papageridis, K.; Goula, M. A.; Reina, T. R. *Energies* **2020**, *13*, 132.
164. Li, S.; Liu, B.; Truong, J.; Luo, Z.; Ford, P. C.; Abu-Omar, M. M. *Green Chem.* **2020**, *22*, 7406–7416.
165. Chan, X. J.; Pu, T. C.; Chen, X. Y.; James, A.; Lee, J.; Parise, J. B.; Kim, D. H.; Kim, T. *Catal. Commun.* **2017**, *97*, 65–69.
166. de Souza, P. M.; Rabelo-Neto, R. C.; Borges, L. E. P.; Jacobs, G.; Davis, B. H.; Sooknoi, T.; Resasco, D. E.; Noronha, F. B. *ACS Catal.* **2015**, *5*, 1318–1329.
167. Wang, H. L.; Ruan, H.; Pei, H. S.; Wang, H. M.; Chen, X. W.; Tucker, M. P.; Cort, J. R.; Yang, B. *Green Chem.* **2015**, *17*, 5131–5135.
168. Qin, S. F.; Li, B. L.; Luo, Z. C.; Zhao, C. *Green Chem.* **2020**, *22*, 2901–2908.

CHAPTER SIX

Catalytic cleavage of lignin C—O and C—C bonds

Chaofeng Zhang and Feng Wang*
State Key Laboratory of Catalysis, Dalian National Laboratory for Clean Energy, Dalian Institute of Chemical Physics, Chinese Academy of Sciences, Dalian, China
*Corresponding author: e-mail address: wangfeng@dicp.ac.cn

Contents

Abstract

Lignin is a unique and abundant renewable organic carbon resource in nature with an abundant aromatic structure, thus, is being regarded as the potential renewable bank to prepare aromatic chemicals. Therefore, in recent decades, researches on lignin depolymerization and the transformation of its down-stream products have drawn enormous attention from academia to industry worldwide, aiming at harvesting aromatic compounds. Due to the complex lignin linkages among the random polymerized sinapyl alcohol (S-unit), coniferyl alcohol (G-unit) and *p*-coumaric alcohol (H-unit) with various robust C—C and C—O bonds, lignin depolymerization to easy-to-handle aromatic monomers with acceptable conversion and selectivity is still a highly challenging task. In this chapter, focusing on lignin catalytic depolymerization to aromatic chemicals, we herein first provide our understanding about the scientific question and strategic

Advances in Inorganic Chemistry, Volume 77
ISSN 0898-8838
https://doi.org/10.1016/bs.adioch.2020.12.004

foundation process in lignin depolymerization, and then clarify them by mainly presenting our recent studies. Firstly, we introduce our research on direct protolignin depolymerization via a fragmentation–hydrogenolysis process in alcohol solvents. Then, focusing on the catalytic cleavage of lignin C—C and C—O bonds, we shed light on a recapitulative adjacent functional group modification (AFGM) strategy for the conversion of lignin models and apply the established methods in the real lignin conversion via a bottom-up research approach. AFGM strategy begins with the adjacent functional group modification of target C—C or C—O bond to directly decrease the bond dissociation enthalpy (BDE) of targeted bonds or to generate new substrate sites to introduce the cleavage reagent for further conversion. Furthermore, the critical point for lignin depolymerization to aromatics is not only promoting the lignin conversion, but also restraining the fragments recondensation, which requires the foundation of efficient strategies taking into account the effects of lignin structure change during pre-abstraction and reaction, catalyst character, and reaction conditions. In addition, our perspectives for lignin utilization are that the products from lignin conversion can be used as monomers for artificial polymerization, or using lignin derivative molecules to synthesize high-value synthetic building blocks for medicine and material, which needs a deeper understanding about lignin chemistry itself, new engineering technologies to intensify the conversion process, and fantastic ideas from different subjects to extend the relative research fields.

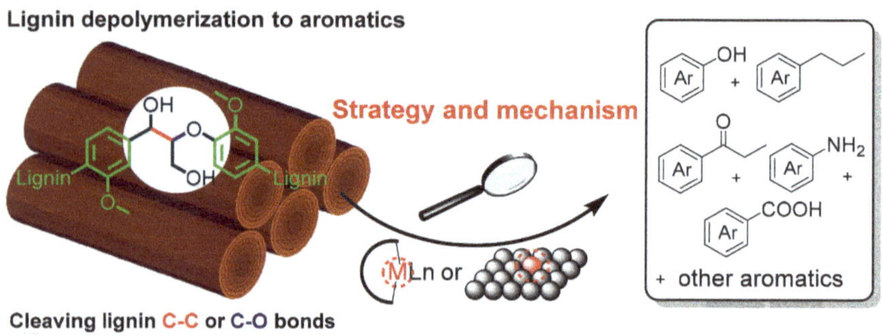

List of abbreviations

AC	active carbon
AFGM	adjacent functional group modification
BDE	bond dissociation enthalpy
^{t}BuONO	tert butyl nitrite
CB	conduction band
DBAD	di-tert-butyl azodicarboxylate
DDQ	2,3-dichloro-5,6-dicyano-1,4-benzoquinone
DFT	density functional theory
EXAFS	extended X-ray absorption fine structure
G-unit	coniferyl alcohol, or lignin phenylpropanoid unit with one –OMe connecting with the aromatic ring
HAADF-STEM	high angle annular dark-field scanning transmission electron microscopy

H-unit	*p*-coumaric alcohol, or lignin phenylpropanoid unit without –OMe connecting with the aromatic ring
LC	lignocellulose
LS	lignosulfonate
LDP	lignin depolymerization
LMCT	ligand-to-metal charge-transfer
MALDI-TOF	matrix-assisted laser desorption/ionization time-of-flight mass spectrometry
Mn	number-average molecular weight
Mw	mass average molecular weight
NHPI	*N*-hydroxyphthalimide
ppu	phenylpropanoid units
PTH	phenothiazine
TEMPO	2,2,6,6-tetramethyl-1-piperidinyloxy
S-unit	sinapyl alcohol, or lignin phenylpropanoid unit with two –OMe connecting with the aromatic ring
UV–vis DRS	ultraviolet-visible diffuse reflectance spectra
VB	valence band
XRD	X-ray diffraction
XPS	X-ray photoelectron spectroscopy

1. Introduction

The world currently is more and more relying on nonrenewable fossil resources. In 2019, the global primary energy consumption reaches to 583.9 EJ (1 EJ = 10^{18} J),[1] and the 84.3% part comes from the traditional fossil resources, coal, natural gas, and crude oil. Besides fuel production and energy output, the fossil resources in billion ton quantities are transformed into various chemicals and materials to support the explosive development of human civilization. On the negative side of the coin, the immoderate utilization of fossil resources released billions of tons of CO_2 into the atmosphere within a short time, exceeding the capacity of the earth's ecological cycle and causing global climate problems like Global Warming and so on. With such background, the increasing awareness of dwindling fossil resources and the ever-increasing need for more chemical feedstock and energy supply has led to extensive research into more efficient and sustainable methods to meet these demands. In order to make the carbon cycle more efficient and controllable, to reduce the dependence on fossil resources, utilizing sustainable carbon resources will be another choice. As one kind of carbon and solar energy storage form, biomass is any organic matter that is renewable over time and is the only renewable organic carbon resource in nature. These are plant-based materials, such as agricultural residues, forestry wastes, and energy crops.[2] Every year, plants on earth can produce about 170 billion tons of lignocellulose from photosynthesis by capturing the low concentration CO_2, immobilizing the carbon and releasing O_2.[3] However, besides the timber, only a small part of biomass carbon resource is utilized, and that is primarily used to generate heat and power. If we can efficiently transform the renewable biomass resource to the desired chemical feedstocks, we need not worry too much about the increasing CO_2 concentration, mainly because the origin of the released CO_2 during the utilization process is from the original CO_2 in the modern-day atmosphere. This would provide a perfect CO_2-neutral cycle, which transforms solar energy, H_2O, and CO_2 into chemical feedstocks and fuels via biomass.

Lignocellulose (LC) with an annual production of around 170 billion tons is the most abundant biomass resource,[3] and its utilization has drawn an enormous amount of research attention from academia to industry worldwide. Lignocellulose comprises three main components hemicellulose, cellulose, and lignin, which build the body of the plants. Among them, hemicellulose (20–30 wt% of LC) and cellulose (40–50 wt% of LC) are the polymers of C_5 and C_6 sugars with certain ether linkages.[4] The lignin (10–35 wt% of LC) is a unique aromatic bank, which is a much more complex polymer and consists of various methoxylated phenylpropanoid units (ppu). The abundant annual output makes lignin an alternative resource to fossil resources for the preparation of aromatic chemicals.

For the generation of lignin, the plants transform a part of carbohydrate from photosynthesis into sinapyl alcohol (S-unit), coniferyl alcohol (G-unit), and *p*-coumaric alcohol (H-unit) via the complex biochemical pathway,[5] which further undergo random polymerization via an enzymatic route (Fig. 1A). According to previous research, the polymerization process is initiated by the enzymatic dehydrogenation of the monolignol phenolic hydroxyl groups by an electron transfer process that yields reactive monolignol radicals and their different resonant structures. With the G-unit as the example (Fig. 1B), the reactive oxygen-centered and carbon-centered monolignol radicals undergo random polymerization with each other, which builds the lignin dilignols with new C—C or C—O bonds generation. Subsequent nucleophilic attack by water, alcohols, or phenolic hydroxyl groups on the benzyl carbon of the quinone methide intermediates will restore the aromaticity of the benzene ring. The resulting aromatic dilignols then undergo further polymerization to generate more complex lignin chains and polymers.[6] Although the exact structure of protolignin is quite complex and differs from

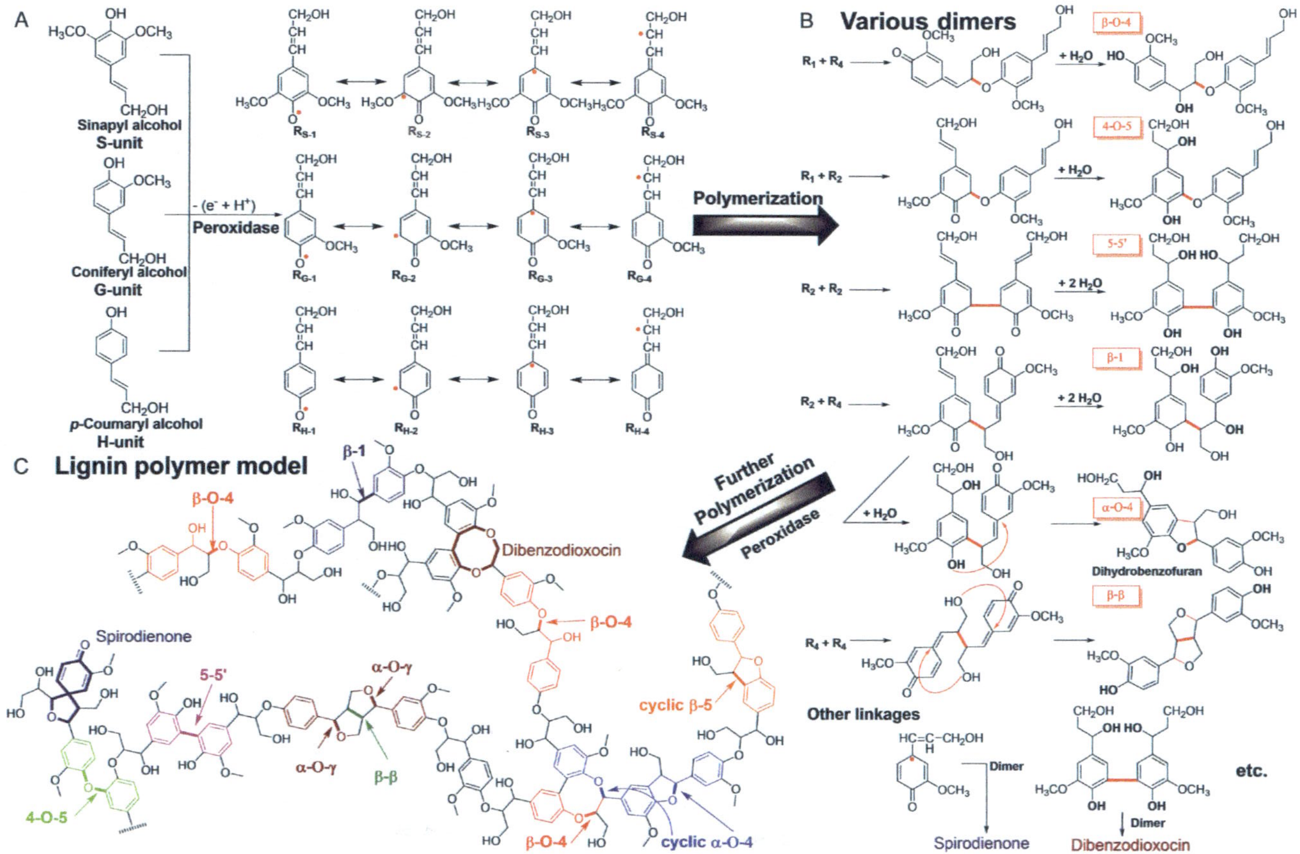

Fig. 1 The formation of lignin polymer from three basic monomers.

Table 1 Types and frequencies of linkages and main functional groups in softwood and hardwood.

	Number (100 ppu)			Number (100 ppu)	
Linkages	**Softwood**	**Hardwood**	**Functional groups**	**Softwood**	**Hardwood**
β-O-4	43–50	50–65	$-OCH_3$	92–96	132–146
5–5′	10–25	4–10	$-OH_{(phenolic)}$	20–28	9–20
β-5	9–12	4–6	$-OH_{(benzyl)}$	16	–
α-O-4	6–8	4–8	$-OH_{(aliphatic)}$	120	–
4-O-5	4	4–7	Carbonyl	20	3–7
β-1	3–7	5–7	Carboxyl	–	11–13
β-β	2–4	3–7			
Others	16	7–8			

plant to plant, researchers have shown that lignin structure includes β-O-4, β-5, α-O-4, β-β, 5-5, 4-O-5, β-1, and other linkages with aromatic structures (Fig. 1C), and their approximate linkage abundance in softwood/hardwood are listed in Table 1.[7]

Although various non-catalytic and catalytic methods have been explored to cleave the various lignin since the 1930s, it remains a big challenge for converting lignin extracts, albeit original lignin, to aromatics selectively. At the beginning of lignin conversion research, the lignin was regarded as a carbon source like oil and coal, and the principles and equipment for oil refinery and coal transformation, such as pyrolysis and gasification processes carried out above 400 °C, were employed in the lignocellulose and lignin utilization. Pyrolysis can produce a certain amount of liquid products, known as bio-oils which are a mixture of phenols, low-carbon alcohol, acid, ester, ether, and hydrocarbon. Gasification can transform some biomass structure to CO, CO_2, H_2, and other gaseous products, but the harsh condition at a high temperature also generates carbon deposits or unconverted carbon polymer from lignin carbon constituents. This can be attributed to the more complex mechanism caused by the disturbing lignin structure with various linkages between the phenylpropane units and the different condensation modes of the lignin fragments under the harsh condition.

In recent decades, with the development and utilization of renewable biomass resources drawing more and more attention from developed economies all over the world, more social resources have been dedicated to this challenging but promising field. Therefore, efficient utilization of biomass with its different constituents, including lignins with different structures from different plants and isolation processes, will require modification of the traditional catalysis methods to give efficient bio-refineries.[8,9]

In this chapter, focusing on lignin depolymerization, we will summarize our recent studies on catalytic lignin conversion to aromatic monomers. We will first give a brief introduction of the lignin conversion from bulk lignin and compare this to the mechanistic analysis of target molecule/linkage transformation and cleavage. This will highlight catalytic strategies, suitable catalysts, solvents, and lignin pretreatments that can enhance the lignin conversion via the selective cleavage of lignin C—C or C—O bonds. In addition to the exploration of the scientific questions about C—C or C—O bond cleavage, we also present our attempts to convert the lignin depolymerization products to high demand polymer monomers and transform the stable lignin fragments to highly functionalized motifs.

2. Scientific questions involved in lignin depolymerization and strategic foundation

Before working on the lignin depolymerization, it is necessary to first clarify the scientific questions involved that are beneficial to the foundation of catalytic strategies for efficient lignin utilization. For the depolymerization of lignin polymer with a complex 3D structure to easily handle aromatic monomers with acceptable conversion and selectivity, we should first focus on the target or substrate structure to be cleaved. Although the exact structure of protolignin is quite complex and different from plant to plant, the target is clear that the key issue for lignin utilization to preparing aromatic chemicals depends on the efficient strategies to cleave the ubiquitous C—C or C—O linkage bonds shown in Table 1, while leaving the aromatic rings intact. The β-O-4 linkage is dominant and accounts for more than half of inter-unit linkages in lignin, and its C–O–Ar and C_α–C_β bonds are mainly the targeted bonds to break during the delignification process. The C—O/C—C bonds in β-5, α-O-4, β-β, 5-5, 4-O-5, and β-1 linkages also need to be cleaved to increase the depolymerization efficiency. To achieve this goal, various non-catalytic and catalytic methods have been explored via a "Direct lignin or isolated lignin conversion" approach or a "from typical models to lignin" approach (bottom-up approach), which place extra emphasis on direct lignin conversion or lignin model conversion, respectively. Although a vast difference exists between studies on model compounds and studies on technical/native lignin because of the complex structure of the latter, the "bottom-up" approach has presented many effective catalytic systems and strategies in the technical/native lignin conversion. Nevertheless, each approach has its advantages and limitations. Hence, both domains should be complementary in finding efficient methods in lignin conversion.

As we discussed above, the key issues for the lignin transformation lie in the development of highly selective and active catalysts to effectively cleave the ubiquitous C—C or C—O linkages, while leaving the aromatic benzene rings unconverted. The next step is to develop effective depolymerization strategies. For lignin depolymerization to aromatic monomers, two factors should be especially concerned: (a) promoting the selective cleavage of the ubiquitous lignin C—O/C—C bonds, increasing the conversion and (b) restraining the lignin fragments from recondensation to more stable multimers to increase the selectivity and monomer yields.

For the first step in lignin depolymerization, different catalytic routes produce different intermediates, and these will lead to variations in the lignin linkage cleavage. The dominating factor may not be the dissociation enthalpy of the target bonds in the original substrate, but one must also consider the activated intermediates. In an ideal route with β-O-4 bond and other linkages cleavage, an activated intermediate is generated after a series of steps with low activation energy, where the target bond has a lower dissociation energy and easily breaks in the subsequent transformations. In 2017, we summarized our methods for β-O-4 and β-1 lignin model conversion as the adjacent functional group modification (AFGM) strategy as illustrated in Fig. 2, which demonstrates changes in the final C—O/C—C bond strengths with different adjacent groups.

The AFGM strategy aims to find the potential lignin conversion routes via the study of the BDE changes of the target linkage bonds upon the formation of an intermediate or the introduction of an active chemical group. AFGM targets at the first substrate transformation to generate an active molecule or fragment intermediate with a weakened C—O/C—C bond, generate an unstable intermediate which can undergo self-decomposition via C—O/C—C bond cleavage with or without the assistance of additives or generate new "substrate sites" for introducing a cleavage reagent for further linkage cleavage. These concepts, combined with appropriate catalytic systems, can make lignin model conversions more efficient and selective. For examples, initial C_α–OH oxidation to C_α=O or dehydroxylation to a C_α-centered radical in β-O-4 linkage can reduce the BDE of C_β–OAr, while initial C_γ–OH oxidation to C_γ=O can weaken the C_α–C_β bond (Fig. 2). Also, although the first C_α–OH oxidation to C_α=O can cause the increase of C_α–C_β BED by 30 kJ mol^{-1}, H• or H^+ can be easily abstracted from C_β–H and the remaining C_β can work as the "substrate site"

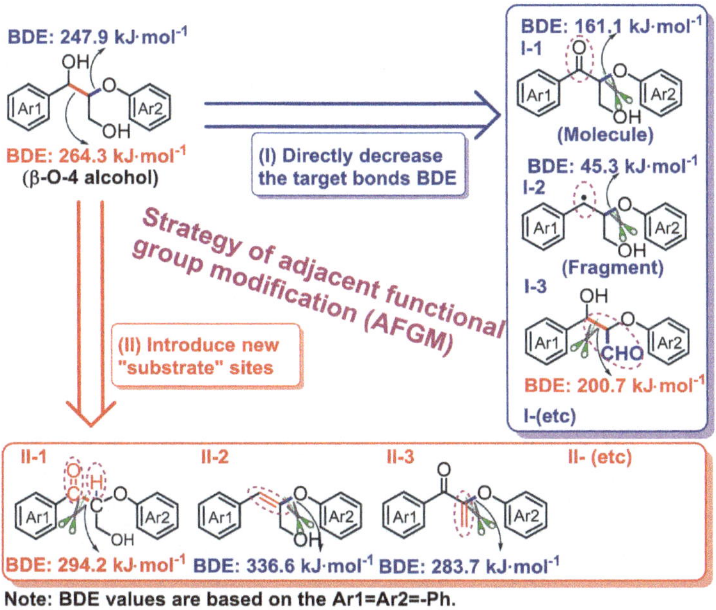

Fig. 2 Adjacent functional group modification (AFGM) strategy for the cleavage of basic lignin C_α–C_β and C_β–OAr bonds in β-O-4 lignin linkage. *The figure was adapted from Zhang, C.F.; Wang, F.* Acc. Chem. Res. ***2020,*** *53, 470.*

to react with active O_2* or other O* species, causing the cleavage of the C_α–C_β bond. The C_α=O can directly react with NH_2OH or peroxide, and the N and O atoms can insert in lignin C—C bonds to generate C(=O)—O and C(=O)—N bonds in the generated ester and amide that could be easily cleaved by the subsequent hydrolysis. Moreover, the more stable C_α=C_β bond of a lignin alkenyl ether can work as the "substrate site" to introduce the proton and promote the hydrolysis of the C_β–OAr bond. Therefore, based on the desired path via certain intermediates, the nature of the catalyst will be explicitly defined, which can improve the efficiency of the catalysts screening and method foundation with specific guidance.

Furthermore, attention must be paid to the effect of lignin structure change upon the pretreatment of lignocellulose, which may result in the lignin linkage transformations and fragment condensation. In Sels' recent review,[9] they point out that new bio-refinery methods need to be directed to the preservation of dominant β-O-4 bonds in lignin. During organic synthesis, one can keep the selectivity of the reaction via the first protection of sensitive groups that do not participate in the conversion. For lignocellulose conversion, the similar "β-O-4 bond protection" mode suggests that one can first protect/transform the active functional groups of lignin at the beginning of pretreatment, which could preserve the β-O-4 structure for further controllable conversion and restrain the recondensation of possible lignin fragments. Recently, the pre-protection methods have drawn much attention to lignocellulose hydrogenation, such as the formaldehyde stabilization method and its derivative methods which involve the initial protection of C_α–OH and C_γ–OH in the β-O-4 structure.[10] Some etherification/esterification pretreatment at C_α–OH and C_γ–OH in the β-O-4 structure with protonic polar solvents or reagents can weaken or replace hydrogen bonds to promote lignin release from lignocellulose, and the etherification/esterification with alcohol/acid can further increase the dissolving capacity of lignin for the following depolymerization. Furthermore, the first oxidation of the C_α–OH to C_α=O could not only restrain the generation of the C_α carbocation which can cause the lignin fragments condensation, but also can promote the hydrogenation cleavage of β-O-4 linkage by weakening the C_β–OAr bond. In addition, tandem depolymerization–stabilization of native lignin depolymerization fragments/monomers is another method to prevent the unwanted condensation. These cascaded models usually combine two different methods to achieve the in situ lignin fragment depolymerization to monomers and instant monomers stabilization or transformation. Compared with the single transformation method, the "Cascaded model," represented by the acidolysis + acetal formation, acidolysis + hydrogenation, acidolysis + decarbonylation,[11] and alkali catalysis + hydrogenation/oxidation methods, can restrain the recondensation of fragments and make the conversion process more controllable.

In Section 3, based on the concepts from the AFGM strategy, directly decreasing the BDE of targeted C—C/C—O bonds or generating new substrate sites to introduce cleavage reagent for further conversion, meanwhile restraining the condensation via carrying out the conversion under a mild condition by introducing photo-catalysis,[12] we will summarize our recent studies on lignin catalytic depolymerization to aromatic monomers, which will highlight the effects of lignin structure, pretreatments, catalyst characters, and reaction conditions on the efficiency of various strategies.

3. Recent lignin depolymerization and utilization strategies

3.1 Alcohol solvents promote lignin depolymerization via the first solvolysis fragmentation

At the beginning of the 21st century, scientists and engineers began researching ways to utilize the lignocellulose to prepare chemicals and fuels. The cellulose and hemicellulose, with relatively simple structures and weak ether bond linkages, were firstly extensively studied and transformed into different biomass-based chemicals and materials. As a comparison, the lignin in lignocellulose with a relatively complex structure and robust linkages has usually been pre-removed or left behind. Then, in 2010–2013, lignin depolymerization (LDP) to aromatics became an attractive approach for lignin valorization. Methods included acid/base catalytic hydrolysis, oxidation, reduction, and hydrogenation were introduced for the LDP. During this research boom, we initiated our LDP research with hydrogenation methods, which was mainly based on the consideration that the reduction/hydrogenation methods could selectively cleave lignin ether linkages into phenols and partially avoid the phenols condensation to oligomers and provide a higher monomer yield.

Given the facts that the hydrogenation strategy needed extra H_2 and that alcohol can work as a solvent and hydrogen donor, we then carried out the study of protolignin alcoholysis over different hydrogenation catalysts. We found that the Ni/AC catalyst prepared by carbothermic reduction of Ni precursor with active carbon (AC) could efficiently catalyze the cleavage of the C_β–OAr bond in birch lignin with a high yield of aromatic rings.[13] The best selectivity toward monomeric phenols at 200 °C was 90% with a 54% birch lignin conversion in methanol (Scheme 1). After the reaction, the reacted mixture was separated into solid and liquid layers. Due to magnetic attraction, the nickel catalyst can be easily separated by a magnetic bar, indicating its facile separation process. Also, the recycled Ni/AC catalyst could be reused for another 4 times and the lignin conversion remained at around 50% for all use, indicating good stability and reusability of the Ni/AC catalyst.

Then, by monitoring the molecular weight change of the dissolved lignin fragments with MALDI-TOF in a different alcohol solvent, we proposed a new fragmentation–hydrogenolysis mechanism for monomeric phenols generation from the reaction between the solid lignin and solid nickel catalyst. By treating birch powders in methanol without Ni catalyst at 200 °C for 30 min, a typical lignin fragment $m/z = 1511.5$ ($M_w = 1309$, $Mn = 1292$) generated, which was calculated to contain seven phenylpropyl rings (Fig. 3A). The ethanol-soluble lignin had a similar pattern of molecular weight distribution ($M_w = 1406$, $Mn = 1309$) compared with the methanol-soluble lignin. The difference was that M_w of the ethanol-soluble lignin is 97 m/z larger than the methanol-soluble sample, approximately equal to seven $-CH_2-$ groups. The difference suggested that each phenylpropyl unit of the fragment might consist of one alkoxyl group from the solvent in the fragmentation step (Fig. 3A). On the basis of the results, we proposed a fragmentation–hydrogenolysis process for protolignin depolymerization into monomeric

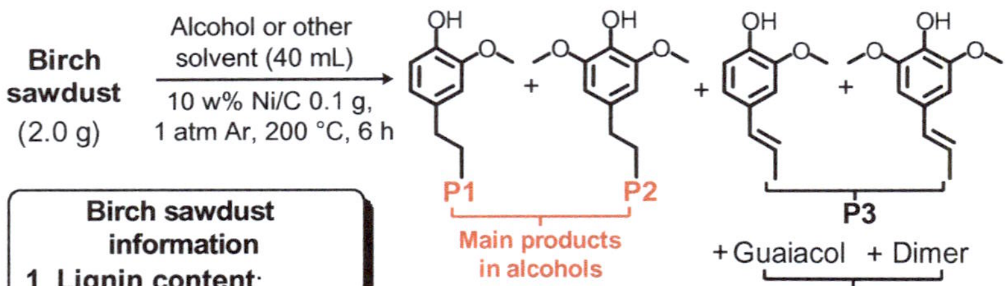

Birch sawdust information

1. **Lignin content**: 19 w% (TAPPI Method)
2. **Ether linkages in birch lignin:** β-O-4 60%; α-O-4 6-8%; 4-O-5 6.5%; Glyceraldehyde-2-aryl ether 2%.
3. **Theoretical yield of phenol monomers from C-OAr bonds cleavage: 74.5~76.5%**

Conversion Results:

Solvent	Lignin Conv. (%)	Products Sel. (%) P1+P2	P3	Others
CH_3OH	54	89	1	10
C_2H_5OH	48	95	3	2
EG	50	97	1	2
Dioxane	15	6	93	<1
Cyclohexane	<1	-	-	-
CH_3OH	62 (220 °C)	80	18	2

$$\text{Lignin Conv.} = \frac{\text{The weight of total detected products}}{\text{The weight of lignin in added birch sawdust}} \times 100\%$$

$$\text{Selec.}(i) = \frac{\text{The intergrated GC area of product } (i)}{\text{The intergrated GC area of all detected product}} \times 100\%$$

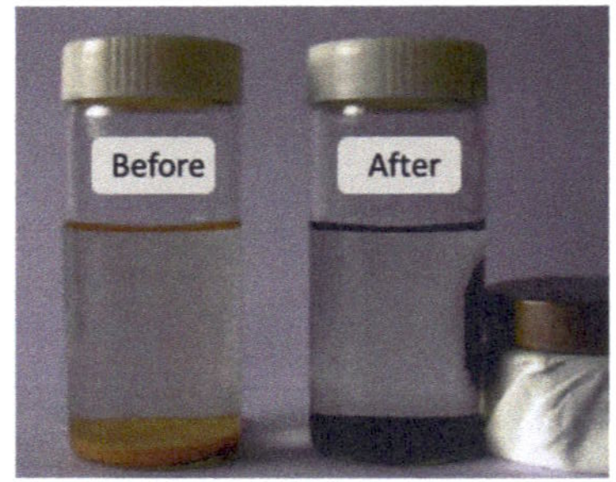

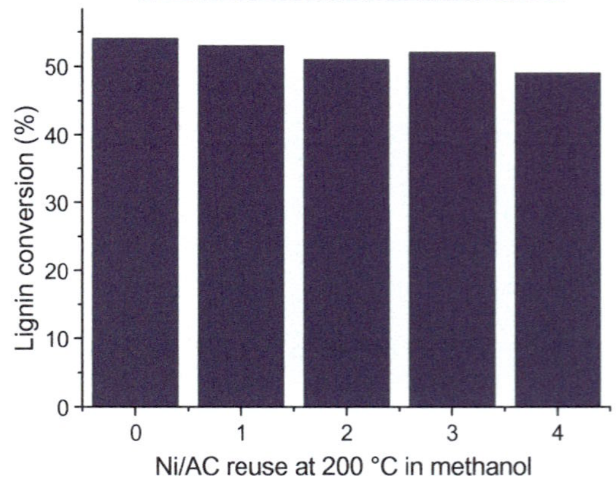

Scheme 1 Lignin depolymerization in methanol with Ni/AC and the reuse experiment in methanol at 200 °C. *The data were abstracted from Song, Q.; Wang, F.; Cai, J.; Wang, Y.; Zhang, J.; Yu, W.; Xu, J.* Energy Environ. Sci. ***2013,*** *6, 994.*

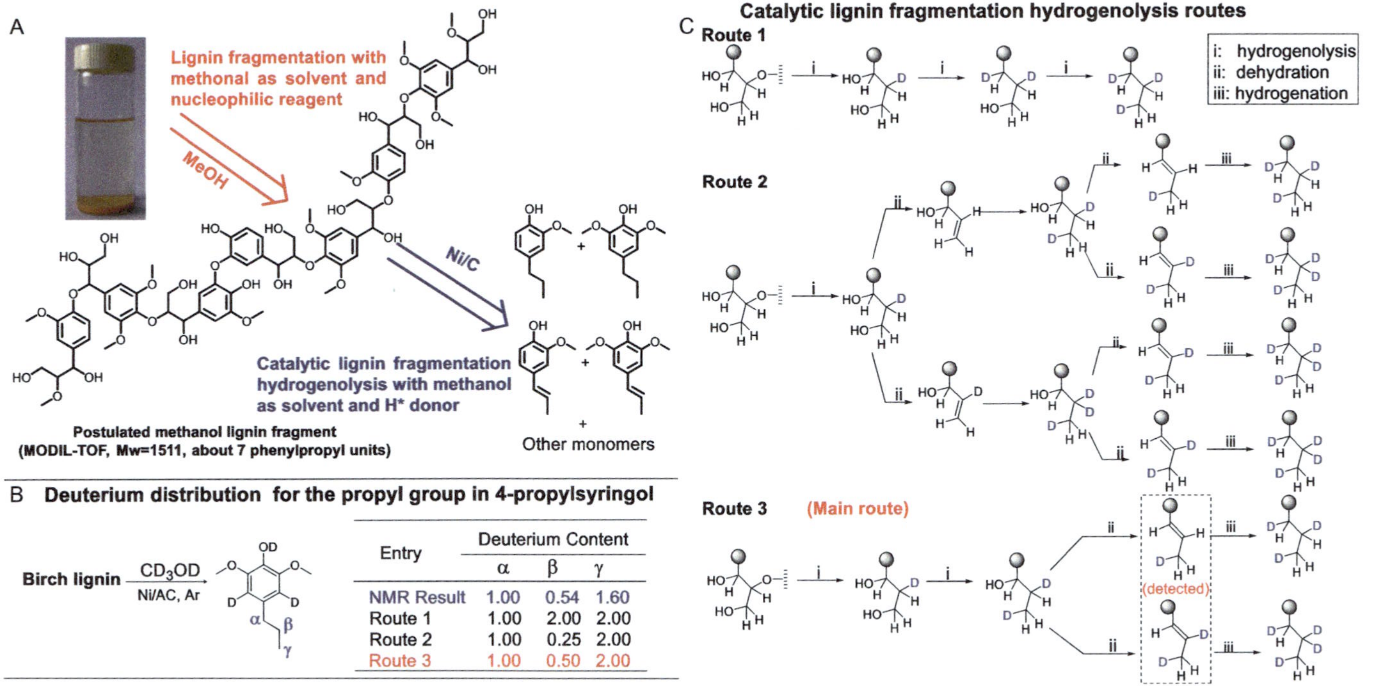

Entry	Deuterium Content α	β	γ
NMR Result	1.00	0.54	1.60
Route 1	1.00	2.00	2.00
Route 2	1.00	0.25	2.00
Route 3	1.00	0.50	2.00

Fig. 3 The proposed fragmentation–hydrogenolysis process with a postulated fragment $m/z=1511$ in methanol (A) and deuterium distribution experiments for determining the main route of further transformation of the soluble lignin fragment (B and C). *The figure was adapted from Song, Q.; Wang, F.; Cai, J.; Wang, Y.; Zhang, J.; Yu, W.; Xu, J.* Energy Environ. Sci. ***2013,*** *6, 994.*

phenols in alcoholic solvents, during which alcohols, such as methanol, ethanol, and ethylene glycol, not only worked as solvent and hydrogen donor, but also acted as the nucleophilic reagent for cleaving ether linkages to achieve the preliminary fragmentation of birch wood into smaller and soluble species, consisting of several aromatic rings. Then, as illustrated by the deuterium distribution experiments (Fig. 3A and C), the unique cooperation between the alcohol solvent and Ni/AC catalyst centers, such as hydrogenation and dehydration sites, could realize the further hydrogenative cleavage of C_β–OAr ether bond and dehydroxylation of C_α–OH and C_γ–OH in the soluble fragments, providing phenylpropane phenols as the final monomer products via Route 3.

It should be also pointed out that the carbothermic reduction with carbon element modifying the Ni particle made the obtained Ni/AC catalyst quite different from other Ni/support, NiM/support catalysts, and noble M/support catalysts (M = Ru, Rh, Pd, Au, Ag, Pt) prepared from H_2 reduction. Even under extra H_2, the Ni/AC prepared from carbothermic reduction could selectively catalyze the cleavage of ether bonds in the lignin β-O-4 model with both high conversion and high yield of aromatic rings (Scheme 2A).[14] Also, the heterogeneous Ni/AC catalyst was robust in catalytic desulphurization without sulfur poisoning, which was a potential and economic catalyst for the hydrogenation depolymerization of lignosulfonate (LS), a technical lignin product from the production of wood pulp using sulfite pulping. Our further studies showed that the heterogeneous Ni/AC catalyst could efficiently catalyze LS hydrogenolysis to phenols without arene hydrogenation.[15] A 91% LS conversion was obtained in ethylene glycol (EG) solvent within 6 h at 200 °C, and the main monomeric products were 4-propyl-guaiacol (PA) and 4-ethyl-guaiacol (EA) (Scheme 2B). On the basis of the mechanistic studies, the activated hydrogen H* over Ni(0) sites in Ni/AC selectively broke caryl–O–calkyl bonds to phenols and methylene structure (–CH_2–), and cleaved hydroxyl groups at side chains to –CH_3 or –CH_2–, producing C_1–C_3 alkane-substituted guaiacols as major products. Furthermore, carbon-sulfur bonds in LS structures could also be reductively cleaved on Ni(0) sites to –CH_2–, and the generated metastable nickel sulfides on Ni/AC surface could be further reduced with active H* to H_2S gas and active Ni(0) metal center for the next catalytic cycle.

During the above process, the promotion effects of alcohol solvent for the lignin depolymerization is as important as the Ni-based catalyst. As a traditional knowledge point, the protonic polar solvents (alcohols) exhibit a higher dissolution capacity than do nonpolar solvents, because the solvents

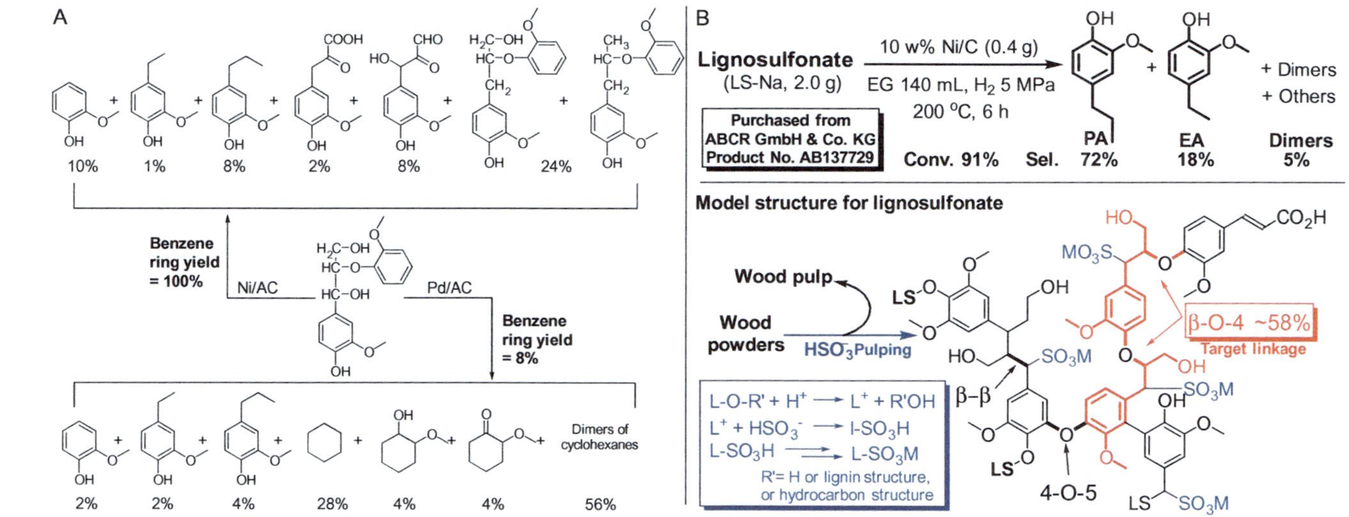

Scheme 2 The conversion of lignin β-O-4 model compound over Ni/AC and Pd/AC (A) and catalytic lignosulfonate hydrogenation to phenols over Ni/AC (B).

with more hydroxyl groups are better for lignin dissolution. The hydroxyl groups can weaken or replace hydrogen bonds to promote lignin release from lignocellulose, and etherification of the lignin fragment with alcohol can further increase the dissolution capacity. In addition, alcohol solvents can dissolve more H_2, lower H_2S concentration, and can work as the hydrogen donor to promote lignin fragments reduction and catalyst in situ regeneration (NiS_x to Ni over the Ni/AC catalyst). Besides, the detailed fragmentation studies in alcohol solvents can provide some guidance in the efficient preparation of the standard lignin fragments with narrow mass distribution and simpler structure than the protolignin. Furthermore, understanding the fragmentation–hydrogenolysis process can provide a solution for the separation difficulty between heterogeneous catalyst particles without magnetic property and carbohydrate solid residue after lignin conversion, such as separating the solvent fragmentation and following catalytic conversion into two processes or trapping the solid catalysts in a basket with strainers.

3.2 Promoting lignin depolymerization by weakening the target linkage C—O and C—C bonds via initial C_αH-OH/$C_\gamma H_2$-OH transformation to carbonyl structure

As demonstrated above, solid lignin polymer can be first dissolved and then be ruptured into smaller soluble fragments via solvolysis. The following key issue is to explore efficient methods to further transform lignin fragments into aromatic monomers. In addition to studying direct lignin conversion with complex or unclear process analysis, another "bottom-up" approach, focusing on the mechanistic study of typical C—O or C—C bond cleavage in lignin models on the molecular level, has presented many effective catalytic systems and strategies in lignin conversion. Among these strategies, the direct idea to efficiently cleave lignin C—O and C—C bond linkages are those which focus on weakening the target linkage C—O and C—C bonds.

In 2010, the Bergman group[16] developed a ruthenium-catalyzed redox neutral method to selectively cleave the C_β–OAr bond in 2-aryloxy-1-arylethanols and β-O-4 polymer to $ArCOCH_3$ and ArOH. The reaction began with a well-known Ru-catalyzed dehydrogenative equilibrium between benzylic alcohol and the corresponding aryl ketone without C_β–OAr bond cleavage. The catalyst Ru^0 center then inserted into the C_β–OAr bond to give a Ru-alkoxide, and reductive elimination with activated H_2 at the Ru center provided aryl ketone and phenol. The net result is C_β–OAr bond hydrogenolysis. Subsequently, the Stephenson group[17] showed that the photo-catalyst $[Ir(ppy)_2(dtbbpy)]PF_6$ could achieve the reductive cleavage of C_β–OAr bond within β-O-4-ketones with a C_α=O, and the generated fragment could be reduced by HCOOH. Westwood and co-workers[18] reported that the pre-oxidized β-O-4 linkages prepared from chemo-selective oxidation of C_αH–OH to C_α=O with DDQ/^{t}BuONO/O_2 system could be cleaved by Zn/NH_4Cl. As shown in Fig. 2, the key point for these strategies is that the first oxidation of C_α–OH to C_α=O can reduce the BDE of C_β–OAr bond by about 90 kJ mol^{-1}. Therefore, the C_β–OAr bond coupled to C_α=O can be cleaved more easily, and the resulting ArCOCHR• (R=H or CH_2OH) and phenolic radical can be further reduced to ketone and phenol.

Given the final lignin utilization target is to prepare aromatic chemicals that can compete economically with the current routes from fossil resources, the depolymerization process costs need to be considered. Therefore, low-cost and/or robust catalysts are a better choice. From this viewpoint, compared with homogeneous catalysts, especially with a noble metal center, heterogeneous catalysts have some relative advantages in lignin depolymerization utilization. We continued our LDP research with heterogeneous catalysts by first testing the catalytic performance of the Pd/C in C_β–OAr bond cleavage within β-O-4 lignol via the self-hydrogen transfer process.[19] We further performed density functional theory (DFT) calculations and experiments to investigate the cleavage mechanism of C_β–OAr bond in 2-phenoxy-1-phenylethanol over Pd(111) surface. These calculations confirmed the dehydrogenation–hydrogenation process, which was different from the reported mechanism of homogenous Ru-based catalysts[20] and the mechanism of heterogeneous Pd-based catalysts.[21–23] We found the favorable reaction pathway carried out on Pd(111) as illustrated in Scheme 3: (i) the β-O-4 lignol **2a** was dehydrogenated first to generate its corresponding ketone **2e** via a carbon radical intermediate **2c**; (ii) **2e** continued to get dehydrogenated on the C_β to **2h** by a first equilibrated keto-enol tautomerization to its enol form **2f** and then –OH dehydrogenation; (iii) the cleavage of C_β–OAr ether bond in intermediate **2h** happened afterward, and the intermediates **1j** and **1d** were hydrogenated to ketone and phenol.

Although the self-hydrogen transfer process is an atom economical method for β-O-4 lignin ether bond cleavage, the reaction temperature of the reported thermal catalytic process is always above 170 °C, which can cause lignin fragments recondensation and over-hydrogenation process with the active noble metal catalyst. To avoid high reaction temperatures, the photocatalytic strategies at a mild condition should be a good choice to explore. Besides the thermal Pd-catalyzed self-hydrogen transfer

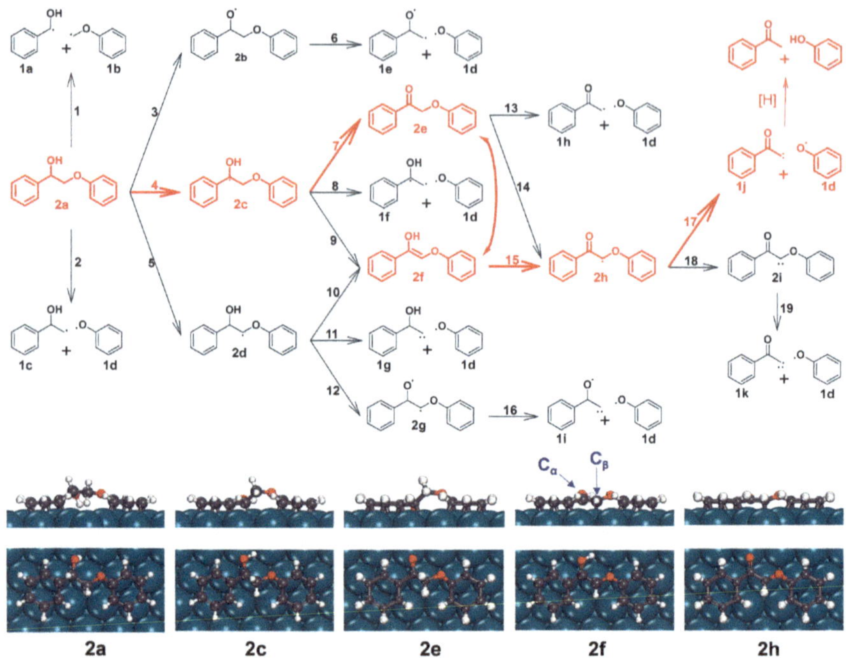

Scheme 3 Reaction network for the β-O-4 bond cleavage of 2-phenoxy-1-phenylethanol over Pd (111) surface and side (upper panel) and top views (lower panel) of the adsorption structures of surface intermediates. *The figure was adapted from Lu, J.; Wang, M.; Zhang, X.; Heyden, A.; Wang, F.* ACS Catal. ***2016,*** 6, *5589.*

hydrogenolysis process, we then reported a photocatalytic protocol with a heterogeneous $ZnIn_2S_4$ catalyst (Scheme 4).[24] In this process, alcoholic groups ($C_\alpha H$–OH) of lignin acted as hydrogen donors. The proliferation of phenolic products to dark substances was suppressed under visible light illumination (450 nm LED) at low temperature (below 50 °C); the formation of a light and transparent reaction solution allowed visible light to be absorbed by the catalyst $ZnIn_2S_4$. With this strategy, 71%–91% yields of phenols in the conversion of lignin β-O-4 models were achieved. Mechanistic studies revealed that the $C_\alpha H$–OH group of β-O-4 linkage was initially dehydrogenated on $ZnIn_2S_4$ holes (h^+) to a ketone intermediate without C_β–OAr bond cleavage. The electron and proton formed a "hydrogen pool" with active H* rather than H_2 at the CB structure, then the C_β–OAr bond of ketone intermediate was subsequently cleaved into two monomers by active H*. Thus, the dehydrogenation and hydrogenolysis reactions are integrated into one-pot with lignin itself as a hydrogen donor. This study shows a promising way of supplying phenolic compounds by

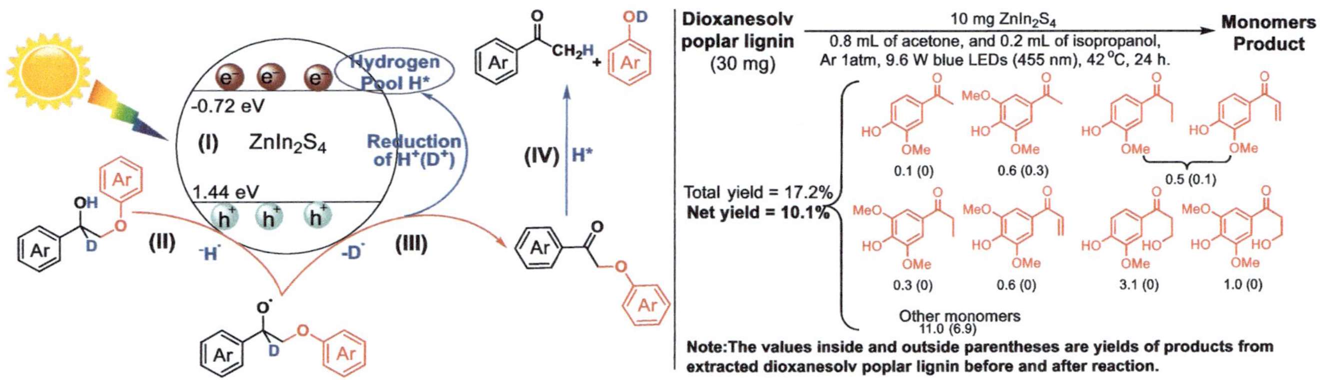

Scheme 4 Photocatalytic self-hydrogen transfer hydrogenolysis of lignin fragmentation. *The figure was adapted from Luo, N.C.; Wang, M.; Li, H.J.; Zhang, J.; Hou, T.T.; Chen, H.J.; Zhang, X.C.; Lu, J.M.; Wang, F.* ACS Catal. ***2017,*** *7, 4571.*

taking advantage of both renewable biomass feedstocks and photo-energy. Where after, $ZnIn_2S_4$ was used to depolymerize dioxanesolv poplar lignin under the photocatalytic transfer hydrogenolysis strategy, providing a ~10% aromatic monomers yield.

Given the first oxidation of C_α–OH to C_α=O can reduce the BDE of C_β–OAr bond, the semiconductor, with appropriate conduction band (CB) potential, could achieve the reductive cleavage of the pre-oxidized β-O-4 linkage more easily with some sacrificial agent at valence band (VB) under a mild condition. After catalysts screening, besides the $ZnIn_2S_4$, we found TiO_2 with abundant surface Ti^{3+} could efficiently catalyze the photo-cleavage of C_β–OAr bond in β-O-4 lignin ketones (Scheme 5A).[25] The oxidation–hydrogenolysis tandem reaction could be accomplished in one-pot to offer ketones and phenols via a dual light wavelength switching strategy from lignin models. With this strategy, 93%–99% yields of phenols in the conversion of lignin β-O-4-ketone models were achieved. In detail, the overall conversion of the β-O-4 model to monomer products began with photo-oxidation of C_α–OH within β-O-4 linkage to C_α=O with 455 nm light over a Pd/$ZnIn_2S_4$ catalyst with O_2. Then, the 365 nm illumination induced the generation of holes (h^+) and electrons (e^-) on TiO_2. Holes of valence band oxidized sacrificial reagent ethanol while electrons reduced surface Ti^{4+} to Ti^{3+}. EPR characterizations indicated the adsorption of β-O-4 ketone to the in situ formed Ti^{3+} weakened the C_β–OAr bond and finally resulted in the ether bond cleavage to phenol and a ketone.

To accelerate the photocatalytic reductive cleavage of the C_β–OAr bond within pre-oxidized β-O-4 linkage, we then loaded some noble metal on the surface TiO_2, targeting at cleaving the C_β–OAr bond with active H* which generates from the combination of H^+ and photo-induced CB electron of the semiconductor. Besides the catalytic cleavage of the C_β–OAr bond to ketones and phenols, our further tests with an optimized Pd/TiO_2 system could efficiently catalyze the photocatalytic deoxygenation of aromatic ketones to alkyl arenes at room temperature,[26] by utilizing alcohols as green hydrogen donors, and no other additive was needed. The absence of hydrogen evolution during this transformation suggested a mechanism of direct hydrogen transfer from alcohol. Control experiments with additives indicated the role of acid sites in this transfer hydrogenolysis, and further characterization indicated a combination of hydrogen transfer sites and acidic sites were responsible for the efficient deoxygenation. More specifically (Scheme 5B), under irradiation of 365 nm light, the photogenerated holes on TiO_2 oxidized the ethanol to acetaldehyde and proton.

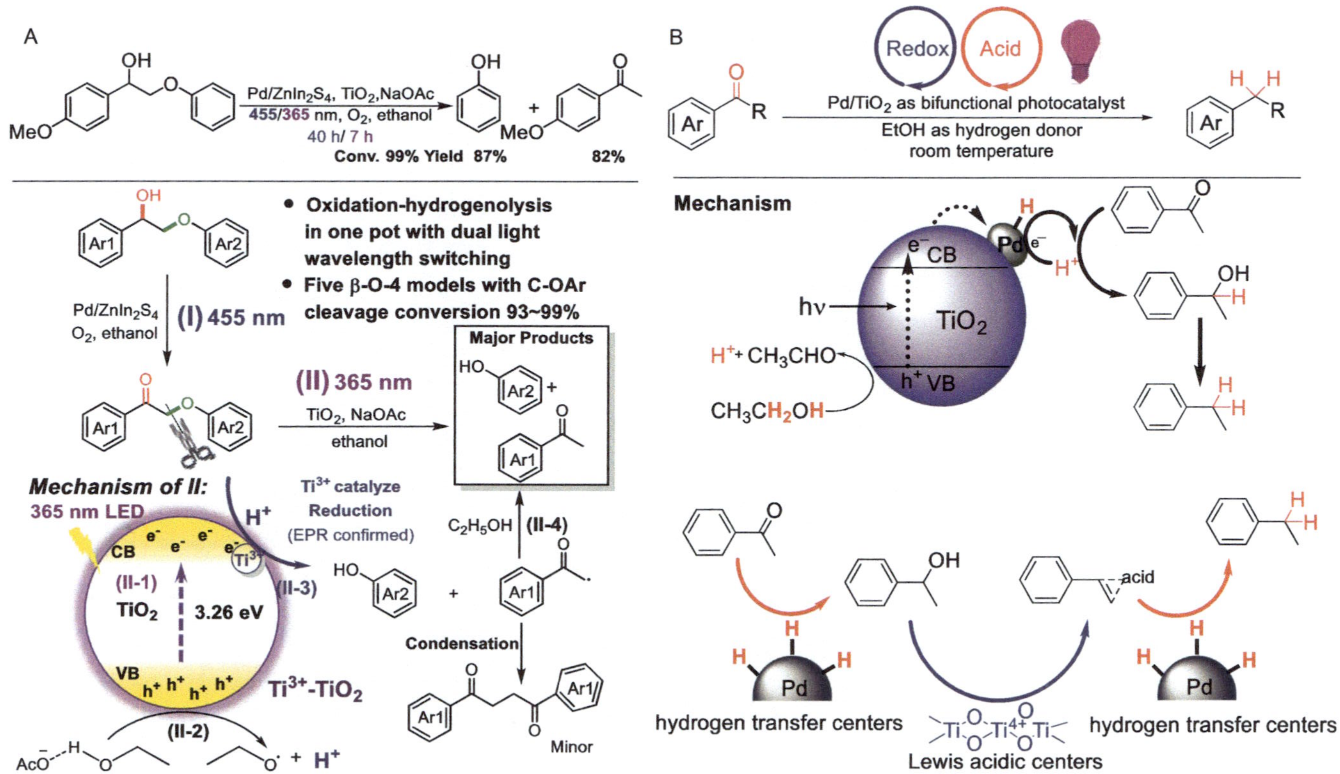

Scheme 5 Photocatalytic C_{β}–OAr bond cleavage of β-O-4 ketones over the Ti^{3+}-TiO_2 (A) and further deoxygenation of aromatic ketones to alkyl arenes over the Pd/TiO_2. *The figure was adapted from Luo, N.C.; Wang, M.; Li, H.J.; Zhang, J.; Liu, H.F.; Wang, F.* ACS Catal. ***2016,*** 6, *7716; Li, H.; Gao, Z.; Lei, L.; Liu, H.; Han, J.; Hong, F.; Luo, N.; Wang, F.* Green Chem. ***2020,*** 22, *3802.*

The photo-generated electrons on TiO_2 migrated to Pd^0 particles and reduced the proton to activated hydrogen species (Pd—H). Then, the ketone substrate was reduced to an alcohol intermediate by Pd—H. In the presence of abundant Lewis acidic sites on TiO_2, the alcohol intermediate was dehydrated and further hydrogenated to ethylbenzene by Pd—H.

Some lignin ether linkages (α-O-4, 4-O-5, etc.) do not have an adjacent hydroxyl group to provide the hydrogen for bond cleavage via the self-hydrogen transfer hydrogenolysis; however, converting the protolignin via the oxidation–hydrogenation strategy is still a feasible method. As we discussed above, the first C_αH–OH oxidation in β-O-4 linkages not only weakens the target ether bonds, it also restrains the lignin fragments condensations caused by the C_α carbocation intermediate to generate oligomers with new C—C bonds (Fig. 4A–D).[10,11,27–29] To achieve this idea, the following hydrogenation system should prefer to catalyze the cleavage of C_β–OAr rather than hydrogenation of the C_α=O bond back to alcohol or of the aromatic rings. After the catalyst screening, we found $NiMoS_x$, prepared by calcination of the precipitate of the reaction between $(NH_4)_2MoS_4$ and $NiCl_2$, was an appropriate hydrogenation catalyst for the selective hydrogenation of pre-oxidized β-O-4 linkage with a C_α=O.[30] On the basis of the XRD characterization, MoS_2, NiS, and $Ni_{0.7}Mo_3S_4$ phases existed in the $NiMoS_x$ catalyst. We further tested the catalytic performances of MoS_2 and NiS. When MoS_2 was used as the catalyst for the hydrogenation of β-O-4-ketone, the main products were target aromatic products with C_β–OAr bond cleavage (71%). A part of the β-O-4-ketone was transformed to β-O-4-alcohol (7%) with a C_αH–OH but without C_β–OAr bond cleavage, which could further undergo hydrogenolysis to β-O-4-H (18%) with a $C_\alpha H_2$, and etherification to β-O-4-E with a C_αH–OR (3%). When NiS was used as the catalyst, the β-O-4-ketone conversion was $<2\%$. The low activity indicated that the NiS_x species might only be the catalytic promoter to the $NiMoS_x$ catalyst. DFT calculations and further characterizations were then carried out to check the reason. DFT calculations of β-O-4 compound adsorption models (Scheme 6E–G) indicated the hydrogenative Mo sites on MoS_2 were in a linear arrangement at the edge of this 2D-layered material. Besides the potential of Mo sites to cleave C_β–OAr bonds, few Mo sites were working as the hydrogenation centers around the aromatic ring, which was different from the face-face adsorption models over bulk Pd(111) and Ni(111). Given that $Ni_{0.7}Mo_3S_4$ has a cubane-type Mo cluster structure similar to MoS_2, if its edge Mo atoms are partly replaced by the low active Ni^{2+}, the potential side reaction sites of the hydrogenation of C_α=O can be reduced,

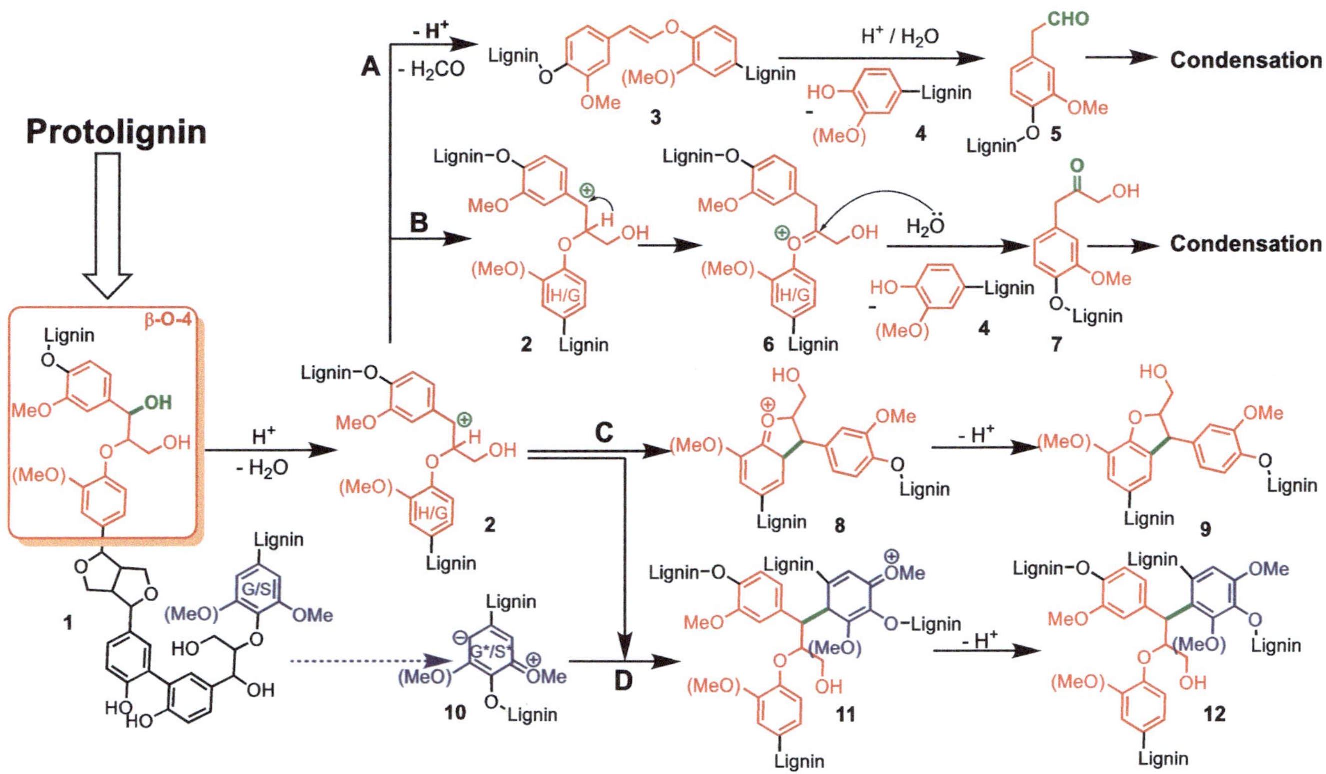

Fig. 4 Lignin fragments condensations or new stable C—C bond generation caused by C_α carbocation. *The figure was adapted from Zhang, C.F.; Li, H.J.; Lu, J.M.; Zhang, X.C.; MacArthur, K.E.; Heggen, M.; Wang, F.* ACS Catal. ***2017,*** *7, 3419.*

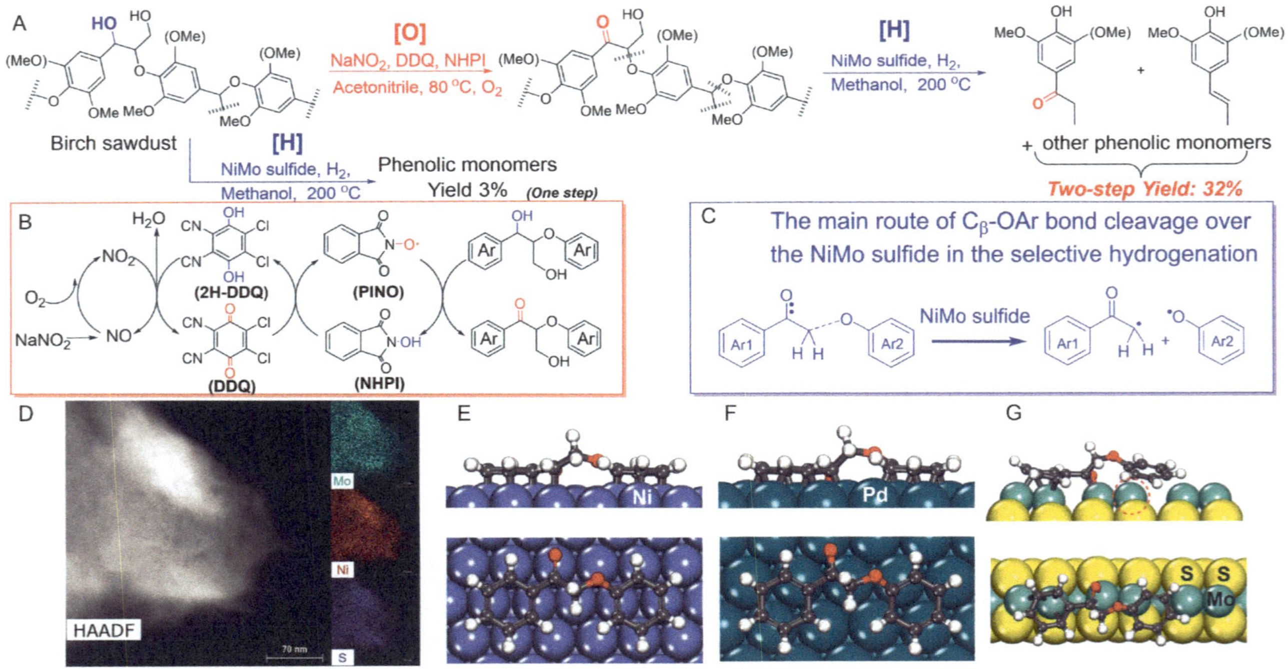

Scheme 6 The conversion of birch sawdust to aromatic monomers via an oxidation–hydrogenation strategy (A–C); DFT adsorption models of $PhCOCH_2OPh$ on Pd(111), Ni(111), and MoS_2 (D–G). *The figure was adapted from Zhang, C.F.; Li, H.J.; Lu, J.M.; Zhang, X.C.; MacArthur, K.E.; Heggen, M.; Wang, F.* ACS Catal. ***2017,*** *7, 3419.*

and the yield of aromatic products generated from the cleavage of C_β–OAr bond can increase. As shown in Scheme 6D, the EDX images show that Ni and Mo atoms are equally distributed at some zones of $NiMoS_x$, which might indicate the hydrogenative Mo sites in NiMo sulfide were not in a good linear arrangement like MoS_2 but interrupted by the Ni^{2+}. Therefore, by partly replacing edge Mo atoms with NiS structures having low hydrogenation activity, the $NiMoS_x$ could further restrain the hydrogenation of C_α=O and aromatic rings while improving C_β–OAr bond cleavage, leading to an increase of monomers yield (Scheme 6C). Coordinating this $NiMoS_x$ hydrogenation system with an O_2/$NaNO_2$/DDQ/NHPI system to selectively oxidize β-O-4 C_αH–OH to C_α=O (Scheme 6B),[31] we achieved the birch sawdust depolymerization with a 32 wt% phenolic monomers yield, which was 10 times higher than the direct hydrogenation (Scheme 6A).

Given the effect of hydrogenation catalyst modification by heteroatoms, we then prepared a carbon modified Ni catalyst from the carbothermal reduction of Ni-doped layered double hydroxides (NiMgAl-LDH) with inserted lignosulfonate (LS),[32] during which an LDH-LS composite was transformed into a MgAl-oxide sheet, and Ni cations were reduced by the carbon to well-dispersed nanoparticles (Fig. 5A). Further XPS, EXAFS, and DFT characterization showed that the high-temperature carbothermal reduction could facilitate a small number of carbon atoms residing in the Ni surface. This greatly modified the nickel 3d electronic structure, giving a Ni catalyst (Ni/MgAlO-C) that preferentially catalyzes hydrogenolysis of C_β–OAr bonds in pre-oxidized lignin models rather than the hydrogenation of the C_α=O bonds or aromatic rings. This strategy was then focused on real lignin as a substrate (Fig. 5B). Birch lignin with 43% content of β-O-4 linkages was first oxidized by 2,3-dichloro-5,6-dicyano-1,4-benzoquinone (DDQ)/^{t}BuONO. Then, the oxidized birch lignin was further subjected to hydrogenolysis over the carbothermal reduced Ni/MgAlO-C and over a control catalyst Ni/MgAlO, which was obtained by further treating Ni/MgAlO-C in the H_2 flow to remove the carbon residing in the Ni nanoparticles. With Ni/MgAlO as the catalyst, the yield of aromatic monomers was only 5 wt%. Comparatively, the enhanced performance was observed with the Ni/MgAlO-C catalyst, which gave a mixture of aromatic monomers in a 22 wt% yield.

As discussed above, the hydrogen transfer, self-hydrogenation transfer, and oxidation–hydrogenation strategies can reduce the BDE of the target C_β–OAr bonds by initial selective oxidation of the C_α–OH to C_α=O. In order to avoid the C_α=O hydrogenation back to C_α–OH,

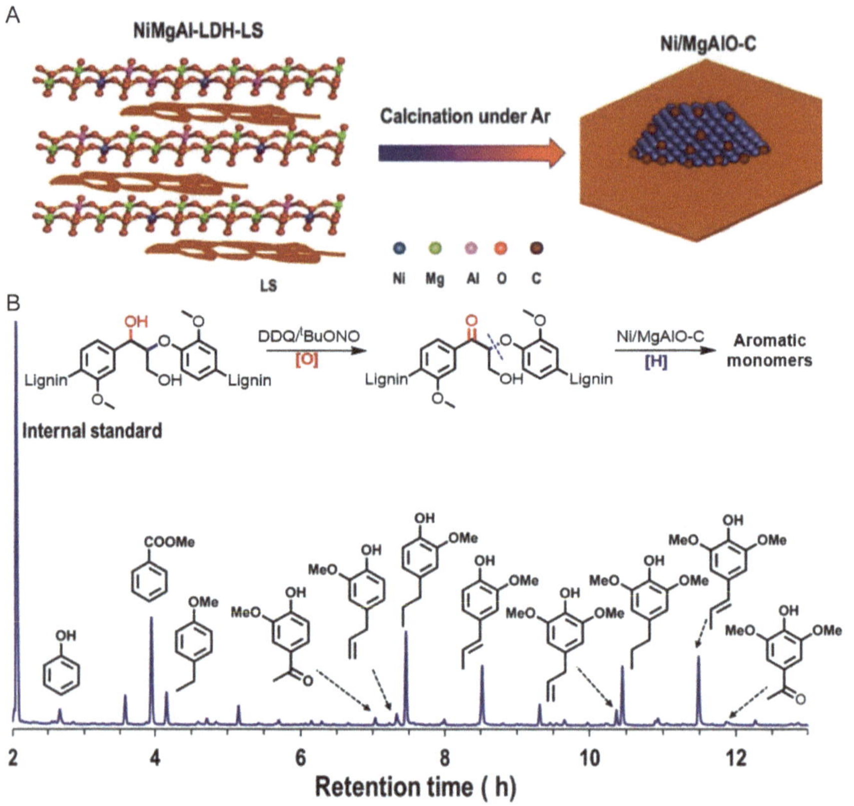

Fig. 5 Carbon modification of nickel catalyst for depolymerization of oxidized lignin to aromatics. *The figure was adapted from Wang, M.; Zhang, X.; Li, H.; Lu, J.; Liu, M.; Wang, F.* ACS Catal. ***2018,*** 8, *1614.*

a weakened or poisoned catalyst, not an active hydrogenation catalyst like Pd/C was employed in the second step for $C_β$–OAr bond reductive cleavage. This approach made the lignin hydrogenolysis more selective at a mild condition. Furthermore, the $C_α$–OH oxidation apparently restrains the lignin fragment recondensation caused by the $C_α$ carbocation, which has been discussed in Fig. 4. Although the strategies involving initial $C_α$H–OH etherification with formaldehyde or alcohol[10,33–35] or $C_α$H–OH transformation to $C_αH_2$ can also restrain the generation of $C_α$ carbocation, the $C_β$–OAr bond in the modified structure is a relatively robust bond. The direct cleavage of the $C_β$–OAr bond, connecting with a $C_α$H–OR or $C_αH_2$, need active catalysts like Ru/C or Pd/C not over a weakened catalyst at a high temperature. Therefore, the oxidation (dehydrogenation)–hydrogenation strategy, containing first C_αH–OH transformation to C_α=O, should also be an efficient and elaborate method. Furthermore, the first C_α–OH oxidation to C_α=O in the β-O-4 linkage can change the substrate reaction site. In 2014, the Stahl group reported a modified lignin hydrolysis strategy,[36] the pre-oxidation of C_α–OH to C_α=O made the intermediate C_γ=C_β–OAr could replace the traditional intermediate C_α=C_β–OAr for hydrolysis, which could restrain the recondensation by avoiding the generation of C_α carbocations, leading to a higher aromatic monomers yield.

Another way to promote the lignin decomposition by decreasing BDE of the target lignin bonds is the selective dehydrogenation or oxidation of –$C_\gamma H_2$OH to aldehyde –C_γHO[37–40] or ester –C_γOOR[41] while keeping the structure of C_α–OH in a β-O-4 or β-1 linkage (Fig. 2). This can weaken the C_α–C_β, inducing the generation of corresponding aromatic aldehydes via retrograde aldol condensation.

3.3 Promoting lignin depolymerization by weakening the target linkage C—O and C—C bonds via fragment intermediate analysis

The above methods just regard the molecule as the key intermediate for the lignin depolymerization. Given that lignin conversion always leads to fragments (radical, cation, or anion), analyzing the pathways from substrate molecules to a potential fragment intermediate may lead to more efficient strategies.

In 2016, we carried out the hydrogenation study of lignin β-O-4 alcohols models over a NiMo sulfide catalyst, which gave unique styrene and ether as the products from cleavage of the C_β–OAr bond.[42] In addition, the competitive conversion of β-O-4 model compounds with C_αH–OH and C_α=O group in a single reactor showed that NiMo sulfide preferred to catalyze the hydrogenative cleavage of C_β–OAr bond connected with a C_αH–OH structure rather than catalyze the cleavage of the weaker C_β–OAr bond connected with a C_α=O structure. This observation is contrary to the relative BDEs and indicated a different reaction mechanism than the above studies. In this case, we proposed an efficient dehydroxylation–hydrogenation strategy for the C_β–OAr bond cleavage (Scheme 7), which did not proceed via C_β–OAr bond direct cleavage or C_α–OH dehydrogenation to C_α=O before the subsequent transformation. The acid site on the NiMo catalyst played a critical role in the lignin model hydrogenolysis cleavage but showed low efficiency in lignin model hydrolysis cleavage without H_2. Further control experiments with radical trap with TEMPO

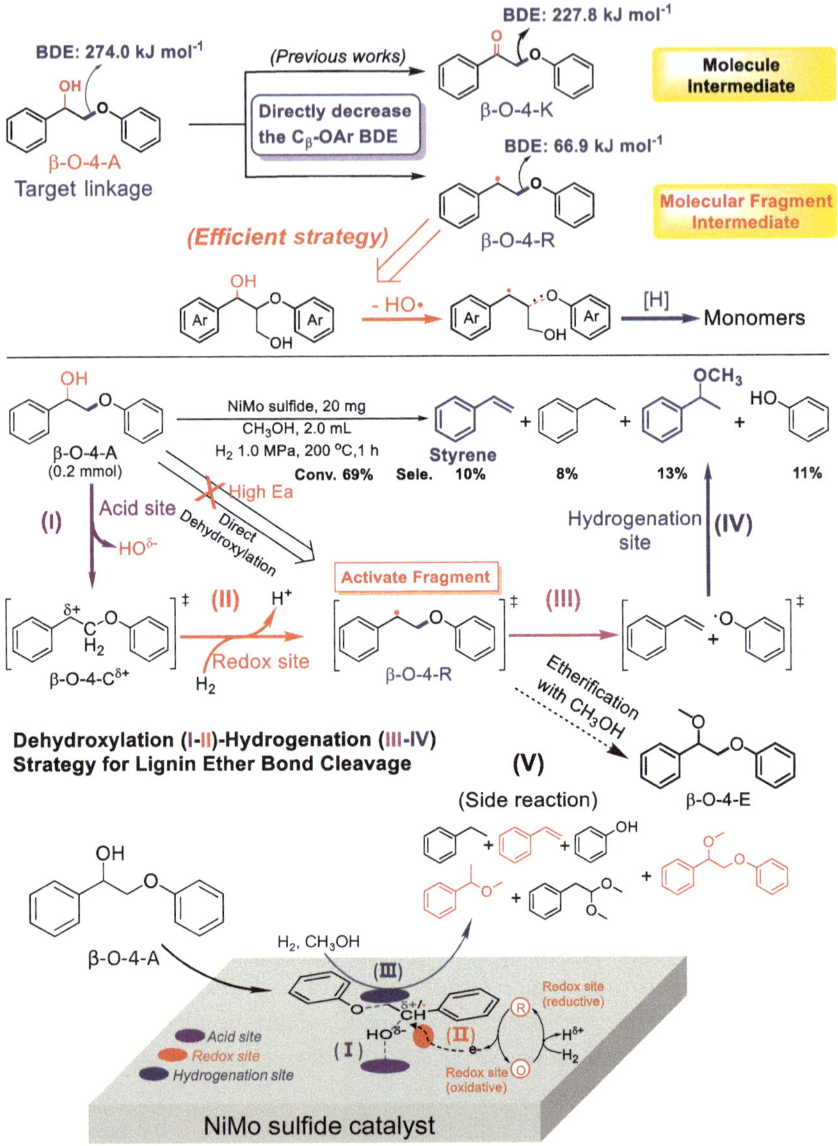

Scheme 7 The dehydroxylation–hydrogenation strategy for β-O-4 ether bond cleavage. *The figure was adapted from Zhang, C.; Lu, J.; Zhang, X.; MacArthur, K.; Heggen, M.; Li, H.; Wang, F.* Green Chem. ***2016,*** 18, *6545.*

(2,2,6,6-tetramethyl-1-piperidinyloxy) suggested that a critical radical intermediate was involved in the process. Further mechanistic study, including the generation analysis of styrene and ether products, and DFT calculations, suggested that the β-O-4-A model PhCH(*OH*)CH_2OPh first lost a hydroxyl group (C_α–OH) at medium strong acid sites of NiMo catalyst. Then the generated $PhCH^{\delta+}CH_2OPh$ transformed to $PhCH{\bullet}CH_2OPh$ (β-O-4-R) by obtaining an electron from the NiMo catalyst in a redox cycle. The DFT calculations suggested that dehydroxylation of C_α–OH to a β-O-4-R radical would greatly decrease the C_β–OAr BDE from 274.0 to 66.9 kJ mol^{-1}. This would facilitate C_β–OAr cleavage, leading to styrene, ethylbenzene, phenol, and various ethers as product under a hydrogenation condition in methanol. The cooperation between the acid-redox sites and hydrogenation sites played a crucial role in this proposed route.

Wang and co-workers[43–45] have also reported some photocatalytic lignin transformation methods with radical intermediates. In one of these, the photo-catalysts CdS or $Zn_4In_2S_7$ would selectively abstract the H from the carbon of C_αH–OH and generate a C_α carbon radical ($PhC{\bullet}(OH)CH_2OPh$) (Scheme 8B). The result is a decreased BDE of the C_β–OAr bond from 183.9 to 32.6 kJ mol^{-1} and induce the self-cleavage of C_β–OAr bond (Scheme 8B). Furthermore, before the above methods with a C_α carbon radical intermediate, Toste and co-workers[46] reported a (V^V=O)-catalyzed non-oxidative mechanism for β-O-4 C_β–OAr bond cleavage under N_2, during which the V^V=O could abstract the H from C_α carbon of the coordinate β-O-4 alkoxide via a V^V–OC_α structure (Scheme 8D upper, M = (V^V=O)). Then, the generated C_α carbon radical induced the cleavage of the C_β–OAr bond. Besides the fragment intermediate about the C_α carbon radical, Goldman and co-workers[47] reported an iridium-catalyzed dehydroaryloxylation mechanism for the C_β–OAr cleavage, during which the Ir center can directly and selectively insert into the C_α–H bond of $Ar_1CH_2CH_2$–OAr_2 to a C_α–M intermediate (Scheme 8C, M = Ir), which further induced the cleavage of C_β–OAr bond to give the phenol Ar_2OH and the olefin $Ar_1CH{=}CH_2$.

The generation of the other radicals or fragments could also promote the cleavage of the β-O-4 linkage. In 2010, Baker and co-workers[48] reported a different (V^V=O)-catalyzed oxidative mechanism for lignin models conversion. A similar intermediate with a V^V–OC_α structure which was involved in the (V^V=O)-catalyzed mechanism reported by Toste[46] could induce the direct cleavage of the C_α–C_β bond under O_2 condition. Based on product analysis, the single electron transferred between the V^V and oxygen in V^V–OC_α produced an active $ArCH(O{\bullet})CH_2OAr$ (Scheme 8D under, M = (V^V=O)), which further induced the selective C_α–C_β bond cleavage. Subsequently, photo-catalysis was introduced in the (V^V=O)-catalyzed mechanism, Soo and co-workers[49] reported that the visible light irradiation could trigger the single electron transfer between the substrate

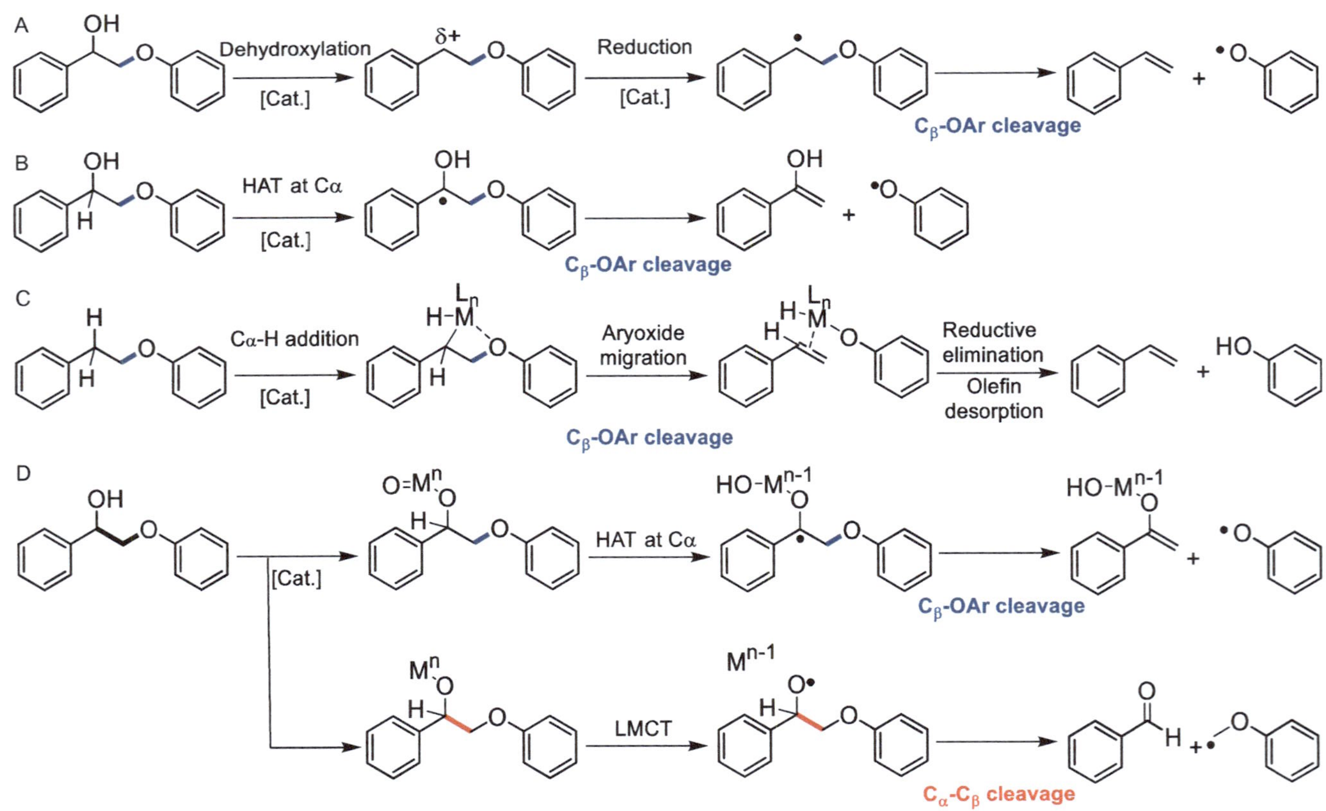

Scheme 8 The lignin depolymerization strategies with β-O-4 fragments generated from C_α–OH firstly activation as the critical intermediate for C_β–OAr and C_α–C_β bonds cleavage.

and the catalyst with a V^V–OC_α structure, leading to an increase in the monomers generated from the decomposition of β-O-4 intermediate ArCH(O•)CH_2OAr (Scheme 8D under, M=(V^V=O)). Similarly, using vanadium catalysts VO(O^iPr)$_3$ under visible light, we recently[50] achieved the selective cleavage of C_α–C_β bonds in β-1 and β-O-4 interlinkages in lignin models and extracts by an oxidative protocol (Scheme 8D under, M=(V^V=O)). For the generation of radical with C_α–O• in β-1 and β-O-4 linkages, besides the V^V species, Zhang and co-workers[51] reported a Ce(III)-catalyzed redox mechanism with DBAD (di-tert-butyl azodicarboxylate) as the oxidant. The $Ce^{III}Cl_n$ firstly coordinated with the lignin model through the deprotonation of C_αO–H, which was further oxidized by DBAD to a $Ce^{IV}Cl_n$/lignin species with a Ce^{IV}–OC_α structure (Scheme 8D, M=Ce^{IV}). The homolysis of $Ce^{IV}Cl_n$/lignin species (Ce^{IV}–OC_α), caused by the photo-induced ligand-to-metal charge-transfer mechanism (LMCT), afforded the key alkoxy radical intermediate ArCH(O•)CH_2OAr along with the release of $Ce^{III}Cl_3$ for the next catalytic cycle. The ArCH(O•)CH_2OAr further decomposed to ArCHO and ArOCH_2•. The ArOCH_2• radical in the Ce-catalyzed process was quenched by the DBAD or DBAD −• species and provided N-containing aromatics.

3.4 Promoting lignin polymer depolymerization by introducing a cleavage reagent at the lignin "substrate site"

Another efficient and selective strategy to catalyze the depolymerization of the lignin is to generate a new "substrate site" for introducing cleavage reagent on the modified lignin linkages for further depolymerization. To explain this idea, consider cutting a robust tree stump into pieces as an example. Compared with the direct splitting the stump with an ax, the method of inserting a wooden wedge into the first made gap in the stump is a little tedious but less effort. In this aspect used in the lignin transformation, Stahl and co-workers[52] found that the O_2/AcNH-TEMPO/HCl/HNO_3 system can realize chemo-selective aerobic oxidation of secondary benzylic alcohols within lignin model compounds to ketones. In the subsequent transformation of pre-oxidized β-O-4 linkages, they achieved the selective C_α–C_β bond cleavage with KOH/H_2O_2.[52] Although the first oxidation of the C_α–OH to ketone can increase the BDE of the C_α–C_β bond (Fig. 2), the newly introduced C_α=O can work as the substrate site to introduce the peroxide to achieve the C_α–C_β or C_{Ar}–C_α bonds cleavage via Baeyer-Villiger oxidation or Darkin reaction with following hydrolysis.

Subsequently, several other oxidation-BV strategies were found for cleaving lignin C—C bonds.[53–55] Besides inserting an O atom, the C_α=O can also work as the substrate site for reaction with NH_2OH to achieve the C_α–C_β or C_{Ar}–C_α bonds cleavage via Beckmann rearrangement and subsequent hydrolysis. We have introduced NH_2OH into the depolymerization of oxidized lignin to N-containing monomers. Different from the following works with target amides as the products,[56] we reported a protocol to convert pre-oxidized lignin to isoxazole and aromatic nitrile with NH_2OH salt and MgO (Scheme 9).[57] Mechanistic studies indicated the product difference was caused by the *cis–trans* isomerism of oximation intermediates with different additives. MgO can promote the generation of isoxazole motif by condensation of β-hydroxyl ketone, containing a $C_\gamma H_2OH$, with NH_2OH (Route 1). Aromatic nitriles and esters were obtained via Beckmann

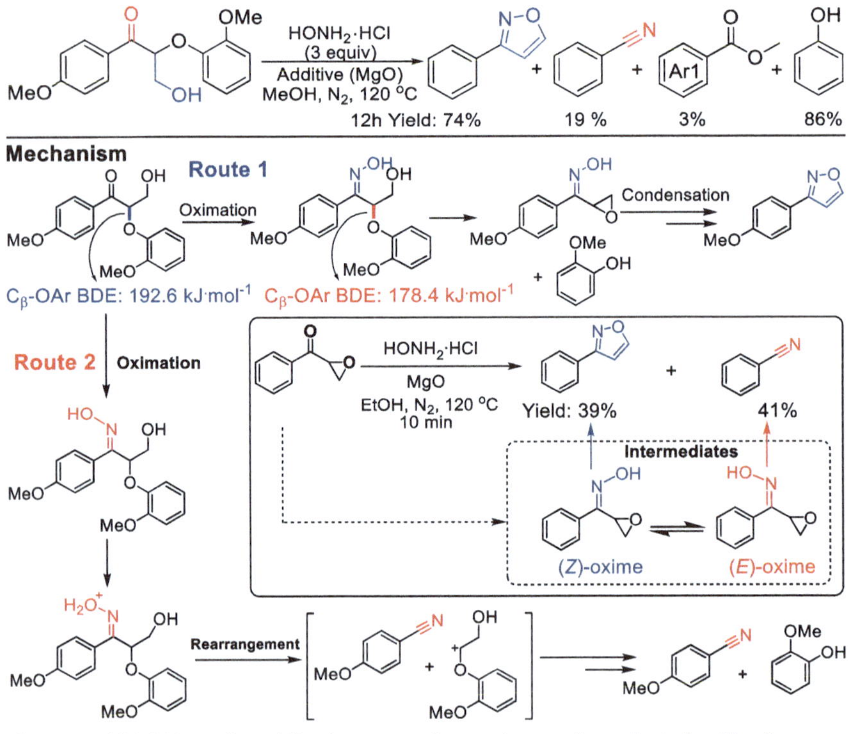

Scheme 9 NH_2OH-mediated lignin conversion to isoxazole and nitrile. *The figure was adapted from Li, H.; Wang, M.; Liu, H.; Luo, N.; Lu, J.; Zhang, C.; Wang, F.* ACS Sustain. Chem. Eng. ***2018,*** 6, *3748.*

rearrangement or an acidolysis reaction depending on the additives (Route 2). Furthermore, this NH_2OH-mediated strategy could provide a 7.6% yield of aromatic isoxazole, nitrile, and ester monomers from pre-oxidized lignin.

Besides cleaving the C—C or Ar—C bonds via inserting an N species, we recently found that introducing an iminyl radical at pre-oxidized lignin C_α could directly cleave the aromatic lignin C_βO–Ar bond (Scheme 10).[58] In detail, we firstly modified the α-alcohol β-aryl ether structure to the α-oximated β-aryl ether structure in lignin. Then, a phenothiazine (PTH), working as the photo-catalyst, could catalyze the decomposition of the α-oximated β-aryl ethers to iminyl radicals. The iminyl radical then induced the aryl migration from O atom to N atom via spiro cyclization. DFT calculations suggested that ortho-methoxy could facilitate the spiro cyclization due

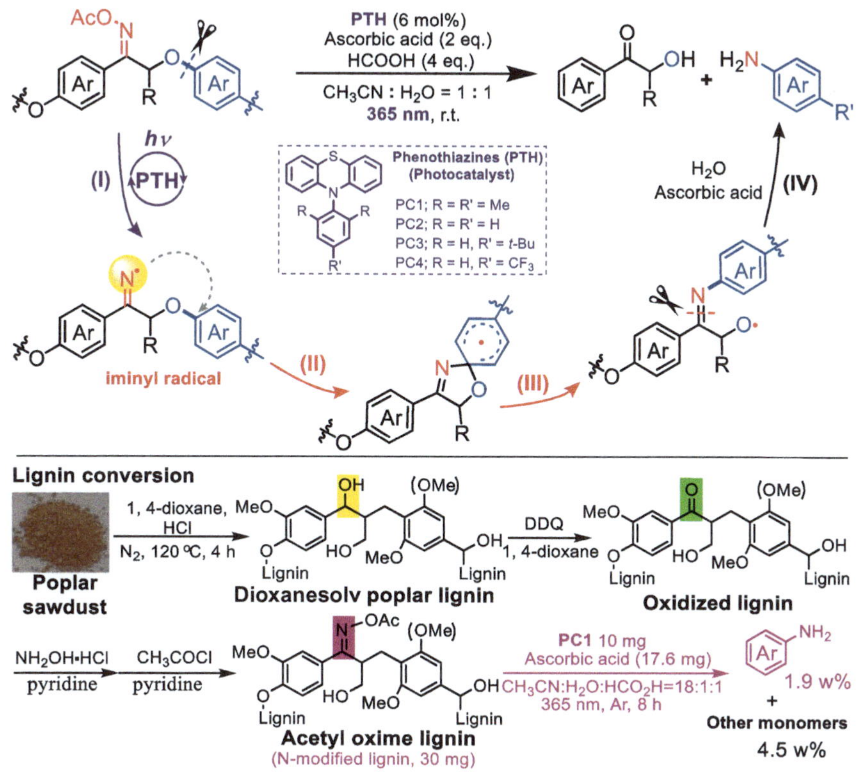

Scheme 10 Introducing N-radical at C_α mediated lignin aryl ether cleavage to anilines. *The figure was adapted from Li, H.J.; Bunrit, A.; Lu, J.M.; Gao, Z.Y.; Luo, N.C.; Liu, H.F.; Wang, F.* ACS Catal. ***2019,*** *9, 8843.*

to its steric effect. After aryl migration, further imine intermediate hydrolysis and O-radical reduction released the primary arylamine and α-hydroxy ketone as the non-phenolic aromatics. The application of this method on N-modified lignin provided a 1.9% yield of primary arylamine monomers and a 2.7% yield of phenol monomers. The moderate yields of desired arylamine monomers could be ascribed to the low content of α-oximated β-aryl ethers after N-modification (21 per 100 aromatic units) and more restricted polymerized structures in real lignin than free dimer models.

In addition, besides working as the "substrate site" to introduce the nucleophilic cleavage reagent like peroxide and NH_2OH, the C_β–H connecting with a C_α=O can be easily abstracted as H• or H^+ and the rest C_β can work as the "substrate site" to react with active O_2* or other O* species, causing the cleavage of lignin C_α–C_β bond. In 2016, we reported a two-step oxidation strategy for lignin C_α–C_β bond cleavage.[59] First, lignin β-O-4 alcohol was oxidized to β-O-4 ketone over an $O_2/VOSO_4$/TEMPO system. Then, the C_α–C_β bond of β-O-4 ketone was selectively cleaved to acids and phenols by oxidation over a Cu^{2+}/1,10-phenanthroline catalyst (Scheme 11A). Our kinetic analysis indicated the C_β–H abstraction was the rate-determining step for C—C bond cleavage in ketone. We then introduced a Lewis acid to assist the activation of the C_β–H for lignin C—C bonds oxidative cleavage. We found that $Cu(OAc)_2/BF_3{\cdot}OEt_2$ can efficiently catalyze the cleavage of C_α–C_β bond in β-O-4 ketone to esters and phenols, and cleave C—C bonds in pre-oxidized β-1 ketone to produce esters and aromatic aldehydes (Scheme 11B).[60] Mechanistic studies suggested the C_β radical after C_β–H abstraction reacted with active O_2 to give an oxygen-centered radical, which led to the cleavage of C_α–C_β bond. Besides Lewis acid assistance, we further found the KOH assisted $O_2/Cu(OAc)_2$/1,10-phenanthroline system could achieve oxidative cleavage of lignin model C—C bond to acids in one-step via the C_α ketone intermediate (Scheme 11C).[61,62] Furthermore, we recently reported a method of amines/ammonia-mediated bonds cleavage in oxidized lignin β-O-4 models with Cu catalyst at room temperature, demonstrating the effect of amine source on the selectivity of products (Scheme 11D). For primary and secondary aliphatic amines, lignin ketone models underwent oxidative C_α–C_β bond cleavage and C_α–N bond formation to generate aromatic amides. For ammonia, the competition between oxygen and ammonia determined the selectivity between C_α–N and C_β–N bond formation, generating amides and α-keto amides, respectively. For tertiary amines, the lignin models underwent oxidative C_α–C_β bond cleavage to benzoic acids.

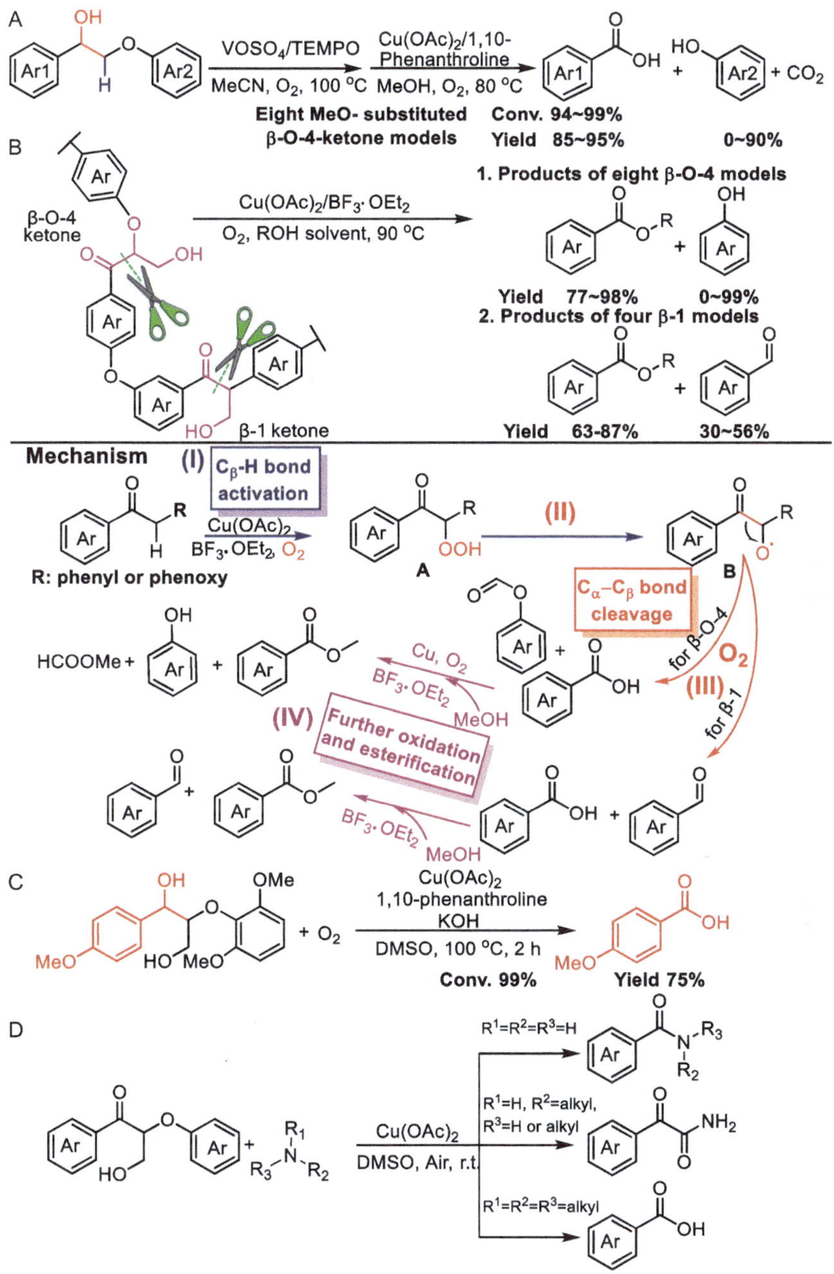

Scheme 11 Cu-catalyzed C_α–C_β bond cleavage in lignin β-O-4 and β-1 linkages.

Control experiments indicated that amines acted as nucleophiles attacking at C_α or C_β position of the oxidized β-O-4 linkage to be cleaved. These results represented an example that the breakage of the oxidized lignin model can be regulated by amines with copper catalysts.

As mentioned above, the thermal catalytic oxidation methods used in lignin C—C bond cleavage usually begin with the oxidation of C_α–OH to ketone, which can promote the activation of C_β–H and following introduction of an O* species. Alternatively, some thermal processes begin with the selective oxidation at C_γ to aldehyde, which can promote the C—C bond cleavage via the retrograde aldol condensation process. As another route, some photo-catalysis processes can achieve the direct activation of the C_β–H bond to generate a substrate site, which can accept the O* species for further C—C bond cleavage.

In 2017, we found that a hybrid CuO_x/ceria/TiO_2(anatase) catalyst could efficiently catalyze the photo-cleavage of β-1 C—C bond under visible light (455 nm), leading to high yields of substituted benzaldehydes.[63] Mechanistic studies suggested the photo-generated holes abstracted hydrogen from C_β–H to generate a C_β-centered radical (Scheme 12A mechanism). Then, the C_β-centered radical combined with an O_2 before forming an unstable peroxide intermediate with photo-generated electron and H^+. The peroxide then underwent C_α–C_β cleavage with the elimination of H_2O to form benzaldehydes through a six-membered ring transition state. The photo-catalyst characterizations played a critical role in controlling the reaction routes. Characterization by high angle annular dark-field scanning transmission electron microscopy (HAADF-STEM) and energy-dispersive X-ray spectroscopy element (EDX) mapping revealed that the CuO_x clusters are highly dispersed on the exposed anatase surface as well as on the nanosized ceria domains. In-depth investigations by Raman and ultraviolet-visible diffuse reflectance spectra (UV–vis DRS), together with density functional theory (DFT) calculations, further verified that the CuO_x clusters present on the ceria domains increase the concentration of surface defects (Ce^{3+} ions and oxygen vacancies) and accordingly improve the photocatalytic activity. The CuO_x clusters decorating the anatase TiO_2 suppress the side reaction (oxydehydrogenation of C_α–OH to C_α=O without C—C bond cleavage) because of an upward shift in the valence band (VB) edge of anatase.

In our following work, we reported another photocatalytic oxidative strategy for the C—C bond cleavage of lignin β-O-4 and β-1 linkages, using a mesoporous graphitic carbon nitride catalyst (mpg-C_3N_4).[64] Identifications by solid-state NMR techniques and density functional theory (DFT) calculations indicated that π–π stacking interactions are most likely present between

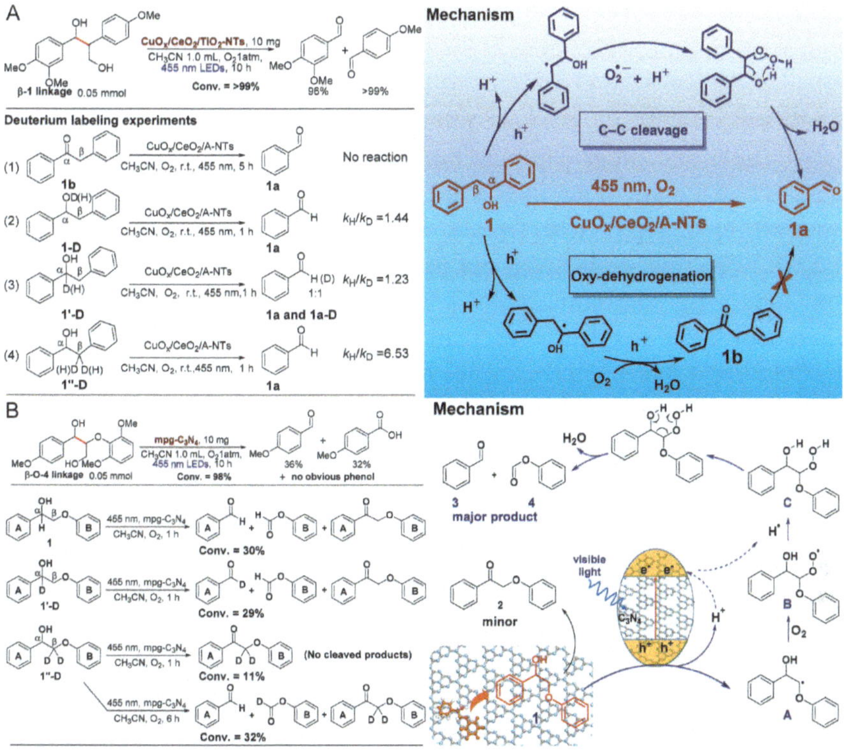

Scheme 12 Photocatalytic oxidative cleavage of lignin C—C bond. *The figure was adapted from Hou, T.T.; Luo, N.C.; Li, H.J.; Heggen, M.; Lu, J.M.; Wang, Y.H.; Wang, F.* ACS Catal. ***2017,*** 7, *3850; Liu, H.; Li, H.; Lu, J.; Zeng, S.; Wang, M.; Luo, N.; Xu, S.; Wang, F.* ACS Catal. ***2018,*** 8, *4761.*

the flexible carbon nitride surface and lignin model molecule. In addition, the low charge recombination efficiency and high specific surface area (206.5 $m^2 g^{-1}$) of the catalyst contributed to its high catalytic activity. Mechanistic investigations revealed that the photo-generated holes, as the main active species, triggered the C—C bond oxidative cleavage. The β-O-4 process was similar to the β-1 linkage cleavage discussed above (Scheme 12B mechanism), via a direct C_β–H abstraction to the C_β-centered radical, but the major products were aromatic aldehyde and phenyl formate.

3.5 Catalytic lignin depolymerization to give narrow monomer distributions and valuable chemicals

Achieving the transformation of protolignin polymer to monomers like phenol[65,66] or other aromatic acids,[67,68] that can be used to synthesize artificial polymers, has the potential to add significant value to lignin depolymerization. Already, some catalytic systems can efficiently produce aromatic monomers via lignin C—O/C—C bonds cleavage. For example, certain hydrogenation processes can facilitate >60% lignin conversion, but these methods generally produce phenol mixtures with C_1–C_3 alkyl tails on the para sites and other complex monomers like aromatic acids and aldehydes. To make use of lignin as a renewable feedstock, new catalytic depolymerization strategies with narrow monomer distributions are necessary. In 2018, we achieved the one-pot preparation of alkyl group-free phenols from lignin, by combining C_α–C_β bond oxidative cleavage to acids and subsequent decarboxylation (Scheme 13).[66] The combination of initial C_β–OAr bond hydrogenolysis over Ru/CeO_2 to release phenolic groups, C_α—C_β bond oxidative cleavage, and subsequent decarboxylation to cleave caryl–C_α with $CuCl_2$, led to the selective conversion of separated poplar lignin into phenol (only Ph–OH) with up to a 13 wt% yield. One reason for the low phenol yield is that the current system preferently transforms the H-type lignin linkages to phenols. For example, an H-type β-O-4 synthetic polymer was converted to phenol in a 75% yield, while G-type and S-type lignin synthetic polymers gave poor phenol yields. Therefore, more modifications need to be explored for the structure universality. Besides, in the wood powders conversion, besides lignin, the cellulose was mostly converted to levulinic acid and γ-valerolactone with above method (Scheme 13).

Besides converting lignin to large unmet need for artificial polymers, transforming stable lignin fragment to highly functionalized and valuable motifs should be another choice.[69] During catalytic lignin depolymerization, certain ether and C—C linkages can be cleaved at a relatively mild condition, but other more stable lignin linkages are less reactive. One example is the biaryl 5–5 linkage. Taking into account the difficulty and low benefit from biaryl C—C bond cleavage to monomers or complete hydrogenation to alkane as fuel, the idea of using lignin biaryl structure derivatives to generate high-value synthetic building blocks is challenging but attractive. In 2019, we developed a photo-induced dearomatization method for spironolactones preparation from biaryl compounds (Scheme 14).[70] The reaction could be performed under aerobic conditions with an acridinium photo-catalyst. A complementary methodology in which commercial 2,3-dichloro-5,6-dicyano-1,4-benzoquinone (DDQ) was used as a photo oxidant to generate the spironolactones under aqueous conditions was shown to tolerate anaerobic synthetic conditions. Scheme 14 displays proposed mechanisms for reaction pathways under aerobic or anaerobic conditions. The first step is formation of key carboxyl radical **A**, which binds to one aromatic ring close to the Ar–Ar bond to induced dearomatization by an

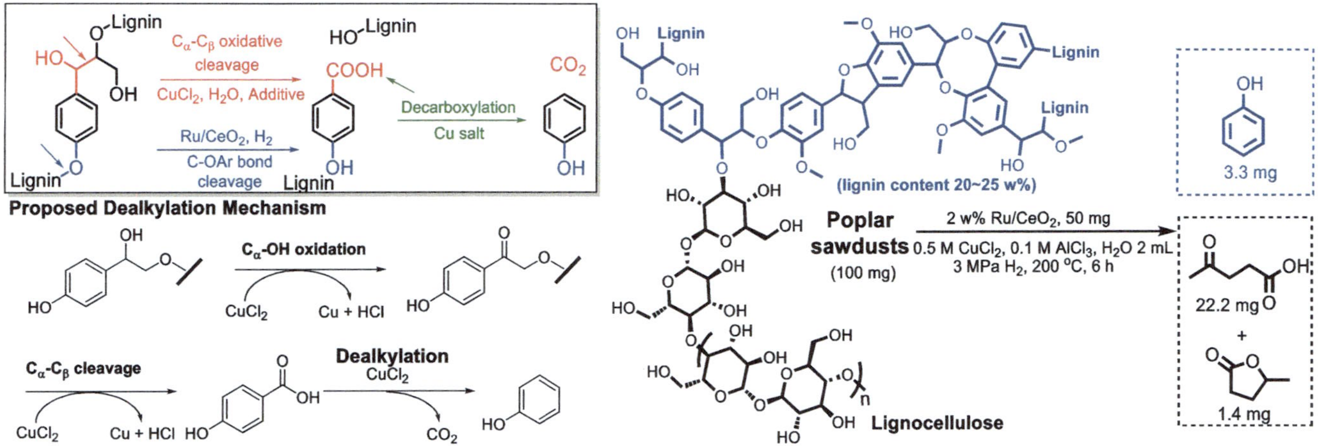

Scheme 13 Lignin dealkylation to phenol via hydrogenation-oxidation-decarboxylation strategy. *The figure was adapted from Wang, M.; Liu, M.; Li, H.; Zhao, Z.; Zhang, X.; Wang, F.* ACS Catal. ***2018,*** *8, 6837.*

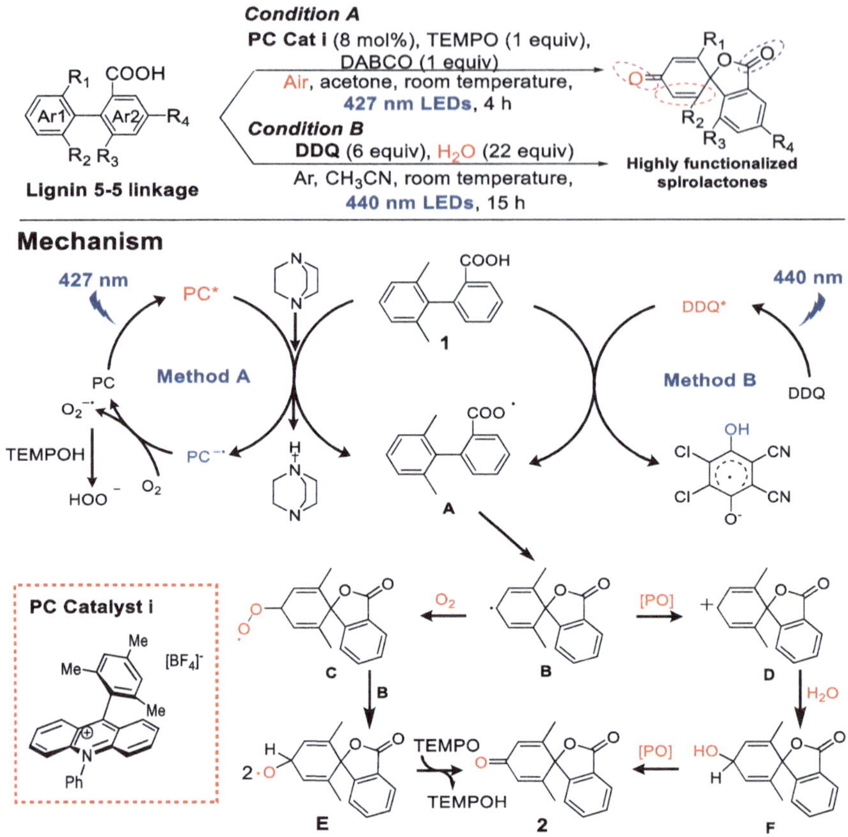

Scheme 14 Photocatalytic spironolactone generation from biaryl compounds. *The figure was adapted from Li, H.; Subbotina, E.; Bunrit, A.; Wang, F.; Samec, J.S.M.* Chem. Sci. ***2019,*** 10, *3681.*

intramolecular cyclization to form intermediate **B**. From this point, the two pathways diverged. Under aerobic Condition I, the nucleophilic attack on intermediate **B** by O_2 could deliver peroxy radical **C**, followed by reaction between **C** and **B** to radical **E**, followed by hydrogen atom transfer in the presence of TEMPO to give spironolactone product **2** via. In the case of Condition II, intermediate **B** underwent oxidation to give cation **D**, which then trapped water to **F**. **F** could undergo further oxidation to give the final spironolactone product.

4. Conclusions and perspectives

In this chapter, we first introduced our research projects on the Ni-catalyzed depolymerization of protolignin and technical lignin to aromatic chemicals. This work highlights the unique effect of the alcohol solvents on lignin depolymerization and presents a fragmentation–hydrogenolysis mechanism for lignin conversion. Then, based on our results, we gave a brief introduction of the lignin conversion from target molecule/linkage transformation analysis. This highlighted strategies where lignin pretreatments in cooperation with corresponding catalysts can enhance the efficiency of lignin depolymerization. Focusing on the catalytic transformation of lignin fragment models and based on our understanding about the lignin conversion from our projects, we shed light on a strategy that begins with the adjacent functional group modification (AFGM) to directly decrease the bond dissociation enthalpy (BDE) of targeted C—C/C—O bonds or to generate new substrate sites to introduce a cleavage reagent for further conversion. These concepts, reducing the target bond BDE and introducing a new "substrate site," can provide some guidance for determining the possible reaction routes, which can facilitate the catalysts screening or mechanistic study. We then demonstrate these two concepts by introducing our recent research projects.

For the perspective of further lignin catalytic utilization, we hold the following points: (a) scaled lignin transformation needs a standard substrate, which depends on the elaborate pretreatment or lignin abstracting methods that can keep more β-O-4 linkages and pre-cut lignin into stable fragments for further catalytic depolymerization and (b) catalytic strategies and methods should make lignin depolymerization profitable, which depends on how to increase the lignin conversion and monomers yields with robust and low-cost catalysts after strategical optimization. In this respect, the target lignin depolymerization products can focus on the high demand monomers for artificial polymerization or low demand, but high valued medical chemicals, functional materials, highly functionalized motifs, and so on; (c) the innovations about the reactors and new technology-process intensification methods can combine the advantages of thermal catalysis, photo-catalysis, electro-catalysis, and even enzymatic catalysis, biotechnology, genetic engineering, which may make lignin conversion efficient under a mild condition.

In the end, although the lignin conversion can be traced back to the 1930s and it is still a challenging topic by now, the process in lignin utilization is still like that of a forest fire; faster when fanned by the winds of new ideas, brighter when passing through other areas of science, which will finally provide a successful solution for human society in the demand of sustainable chemicals and energy.

Acknowledgments

This work was supported by the National Natural Science Foundation of China (21721004, 21991094, and 21802137), the Ministry of Science and Technology of the People's Republic of China (2018YFE0118100), the CAS-NSTDA Joint Research Project (GJHZ2075) and Science and Technology Foundation of Liaoning (2019-BS-245).

References

1. BP Statistical Review. *BP Statistical Review of World Energy 2020*; BP p.l.c: London, 2020.
2. Li, C.; Zhao, X.; Wang, A.; Huber, G. W.; Zhang, T. *Chem. Rev.* **2015**, *115*, 11559.
3. Amidon, T. E.; Liu, S. *Biotechnol. Adv.* **2009**, *27*, 542.
4. Carpenter, D.; Westover, T. L.; Czernik, S.; Jablonski, W. *Green Chem.* **2014**, *16*, 384.
5. Zhang, C. F.; Wang, F. *Acc. Chem. Res.* **2020**, *53*, 470.
6. Chakar, F. S.; Ragauskas, A. J. *Ind. Crop Prod.* **2004**, *20*, 131.
7. Pinto, P. C. R.; da Silva, E. A. B.; Rodrigues, A. E. *Ind. Eng. Chem. Res.* **2011**, *50*, 741.
8. Zakzeski, J.; Bruijnincx, P. C. A.; Jongerius, A. L.; Weckhuysen, B. M. *Chem. Rev.* **2010**, *110*, 3552.
9. Renders, T.; Van den Bosch, S.; Koelewijn, S. F.; Schutyser, W.; Sels, B. F. *Energy Environ. Sci.* **2017**, *10*, 1551.
10. Shuai, L.; Amiri, M. T.; Questell-Santiago, Y. M.; Héroguel, F.; Li, Y.; Kim, H.; Meilan, R.; Chapple, C.; Ralph, J.; Luterbacher, J. S. *Science* **2016**, *354*, 329.
11. Deuss, P. J.; Scott, M.; Tran, F.; Westwood, N. J.; de Vries, J. G.; Barta, K. *J. Am. Chem. Soc.* **2015**, *137*, 7456.
12. Wu, X.; Luo, N.; Xie, S.; Zhang, H.; Zhang, Q.; Wang, F.; Wang, Y. *Chem. Soc. Rev.* **2020**, *49*, 6198.
13. Song, Q.; Wang, F.; Cai, J.; Wang, Y.; Zhang, J.; Yu, W.; Xu, J. *Energy Environ. Sci.* **2013**, *6*, 994.
14. Song, Q.; Cai, J.; Zhang, J.; Yu, W.; Wang, F.; Xu, J. *Chin. J. Catal.* **2013**, *34*, 651.
15. Song, Q.; Wang, F.; Xu, J. *Chem. Commun.* **2012**, *48*, 7019.
16. Nichols, J. M.; Bishop, L. M.; Bergman, R. G.; Ellman, J. A. *J. Am. Chem. Soc.* **2010**, *132*, 12554.
17. Nguyen, J. D.; Matsuura, B. S.; Stephenson, C. R. *J. Am. Chem. Soc.* **2014**, *136*, 1218.
18. Lancefield, C. S.; Ojo, O. S.; Tran, F.; Westwood, N. J. *Angew. Chem. Int. Ed.* **2015**, *54*, 258.
19. Lu, J.; Wang, M.; Zhang, X.; Heyden, A.; Wang, F. *ACS Catal.* **2016**, *6*, 5589.
20. Chmely, S. C.; Kim, S.; Ciesielski, P. N.; Jimenez-Oses, G.; Paton, R. S.; Beckham, G. T. *ACS Catal.* **2013**, *3*, 963.
21. Zhou, X.; Mitra, J.; Rauchfuss, T. B. *ChemSusChem* **2014**, *7*, 1623.
22. Srifa, P.; Galkin, M. V.; Samec, J. S. M.; Hermansson, K.; Broqvist, P. *J. Phys. Chem. C* **2016**, *120*, 23469.
23. Galkin, M. V.; Dahlstrand, C.; Samec, J. S. M. *ChemSusChem* **2015**, *8*, 2187.
24. Luo, N. C.; Wang, M.; Li, H. J.; Zhang, J.; Hou, T. T.; Chen, H. J.; Zhang, X. C.; Lu, J. M.; Wang, F. *ACS Catal.* **2017**, *7*, 4571.
25. Luo, N. C.; Wang, M.; Li, H. J.; Zhang, J.; Liu, H. F.; Wang, F. *ACS Catal.* **2016**, *6*, 7716.
26. Li, H.; Gao, Z.; Lei, L.; Liu, H.; Han, J.; Hong, F.; Luo, N.; Wang, F. *Green Chem.* **2020**, *22*, 3802.
27. Sturgeon, M. R.; Kim, S.; Lawrence, K.; Paton, R. S.; Chmely, S. C.; Nimlos, M.; Foust, T. D.; Beckham, G. T. *ACS Sustain. Chem. Eng.* **2014**, *2*, 472.

28. Lahive, C. W.; Deuss, P. J.; Lancefield, C. S.; Sun, Z.; Cordes, D. B.; Young, C. M.; Tran, F.; Slawin, A. M.; de Vries, J. G.; Kamer, P. C.; Westwood, N. J.; Barta, K. *J. Am. Chem. Soc.* **2016**, *138*, 8900.
29. Miles-Barrett, D. M.; Neal, A. R.; Hand, C.; Montgomery, J. R. D.; Panovic, I.; Ojo, O. S.; Lancefield, C. S.; Cordes, D. B.; Slawin, A. M. Z.; Lebl, T.; Westwood, N. J. *Org. Biomol. Chem.* **2016**, *14*, 10023.
30. Zhang, C. F.; Li, H. J.; Lu, J. M.; Zhang, X. C.; MacArthur, K. E.; Heggen, M.; Wang, F. *ACS Catal.* **2017**, 7, 3419.
31. Zhou, L. P.; Zhang, C. F.; Fang, T.; Zhang, B. B.; Wang, Y.; Yang, X. M.; Zhang, W.; Xu, J. *Chin. J. Catal.* **2011**, *32*, 118.
32. Wang, M.; Zhang, X.; Li, H.; Lu, J.; Liu, M.; Wang, F. *ACS Catal.* **2018**, *8*, 1614.
33. Chen, J.; Lu, F.; Si, X.; Nie, X.; Chen, J.; Lu, R.; Xu, J. *ChemSusChem* **2016**, *9*, 3353.
34. Zhu, G.; Qiu, X.; Zhao, Y.; Qian, Y.; Pang, Y.; Ouyang, X. *Bioresour. Technol.* **2016**, *218*, 718.
35. Barrett, J. A.; Gao, Y.; Bernt, C. M.; Chui, M.; Tran, A. T.; Foston, M. B.; Ford, P. C. *ACS Sustain. Chem. Eng.* **2016**, *4*, 6877.
36. Rahimi, A.; Ulbrich, A.; Coon, J. J.; Stahl, S. S. *Nature* **2014**, *515*, 249.
37. vom Stein, T.; den Hartog, T.; Buendia, J.; Stoychev, S.; Mottweiler, J.; Bolm, C.; Klankermayer, J.; Leitner, W. *Angew. Chem. Int. Ed.* **2015**, *54*, 5859.
38. Lancefield, C. S.; Teunissen, L. W.; Weckhuysen, B. M.; Bruijnincx, P. C. A. *Green Chem.* **2018**, *20*, 3214.
39. Dabral, S.; Hernandez, J. G.; Kamer, P. C. J.; Bolm, C. *ChemSusChem* **2017**, *10*, 2707.
40. Zhou, Z.-Z.; Liu, M.; Li, C.-J. *ACS Catal.* **2017**, 7, 3344.
41. Rafiee, M.; Alherech, M.; Karlen, S. D.; Stahl, S. S. *J. Am. Chem. Soc.* **2019**, *141*, 15266.
42. Zhang, C.; Lu, J.; Zhang, X.; MacArthur, K.; Heggen, M.; Li, H.; Wang, F. *Green Chem.* **2016**, *18*, 6545.
43. Wu, X. J.; Fan, X. T.; Xie, S. J.; Lin, J. C.; Cheng, J.; Zhang, Q. H.; Chen, L. Y.; Wang, Y. *Nat. Catal.* **2018**, *1*, 772.
44. Lin, J.; Wu, X.; Xie, S.; Chen, L.; Zhang, Q.; Deng, W.; Wang, Y. *ChemSusChem* **2019**, *12*, 5023.
45. Wu, X. J.; Xie, S. J.; Liu, C. X.; Zhou, C.; Lin, J. C.; Kang, J. C.; Zhang, Q. H.; Wang, Z. H.; Wang, Y. *ACS Catal.* **2019**, *9*, 8443.
46. Son, S.; Toste, F. D. *Angew. Chem. Int. Ed.* **2010**, *49*, 3791.
47. Haibach, M. C.; Lease, N.; Goldman, A. S. *Angew. Chem. Int. Ed.* **2014**, *53*, 10160.
48. Hanson, S. K.; Baker, R. T.; Gordon, J. C.; Scott, B. L.; Thorn, D. L. *Inorg. Chem.* **2010**, *49*, 5611.
49. Gazi, S.; Ng, W. K. H.; Ganguly, R.; Moeljadi, A. M. P.; Hirao, H.; Soo, H. S. *Chem. Sci.* **2015**, *6*, 7130.
50. Liu, H. F.; Li, H. J.; Luo, N. C.; Wang, F. *ACS Catal.* **2020**, *10*, 632.
51. Wang, Y.; He, J.; Zhang, Y. *CCS Chem.* **2020**, *2*, 107.
52. Rahimi, A.; Azarpira, A.; Kim, H.; Ralph, J.; Stahl, S. S. *J. Am. Chem. Soc.* **2013**, *135*, 6415.
53. Patil, N. D.; Yao, S. G.; Meier, M. S.; Mobley, J. K.; Crocker, M. *Org. Biomol. Chem.* **2015**, *13*, 3243.
54. Jennings, J. A.; Parkin, S.; Munson, E.; Delaney, S. P.; Calahan, J. L.; Isaacs, M.; Hong, K.; Crocker, M. *RSC Adv.* **2017**, 7, 25987.
55. Wang, Y. L.; Wang, Q. Y.; He, J. H.; Zhang, Y. T. *Green Chem.* **2017**, *19*, 3135.
56. Wang, Y.; Du, Y.; He, J.; Zhang, Y. *Green Chem.* **2018**, *20*, 3318.
57. Li, H.; Wang, M.; Liu, H.; Luo, N.; Lu, J.; Zhang, C.; Wang, F. *ACS Sustain. Chem. Eng.* **2018**, *6*, 3748.
58. Li, H. J.; Bunrit, A.; Lu, J. M.; Gao, Z. Y.; Luo, N. C.; Liu, H. F.; Wang, F. *ACS Catal.* **2019**, *9*, 8843.

59. Wang, M.; Lu, J. M.; Zhang, X. C.; Li, L. H.; Li, H. J.; Luo, N. C.; Wang, F. *ACS Catal.* **2016**, *6*, 6086.
60. Wang, M.; Li, L. H.; Lu, J. M.; Li, H. J.; Zhang, X. C.; Liu, H. F.; Luo, N. C.; Wang, F. *Green Chem.* **2017**, *19*, 702.
61. Wang, M.; Lu, J. M.; Li, L. H.; Li, H. J.; Liu, H. F.; Wang, F. *J. Catal.* **2017**, *348*, 160.
62. Liu, H.; Wang, M.; Li, H.; Luo, N.; Xu, S.; Wang, F. *J. Catal.* **2017**, *346*, 170.
63. Hou, T.; Luo, N.; Li, H.; Heggen, M.; Lu, J.; Wang, Y.; Wang, F. *ACS Catal.* **2017**, 7, 3850.
64. Liu, H.; Li, H.; Lu, J.; Zeng, S.; Wang, M.; Luo, N.; Xu, S.; Wang, F. *ACS Catal.* **2018**, *8*, 4761.
65. Liao, Y. H.; Zhong, R. Y.; Makshina, E.; d'Halluin, M.; van Limbergen, Y.; Verboekend, D.; Sels, B. F. *ACS Catal.* **2018**, *8*, 7861.
66. Wang, M.; Liu, M.; Li, H.; Zhao, Z.; Zhang, X.; Wang, F. *ACS Catal.* **2018**, *8*, 6837.
67. Bai, Z. Y.; Phuan, W. C.; Ding, J.; Heng, T. H.; Luo, J. Z.; Zhu, Y. H. *ACS Catal.* **2016**, *6*, 6141.
68. Song, S.; Zhang, J. G.; Gozaydin, G.; Yan, N. *Angew. Chem. Int. Ed.* **2019**, *58*, 4934.
69. Li, H.; Bunrit, A.; Li, N.; Wang, F. *Chem. Soc. Rev.* **2020**, *49*, 3748.
70. Li, H.; Subbotina, E.; Bunrit, A.; Wang, F.; Samec, J. S. M. *Chem. Sci.* **2019**, *10*, 3681.

CHAPTER SEVEN

Design of task-specific metal phosphides for the sustainable manufacture of advanced biofuels

Michael Dierks[a], Zhenweng Cao[a,b], and Roberto Rinaldi[c,*]
[a]Max-Planck-Institut für Kohlenforschung, Mülheim an der Ruhr, Germany
[b]Qingdao Institute of Bioenergy and Bioprocess Technology, Chinese Academy of Sciences, Qingdao, China
[c]Imperial College London, Department of Chemical Engineering, South Kensington Campus, London, United Kingdom
[*]Corresponding author: e-mail address: r.rinaldi1@ic.ac.uk

Contents

Abstract

Hydrodeoxygenation (HDO) of biomass waste streams constitutes a broad class of catalytic processes crucial to the sustainable manufacture of advanced biofuels. In this context, metal phosphides have received increasing attention as catalysts for HDO of vegetable oil, animal fats, bioderived phenolic streams (e.g., lignins, their depolymerization products, and pyrolysis oil). Metal phosphides present highly tunable surface chemistry accessible via uncomplicated synthetic protocols based on low-cost precursors. Herein, a critical analysis of the sustainable manufacture of advanced biofuels is provided. Emphasis is placed on nickel phosphides as catalysts with unique properties for the HDO of bioderived phenolic. The task-specific design of nickel phosphides is centered on optimizing catalyst activity, product distribution, product selectivity (aromatic vs. aliphatic hydrocarbons). In the manufacture of advanced biofuels, catalyst deactivation constitutes a primary challenge. We briefly discuss mechanisms underlining catalyst deactivation and protective effects of surface modification on the nickel phosphide stability.

Advances in Inorganic Chemistry, Volume 77
ISSN 0898-8838
https://doi.org/10.1016/bs.adioch.2021.02.002

1. Introduction

Sustainable manufacture of advanced biofuels provides drop-in replacements of fossil fuels for aviation, shipping, and heavy freight trucks.[1] Thereby, only minor changes in current infrastructure for fuel distribution or the transport fleet are required.[1] In the long-haul heavy-duty transport sector, advanced biofuels will be needed for decades to meet the long-term climate targets of the Paris Agreement.[1,2] Moreover, to achieve the 2030 UN Sustainability Goals, the sustainable manufacture of advanced biofuels must be based on sustainable agricultural intensification,[3] to increase biofuel yields per hectare and, thus, alleviate the pressure on the conversion of non-agricultural lands into energy crops.[3] Such an essential requirement is met by utilizing lignocellulosic biomass waste, which constitutes "any organic material apart from the primary material for which the plants were originally grown," as defined by Tuck et al.[4]

The sugarcane industry is a case in point useful to recognize the untapped potential of lignocellulosic biomass waste for the manufacture of advanced biofuels.[5] In 2018, the world sugarcane industry produced sugar (141 Mt) and fuel ethanol (24 Mt) in addition to over 540 Mt of bagasse as a waste stream.[6,7] For comparison, the amount of bagasse exceeds by about 50% in weight the annual world production of plastics (i.e., approximately 359 Mt in 2018).[8] The bagasse waste is employed as a low-energy solid fuel to power the industrial operations.[6] In recent years, improvements in sugarcane mills' energy efficiency enabled part of the power to be sold to the energy grid as renewable energy.[9,10] In Brazil, electricity cogeneration transformed 181 Mt of bagasse into 36.8 TWh of bioenergy, supplying approximately 10% of the electricity demand in 2020.[11,12] However, the full potential is estimated at 20% of Brazil's electricity demand.[10]

The use of sugarcane bagasse and other lignocellulosic biomass wastes from first-generation ethanol plants (e.g., maize stalk in the United States) could further reduce net greenhouse gas emissions, mitigating climate change.[2,4,13] However, lignocellulosic biomass wastes' complex structures are difficult to deconstruct into "easy-to-process" fractions.[14–19] Accordingly, catalytic processing requires high-severity conditions, which reduce the net flows of carbon and energy (to the synthetic biofuel).[20–22] Moreover, the term "waste" does not imply that the material is worthless.[4] As of 2021, owing to the high electricity prices in Brazil, electricity cogeneration remains the most economically attractive option for sugarcane bagasse's utilization.[10]

However, the economic scenario may dramatically change in few years with the fast increasing role of more affordable renewable electricity in the power grid (e.g., wind power).[23]

2. Manufacture of advanced biofuels from lignin streams

For the sustainable manufacture of advanced fuels, lignin is seldom considered a feedstock more attractive than cellulose and hemicellulose, which together correspond to 75% lignocellulose composition by weight.[24,25] In advanced biotechnological biorefineries, "cellulosic" ethanol production relies upon pretreatment steps to facilitate enzymatic hydrolysis of cellulose and hemicellulose.[17,26] The pretreatments are optimized to improve the yields of fermentable sugars and their quality for downstream fermentation (by limiting yeast inhibitors' formation, e.g., furfurals and other aromatic compounds).[27] Notably, advanced biotechnological biorefinery operations have not been designed to control lignin degradation via condensation reactions in their multiple lignocellulose pretreatment stages.[14,17,19] Lignin degradation decreases the structural uniformity and lowers the reactivity toward catalytic processes for depolymerization and hydrotreatment.[14,16,24,28–34] Nonetheless, a techno-economic analysis suggested a production cost of 0.3 € L^{-1} of high-octane reformulated fuel from lignin.[35] Such analysis considers the lignin biofuels to be produced from base-catalyzed lignin depolymerization and subsequent catalytic hydroprocessing processes,[35] affording compounds characterized by higher and more varied molecular weights ($C_{6–9+}$) than biofuels derived from the carbohydrate fraction of biomass (primarily restricted to C_2–C_6,[36–38] and more recently expanded to heavier products).[39–42]

To unlock the utilization of lignocellulosic biomass to its fullest, the degradation of lignin in the fractionation processes is mandatory in tackling lignin valorization challenges into chemicals biofuels.[14,43,44] As such, several lignocellulosic approaches to biomass fractionation have emerged, considering the control over the lignin species' reactivity released by solvolysis.[14,16,18,43,44] These approaches are now defined under the umbrella term "Lignin-first Biorefining." They utilize active stabilization approaches (e.g., reductive processes or capping agents) to protect lignin species released from native lignocellulosic biomass from condensation reactions.[19] These reactions are responsible for the generation of recalcitrant lignin polymers. Notably, lignin-first biorefining processes are not detrimental to the cellulose and hemicellulose valorization since high-quality pulps can also be isolated.[45] Therefore, lignin-first biorefining processes can also be described as a process for the integral utilization of lignocellulosic feedstocks. In this case, the depolymerized lignin streams provide a platform to produce several valuable materials, including advanced biofuels.[24,30,46–48]

By arranging the biopolymers (lignin, cellulose, and hemicellulose) and fossil fuels (crude oil, gasoline, diesel, kerosene, and methane) by their H/C and O/C coordinates in a van Krevelen plot (Fig. 1),[49] it becomes clear that the catalytic conversion of biopolymers into liquid biofuels demands a decrease in the O/C ratio while increasing H/C. Importantly, although fossil fuels present O/C equal to zero, full deoxygenation is not required for the production of biofuels.[50,51] In fact, oxygenates are added to fossil fuels to improve the combustion properties in the engine (e.g., ethanol in gasoline; fatty acid methyl ester, that is, biodiesel blended in diesel).

Considering the H-atoms' fate in the O-content removal, the manufacture of advanced biofuels from lignins appears to be more efficient than cellulosic biofuels. Let us examine the conversion of C–OH into CH_2 group,

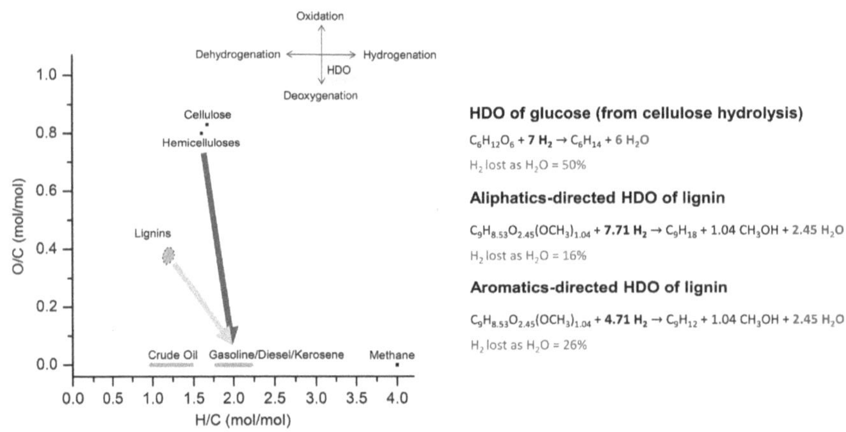

Fig. 1 (*Left*) Van Krevelen diagram arranging biopolymers (lignin, cellulose, and hemicellulose) and fossil fuels (crude oil, gasoline, diesel, kerosene, and methane) according to their coordinates H/C and O/C (mol/mol). The inset indicates how the direction along which H/C and O/C shifts via oxidation, deoxygenation, dehydrogenation and hydrogenation (as single processes), and the space covered by hydrodeoxygenation (HDO); (*Right*) Comparison of the net loss of H_2 equivalent per O-atom eliminated as H_2O, for hypothetical chemical equations, to produce target products (*indicated between parentheses*) via HDO of glucose (*n*-hexane), aliphatics-directed HDO of lignin (*n*-propylcyclohexane), and aromatic-directed HDO of lignin (*n*-propylbenzene).

which occurs via either C—OH hydrogenolysis or a tandem process combining water elimination (forming a C=C bond) followed by hydrogenation (of the C=C bond). This conversion decreases O/C but has no net effect on the H/C mole ratio. Nevertheless, there is a substantial net loss of one H atom per O atom eliminated as H_2O. By contrast, monomeric lignin units present methoxyl groups on the aromatic ring. In this case, part of the O-content is eliminated as methanol, which, as found for other alcohols,[33,44,52] can be utilized as an internal H-source for depolymerization and hydrodeoxygenation (HDO) processes.[53–55]

Despite being H_2 intensive, the full saturation of lignin's aromatic ring constitutes a sensible approach in a scenario in which renewable H_2 is broadly available. Such a hydrogenation process incorporates both atoms of H_2 into the product (fuel). Therefore, the chemical energy of H_2 remains stored in the aliphatic hydrocarbon to a large extent. As a matter of fact, the hydrogenation of benzene to cyclohexane increases the heat evolved from combustion from 3273 kJ/mol (benzene) to 3920 kJ/mol (cyclohexane) at 298 K under constant pressure.[56]

Let us further examine the net loss of H_2 equivalent per O-atom eliminated as H_2O in HDO processes (Fig. 1). In the hypothetical conversion of glucose to *n*-hexane, the elimination of each alcohol group leads to only one H-atom added to the product fuel, since the other H-atom is lost as H_2O, decreasing the O/C mole ratio. However, much better H_2 utilization is achieved in the aliphatic-directed HDO of lignin (in which 16% H_2 is lost as H_2O) and, to a lesser extent, for aromatic-directed HDO of lignin (26% H_2 is lost as H_2O).

The manufacture of advanced biofuels based on catalytic hydroprocessing is undoubtedly an H_2 intensive process.[57] This fact impacts the contribution of advanced biofuels to global warming mitigation as H_2 is obtained primarily from the reforming of non-renewable carbonaceous feedstocks. A high H_2 demand currently implies a high CO_2 emission (e.g., 11.88 kg CO_2/kg H_2 obtained from natural gas reforming; approximately 25 kg CO_2/kg H_2 via coal gasification).[57] Therefore, lignin fuels' sustainability relies upon renewable H_2 (e.g., from biomass gasification or water electrolysis in the best case).[24,57] In a future scenario of an abundant supply of renewable H_2, the manufacture of advanced biofuels via aliphatic-directed HDO of lignin could provide efficient chemical storage of H_2 in the form of energy-dense hydrocarbons, also delivering a cost-efficient solution for the otherwise cost-intensive H_2 storage and expensive logistics of H_2 as renewable fuel by its own right.[57]

3. Metal phosphides as hydrotreating catalysts

Metal phosphides are accessible via uncomplicated synthetic protocols based on low-cost precursors.[58,59] Since the late 1950s, it is known that metal (M) phosphides present hydrogenation activity. Such a unique feature indicates that the M sites exhibit a metallic nature.[60,61] In a broader sense, metal phosphides can be classified as bifunctional catalysts.[62–65] This classification is supported by several key features of metal phosphides' surface chemistry, which confer acidity to the surface along with the hydrogenating properties associated with metal sites.

In a detailed study on the structural characterization of Ni_2P catalysts, Oyama and Lee[66] provided in-depth insight into the Ni sites' roles in the hydrodesulfurization reactions of 4,6-dimethyldibenzothiophene. As a result of its hexagonal structure, Ni_2P presents two types of Ni sites, denoted as Ni(1) and Ni(2) in Fig. 2.[66] Surrounded by four nearest-neighbor P atoms, Ni(1) site exhibits a quasi-tetrahedral geometry, while, surrounded by five nearest-neighbor P atoms, Ni(2) site shows square pyramidal geometry. From the differences in the coordination environment, it is plausible to expect different levels of charge transfer(from Ni to P) to make Ni(1) and Ni(2) sites play different roles in hydrotreating processes. In a seminal work, Oyama and Lee provided compelling evidence supporting this working hypothesis.[66] They suggested that direct desulfurization of 4,6-dimethyldibenzothiophene takes place on Ni(1) sites, while hydrogenation on Ni(2) surface sites.[66] More recently, Teixeira Da Silva et al. studied the HDO of cresols in the

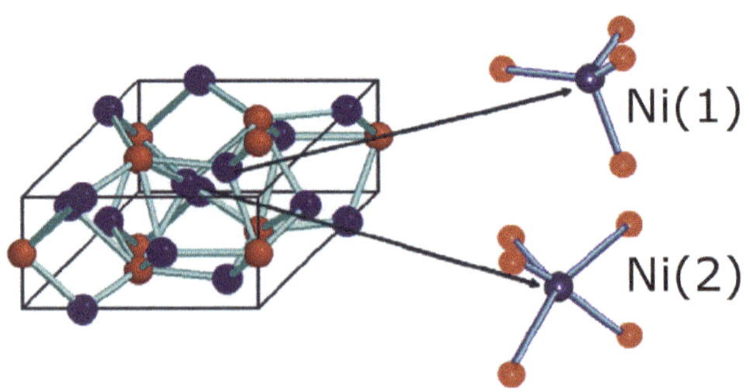

Fig. 2 Hexagonal crystal structure of Ni_2P creates two different Ni sites. As indicated, Ni(1) site exhibits a quasi-tetrahedral geometry, while Ni(2) site presents a square pyramidal geometry (five nearest-neighbor P atoms). *Reproduced with permission from Oyama, S. T.; Lee, Y. K.* J. Catal. ***2008,*** 258 *(2), 393–400.*

presence of Ni_2P/SiO_2.[67] They proposed the participation of different active sites and adsorption modes to explain the product selectivity and distribution.

For HDO processes, acid sites are essential to several reactions. As a case in point, Ni_2P also presents Lewis acidity owing to the charge transfer from Ni to P.[68,69] Moreover, P atoms exposed on the surface can be converted into P-OH groups, acting as Brønsted acid sites. P–OH groups are suggested to play a critical role in hydrogen spillover.[64,70,71] Moreover, there is evidence that P-OH surface groups could interact preferentially with water, protecting the bulk structure of Ni_2P from oxidation.[68,70]

4. General trends of metal phosphides as hydrotreating catalysts

Metal phosphide catalysts have gained increasing attention in various hydrotreating processes, e.g., hydrodesulfurization,[59,72–78] hydrotreatment of vegetable oils and animal fats rendering linear alkanes as second-generation biodiesel,[64,79–89] HDO of variety of biobased phenolic compounds in pyrolysis bio-oil,[62,90–94] and more recently lignin-streams from lignin-first biorefining.[24,95] This section provides an overview of the application of metal phosphides, in special Ni_2P, in catalytic hydrotreating of lignin models and lignin-derived streams. Scheme 1 presents the typical classes of hydrotreating processes on lignin-derived streams as a general guide for the readers.[89,96–105] Table 1 presents an overview of metal phosphides explored in the HDO of lignin model compounds (entries 1–20), technical lignins (entries 21–22), and pyrolysis bio-oils (entry 23).

Some general trends highlighting the potential of metal phosphides, particularly Ni_2P, can be extracted from Table 1. However, a direct quantitative comparison among the examples is not possible since reactor setup and reaction conditions broadly vary. Nonetheless, by cross-examining several studies, the Ni_2P phase was identified as the most active for HDO of guaiacol (Table 1, entry 2).[91] Furthermore, a comparison of the catalytic activity among the different nickel phosphide phases revealed that Ni_3P/SiO_2 and $Ni_{12}P_5/SiO_2$ presented an apparent activity higher than Ni_2P/SiO_2 (Table 1, entry 8).[112] However, in guaiacol conversion, Ni_2P/SiO_2 showed a higher selectivity toward direct deoxygenation (yielding anisole) and demethoxylation (producing phenol), while Ni_3P/SiO_2 and $Ni_{12}P_5/SiO_2$ showed enhanced demethylation activity, yielding catechol (which is a coke precursor).

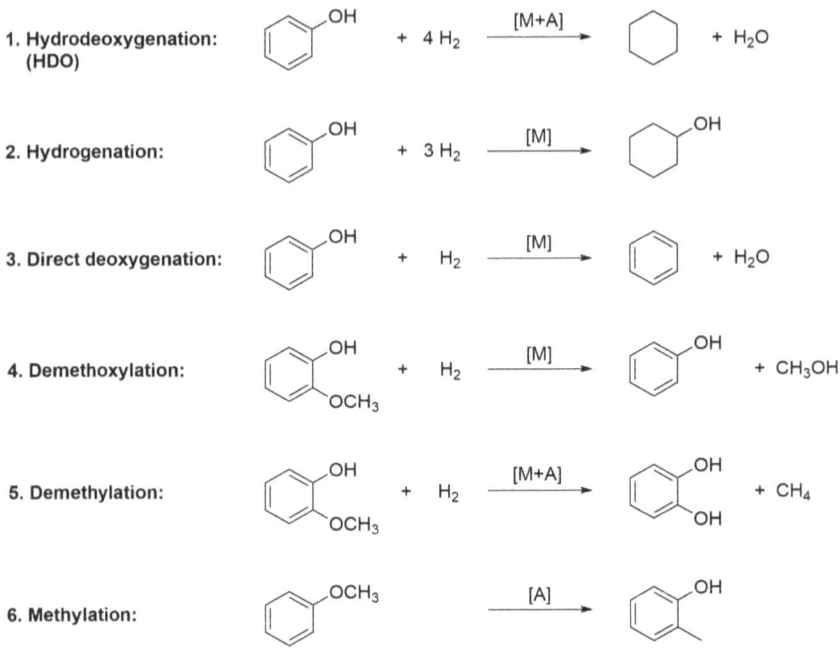

Scheme 1 Model reactions on oxygenated aromatics exemplifying typical classes of hydrotreating processes on lignin-derived streams. The nature of the catalytic site(s) is indicated by [A] for acid sites, [M] for metal sites, [M+A] bifunctional catalysts presenting both acid and metal sites.

Table 1 List of examples for metal phosphide catalysts applied to HDO reactions of lignin model compounds (*entries 1–20*), technical lignins (*entries 21–22*), and pyrolysis bio-oils (*entry 23*).

Entry	Catalyst	Substrate	Products	References
1	Ni_2P, MoP, NiMoP supported on SiO_2	Anisole	With increased Ni content, mainly cyclohexane	106
2	Ni_2P, WP, MoP, Co_2P, Fe_2P supported on SiO_2	Guaiacol	Mainly aromatic products (Ni_2P/SiO_2 found as the most active catalyst)	91
3	Unsupported MoP	4-Methylphenol	Mainly toluene and methylcyclohexane	107
4	Unsupported FeMoP	Phenol, anisole	Mainly aromatic products	108

Table 1 List of examples for metal phosphide catalysts applied to HDO reactions of lignin model compounds (*entries 1–20*), technical lignins (*entries 21–22*), and pyrolysis bio-oils (*entry 23*).—cont'd

Entry	Catalyst	Substrate	Products	References
5	Ni_2P/SiO_2, Ni_2P/ZrO_2, Ni_2P/Al_2O_3	Guaiacol	With Ni_2P/SiO_2: phenol and benzene; other catalysts: catechol and coke	[109]
6	MoP/TiO_2	Phenol	Mainly benzene	[110]
7	Ni_2P/SiO_2	Guaiacol	Benzene or cyclohexane (depending on the reaction conditions)	[111]
8	Ni_3P/SiO_2, $Ni_{12}P_5/SiO_2$, Ni_2P/SiO_2	Guaiacol	With Ni_2P/SiO_2: benzene; other catalysts: catechol and coke	[112]
9	Ni_2P/SiO_2, CoP/SiO_2	Dibenzofuran	Aliphatic products	[113]
10	Ni_2P/SiO_2, Ni_2P/ZrO_2, Ni_2P/AC	Guaiacol	Mainly cyclohexane	[114]
11	Ni_2P/SiO_2, Ni_2P/ZSM-5, Ni_2P/USY, Ni_2P/ASA	Guaiacol	With Ni_2P/SiO_2: benzene; other catalysts: catechol	[115]
12	Ni_2P/SiO_2	Phenol	Mainly benzene and cyclohexane	[116]
13	Ni_2P, Co_2P, MoP supported on Al-SBA-15, γ-Al_2O_3 or CMK-3	Phenol	With Ni_2P/Al-SBA-15: mainly cyclohexane; other catalysts: oxygenated products	[117]
14	Ni_2P/SiO_2 promoted by Pt, Rh, Ir, or Ru	Dibenzofuran	Mainly bicyclohexane	[118]
15	Ni_2P/SiO_2, Rh_2P/SiO_2	Guaiacol	Mainly benzene Rh_2P/SiO_2 shows better activity	[119]

Continued

Table 1 List of examples for metal phosphide catalysts applied to HDO reactions of lignin model compounds (*entries 1–20*), technical lignins (*entries 21–22*), and pyrolysis bio-oils (*entry 23*).—cont'd

Entry	Catalyst	Substrate	Products	References
16	Unsupported FeMoP	Phenol	Benzene or cyclohexane depending on the preparation method	71
17	$Ni_2P/Al_2O_3@TiO_2$, Ni_2P/Al_2O_3	Benzofuran	Mostly aliphatic products	120
18	Hydrophobic reduced graphene oxide supported Ni-B-P-O and Co-B-P-O catalysts	Cresol	Mostly aliphatic products	121
19	Ni_2P/SiO_2	Cresol	Aliphatics or aromatics	67
20	Ni_2P/SiO_2	Dibenzofuran	Bicyclohexane	122
21	WP, WP/SiO_2, WP/AC, Ni-WP/AC, Fe-WP/AC	Alkaline lignin	Phenols (yield: 67.0 mg/g lignin)	123
22	$CuMgAlO_x$ combined with Ni_2P/SiO_2	Alkaline lignin	53% Lignin monomer yield at a deoxygenation degree of 83% HDO	124
23	Ni_2P/SiO_2	2-MTHF and pyrolysis vapors	Moderate oxygen removal (50%) from pyrolysis vapors	90

Li et al. investigated bimetallic $NiMoP/SiO_2$ for HDO of anisole.[106] They reported that both catalytic activity and saturated products' selectivity improve with increasing Ni content, similar to NiMoS catalysts (Table 1, entry 1). Rensel et al. also studied a bimetallic metal phosphide (unsupported FeMoP), suggesting that aromatics' selectivity depends on catalyst acidity.[108] Comparing metal phosphide and a commercial $CoMoS/Al_2O_3$ indicated the metal phosphide catalysts to be more active than transition metal sulfide catalysts (Table 1, entry 1–2 and 19).[67,91,106]

Some studies have been performed on technical lignins or biomass feedstocks (Table 1, entry 21–22). Ma et al. investigated the conversion of alkaline lignin into phenols using different supported WP catalysts in the presence of hydrogen.[123] In this instance, they found WP/AC (AC: activated carbon) to provide a phenols yield of 67.0 mg/g lignin (Table 1, entry 21).

Korányi et al.[124] reported a combination of $CuMgAlO_x$ and Ni_2P/SiO_2 to convert alkaline lignin in supercritical ethanol (which also served as a hydrogen source for the process, Table 1, entry 22). In this catalyst system, $CuMgAlO_x$ provides the hydrogen from ethanol and depolymerizes the lignin to intermediates of low molecular weight,[53] Ni_2P/SiO_2 promoted the intermediates' HDO, resulting in a lignin monomer yield of 53 wt% at an HDO degree of 83 wt%.[124]

Koike et al. investigated the upgrading of biomass pyrolysis vapors in the presence of Ni_2P/SiO_2 (Table 1, entry 23).[90] For this purpose, a two-stage system was used, in which biomass was pyrolyzed using a fluidized bed reactor, and then the pyrolysis vapors were upgraded in a fluidized bed reactor in the presence of NiP_2/SiO_2. The process generated phenols as the main products. In this process, Ni_2P/SiO_2 outperformed Ni/SiO_2, Pd/C, and acid catalysts.

5. Advanced biofuels from lignin streams

For the manufacture of advanced biofuels from lignin-derived streams, product selectivity and product distribution control are crucial. In this context, the sustainable manufacture of advanced biofuels would benefit from HDO catalysts capable of switching the selectivity from aliphatics to arenes. Such flexibility in process operation allows for meeting different regional and temporal demands for hydrogen storage/consumption in the advanced biorefinery site.[24,30,46,48]

In the HDO of lignin oils from lignin-first biorefining, we demonstrated that the product selectivity could be switched from saturated hydrocarbons to arenes by adjusting H_2 pressure and temperature (Fig. 3).[24]

For both reactions, colorless transparent product mixtures were obtained. No lignin-derived insoluble char was formed. HDO of lignin oil carried out under 5 MPa H_2 at 573 K generated a product mixture containing saturated cycloalkanes (Fig. 3A). In contrast, the HDO process performed under 0.5–1 MPa H_2 at 623 K rendered aromatics as the primary products along with saturated cycloalkanes (Fig. 3C). Building upon our initial results, Cao et al. reported a study combining experiment and theory to identify the mechanisms underlining the switchability of product

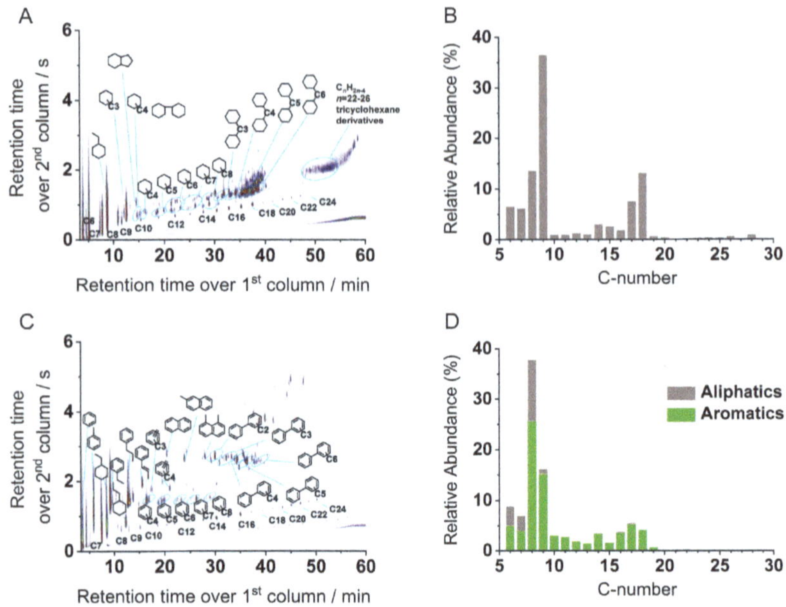

Fig. 3 Characteristic bio-hydrocarbons and product distributions obtained from HDO of lignin oil in the presence of a phosphidated Ni/SiO_2 catalyst. (A) and (C) GC × GC − MS (FID) traces highlighting volatile products obtained from the HDO of poplar lignin oil. As estimated by thermogravimetric analysis, up to 90% of products are volatile at 300 °C, the gas chromatograph's injector temperature. (B) and (D) Distribution of products according to their C-atoms. Reaction conditions: (A) and (B) Aliphatics-directed HDO: 5 MPa H_2, 573 K; (C) and (D) Aromatics-directed HDO: 0.5–1 MPa H_2, 623 K. *Reproduced with permission from Cao, Z.; Dierks, M.; Clough, M. T.; Daltro de Castro, I. B.; Rinaldi, R.* Joule ***2018,*** 2 *(6), 1118–1133.*

distribution in the presence of Ni_2P catalysts.[46] They found that the selectivity is controlled by the availability of hydrogen on the catalyst surface. DFT calculations showed that the activation barriers strongly depend on the surface hydrogen coverage. They found that when H_2 coverage increases, all activation barriers decrease; however, their relative values change. Therefore, this feature enables directing the HDO selectivity to aliphatics or aromatic hydrocarbons, depending on the H_2 pressure.[46]

Notably, as presented in Fig. 3B and D, both HDO processes produced two well-separated cuts of hydrocarbons (C_6–C_{10} and C_{14}–C_{20}), which could directly be utilized as drop-in biofuels. To understand this remarkable product distribution profile, one should consider that the fractionation of lignocellulosic biomass wastes by lignin-first biorefining preserves the lignin species from condensation. In this manner, the depolymerized lignin stream is generated with a high structural uniformity (herein defined as the similarity in contiguous carbon backbones, which are terminated by an ether linkage). Therefore, in the depolymerization of lignin oligomers followed by HDO, provided no cracking or alkylation takes place, there should be no change in the resulting units' carbon backbones. In this manner, carbon backbones are preserved in the stream throughout the HDO process, producing well-separated hydrocarbon cuts.

By rendering two well-separated cuts of branched hydrocarbons (gasoline: C_6–C_{10}) and (kerosene/diesel: C_{14}–C_{20}), the manufacture of lignin fuels provides a critical technological advantage over the traditional Fischer–Tropsch (FT) synthesis of linear hydrocarbons, as summarized in Fig. 4. In industrial practice, FT synthesis forms large amounts of wax and paraffin. Thereby, for adjusting FT product distribution to meet the market demand for high-performance fuels, expensive hydrotreatment units are required (i.e., hydrocracking and hydroisomerization). This requirement is relaxed by a task-specific HDO catalyst able to preserve the lignin streams' uniformity as close as obtained from lignin-first biorefining.

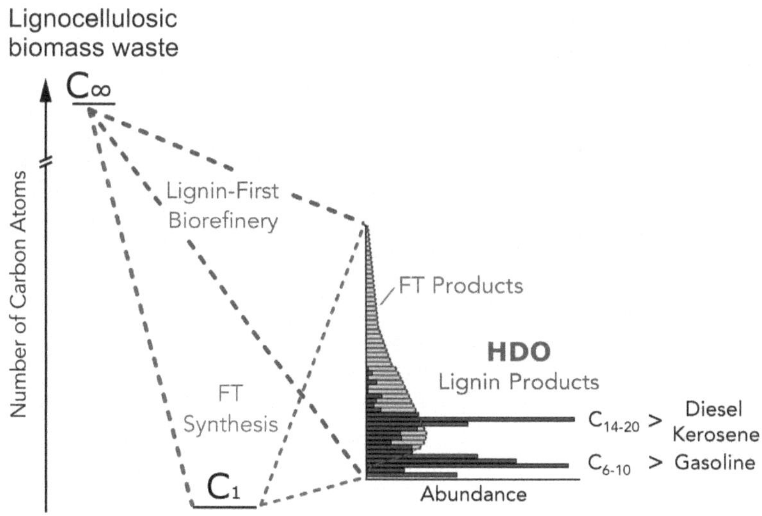

Fig. 4 Comparison between HDO lignin products (well-separated cuts of hydrocarbons: C_6–C_{10} and C_{14}–C_{20}) and Fischer-Tropsch (FT) products (following the Anderson–Flory–Schultz probability distribution for chain growth, which substantial formation of wax and paraffin, $C_{20–40}$, at a high chain growth probability).

6. Mechanisms underlining the catalyst deactivation

The limited stability of metal phosphide catalysts under HDO conditions constitutes a significant challenge for their practical application.[100–105] The most critical catalyst deactivation routes are (1) *coking* and (2) *metal phosphide oxidation*. In this section, the fundamental aspects of metal phosphide deactivation are discussed.

By comparing different nickel phosphide phases in the conversion of guaiacol, Wu et al. found that the less acidic Ni_2P/SiO_2 catalyst was less prone to coke deactivation than more acidic counterparts, $Ni_{12}P_5/SiO_2$ and Ni_3P/SiO_2 (Table 1, entry 8).[112] This result suggests that strong Brønsted acid sites participate in the coke formation (e.g., P-OH groups on the metal phosphide phase). Nonetheless, the Lewis acid sites on ZrO_2 and Al_2O_3 are also known to further facilitate the coke formation.[109,125] In the HDO of guaiacol, acid sites in Ni_2P/ZrO_2 and Ni_2P/Al_2O_3 promoted guaiacol demethylation to catechol, which was identified as a coke precursor and, thus, the factor responsible for catalyst deactivation (Table 1, entry 5).[109] In contrast, the relatively stable Ni_2P/SiO_2 promoted direct deoxygenation and demethoxylation reactions (Table 1, entries 5 and 11).[109,115] Therefore, to alleviate the deactivation problems associated with coke formation, neutral supports should be utilized in the catalyst synthesis (SiO_2).[91]

Water can oxidize metal phosphides deactivating these catalysts.[70,106,114,120,126] Li et al. showed the catalyst deactivation to depend on the metal oxophilicity in the metal phosphide phase. Thus, Ni_2P/SiO_2 was more resistant than MoP/SiO_2 against oxidation (Table 1, entry 1).[106] Similarly, Infantes-Molina et al. found that Ni_2P/SiO_2 was more stable than CoP/SiO_2 in dibenzofuran HDO (Table 1, entry 9).[113] Ni_2P/SiO_2 catalyst presented an increased resistance against oxidation compared to the CoP/SiO_2 catalyst. The authors proposed that the higher population of P surface sites contributed to the higher stability of Ni_2P/SiO_2 relative to CoP/SiO_2. Cecilia et al. also reported the stability of Ni_2P/SiO_2 in the HDO of dibenzofuran to improve with the increase in the P:Ni ratio employed in catalyst synthesis (Table 1, entry 20). The mechanism proposed to protect the metal against oxidation involves P sites and water reacting to form P-OH species. The oxyphosphide species are proposed to interact preferentially with water, thus protecting the active phase from further oxidation.[122] This explanation is supported by previous studies on gas-shift reactions catalyzed by Ni_2P.[68,70]

Cecilia et al. found noble metals increased the stability of Ni_2P/SiO_2 (Table 1, entry 14).[118] Compared to Ni_2P/SiO_2 catalyst, $Ni_{1.8}Rh_{0.2}P/SiO_2$, $Ni_{1.8}Ru_{0.2}P/SiO_2$, and $Ni_{1.8}Ir_{0.2}P/SiO_2$ showed an improved activity and stability attributed to the increase in the occupation of the catalyst surface by hydrogen.

Moon et al. observed a strong deactivation of Ni_2P/SiO_2 during guaiacol HDO (Table 1, entry 10).[114] The authors compared Ni_2P/ZrO_2, Ni_2P/AC (AC: activated carbon), and Ni_2P/SiO_2. Ni_2P/SiO_2 catalyst showed a superior activity, but already in the second run, the activity dropped significantly, which was not the case for Ni_2P/ZrO_2 and Ni_2P/AC. Ni K-edge X-ray absorption near edge structure spectroscopy revealed that Ni_2P/SiO_2 underwent oxidation in the presence of phenol and water at 573 K, while Ni_2P/ZrO_2 displayed no changes. This observation appears to be related to SiO_2 hydrophilicity, which facilitates the hydration of the surface, allowing for the active phase oxidation. Consequently, it was proposed that using less hydrophilic supports would be beneficial for the stability of Ni_2P catalysts in HDO reactions.

Song et al. reported benzofuran HDO in the presence of a Ni_2P/Al_2O_3 catalyst modified with a hydrophobic TiO_2 shell. Increased stability and activity for the TiO_2 modified catalyst were achieved. The authors identified the hydrophobic TiO_2 shell as the factor responsible for the better catalyst performance (Table 1, entry 17).[120] In line with these results, Wang et al. compared Ni-B-P-O and Co-B-P-O supported on hydrophobic graphene oxide, prepared by reducing oxygen-containing functional groups on the catalyst surface. The hydrophobic catalysts exhibited a better selectivity to deoxygenated products in HDO reactions performed in the presence of water (Table 1, entry 18).

Recently, Dierks et al. presented a systematic study of the impact of support polarity on the activity and stability of the Ni_2P catalyst phase in the HDO of guaiacol.[47] Ni_2P supported on silica was chosen as a model system since it is known to show good HDO activity, yet due to silica's hydrophilic character, these catalysts show strong deactivation by oxidation processes involving water.[106,114] Model catalysts were prepared using a sol–gel approach. The polarity of the final catalyst can be adjusted by copolymerization of tetraethyl orthosilicate (TEOS) with different types and amounts of organosilanes in the presence of nickel nitrate and citric acid. After thermal treatment under an inert atmosphere, the obtained Ni/SiO_2 catalysts were converted into the corresponding Ni_2P phase via phosphidation with PH_3 generated by thermal decomposition of NaH_2PO_2. Recycling experiments performed using guaiacol as a substrate revealed that it is possible to stabilize the catalyst against water deactivation by copolymerizing TEOS with 50 mol% methyltriethoxysilane. The improved catalyst stability correlated well with its surface polarity properties, as determined by inverse gas chromatography.

7. Outlook

Metal phosphides are unique for HDO of lignin-derived streams. Among the different metal phosphide phases, Ni_2P displayed the most promising performance in terms of activity and stability. However, Ni_2P supported on acidic supports (e.g., ZrO_2, Al_2O_3) suffer from deactivation due to coke deposition. This problem can be circumvented using less acidic supports (SiO_2). Nonetheless, oxidation of the active phase remains an issue that needs to be solved. Therefore, we identify the synthetic methodologies improving metal phosphides' stability by utilizing hydrophobic supports[47,93,120,121] or organic phosphines as task-specific phosphidation agents producing water-tolerant phosphidated metal catalysts,[24,127–130] as timely research frontiers to advance fundamental knowledge on metal phosphides' chemistry toward task-specific applications.

The uniqueness of metal phosphides in controlling product selectivity and product distribution plays a central role in the sustainable manufacture of advanced biofuels from lignin-derived streams. In the presence of Ni_2P/SiO_2 or phosphidated Ni/SiO_2, switchable selectivity from aromatics to aliphatics bio-hydrocarbons from biophenols is attainable, choosing the HDO conditions (temperature and H_2 pressure).[24,46] Lignocellulosic biomass fractionation via lignin-first biorefining provided a depolymerized lignin stream of high structural uniformity.[14,18,19,43,44] In this manner, by utilizing catalysts presenting low cracking and alkylation activities, the HDO processes preserve the initial stream's carbon backbones to a great extent. As such, both aromatics- and aliphatics-directed HDO processes produced two well-separated cuts of branched hydrocarbons, which could serve as drop-in replacements of gasoline (C_6–C_{10}) and kerosene/diesel (C_{14}–C_{20}).[24] In this manner, the HDO of lignin could well hold the key to process intensification while offering a promising pathway toward sustainable manufacture of advanced fuels via better utilization of natural resources.

Acknowledgments

The authors are grateful to the CASCATBEL project framework funded by the European Commission (Grant Agreement No. 604307). The authors are indebted to Prof. Ferdi Schüth for insightful discussions on materials chemistry and supportive mentorship. R.R. acknowledges the ERC Consolidator Grant (LIGNINFIRST, Project Number: 725762).

References

1. Oksanen, S.; Gielen, D.; Kang, S.; Leme, R.; Masuyama, T. *Advanced Biofuels: What Holds Them Back?*; International Renewable Energy Agency (IRENA): Abu Dhabi, 2019.
2. Field, J. L.; Richard, T. L.; Smithwick, E. A. H.; Cai, H.; Laser, M. S.; LeBauer, D. S.; Long, S. P.; Paustian, K.; Qin, Z.; Sheehan, J. J.; Smith, P.; Wang, M. Q.; Lynd, L. R. *Proc. Natl. Acad. Sci. U. S. A.* **2020**, *117* (36), 21968–21977.
3. Pretty, J.; Bharucha, Z. P. *Ann. Bot.* **2014**, *114* (8), 1571–1596.
4. Tuck, C. O.; Pérez, E.; Horváth, I. T.; Sheldon, R. A.; Poliakoff, M. *Science* **2012**, *337* (6095), 695–699.
5. Barros, S. *Global Agricultural Information Network (GAIN) Report*; 2019; pp. 1–30.
6. Bezerra, T. L.; Ragauskas, A. J. *Biofuels Bioprod. Biorefin.* **2016**, *10* (5), 634–647.
7. *Agricultural Outlook 2020–2029*; 2020.
8. https://www.plasticseurope.org/.
9. Guerra, J.; Coleta, J.; Arruda, L.; Silva, G.; Kulay, L. *Int. J. Life. Cycle Assess.* **2014**, *19*, 814–825.
10. Turdera, M. V.; Garcia, M. D. S. *Low Carbon Transition—Technical, Economic and Policy Assessment*; InTech, 2018. https://doi.org/10.5772/intechopen.76251.
11. Hofsetz, K.; Silva, M. A. *Biomass Bioenergy* **2012**, *46*, 564–573.
12. Dantas, G. A.; Legey, L. F. L.; Mazzone, A. *Renew. Sustain. Energy Rev.* **2013**, *21*, 356–364.
13. Wang, H.; Tucker, M.; Ji, Y. *J. Appl. Chem.* **2013**, *2013*, 1–9.
14. Rinaldi, R.; Jastrzebski, R.; Clough, M. T.; Ralph, J.; Kennema, M.; Bruijnincx, P. C. A.; Weckhuysen, B. M. *Angew. Chem. Int. Ed.* **2016**, *55* (29), 8164–8215.
15. Bruijnincx, P. C. A.; Rinaldi, R.; Weckhuysen, B. M. *Green Chem.* **2015**, *17* (11), 4860–4861.
16. Rinaldi, R. *Angew. Chem. Int. Ed.* **2014**, *53* (33), 8559–8560.
17. Ragauskas, A. J.; Beckham, G. T.; Biddy, M. J.; Chandra, R.; Chen, F.; Davis, M. F.; Davison, B. H.; Dixon, R. A.; Gilna, P.; Keller, M.; et al. *Science* **2014**, *344* (6185). https://doi.org/10.1126/science.1246843.
18. Schutyser, W.; Renders, T.; Van den Bosch, S.; Koelewijn, S. F.; Beckham, G. T.; Sels, B. F. *Chem. Soc. Rev.* **2018**, *47* (3), 852–908.
19. Questell-Santiago, Y. M.; Galkin, M. V.; Barta, K.; Luterbacher, J. S. *Nat. Rev. Chem.* **2020**, *4* (6), 311–330.
20. Huber, G. W.; Iborra, S.; Corma, A. *Chem. Rev.* **2006**, *106*, 4044–4098.
21. Huber, G. W.; Corma, A. *Angew. Chem. Int. Ed.* **2007**, *46* (38), 7184–7201.
22. Fasahati, P.; Maravelias, C. T. *Joule* **2018**, *2* (10), 1915–1919.
23. Lucena, J. D. A. Y.; Lucena, K.Â. A. *Clean Energy* **2019**, *3* (2), 69–84.
24. Cao, Z.; Dierks, M.; Clough, M. T.; Daltro de Castro, I. B.; Rinaldi, R. *Joule* **2018**, *2* (6), 1118–1133.
25. Arapova, O. V.; Chistyakov, A. V.; Tsodikov, M. V.; Moiseev, I. I. *Pet. Chem.* **2020**, *60* (3), 227–243.
26. Galbe, M.; Wallberg, O. *Biotechnol. Biofuels* **2019**, *12* (1), 294.

27. Singh, S.; Cheng, G.; Sathitsuksanoh, N.; Wu, D.; Varanasi, P.; George, A.; Balan, V.; Gao, X.; Kumar, R.; Dale, B. E.; et al. *Front. Energy Res.* **2015**, *2* (62). https://doi.org/10.3389/fenrg.2014.00062.
28. Calvaruso, G.; Clough, M. T.; Rechulski, M. D. K.; Rinaldi, R. *ChemCatChem* **2017**, *9* (14), 2691–2700.
29. Calvaruso, G.; Clough, M. T.; Rinaldi, R. *Green Chem.* **2017**, *19* (12), 2803–2811.
30. Cao, Z.; Engelhardt, J.; Dierks, M.; Clough, M. T.; Wang, G. H.; Heracleous, E.; Lappas, A.; Rinaldi, R.; Schüth, F. *Angew. Chem. Int. Ed.* **2017**, *56* (9), 2334–2339.
31. Ji, N.; Wang, X.; Weidenthaler, C.; Spliethoff, B.; Rinaldi, R. *ChemCatChem* **2015**, *7* (6), 960–966.
32. Wang, X.; Rinaldi, R. *Angew. Chem. Int. Ed.* **2013**, *52* (44), 11499–11503.
33. Wang, X.; Rinaldi, R. *Energ. Environ. Sci.* **2012**, *5* (8), 8244–8260.
34. Wang, X.; Rinaldi, R. *ChemSusChem* **2012**, *5* (8), 1455–1466.
35. Johnson, D. K.; Chornet, E.; Zmierczak, W.; Shabtai, J. *Fuel Chem. Div. Prepr.* **2002**, *47* (1), 380e381.
36. Barrett, J. A.; Jones, Z. R.; Stickelmaier, C.; Schopp, N.; Ford, P. C. *ACS Sustain. Chem. Eng.* **2018**, *6* (11), 15119–15126.
37. Liu, S.; Tamura, M.; Nakagawa, Y.; Tomishige, K. *ACS Sustain. Chem. Eng.* **2014**, *2* (7), 1819–1827.
38. Op de Beeck, B.; Dusselier, M.; Geboers, J.; Holsbeek, J.; Morré, E.; Oswald, S.; Giebeler, L.; Sels, B. F. *Energ. Environ. Sci.* **2015**, *8* (1), 230–240.
39. Huber, G. W.; Cortright, R. D.; Dumesic, J. A. *Angew. Chem. Int. Ed.* **2004**, *43* (12), 1549–1551.
40. Zakzeski, J.; Bruijinincx, P. C. A.; Jongerius, A. L.; Weckhuysen, B. M. *Chem. Rev.* **2010**, *110*, 3552–3599.
41. Liu, Y.; Li, G.; Hu, Y.; Wang, A.; Lu, F.; Zou, J.-J.; Cong, Y.; Li, N.; Zhang, T. *Joule* **2019**, *3* (4), 1028–1036.
42. Liu, Y.; Wang, R.; Qi, H.; Liu, X.; Li, G.; Wang, A.; Wang, X.; Zhang, T.; Li, N. *Nat. Commun.* **2021**, *12*. https://doi.org/10.1038/s41467-020-20264-3.
43. Abu-Omar, M. M.; Barta, K.; Beckham, G. T.; Luterbacher, J. S.; Ralph, J.; Rinaldi, R.; Román-Leshkov, Y.; Samec, J. S. M.; Sels, B. F.; Wang, F. *Energ. Environ. Sci.* **2021**, *14* (1), 262–292.
44. Ferrini, P.; Rinaldi, R. *Angew. Chem. Int. Ed.* **2014**, *53* (33), 8634–8639.
45. Ferrini, P.; Rezende, C. A.; Rinaldi, R. *ChemSusChem* **2016**, *9* (22), 3171–3180.
46. Cao, Z.; Xu, Y.; Lyu, P.; Dierks, M.; Morales-García, Á.; Schrader, W.; Nachtigall, P.; Schüth, F. *ChemSusChem* **2021**, *14* (1), 373–378.
47. Dierks, M.; Cao, Z.; Manayil, J. C.; Akilavasan, J.; Wilson, K.; Schüth, F.; Rinaldi, R. *ChemCatChem* **2018**, *10* (10), 2219–2231.
48. Wang, G. H.; Cao, Z.; Gu, D.; Pfänder, N.; Swertz, A. C.; Spliethoff, B.; Bongard, H. J.; Weidenthaler, C.; Schmidt, W.; Rinaldi, R.; et al. *Angew. Chem. Int. Ed.* **2016**, *55* (31), 8850–8855.
49. Rinaldi, R.; Schüth, F. *Energ. Environ. Sci.* **2009**, *2* (6), 610–626.
50. Leitner, W.; Klankermayer, J.; Pischinger, S.; Pitsch, H.; Kohse-Höinghaus, K. *Angew. Chem. Int. Ed.* **2017**, *56* (20), 5412–5452.
51. Boot, M. D.; Tian, M.; Hensen, E. J. M.; Mani Sarathy, S. *Prog. Energy Combust. Sci.* **2017**, *60*, 1–25.
52. Cheng, C.; Truong, J.; Barrett, J. A.; Shen, D.; Abu-Omar, M. M.; Ford, P. C. *ACS Sustain. Chem. Eng.* **2020**, *8* (2), 1023–1030.
53. Barta, K.; Ford, P. C. *Acc. Chem. Res.* **2014**, *47* (5), 1503–1512.
54. Barta, K.; Matson, T. D.; Fettig, M. L.; Scott, S. L.; Iretskii, A. V.; Ford, P. C. *Green Chem.* **2010**, *12* (9), 1640–1647.
55. Song, Q.; Wang, F.; Cai, J.; Wang, Y.; Zhang, J.; Yu, W.; Xu, J. *Energ. Environ. Sci.* **2013**, *6* (3), 994–1007.

56. https://webbook.nist.gov/.
57. Schüth, F. In *Catalytic Hydrogenation for Biomass Valorization*; Rinaldi, R., Ed.; RSC, 2014. https://doi.org/10.1039/9781782620099-0000.
58. Prins, R.; Bussell, M. E. *Catal. Lett.* **2012**, *142* (12), 1413–1436.
59. Oyama, S. T.; Gott, T.; Zhao, H. Y.; Lee, Y. K. *Catal. Today* **2009**, *143* (1–2), 94–107.
60. Sweeny, N. P.; Rohrer, C. S.; Brown, O. W. *J. Am. Chem. Soc.* **1958**, *80* (4), 799–800.
61. Muetterties, E. L.; Sauer, J. C. *J. Am. Chem. Soc.* **1974**, *96* (11), 3410–3415.
62. Robinson, A. M.; Hensley, J. E.; Medlin, J. W. *ACS Catal.* **2016**, *6* (8), 5026–5043.
63. Lee, Y.-K.; Oyama, S. T. *J. Catal.* **2006**, *239* (2), 376–389.
64. Chen, J. X.; Shi, H.; Li, L.; Li, K. L. *Appl. Catal. Environ.* **2014**, *144*, 870–884.
65. Liu, X.; Chen, J.; Zhang, J. *Ind. Chem. Eng. Chem. Res.* **2008**, *47* (15), 5362–5368.
66. Oyama, S. T.; Lee, Y. K. *J. Catal.* **2008**, *258* (2), 393–400.
67. Gonçalves, V. O. O.; de Souza, P. M.; da Silva, V. T.; Noronha, F. B.; Richard, F. *Appl. Catal. Environ.* **2017**, *205*, 357–367.
68. Liu, P.; Rodriguez, J. A.; Takahashi, Y.; Nakamura, K. *J. Catal.* **2009**, *262* (2), 294–303.
69. Liu, P.; Rodriguez, J. A.; Asakura, T.; Gomes, J.; Nakamura, K. *J. Phys. Chem. B* **2005**, *109* (10), 4575–4583.
70. Ruddy, D. A.; Schaidle, J. A.; Ferrell, J. R.; Wang, J.; Moens, L.; Hensley, J. E. *Green Chem.* **2014**, *16* (2), 454–490.
71. Rensel, D. J.; Kim, J.; Bonita, Y.; Hicks, J. C. *Appl. Catal. A Gen.* **2016**, *524*, 85–93.
72. Oyama, S. T.; Wang, X.; Lee, Y. K.; Bando, K.; Requejo, F. G. *J. Catal.* **2002**, *210* (1), 207–217.
73. Oyama, S. T.; Wang, X.; Lee, Y. K.; Chun, W. J. *J. Catal.* **2004**, *221* (2), 263–273.
74. Cecilia, J. A.; Infantes-Molina, A.; Rodriguez-Castellon, E.; Jimenez-Lopez, A. *Appl. Catal. Environ.* **2009**, *92* (1–2), 100–113.
75. Kanda, Y.; Temma, C.; Nakata, K.; Kobayashi, T.; Sugioka, M.; Uemichi, Y. *Appl. Catal. A Gen.* **2010**, *386* (1), 171–178.
76. Song, L. M.; Zhang, S. J.; Wei, Q. W. *Catal. Commun.* **2011**, *12* (12), 1157–1160.
77. Kanda, Y.; Temma, C.; Sawada, A.; Sugioka, M.; Uemichi, Y. *Appl. Catal. A Gen.* **2014**, *475*, 410–419.
78. Sawada, A.; Kanda, Y.; Sugioka, M.; Uemichi, Y. *Catal. Commun.* **2014**, *56*, 60–64.
79. Roberto Gomes, J.; Bezergianni, S.; Zotin, J. L.; Sousa-Aguiar, E. F. *Catalytic Hydrogenation for Biomass Valorization*; The Royal Society of Chemistry, 2015. https://doi.org/10.1039/9781782620099-00204.
80. Pimenta, J.; Barreto, R. D. T.; dos Santos, O. A. A.; Jorge, L. M. D. *Environ. Prog. Sustain. Energy* **2020**, *39* (5). https://doi.org/10.1002/ep.13450.
81. Camargo, M. D.; Pimenta, J.; Camargo, M. D.; Arroyo, P. A. *Fuel* **2020**, *281*. https://doi.org/10.1016/j.fuel.2020.118719.
82. Zhang, Z. G.; Bi, G. C.; Zhang, H. D.; Zhang, A. P.; Li, X.; Xie, J. *Fuel* **2019**, *247*, 26–35.
83. Phimsen, S.; Kiatkittipong, W.; Yamada, H.; Tagawa, T.; Kiatkittipong, K.; Laosiripojana, N.; Assabumrungrat, S. *Energ. Conver. Manage.* **2017**, *151*, 324–333.
84. Nie, Z. Y.; Zhang, Z. N.; Chen, J. X. *Appl. Surf. Sci.* **2017**, *420*, 511–522.
85. Liu, X. G.; Li, Z. Y.; Zhang, B. Q.; Hu, M. C. *Fuel* **2017**, *204*, 144–151.
86. Kordulis, C.; Bourikas, K.; Gousi, M.; Kordouli, E.; Lycourghiotis, A. *Appl. Catal. Environ.* **2016**, *181*, 156–196.
87. Arun, N.; Sharma, R. V.; Dalai, A. K. *Renew. Sustain. Energy Rev.* **2015**, *48*, 240–255.
88. Zeng, Y.; Zhao, B. H.; Zhu, L. F.; Tong, D. M.; Hu, C. W. *RSC Adv.* **2013**, *3* (27), 10806–10816.
89. Furimsky, E. *Catal. Today* **2013**, *217*, 13–56.
90. Koike, N.; Hosokai, S.; Takagaki, A.; Nishimura, S.; Kikuchi, R.; Ebitani, K.; Suzuki, Y.; Oyama, S. T. *J. Catal.* **2016**, *333*, 115–126.

91. Zhao, H. Y.; Li, D.; Bui, P.; Oyama, S. T. *Appl. Catal. A Gen.* **2011**, *391* (1–2), 305–310.
92. Inocêncio, C. V. M.; de Souza, P. M.; Rabelo-Neto, R. C.; da Silva, V. T.; Noronha, F. B. *Catal. Today* **2020**. https://doi.org/10.1016/j.cattod.2020.07.077.
93. Leal Mendes, F.; Teixeira da Silva, V.; Edral Pacheco, M.; de Rezende Pinho, A.; Assumpção Henriques, C. *Fuel* **2020**, *264*. https://doi.org/10.1016/j.fuel.2019.116764.
94. Mendes, F. L.; da Silva, V. T.; Pacheco, M. E.; Toniolo, F. S.; Henriques, C. A. *Fuel* **2019**, *241*, 686–694.
95. Samec, J. S. M. *Chem* **2018**, *4* (6), 1199–1200.
96. Zacher, A. H.; Olarte, M. V.; Santosa, D. M.; Elliott, D. C.; Jones, S. B. *Green Chem.* **2014**, *16* (2), 491–515.
97. Mortensen, P. M.; Grunwaldt, J. D.; Jensen, P. A.; Knudsen, K. G.; Jensen, A. D. *Appl. Catal. A Gen.* **2011**, *407* (1–2), 1–19.
98. Adjaye, J. D.; Bakhshi, N. *Fuel Process. Technol.* **1995**, *45* (3), 161–183.
99. Wildschut, J.; Mahfud, F. H.; Venderbosch, R. H.; Heeres, H. J. *Ind. Chem. Eng. Chem. Res.* **2009**, *48* (23), 10324–10334.
100. Zhang, J. H.; Sun, J. M.; Wang, Y. *Green Chem.* **2020**, *22* (4), 1072–1098.
101. Smith, K. J. *Curr. Opin. Green Sustain. Chem.* **2020**, *22*, 47–53.
102. Golubeva, M. A.; Zakharyan, E. M.; Maximov, A. L. *Pet. Chem.* **2020**, *60* (10), 1109–1128.
103. Han, Y. L.; Gholizadeh, M.; Tran, C. C.; Kaliaguine, S.; Li, C. Z.; Olarte, M.; Garcia-Perez, M. *Fuel Process. Technol.* **2019**, *195*. https://doi.org/10.1016/j.fuproc.2019.106140.
104. Si, Z.; Zhang, X. H.; Wang, C. G.; Ma, L. L.; Dong, R. J. *Catalysts* **2017**, 7 (6). https://doi.org/10.3390/catal7060169.
105. Bussell, M. E. *React. Chem. Eng.* **2017**, *2* (5), 628–635.
106. Li, K. L.; Wang, R. J.; Chen, J. X. *Energy Fuel* **2011**, *25* (3), 854–863.
107. Whiffen, V. M. L.; Smith, K. J.; Straus, S. K. *Appl. Catal. A-Gen.* **2012**, *419*, 111–125.
108. Rensel, D. J.; Rouvimov, S.; Gin, M. E.; Hicks, J. C. *J. Catal.* **2013**, *305*, 256–263.
109. Wu, S. K.; Lai, P. C.; Lin, Y. C.; Wan, H. P.; Lee, H. T.; Chang, Y. H. *ACS Sustain Chem. Eng.* **2013**, *1* (3), 349–358.
110. Boullosa-Eiras, S.; Lodeng, R.; Bergem, H.; Stocker, M.; Hannevold, L.; Blekkana, E. A. *Catal. Today* **2014**, *223*, 44–53.
111. Moon, J. S.; Kim, E. G.; Lee, Y. K. *J. Catal.* **2014**, *311*, 144–152.
112. Wu, S. K.; Lai, P. C.; Lin, Y. C. *Catal. Lett.* **2014**, *144* (5), 878–889.
113. Infantes-Molina, A.; Gralberg, E.; Cecilia, J. A.; Finocchio, E.; Rodriguez-Castellon, E. *Cat. Sci. Technol.* **2015**, *5* (6), 3403–3415.
114. Moon, J. S.; Lee, Y. K. *Top. Catal.* **2015**, *58* (4–6), 211–218.
115. Oyama, S. T.; Onkawa, T.; Takagaki, A.; Kikuchi, R.; Hosokai, S.; Suzuki, Y.; Bando, K. K. *Top. Catal.* **2015**, *58* (4–6), 201–210.
116. Yan, K. Q.; Li, Y. H.; Zhang, X.; Yang, X.; Zhang, N. W.; Zheng, J. B.; Chen, B. H.; Smith, K. *J. Int. J. Hydrogen Energy* **2015**, *40* (46), 16137–16146.
117. Berenguer, A.; Sankaranarayanan, T. M.; Gomez, G.; Moreno, I.; Coronado, J. M.; Pizarro, P.; Serrano, D. P. *Green Chem.* **2016**, *18* (7), 1938–1951.
118. Cecilia, J. A.; Infantes-Molina, A.; Sanmartin-Donoso, J.; Rodriguez-Aguado, E.; Ballesteros-Plata, D.; Rodriguez-Castellon, E. *Cat. Sci. Technol.* **2016**, *6* (19), 7323–7333.
119. Griffin, M. B.; Baddour, F. G.; Habas, S. E.; Ruddy, D. A.; Schaidle, J. A. *Top. Catal.* **2016**, *59* (1), 124–137.
120. Song, H.; Gong, J.; Song, H.-L.; Li, F.; Zhang, J.; Chen, Y.-G. *Catal. Commun.* **2016**, *85*, 1–4.

121. Wang, W. Y.; Liu, P. L.; Wu, K.; Tan, S.; Li, W. S.; Yang, Y. Q. *Green Chem.* **2016**, *18* (4), 984–988.
122. Cecilia, J. A.; Infantes-Molina, A.; Rodriguez-Castellon, E.; Jimenez-Lopez, A.; Oyama, S. T. *Appl. Catal. Environ.* **2013**, *136*, 140–149.
123. Ma, X. L.; Tian, Y.; Hao, W. Y.; Ma, R.; Li, Y. D. *Appl. Catal. A Gen.* **2014**, *481*, 64–70.
124. Koranyi, T. I.; Huang, X. M.; Coumans, A. E.; Hensen, E. J. M. *ACS Sustain Chem. Eng.* **2017**, *5* (4), 3535–3543.
125. Bredenberg, J. B.; Sarbak, Z. *J. Chem. Technol. Biotechnol.* **1988**, *42* (3), 221–234.
126. Laurent, E.; Delmon, B. *J. Catal.* **1994**, *146* (1), 281–291.
127. Henkes, A. E.; Vasquez, Y.; Schaak, R. E. *J. Am. Chem. Soc.* **2007**, *129* (7), 1896.
128. Henkes, A. E.; Schaak, R. E. *Chem. Mater.* **2007**, *19* (17), 4234–4242.
129. Rinaldi, R.; Porcari, A. D. M.; Rocha, T. C. R.; Cassinelli, W. H.; Ribeiro, R. U.; Bueno, J. M. C.; Zanchet, D. *J. Mol. Catal. A Chem.* **2009**, *301* (1–2), 11–17.
130. Carenco, S.; Liu, Z.; Salmeron, M. *ChemCatChem* **2017**, *9* (12), 2318–2323.

CHAPTER EIGHT

The RCF biorefinery: Building on a chemical platform from lignin

Wouter Arts[†], Dieter Ruijten[†], Korneel Van Aelst[†], Laura Trullemans, and Bert Sels*
Centre for Sustainable Catalysis and Engineering, KU Leuven, Leuven, Belgium
*Corresponding author: e-mail address: bert.sels@kuleuven.be

Contents

Abstract

Lignin valorization will be pivotal for the success of future lignocellulose biorefineries. This chapter focuses on the emerging reductive catalytic fractionation (RCF) technology. In Section 1, the structure of native lignin is briefly discussed and the RCF biorefinery concept is presented. The importance of proper characterization of the resulting lignin fraction as well as their relation to the native lignin structure is highlighted herein. In Section 2, a selection of novel phenolic monomer valorization strategies is discussed including: dimerization toward renewable and safer bisphenols, amination strategies to obtain lignin-derived amines, and several production routes to phenol. In a final part,

[†] These authors contributed equally to this work.

Advances in Inorganic Chemistry, Volume 77
ISSN 0898-8838
https://doi.org/10.1016/bs.adioch.2021.02.006

the importance of a techno-economic and life cycle assessment integrated with a thorough process design for the development of future RCF biorefineries is exemplified by a *lignocellulose-to-phenol* case study.

Abbreviations

4-PG	4-*n*-propylguaiacol
4-PS	4-*n*-propylsyringol
BEP	Break-even Point
BGF	Bisguaiacol F
BPA	Bisphenol A
BPF	Bisphenol F
BPS	Bisphenol S
CAPEX	Capital Investments
CF	Cash Flow
CRLO	Crude Lignin Oil
CUB	Catalytic Upstream Biorefinery
DCM	Dichloromethane
DESs	Deep Eutectic Solvents
DHDC	Dehydrogenative Decarbonylation
EA	Estrogenic Activity
ECHA	European Chemicals Agency
EtOAc	Ethyl Acetate
G	Guaiacyl
GCT	Group Contribution Theory
GWP	Global Warming Potential
H	Hydroxyphenol
HBA	Hydrogen Borrowing Amination
HDO	Hydrodeoxygenation
hERα	Human Estrogen Receptor α
HSQC	Heteronuclear Single Quantum Coherence Spectroscopy
ILs	Ionic Liquids
IRR	Internal Rate of Return
LCA	Life Cycle Assessment
LCI	Life Cycle Inventory
LLE	Liquid-Liquid Extraction
NPV	Net Present Value
OPEX	Operational Costs
PFD	Process Flow Diagram
POH-G	4-*n*-propanolguaiacol
POH-S	4-*n*-propanolsyringol
RCF	Reductive Catalytic Fractionation
$\mathbf{R_{wacc}}$	Weighted Average Cost of Capital
S	Syringyl
SHSs	Switchable Hydrophilicity Solvents
SML	Specific Migration Limit
SVHC	Substance of Very High Concern
TEA	Techno-Economic Assessment
VOCs	Volatile Organic Compounds
Z140-H	H-ZSM-5 zeolite

Introduction

Our society highly depends on the geological reserves of fossil raw materials—gas, coal, and oil—that have formed over millions of years and that are currently exploited massively to generate energy and produce chemicals. The negative impact of these non-renewables makes urgent the search for alternative technologies and feedstock to sustain society's prosperity worldwide.[1–3] Various alternative energy productions with a lower carbon footprint have been developed and implemented such as hydro, solar, wind, geothermal and nuclear power, which can be placed under the umbrella of renewable energy. However, the production of many materials and chemicals relies on organic molecules with a carbon skeleton, and therefore opens the search for renewable carbon feedstock.[4] The world's largest source of renewable carbon is biomass. Plant biomass is formed by plants converting solar energy, H_2O, and CO_2 into organic molecules through photosynthesis and thus can be considered as a renewable resource. These organic molecules can be extracted in a biorefinery and converted into bio-based chemicals. For a biorefinery to be economical competitive, it is essential that most of the biomass carbon can be converted into valuable products.[5]

The most widely available form of inedible plant biomass is lignocellulose. It is the major structural component of plants, mainly composed of three polymers: cellulose, hemicellulose and lignin.[6–8] The ratio of these polymers, located in the cell wall as a complex structure, varies with the type of biomass, although it is generally in the range of 40–60% cellulose, 10–40% hemicellulose and 15–30% lignin (Fig. 1). Both cellulose and hemicellulose are sugar-based polymers, whereas lignin is composed of aromatic building blocks. Hence, lignin is the most abundant source of renewable aromatics on earth. Since aromatics represent a large share of today's chemical building blocks, lignin is a promising future feedstock to replace current fossil derived aromatics.[6,8,9]

Native lignin, a complex polymer, results from a radical polymerization process of three 4-hydroxyphenylpropanoid building blocks or monolignols: *p*-coumaryl alcohol, coniferyl alcohol and sinapyl alcohol. These monolignols, originating from the phenylpropanoïd pathway, are substituted aromatic monomers and differ in the number of methoxy groups on the phenolic moiety being zero (hydroxyphenyl, H), one (guaiacyl, G) or two (syringyl, S).[7,10,11] Their polymerization process is initiated in the plant cell wall by laccase and peroxidase enzymes, resulting in phenoxy radicals. Once generated, radical polymerization is a purely

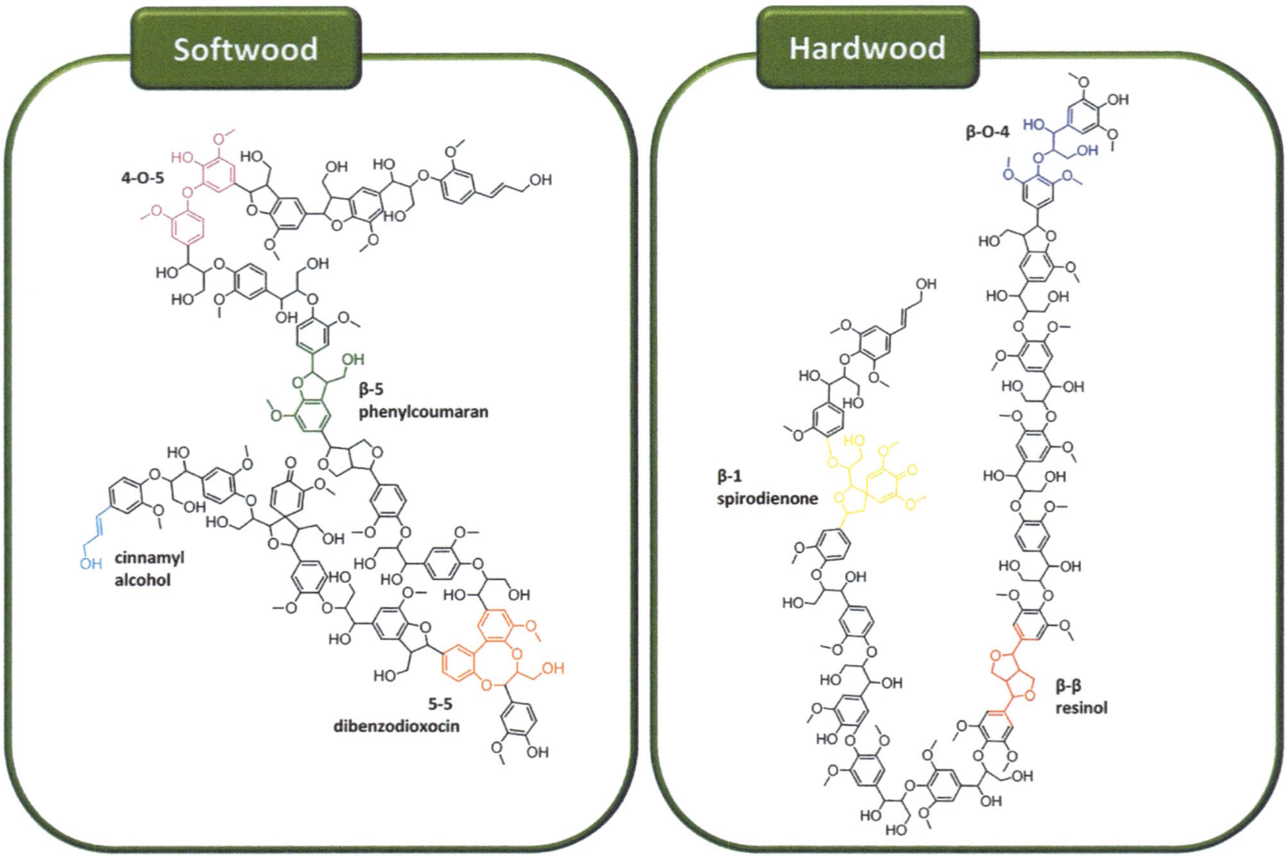

Fig. 1 Softwood (left) and hardwood (right) pristine lignin structures.

chemical process, predominantly occurring through endwise coupling reactions, next to some cross-couple reactions. Some mechanistic principles during this radical lignification have to be obeyed, resulting in a lignin polymer, which has been referred to as a racemic polymer with bewildering complexity.[12]

During this radical polymerization process, various inter-unit linkages can be formed due to the conjugated π-system, including relatively stable carbon-carbon bonds (viz. β-5 phenylcoumaran, β-β resinol, β-1 spirodienone, and 5–5 dibenzodioxocin) and relatively more reactive ether bonds (viz. β-O-4, 4-O-5, α-O-4, α-O-γ, and α-O-α) shown in Fig. 1. The latter three ether linkages always occur in combination with a more resilient C–C inter-unit linkage in native lignin.[12] The relative occurrence of the inter-unit linkages is determined by many factors, the most important one being the plant variety since the abundance of each monolignol is species specific. Generally, hardwoods have a higher amount of syringyl than guaiacyl units, whereas softwood lignin is almost exclusively composed of guaiacyl units. Lignin originating from herbaceous biomass contains these three monolignols next to a considerable amount of hydroxycinnamates.[7,11] These variations have a direct impact on the relative occurrence of inter-unit bonds, thus generating different lignin structures. For instance, due to the higher syringyl content in hardwoods, more 5-positions on the phenolic ring are blocked, impeding the formation of β-5, 5–5, and 4-O-5 linkages. As a result, hardwoods contain larger amounts of β-O-4 and β–β inter-unit ether bonds than softwoods, in which the 5-substituted carbon–carbon inter-unit bonds are predominant. This is a crucial aspect in lignin valorization, since ether bonds are easier to cleave, facilitating lignin extraction from the lignocellulosic matrix as well as depolymerization.[6]

A general challenge of wood fractionation and lignin valorization is the structural variety between and within wood species and the trade-off between delignification and repolymerization.[13] A high delignification yield typically requires harsh process conditions, causing undesired repolymerization reactions of the lignin fragments. Conversely, a high structural preservation often requires mild process conditions, resulting in lower lignin yields.[6] This issue can be tackled by stabilizing the native structures or the reactive intermediates through either protection group chemistry or catalysis, hereby impeding undesired repolymerization while reaching high delignification yields.[6,13,14] This strategy—a *lignin-first* strategy—has been implemented by several technologies: (i) high alcohol organosolv, wherein the α-OH of the β-O-4 linkage is etherified with a higher alcohol such as *n*-butanol, effectively preserving the β-O-4 linkage,[15,16] (ii) acetal formation by reacting the native α-OH and γ-OH of the β-O-4 linkage with short aldehydes, forming a 1,3-dioxane structure, which preserves the β-O-4 linkage,[17–19] (iii) acetal formation by reacting monomeric aldehydes formed during acid-catalyzed lignin depolymerization with external diols,[20–22] or (iv) reductively stabilizing reactive intermediates with a process named Reductive Catalytic Fractionation (RCF).[23]

Reductive catalytic fractionation technology is the primary focus of this chapter. In Section 1, a brief description of the RCF process is given, after which the composition of the obtained RCF lignin oil and some specific analysis methods are discussed, focusing on both monomer and oligomer analysis. Since valorization of the lignin stream is pivotal to obtain an economical beneficial biorefinery, some of the many monomer valorization pathways are highlighted as well in Section 2. The importance of techno-economic and life cycle assessment methods and their relation to an accurate process design are discussed on the basis of a case study: "RCF of *lignocellulose-to-phenol*" in Section 3 (Fig. 2).

1. Part I: Reductive catalytic fractionation

1.1 RCF overview

Reductive catalytic fractionation of lignocellulose was first described in 1940 by Hibbert et al.[24] Back then, RCF was mainly used to obtain hydrogenolysis oils for the structural analysis of native lignin.[25] In recent years, the RCF technology has emerged into the spotlight as an attractive strategy for biomass valorization. Its elegant combination of fractionation, depolymerization, and stabilization allows one to obtain high lignin oil yields combined with high monomeric yields.[14,23,26,27] Recently, several excellent reviews have been published describing the role of the catalyst and process parameters on lignin oil yield and monomers composition,[6,26,28,29] scale-up hurdles,[30] feedstock flexibility[23] and downstream processing.[31] Therefore, we will only briefly summarize the basic RCF strategy and pin-point some specific outlooks and challenges based on the hypothetical softwood lignin depolymerization as shown in Fig. 1.

An overview of the basic lab-scale RCF strategy is presented in Fig. 3. Briefly, a lignocellulosic feedstock is loaded into a stirred batch reactor together with a heterogeneous redox catalyst, a solvent and a hydrogen source. The reactor is heated to 160–250 °C for a reaction time of 1–24 h.

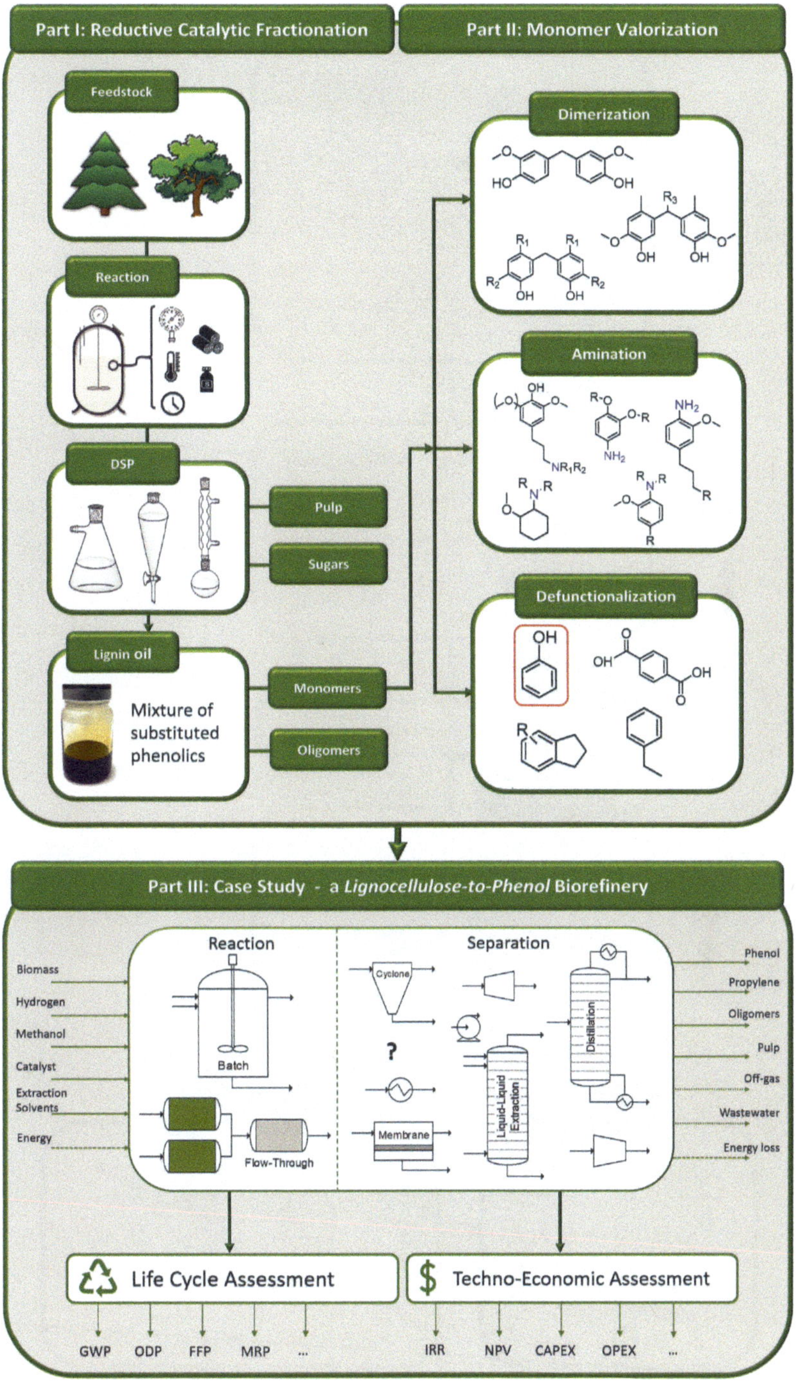

Fig. 2 Schematic overview of this chapter.

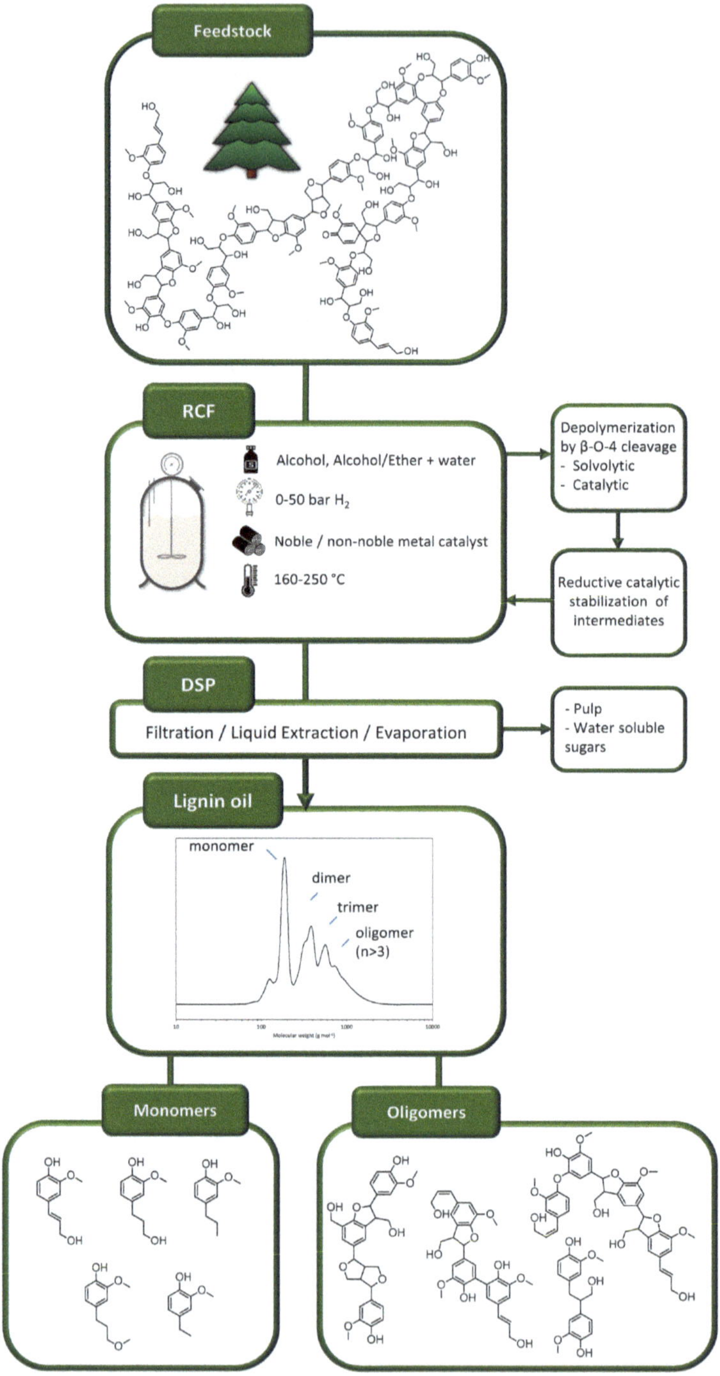

Fig. 3 Lab scale RCF strategy processing softwood into phenolic monomers and oligomers.

During reaction, the following events occur: (i) the ether and ester linkages between lignin and hemicellulose are cleaved together with a fraction of the β-O-4 inter-unit linkages, allowing the release of lignin from the lignocellulosic matrix and solubilization in the solvent, (ii) once solubilized, the lignin fragments diffuse out of the lignocellulose matrix[32] and simultaneously undergo solvolytic deconstruction[33] and catalytic depolymerization[34,35] by cleavage of β-O-4 and α-O-4 inter-unit lignin bonds, generating reactive low M_w intermediates, (iii) these reactive intermediates are reductively stabilized, preventing condensation and generating a stable mixture of substituted aromatic molecules. Interestingly, the role of the solvent and catalyst in the β-O-4 cleavage toward the reactive intermediates (viz. 4-propenol substituted methoxyphenols) is still debated in the literature. The carbohydrate fraction in the lignocellulosic feedstock is largely retained during RCF, resulting in a delignified carbohydrate-enriched pulp.[36,37] However, some hemicellulose is solubilized and ends up in the crude bio-oil mixture, which is obtained after solvent evaporation. Usually, this crude bio-oil mixture is extracted with a water-organic solvent (ethyl acetate (EtOAc))[38] or dichloromethane (DCM)[36] mixture, effectively separating the water-soluble sugars and the organic solvent soluble fraction, viz. the low molecular weight lignin oil.

1.2 Composition RCF lignin

RCF of a hypothetical softwood lignin structure produces a low molecular weight lignin oil as exemplified by the GPC profile in Fig. 3. The final lignin oil composition is determined by several events occurring during RCF. The reductive stabilization by the heterogeneous noble (Pd, Ru, Rh, Pt) or non-noble (Ni, Cu, Fe, Mo) redox catalyst is hereby crucial as it avoids repolymerization of reactive (lignin) intermediates. Hence, this step is regarded as the distinctive parameter between RCF and organosolvolysis. Besides reductive stabilization, the catalyst can selectively alter the aliphatic functionality of the formed phenolic monomers, to mainly 4-*n*-propyl or 4-*n*-propanol chains (Fig. 3, monomers), next to some other minor side-chains, such as 4-ethyl, 4-*n*-propenol or 4-(3-methoxypropyl).[6,26,39]

Crucial for proper catalytic action is a good and quick contact between the solid catalyst, the reactive intermediates, and the hydrogen source. From the biomass perspective, fast diffusion of the intermediates out of the lignocellulose particle ensures optimal catalytic action. Recently, important guidelines with regard to this intra-particle diffusion were formulated following a detailed and experimentally verified CFD model of wood particle lignin extraction.[32] From the catalyst perspective, an optimal contact can be achieved by applying a powder catalyst. However, industrial implementation of a catalyst powder seems impractical given its tendency to retain in the pulp, which complicates recuperation and pulp valorization.[30] To circumvent these problems, (i) ferromagnetic catalysts,[40,41] (ii) catalyst baskets using pelletized catalysts[33,42] or (iii) flow through fixed bed set-ups[43–46] with a separate biomass bed and catalyst bed, have recently been studied. In these set-ups, mesoporous (preformed) catalysts might play an important role, though, they have not been investigated in RCF.[30]

An important distinction has to be made between RCF lignin monomers and oligomers. Most of RCF research has focused on analyzing monomer compositions by, e.g., GC from an analytical as well as a kinetic perspective.[33,43,46,47] These phenolic monomers result from the cleavage of 2 adjacent ether linkages or 1 end-unit and 1 ether linkage. Assuming a complete cleavage of the β-O-4 linkages, dimers and oligomers (essentially two or more covalently linked phenolic groups) are derived from the native C–C inter-unit linkages, shown in hypothetical depolymerized product mixture in Fig. 3. Identifying the molecular structures of these oligomers is important to better understand the RCF chemistry and to enhance the bottom-up development of novel applications (e.g., chemicals, polymers, or materials). The latter is pivotal as oligomer utilization is necessary to claim an economically feasible RCF biorefinery as discussed in Section 3.[5,48]

Thus far, few studies have succeeded to elucidate molecular structures of some components contained in this fraction, yet a complete structural analysis of dimers and oligomers is still needed since this represents a major analytical undertaking.[27] Nevertheless, analysis of the dimer fraction is possible by GC-FID and GC–MS after trimethylsilylation of the phenolic and hydroxyl groups. In the last decade, the research groups of Torr et al.,[49] Roman Leshkov & Beckham et al.,[44,46] and Sels et al.[33,36,39,50–52] reported a large number of dimers, as shown in Fig. 4. Some important qualitative observations can be made, based on these reported structures.

Firstly, the C–C linkages found in native lignin are preserved. These rigid C–C remain intact during RCF processing whereas most ether linkages of the native lignin structure appear to be broken. For instance, the α-O-4 linkage in the β-5 phenylcoumaran structure is cleaved, resulting in an ethyl linkage between the two phenolics. Similar observations can be made for the β-1 linkage. For both β-5 and β-1, the ethyl linkage can be substituted by a hydroxymethyl group. This group provides an additional aliphatic OH-functionality in the dimer, which is an appealing feature for application development. Selectively obtaining this hydroxymethyl substitution has not

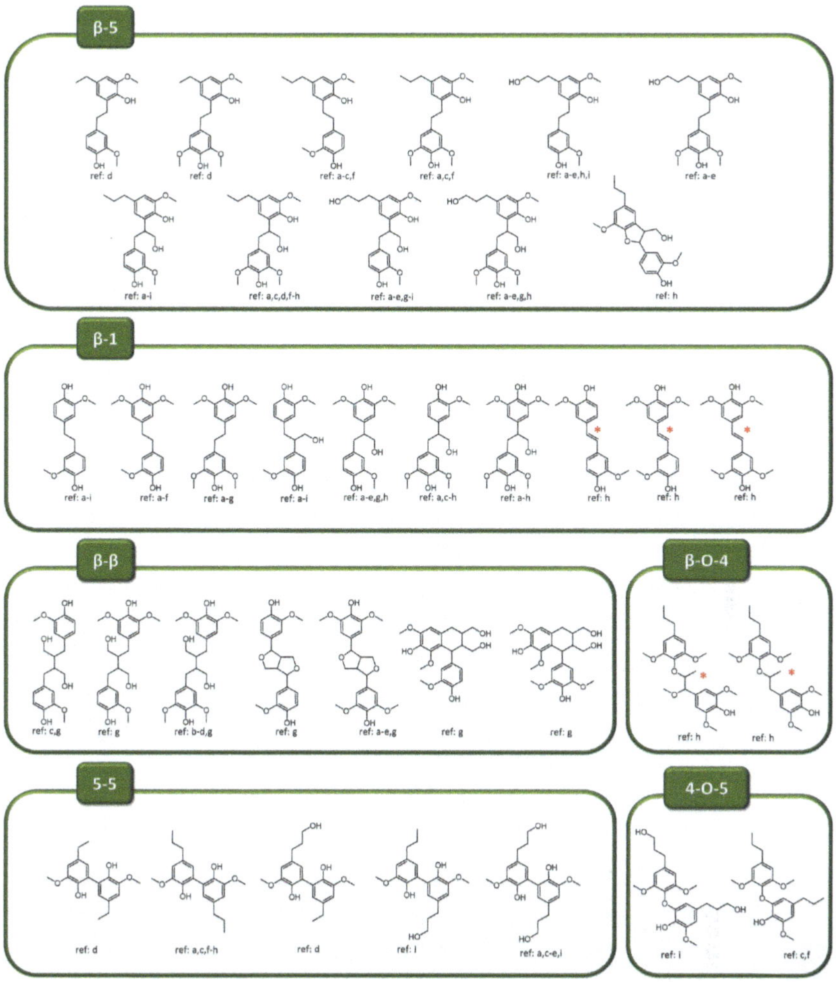

Fig. 4 Reported phenolic dimers in the RCF literature over the past 10 years. Dimers marked with an asterisk are observed at no or low catalyst loadings. *References: a: Van Den Bosch, S.; Renders, T.; Kennis, S.; et al.* Green Chem. ***2017***, 19, *3313–3326; b: Schutyser, W.; Van Den Bosch, S.; Renders, T.; et al.* Green Chem. ***2015***, 17, *5035–5045; c: Van Den Bosch, S.; Schutyser, W.; Koelewijn, S.-F. F.; et al.* Chem. Commun. ***2015***, 51, *13158–13161; d: Renders, T.; Schutyser, W.; Van Den Bosch, S.; et al.* ACS Catal. ***2016***, 6, *2055–2066; e: Renders, T.; Van Den Bosch, S.; Vangeel, T.; et al.* ACS Sustain. Chem. Eng. ***2016***, 4, *6894–6904; f: Van Den Bosch, S.; Schutyser, W.; Vanholme, R.; et al.* Energy Environ. Sci. ***2015***, 8, *1748–1763; g: Anderson, E. M.; Stone, M. L.; Katahira, R.; et al.* Nat. Commun. ***2019***, 10, *1–10; h: Anderson, E. M.; Stone, M. L.; Katahira, R.; et al.* Joule ***2017***, 1, *613–622; i: Torr, K. M.; van de Pas, D. J.; Cazeils, E.; et al.* Bioresour. Technol. ***2011***, 102, *7608–7611.*

yet been explored. Secondly, the β-5, 5–5, 4-O-5 and β-O-4 dimers have functionally similar aliphatic end-groups as the observed aliphatic chains in RCF monomers, viz. mainly *n*-propyl and *n*-propanol. In a qualitative study, Van den Bosch et al. showed that the end-group substitution of dimers can be altered by choosing either a Ru/C or a Pd/C catalyst, analogously to monomers. Obtaining quantitative information on end-group selectivity of dimers remains challenging and can be of interest in future research. Lastly, it is remarkable that one or more native β–β resinol linkages are often observed. This might be explained by the higher bonding dissociation enthalpy[7] of the ether-linkages in the resinol structure, lowering their conversion rate. Obtaining quantitative information on all these particular events in the dimer fraction remains challenging and can be of interest in future research.

The number of distinct dimers is clearly higher than the amount of different monomers. This is the result of the above reported observations namely: (i) preservation of native C–C linkage, (ii) preservation or cleavage of native C–O linkage, and (iii) if the C–O linkage is cleaved, different substitutions are possible, in combination with the interchangeability of S or G phenolic moieties in some cases.

Accordingly, the number of distinct trimers, tetramers, etc., present in the RCF lignin oil, is expected to be even higher. Besides, the mass fraction they represent is much smaller, thereby complicating their analysis. Lignin oligomers with 3 or more aromatic groups are hard to analyze by GC due to, amongst other things, their low volatility, the probability of peak overlap (due the large number of possible molecules), and absence of commercial standards. Yet, more knowledge apart from their presence is necessary as they can constitute over 50% of the entire lignin oil as is exemplified in the hypothetical oligomers mixture of Fig. 3. Several other analytical methods are frequently applied to gain molecular insights on (technical) lignin oligomers such as HPLC, FT-IR and NMR.[53] Particularly NMR is a very valuable tool for the analysis of lignin. Two types of NMR experiments are highlighted herein: ^{31}P NMR and ^{1}H–^{13}C Heteronuclear Single Quantum Coherence (HSQC) spectroscopy.

Quantitatively assessing the functional properties of lignin is important regarding its valorization. In this context, the hydroxyl and phenolic groups present in lignin oligomers are likely the most important. Both their number and nature can be analyzed by ^{31}P NMR. This method relies on the derivatization of both the lignin OH-groups and the internal standard (either cholesterol or *endo*-N-hydroxy-5-norbornene-2,3-dicarboximide) with

2-chloro-3,3,4,4-tetramethyl-1,3,2-dioxophospholane, generating ^{31}P nuclei with characteristic shifts. The resulting NMR spectrum is quickly obtained and only uses a small amount of sample.[53,54] However, ^{31}P NMR is seldom applied in RCF or lignocellulose hydrogenolysis literature with the exception of the research groups of Torr et al. and Sels et al. They found phenolic OH contents between 3.6 and 4.9 mmol OH per gram of lignin oil and aliphatic OH contents between 1.6 and 4.4 mmol OH per gram mol for both softwood and hardwood hydrogenolysis oils.[49,55–58]

Another powerful tool to study lignin molecular structures is ^{1}H–^{13}C HSQC spectroscopy. It has the advantage over regular 1D experiments that overlapping protons directly attached to carbons with different shifts are separated in the carbon dimension, and vice versa for overlapping carbons, thereby greatly improving resolution.[53,59] Several book chapters and reviews on the application of this method on lignin and the related qualitative and quantitative challenges have been published and are not discussed here.[53,59,60]

Many RCF studies use ^{1}H–^{13}C HSQC to study the molecular structure of the entire lignin oil. Van den Bosch et al. proved that the conversion rate of the β-O-4 linkage is similar for catalyst free RCF reactions and Ni-Al_2O_3 catalyzed reactions, indicating that most—if not all—β-O-4 linkages can be solvolytically cleaved over time. They also observed the disappearance of native lignin structures such as β-5 phenylcoumaran and β–β resinol.[33] Similar observations with regard to the disappearance of native inter-unit linkages have been made by other research groups applying the RCF strategy.[37,41,45,46,49,61–64] Important missing links in these ^{1}H–^{13}C HSQC analyses—to fully use the power of the method—are that (i) it is not fully corroborated what the reaction products of the native inter-unit linkages and their corresponding chemical shifts in the NMR spectrum are and (ii) (semi-)quantitative analysis of these spectra is rare, although it is commonly used in technical lignin analysis.[59]

2. Part II: Phenolic monomers valorization

As lignin-first depolymerization methodologies such as RCF become increasingly developed, downstream monomer upgrading will gradually become a center piece in new research. The success of future biorefineries is inextricably linked to optimal valorization of the obtained phenolics. Depending on the depolymerization mixture's complexity and ease of separation into purer streams, an adapted valorization pathway can be chosen.

Mixtures containing few compounds with distinct physicochemical and thermodynamic properties, lend themselves better for purification to pure chemicals, whereas complex mixtures containing many components with similar properties may not be completely separable due to technical or economic reasons (see Section 3). The latter may be the case for phenolic monomers but is definitely true for the oligomeric stream as was discussed in Section 1. This stream may be used for different applications where specific physical and chemical properties are required such as solvents, printing inks[5] or polymers including polyurethanes[64,65] or epoxy resins.[55,56] Monomer modifications preferentially rely on atom efficient routes to generate new functionalities, while maximally preserving the intrinsic moieties of lignin (i.e., functionalization).[66,67] However, for bulk chemicals requiring lower complexity, selectively removing functional groups (i.e., defunctionalization) is necessary. By combining both upgrading forms, the pool of valuable lignin-derivable molecules can be enlarged even further.[6,29]

In this section, a selection of novel up-to-date monomer upgrading strategies including functionalization and defunctionalization routes is reviewed. Firstly, monomer dimerization, i.e., coupling two monomers, is addressed, leading into the field of renewable bisphenol alternatives. Next, nitrogen incorporation by transforming aliphatic, as well as aromatic C—OH bonds toward C-NR_2 is discussed. In addition to these selective functionalizations, defunctionalization strategies (e.g., hydrodeoxygenation) are addressed as valuable techniques toward bulk platform chemicals. The section is wrapped up with a critical outlook on future monomer developments.

2.1 Functionalization

2.1.1 Dimerization

Lignin monomers can be coupled into bio-derived diols, or bisphenols, effectively preserving the phenolic group. This dimerization has showed to be particularly interesting in the quest for a renewable and safer *para-para* bisphenol A (*p,p′*-BPA or BPA, Fig. 5) alternative. *p,p′*-BPA is an indispensable building block for various polymers (and polymer additives) and is produced by the acid-catalyzed hydroxyalkylation of two phenols

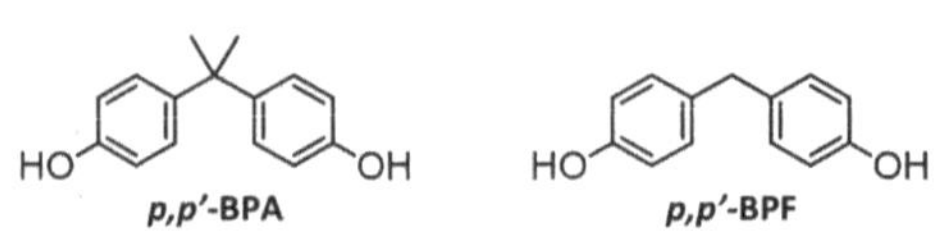

Fig. 5 Molecular structure of para-para bisphenol A (*p,p′*-BPA) and para-para bisphenol F (*p,p′*-BPF).

with acetone using an excess of phenol. The main side products during *p,p′*-BPA synthesis are other regio-isomers, such as *ortho,ortho′*-BPA (*o,o′*-BPA) and *ortho,para′*-BPA (*o,p′*-BPA), which are unsuitable for polymer production.[68] The global estimated market size of *p,p′*-BPA was 7.2 Mt. in 2018, a compound annual growth rate of more than 5% is anticipated for 2020–2025.[69,70] The biggest share of the European BPA production volume is used to manufacture polycarbonates (73%) and epoxy resins (26%).[71] BPA owes its worldwide usage to its stable and rigid molecular structure. The limited degrees of freedom associated with the molecular structure of *p,p′*-BPA are essential, providing high performance properties of the polymers.[72]

Extensive research showed adverse health effects associated to BPA (e.g., cancer,[73,74] obesity[75,76] and diabetes[77]). Due to its structural resemblance to natural steroid hormones, BPA may induce unwanted hormonal activity (cf. endocrine disruption), in particular estrogenic activity (EA).[78,79] Significant levels of free monomeric BPA, caused by incomplete polymerization, (bio)chemical degradation and thermo-mechanical stress on BPA-based polymers, are detected in wastewater, food, human fluids and tissues.[80–86] In 2017, the European Chemicals Agency (ECHA) added BPA to the Candidate List of substances of very high concern (SVHC's), recognizing BPA's endocrine disrupting properties and toxicity to reproduction.[87] Hence, the European Commission lowered the accepted specific migration limit (SML) from 0.6 to 0.05 mg/L and banned BPA in all plastic bottles and food packaging for children under 3 years old.[88]

The molecular structure of BPA should be (re)designed to provide a safe alternative without compromising on its technical performance. The key molecular feature that induces EA, is the free, unhindered phenolic hydroxyl group such as present in the natural estrogen, BPA and current BPA alternatives (i.e., BPF, BPS). Steric shielding and electronic hindrance by intramolecular hydrogen bonds in *ortho* position to the hydroxyl (e.g., *o*-methoxy group(s)), therefore significantly impede the EA.[89–91] This beneficial effect of *o*-methoxy groups on EA was confirmed by Koelewijn et al. (2019) by comparing the petro-based *p,p′*-BPF with its bio-based variant Bisguaiacol F, (*p,p′*-BGF, Fig. 6) in an in vitro human estrogen receptor α (hERα) transactivation assay. The additional *o*-methoxy groups lowered the potency with almost two orders of magnitude, while the efficacy attained only 38% of the original *p,p′*-BPF value.[92] Consequently, the natural design of lignin-derived *ortho*-methoxy-substituted phenols can be preserved, and directly exploited, to gain access to not only renewable, but also safe(r) BPA replacements.[93]

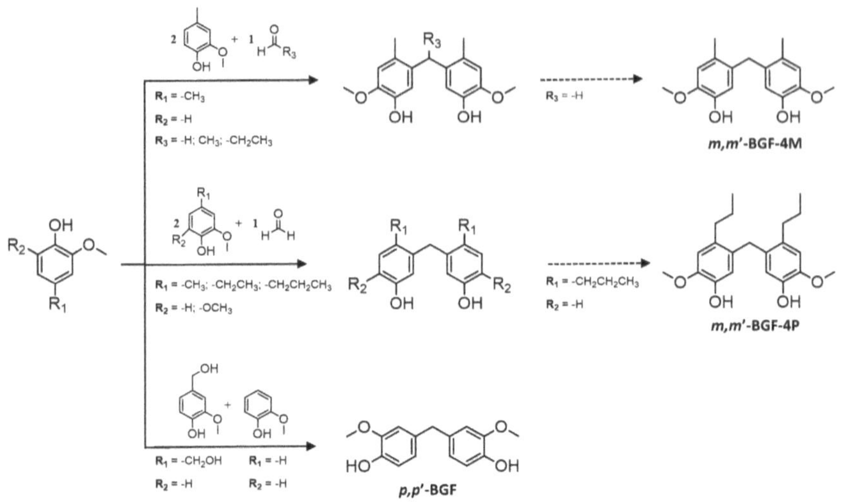

Fig. 6 Coupling reactions of lignin-derived monomers toward bisphenol A alternatives.

In a first attempt to develop renewable bisphenol alternatives, Meylemans et al. presented the hydroxylation reaction of creosol, a reduced form of vanillin, with aldehyde coupling agents like, formaldehyde, acetaldehyde and propionaldehyde, in stoichiometric amounts. Using formaldehyde, a total bisguaiacol yield of 63 mol% was obtained after 16 h reaction in a biphasic system with aqueous hydrochloric acid as catalyst at reflux temperature. A high selectivity (i.e., 97%) toward the *meta*-coupling was obtained, yielding primary 5,5′-methylenebis(4-methylguaiacol), abbreviated *m,m*′-BGF-4M (Fig. 6). The *para*-directing *o*-methoxy group dominates the *ortho*-directing hydroxyl group, therefore favoring alkylation *meta* to the hydroxyl functionality.[94]

Inspired by the work of Meylemans et al. and the safety-by-natural design concept, Koelewijn et al. presented a collection of new bio-derived bisphenols with a substantially lower EA (ca. 10–100 times) using RCF lignin-derived 4-*n*-alkylphenols and formaldehyde.[95,96] For instance, 5,5′-methylenebis(4-*n*-propylguaiacol), abbreviated (*m,m*′-BGF-4P, Fig. 6), was synthesized from 4-*n*-propylguaiacol (4-PG), a major monomeric product from RCF (i.e., 15–20 wt% yield based on total lignin). The hydroxyalkylation reaction of stoichiometric amounts 4-PG and formaldehyde (i.e., 37 wt% solution in water) was catalyzed by aqueous hydrochloric acid in a biphasic system. After 6 h reaction at 100 °C, 94% 4-PG conversion was obtained. The total bisguaiacol yield was 81%, with oligomers as the main side-product. The bisphenol fraction comprises three different regio-isomers: *the o,o′*-isomer (>1 wt%), the *o,m′*-isomer (12 wt%) and the desired *m,m′*-isomer (85 wt%).[95] Because the homogeneous catalyst used for this dimerization of 4-*n*-alkylguaiacols is toxic and corrosive, Ferrini et al. (2017) employed zeolite catalysts instead. They successfully synthesized *m,m′*-BGF-4M with toluene as reaction solvent at 100 °C using a variety of zeolite frameworks (e.g., BEA, MOR, FER, FAU).[97]

A high bisphenol purity (>99%) is required for the synthesis of high MW polymers. All *m,m′*-bisguaiacols were purified via a one-step multi-gram evaporation crystallization in an organic phase system (i.e., acetone/heptane or diethyl ether/heptane), attaining isolated yields of 57–72%. Despite their unique orientation, Meylemans et al. and Koelewijn et al. were able to utilize these novel *m,m′*-bisguaiacols in the synthesis of thermoplastics (e.g., polycarbonate and polyesters) and thermosets (e.g., cyanate ester resins).[94,95]

Bisphenols with a *p,p′*-orientation of the connecting alkyl chain such as in BPA are preferred as this will presumably lead to similar properties. This molecular feature can be obtained by using substrates with (i) a free *para*-position to the hydroxyl group and/or (ii) an intrinsic reactivity for alkylation in the *para*-position. In this regard vanillyl alcohol, or related compounds, proved to be promising substrates for the synthesis of renewable *p,p′*-bisphenol alternatives.[98] Today, the bio-derived bisphenol, which is closest related to the *p,p′*-BPA molecular structure, is the renewable *p,p′*-BPF variant, Bisguaiacol F (*p,p′*-BGF or BGF). The synthesis and application of BGF was invented and patented by the research group of Reno et al.[98] In 2016, Hernandez et al. reported the condensation reaction of vanillyl alcohol and guaiacol to yield *p,p′*-BGF. This reaction was performed in an excess guaiacol using a heterogeneous ion-exchanging resin DOWEX DR2030. After reaction of 12–24 h at 60 °C, a total bisguaiacol yield of 70% was obtained and the main side products showed to be oligomers and other regio-isomers. The bisguaiacol fraction comprised three different isomers: *o,p′*-BGF (3 wt%), *m,p′*-BGF (15 wt%) and *p,p′*-BGF (83 wt%).[99]

In this regard, Koelewijn et al.[92] looked into the valorization of the *m,p*-BGF isomer side-product, and determined its in vitro EA and the influence on the polycarbonate synthesis. Similar to the EA of *p,p′*-BGF, the *m,p′*-BGF isomer showed a potency almost two orders of magnitude lower than BPF and an efficacy of 29% of BPFs' max induction in an in vitro human estrogen receptor α (hERα) transactivation assay. Moreover, *m,p′*-BGF could be incorporated in polycarbonate with an amount up to 25 wt%, without affecting the polymer thermal properties considerably. The BGF-based polycarbonates showed a glass transition temperature (T_g) of 106 °C and a maximal weight loss temperature (T_{dmax}) of 418 °C. This is, respectively, 7% and 6.5% lower than the synthesized petro-based BPF-based polycarbonate.[92]

To investigate the effect of the *o*-methoxy moieties and the isomer bisphenol orientations on the thermomechanical properties of epoxy resins, Nicastro et al. (2019) synthesized bisphenols with a varying amount of *o*-methoxy moieties and isomer orientations (i.e., *p,p′*, *m,p′*, *o,p′* en *o,o′*). All cured bio-derivable resins derived from these bisphenols showed T_g's between 110 and 151 °C, $Td_{5\%}$ ranging from 344 to 368 °C and a room-temperature glassy storage moduli above 2.0 GPa. These results were comparable to the BPA-based resins (i.e., T_g 167 °C; $Td_{5\%}$ 381 °C and, E'_{30} 2.5 GPa) hereby demonstrating the potential of tuning the bio-derived epoxy resins to match desired material specification.[100]

2.1.2 Amination

Beyond dimerization, the possibilities for lignin monomer functionalization are incredibly plentiful and only limited by the researcher's resourcefulness.[29] This section is devoted to catalytic methods providing aminated products starting from lignin-derived monomers. Until recently, research on these functionalizations was rather limited due to the catalytic difficulties and associated lower yields.[31] However, aromatic amines are an indispensable class of molecules in chemical industry, being omnipresent in bio-active molecules and widely used in medicinal, polymer and agrochemistry.[101–104] Incorporating nitrogen into lignin monomers greatly expands their usability so that new markets can be tapped into. Lignin depolymerization products are rich in polar C—OH bonds in the form of primary alcohols and phenolic moieties, which can be exploited as potential amination sites.[105] However, despite their omnipresence and polarity, hydroxyl groups are still relatively unreactive compared to alkyl halides (better leaving groups) or a carbonyl moiety (electrophilic carbon) and thus require more extreme reaction conditions.[106,107] During so-called hydrogen borrowing amination (HBA), the alcohol is converted in situ to a more reactive and highly electrophilic carbonyl intermediate.[107] Nucleophilic amines react readily with this carbonyl during what is in essence a reductive amination. The hydrogen consumed during final imine/enamine hydrogenation, is furnished by the initial alcohol dehydrogenation.[108] This atom efficient reaction route generates water as sole by-product and can be catalyzed by both homogeneous (mostly Ru and Ir, but Fe and Ni are emerging) or heterogeneous catalysts (Ni, Cu and Co) as reviewed by Kempe.[109] Currently, HBA is the most common synthesis route for lower alkylamines (C_1–C_6) on industrial scale.[101] Although reports on bio-based alcohol aminations are plentiful,[110–112] and was recently reviewed by Pelckmans et al.,[107] examples for lignin monomers are scarce. In 2018, Barta et al. were the first to report the synthesis of a primary alkylamine from 4-*n*-propanolguaiacol (POH-G) via a two-step protocol with ammonia and a heterogeneous Ni/SiO_2-Al_2O_3 catalyst.[62] In a first step, they aimed for a selective nitrile synthesis starting from the corresponding alcohol, via HBA with ammonia followed by a consecutive dehydrogenation. Next, a highly selective hydrogenation step yielded the primary amine (Fig. 7, top). By selecting a two-step approach, it was easier to maximize the selectivity of each individual reaction, resulting in an overall higher yield, at the expense of a more complex synthesis. More recently, direct HBA of POH-G and 4-*n*-propanolsyringol (POH-S) was demonstrated by the same group, using a homogeneous Ru catalyst and various amines (Fig. 7, bottom).[38] This is especially remarkable due to the difficulties involved with potentially strongly coordinating moieties such as guaiacol and syringol.[38,113] Here, the catalyst choice was of great importance, as the applied homogeneous Ru complex (so-called Shvo's catalyst) had already proven to be active for this type of reaction and could cope with the demanding lignin-substrates.[114] Additionally, the aminated lignin monomers were processed further to make the biologically active

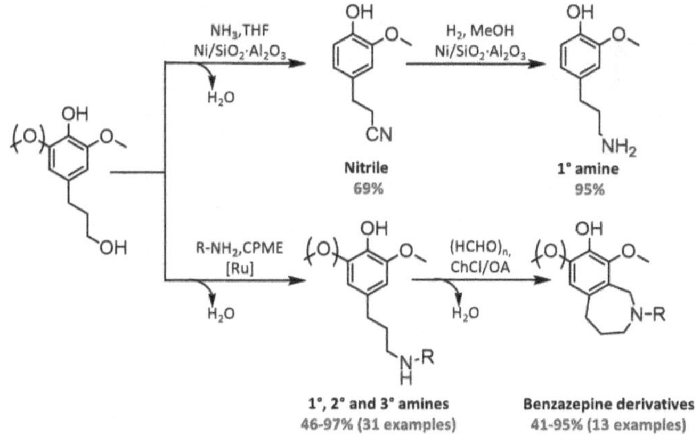

Fig. 7 C—OH amination strategies for lignin-derived monomers toward aromatic alkylamines, nitriles and benzazepine derivates Isolated yields for each step are shown.[38,62] (ChCl/OA = choline chloride/oxalic acid).

tetrahydro-2-benzazepines (Fig. 7). These conventionally petro-based compounds were selectively synthesized from the aminated monomers, during a Mannich-type cyclization in deep eutectic solvents.

Next to targeting the alcohol moiety in lignin monomers, amination of phenolic hydroxyls, yielding substituted anilines or cyclohexylamines can be desirable as well.[115,116] Anilines are important as polymer building blocks in polyurethane, rubber vulcanization accelerators, dyes and pigments, and pharma- and agrochemicals.[115] Gas phase aniline synthesis from phenol and ammonia in the presence of a silica—alumina catalyst is a common industrial production route, next to nitrobenzene hydrogenation.[115,117] The reaction mechanism entails a partial reduction of phenol to cyclohexanone, followed by a reductive amination via an imine intermediate toward the cyclohexylamine. Subsequent dehydrogenation of the saturated amine yields aniline.[118] Gas phase amination of lignin-derived phenolics, in particular alkyl-substituted phenols, has also been studied and is dominated by noble metal supported heterogeneous catalysis at elevated temperatures.[118–121] *o*-Methoxyphenols have not yet been studied as reagents for gas-phase amination—to the best of our knowledge—possibly due to (i) electronic and/or steric effects of the methoxy groups or (ii) unwanted demethylation/demethoxylation.[107]

Amination of phenols toward aniline[122–124] or cyclohexylamine[125–127] can also take place in a liquid phase at milder temperatures compared to gas phase aminations.[107] Although decent yields are obtained starting from (*n*-alkyl)phenol (derived from (*n*-alkyl)guaiacol by demethoxylation[6]), direct catalytic amination of the original lignin-derived *o*-methoxyphenols remains challenging.[29,125] Nevertheless, Li and colleagues achieved to aminate lignin derived alkylphenols as well as guaiacol toward cyclohexylamines[127] and anilines (Fig. 8).[122] For substituted cyclohexylamines, a Pd on carbon catalyst was used with sodium formate (6 equiv.) as the hydrogen source at 120 °C. For substituted anilines, they used the same catalyst but with less sodium formate (1.5 equiv.), a higher temperature (140 °C) and trifluoroacetic acid.[122] Besides guaiacol, various other substituted phenols were coupled with diverse primary and secondary amines and anilines. In another example from Barta et al., several 2-methoxy-4-ethylaniline derivatives were synthesized through a homogeneous Ni catalyzed cross-coupling, starting from 4-ethylguaiacol (Fig. 8).[62] Here, the starting lignin monomer was derivatized with the reactive diethylcarbamyl chloride to allow cross-coupling in the next step. Various amines were tested from which some showed decent yields (e.g., morpholine and anilines), whereas others did

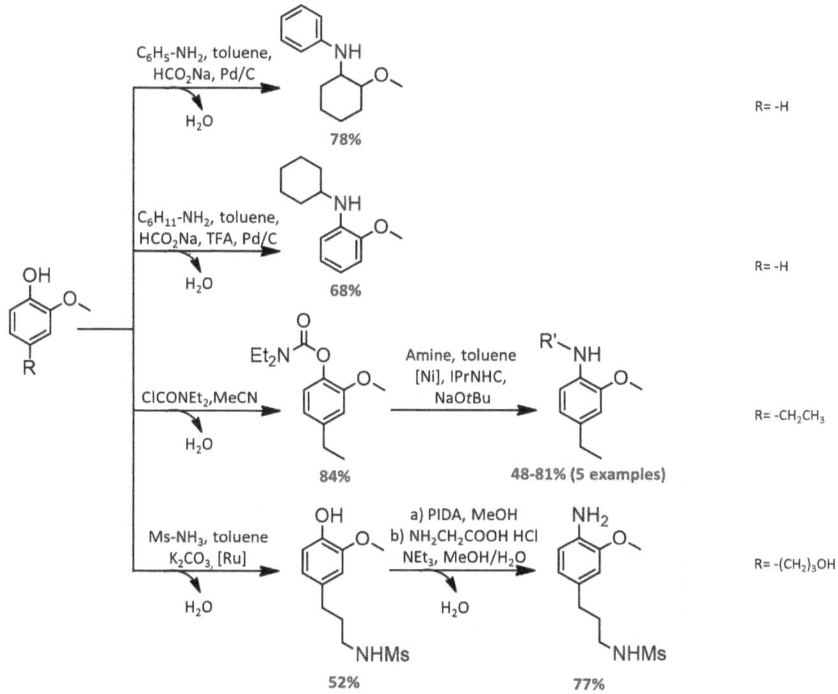

Fig. 8 Phenolic C—OH amination of lignin-derivable aromatics toward anilines, cyclohexylamine and aromatic diamines. Isolated yields for each step are given.[62,122,127]

not react (e.g., aliphatic primary amines). Intriguingly, combined amination of both the phenolic and aliphatic hydroxyl of POH-G was also achieved using a multistep synthesis protocol with a homogeneous Ru catalyst (Fig. 8).[62] As demonstrated above, only few groups have reported successful phenolic C—OH amination with *o*-methoxyphenols. The complex synthesis protocols and inherent structure of the lignin-derived monomers put limits to the applicability of the current amination methods due to lower reactivity or competing side reactions. Overcoming these challenges by developing improved amination strategies is thus an ongoing challenge.

Other approaches not starting from hydroxyl moieties have been reported as well. As a first example, instead of converting the phenolic hydroxyl group directly, its *ortho,para*-directing effect can be exploited for selective amino alkylation during so-called Mannich reactions. This reaction involves the condensation of an amine (mostly secondary), formaldehyde and the substrate (viz. phenol derivatives).[128] As guaiacyl units in lignin have occupied *para* positions and only one available *ortho*-position, the

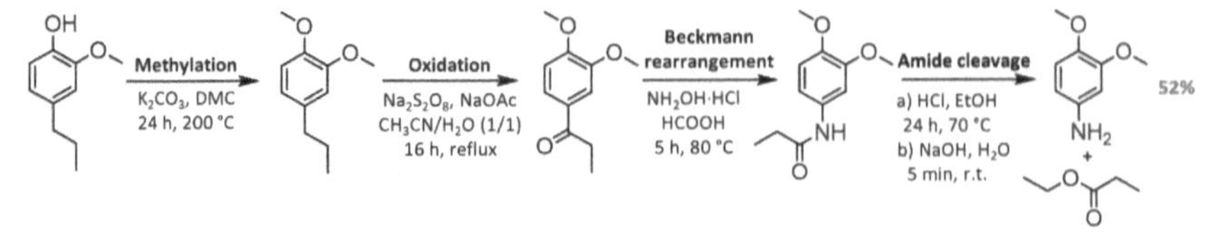

Fig. 9 Synthesis of 3,4-dimethoxyaniline and ethyl propionate from 4-*n*-propylguaiacol. Overall isolated yield is given.[132]

condensation occurs selectively on the aromatic C5. As already discussed, this reaction was used by Barta et al. for the modification of aminated lignin monomers.[38] Additional examples are reported, starting mainly from crude technical lignin samples in order to obtain an amine enriched lignin endproduct.[129–131]

In a second example, the groups of Maes and Sels developed an interesting four-step process where the propyl chain of 4-*n*-propylguaiacol is replaced by an amino group, as shown in Fig. 9.[132] This approach is appealing from an industrial point of view as only cheap bulk reagents and reactants are used and expensive purification techniques, such as column chromatography, are omitted. The new strategy comprises (i) a methylation with environmentally benign dimethyl carbonate, (ii) an oxidation with $Na_2S_2O_4$, (iii) a Beckmann rearrangement of the in situ formed oxime and (iv) an amide cleavage yielding the final amine. Next to 4-*n*-propylguaiacol, this robust methodology can also be applied to 4-*n*-propylcatechol, making 3,4-dialkoxyanilines, which can serve as drop-in chemicals in anticancer drugs synthesis.

2.2 Defunctionalization

Beside the phenolic core, other structural functionalities such as *o*-methoxy groups and *p*-sidechains, are inherent to lignin-derived aromatics. In some specific cases, these unmodified monomeric compounds can be used directly without further modifications. For example vanillin, widely used as vanilla flavoring agent and fragrance is derived from lignocellulose by alkaline oxidative depolymerization on an industrial scale.[133] If exploited well, these additional groups can lead to unique applications for which modified properties, compared to conventional oil-derived aromatics, are essential. Moreover, obtaining similar *n*-alkylphenols from crude oil requires a multistep process starting from phenol (derived from benzene via Hock process[134]), followed by phenol alkylation with alkylating agents (olefins or alcohols) and final purification.[135,136] However, reduced functionality and/or a lower oxygen content is required for many current bulk chemicals (e.g., BTX and phenol) and fuels.[6,137] Defunctionalization of the monomers' phenolic core is generally obtained through hydrodeoxygenation (HDO) and results in less substituted products. The extent of deoxygenation and defunctionalization depends on the reaction parameters, resulting in different end products, namely (*n*-alkyl)phenols, (*n*-alkyl)cyclohexanols, aromatics or aliphatics.[6] For example, Schutyser et al. demonstrated the conversion of lignin-derived phenolic compounds to 4-*n*-alkylcyclohexanols and -cyclohexanones.[138,139] Many dedicated reviews on deoxygenation and defunctionalization exist.[9,137,140,141] Here, some fascinating recently reported lignin monomer upgrading approaches for key chemicals such as phenol, terephthalic acid and other arenes are discussed in more detail, given their tremendous economic importance.[134,142,143] A schematic overview is given in Fig. 10.

2.2.1 Phenol

At the moment of writing, phenol is mainly used to make bisphenol A (40–50% of global output), phenolic resins and polyamide 6. Almost all phenol (<99%) is produced by oxidation of cumene, which is derived from fossil based benzene.[134] Our group recently developed an holistic lignocellulose biorefinery process to make phenol and propylene (monomer fraction), together with bioethanol (carbohydrate pulp) and printing ink (oligomer fraction).[5] As explained in Section 1, RCF of lignocellulose yields an aromatic lignin oil (containing, mono-, di- and oligomers) and a solid carbohydrate pulp. In this new integrated approach, the aromatic monomer fraction, is converted to phenol and propylene (or ethylene).[144] Furthermore, the carbohydrate pulp is converted to bioethanol while the residual aromatic oligomers serve as ingredient for printing ink. The conversion of aromatic monomers to phenol and propylene occurs in two consecutive steps (Fig. 11, top route). In a first upgrading step, gas-phase demethoxylation of pure 4-*n*-propylguaiacol and -syringol with hydrogen using a Ni on silica catalyst renders alkyl phenols with high yields (75–85 mol%) producing water as by-product.[5] For this step, unseparated monomer mixtures derived from pine or birch wood showed yields of 70–77 mol%, stressing the methodology's robustness. In a subsequent dealkylation step, the hydroprocessing products are converted to phenol and olefin with hydrogen over a hierarchical H-ZSM-5 zeolite (Z140-H). Next to an optimum Brønsted-Lewis acidity, the unique tailored pore structure of balanced micro- and mesopores of Z140-H proved to be crucial to achieve

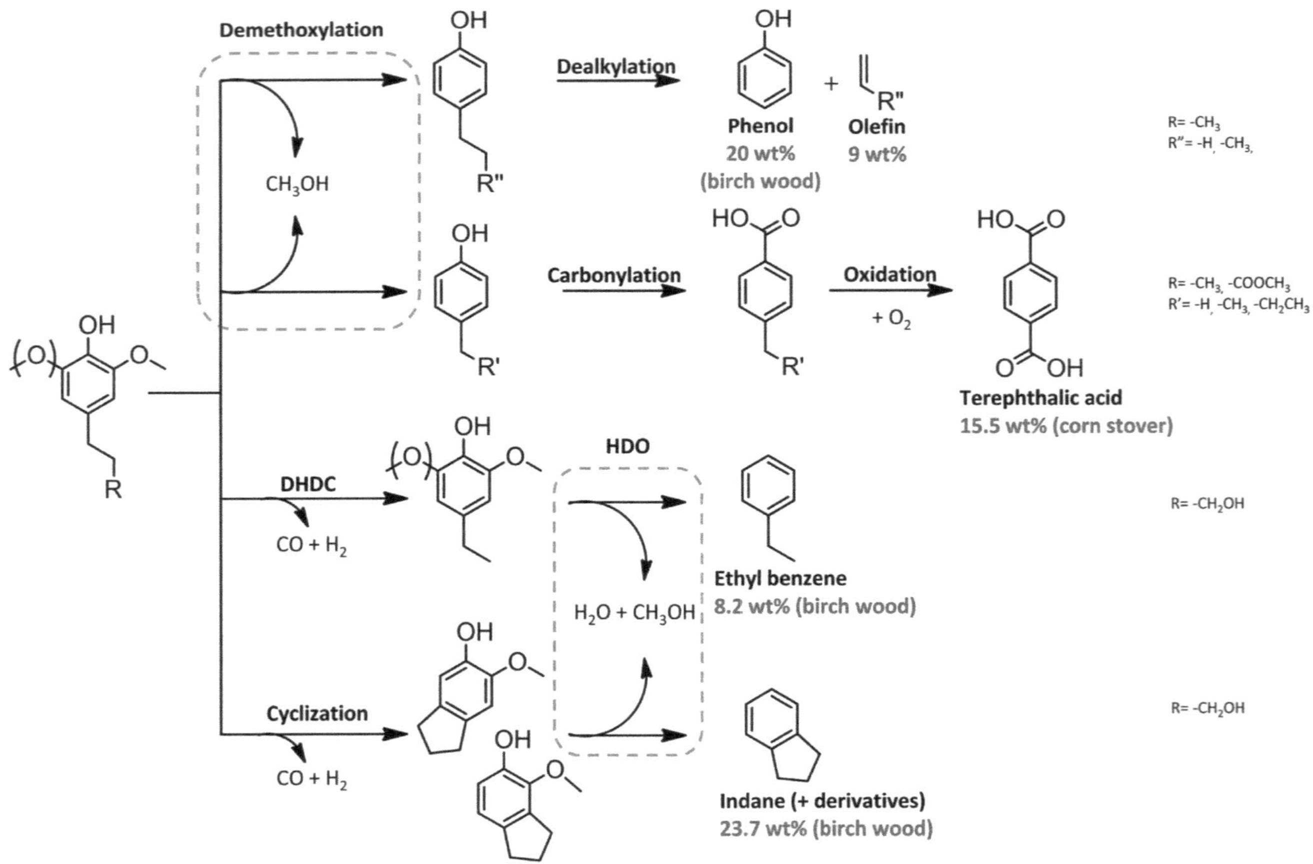

Fig. 10 Catalytic defunctionalization strategies of lignin-derived monomers toward valuable aromatic chemicals. Mass yields were calculated based on the wood's lignin content.

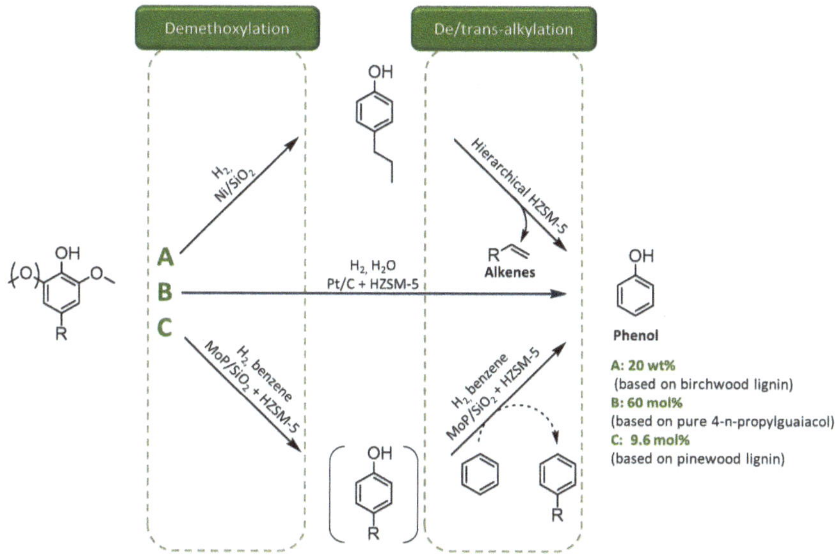

Fig. 11 Transformation of crude lignin-derived 4-*n*-alkylguaiacol and -syringol to phenol via different pathways as reported by (A) Liao et al.,[5] (B) Zhang et al.[145] (only pure substrate) and (C) Ouyang et al.[146] Phenol yields are indicated in green.

a high activity and/or stability. Detailed kinetic studies previously conducted by our group demonstrated that in this zeolite hierarchization, excessive coke formation and site-access restrictions were anticipated to occur when using a commercial ZSM-5 for processing the complex hydroprocessed crude.[109,147] It was also found that phenol condensation, which reduces the catalyst lifetime by pore blocking, was effectively inhibited by co-feeding water into the reactor.[148] The co-production of water in the first hydroprocessing step turned out to be a suitable feature from a process design perspective (see Section 3). Both 4-*n*-propyl- and 4-ethylphenols could be effectively dealkylated using this catalyst.[144,148] Overall, in this integrated biorefinery approach, crude lignin monomer mixtures are converted to bio-phenol and bio-propylene in 20 and 9 wt% yields, respectively, on birch wood lignin weight basis. For pine wood, only yields of 6 and 3 wt% are achieved, respectively, due to the lower degree of delignification and a reduced softwood depolymerization efficiency.[5] Combined with bio-ethanol derived from the carbohydrate pulp and the printing ink application for the oligomers, an impressive 76–78% of the initial carbon mass found in birch wood, was valorized into valuable chemicals. We strongly believe that these integrated biorefinery approaches are essential for the success of sustainable and profitable biorefineries (Section 3).

The Hensen group demonstrated that supported metal phosphides are also efficient catalysts for monomeric 4-*n*-propylguaiacol demethoxylation.[146] In their work, a MoP/SiO_2 catalyst was used for liquid phase demethoxylation in benzene under 90 bar hydrogen pressure, yielding a 88 mol% pure stream of 4-*n*-propylphenol. Instead of dealkylation, H-ZSM-5 catalyzed 4-*n*-propylphenol transalkylation to benzene resulted in the final phenol, along with alkylated benzene compounds (i.e., cumene and propylbenzene). Starting from pinewood, a single-step process was performed with both catalysts combined that achieved an overall phenol yield of 9.6 mol% (based on the initial lignin content). Another interesting single-step 4-*n*-propylguaicol-to-phenol approach was proposed by Yan et al.[145] A combined demethoxylation and dealkylation using a physical mixture of Pt on carbon and H-ZSM-5 for these reactions, respectively, was realized. The yield toward phenol was 60 mol% starting from pure 4-*n*-propylguaiacol.

2.2.2 Terephthalic acid

Terephthalic acid is an industrially important aromatic acid, almost exclusively used as starting material for saturated polyester, mainly poly(ethylene terephthalate) (>90%).[142] It is almost entirely made by oxidation of petro-derived *p*-xylene. Pathways toward biobased *p*-xylene or terephthalic acid mainly focus on carbohydrates such as cellulose or hemicellulose.[149] Nonetheless, Yan and colleagues were the first to report a three-step process from corn stover lignin to terephthalic acid (Fig. 10).[150] Starting from depolymerized RCF lignin oil, it includes (i) a Mo on carbon catalyzed demethoxylation, (ii) a carbonylation of the obtained *n*-alkylphenol with CO by a homogeneous Pd-catalyst, and finally (iii) an oxidation of the 4-*n*-propylbenzoic acid to terephathalic acid by a Co-Mn-Br catalyst under O_2 pressure. The demethoxylation was performed on crude lignin oil, which was obtained from corn stover by RCF in methanol over a Ru on carbon catalyst. The 4-*n*-alkylphenol yield after demethoxylation, starting from a guaiacyl and syringyl monomer mixture, was 65.7 mol% based on the total monomers (16.1 wt% based on lignin content). Interestingly, they found that the Mo catalyst also removed *para*-substituted ester group from phenolic monomers in addition to its demethoxylation activity. Next, CO was inserted using carbonylation of 4-*n*-alkylphenols by a homogeneous Pd catalyst. To increase reactivity, 4-*n*-alkylphenol triflates were formed prior to carbonylation. On average, a 75 mol% yield was obtained for all 4-alkylphenols (methyl, ethyl and *n*-propyl) toward the 4-alkylbenzoic acid. Finally, the oxidation yield of 4-alkylbenzoic acid mixture toward terephthalic acid was 60 mol%. Remarkably, terephthalic acid could be obtained from the reaction mixture by simple filtration and washing with water. Looking at the overall process, the terephthalic acid yield starting from the lignin oil was 30 mol% and corresponds to 15.5 wt% based on corn stover lignin content.[150]

2.2.3 Other aromatic hydrocarbons

More drastic defunctionalization results in aromatic hydrocarbons similar to those obtained by oil refineries. By completely removing oxygen-containing moieties in lignin monomers, while preserving aromaticity, benzene derivatives are accessed. Excellent reviews discussing these defunctionalization already exist,[6,9] therefore we will only briefly focus on two intriguing more recent examples for indane and ethylbenzene.

Indane and its derivatives are used as lubricant and high-quality fuel, and as precursor for pharmaceuticals, fragrances and insecticides.[151] Recently, Wang and colleagues managed their synthesis from native lignin by hydrogenolysis and intramolecular cyclization with a curbed Ru/Nb_2O_5 catalyst (Fig. 10).[152] By weakening ruthenium's hydrogenolysis activity with strongly interacting CH_2Cl_2, the parallel occurring intramolecular cyclization toward indane is promoted. Starting from birch-wood, indane and its derivatives were obtained in 23.7 wt% based on the lignin oil weight.

Another example is ethylbenzene. As main precursor for styrene and styrene based polymers, it is an important bulk chemical for which its lignin-derivability is proven by Wang's group (Fig. 10).[143,153] Through combined dehydrogenative decarbonylation (DHDC) and HDO with a FeO_x modified Ru/Nb_2O_5 catalyst, ethylbenzene was selectively derived from birch lignin oil (propanolphenols). They concluded that FeO_x species lower the metallic Ru content. This enhances decarbonylation by restraining undesired direct hydrogenolysis of the aliphatic C—O bond. Secondly, more oxygen vacancies and low-coordinated Nb^{4+} due to the FeO_x species promoted phenolic oxygen deoxygenation. Overall, a yield of 8.2 wt% based on birch lignin oil was obtained under optimal conditions.

2.3 Critical considerations

Rapidly growing monomer functionalization and defunctionalization methodologies will enlarge the pool of available bulk and high-value lignin-derivable compounds. This evolution will stimulate the development of future biorefineries, as it allows them to offer a more competitive alternative for fossil refineries. However, to enable progress, extension of current chemo-catalytic upgrading routes toward more demanding compounds, and eventually crude depolymerization mixtures, is crucial. More robust catalytic systems, able to cope with impure feed streams and more demanding substrates (e.g., guaiacol, syringol and catechol containing moieties) must be developed. Intense upstreaming purification to yield more "clean" substrate streams can be used as a remedy, albeit often coming at a significant price. Many current upgrading strategies demonstrated on model compounds have a limited substrate scope and are restricted to more reactive but less prevalent functional groups (e.g., benzylic, allylic, halides and carbonyl moieties). Therefore, developing selective routes for less active but more common groups (e.g., alcohols) is a future challenge. For liquid-phase conversions, the solvent choice is limited by the solubility of lignin depolymerization products. For instance, HBA reactions commonly use apolar solvents of the toluene-type, however, they do not dissolve polar lignin compounds very well. Despite these challenges, lignin monomer upgrading will undoubtedly continue to develop rapidly and diversify toward more high value and bulk products. Eventually, the advancements on this aspect will greatly influence the economical attractiveness of the above upcoming holistic biorefineries.

3. Part III: RCF in a sustainable and profitable biorefinery

Implementation of an integrated biorefinery can only be successful if it is able to deliver chemical product(s) through a process which is economically feasible and environmentally sustainable. Techno-economic assessment (TEA) and life cycle assessment (LCA) are two well-known methodologies commonly used to evaluate new and existing processes by calculating typical performance indicators such as, e.g., the net present value (NPV) and global warming potential (GWP), respectively.[154,155] Both analyses depend on (at least) detailed process simulations which include process flow diagrams and mass and energy balances, providing the main process equipment units, their inputs, and outputs as shown in Fig. 12. The reliability of the economic and environmental performance indicators depends highly on the accuracy of the process model hence, care should be taken by policy- and decision-makers not to solely rely on these calculation methods to make decisions, especially for future, non-existing plants which lack accurate process data. In this respect, it is relevant to outline recent

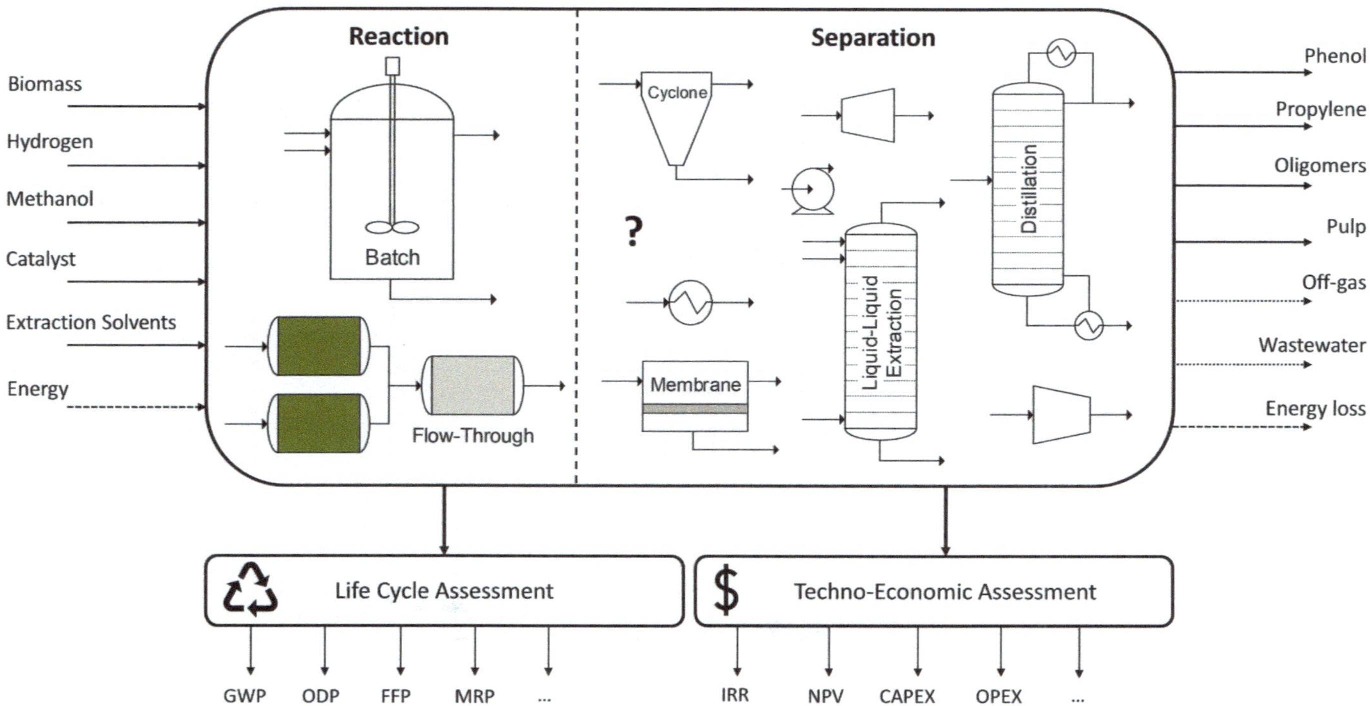

Fig. 12 Integrated process design with life cycle and techno-economic assessment methods.

advances and identify relevant challenges in the field of process design for RCF-based biorefineries and its impact on techno-economic and life-cycle assessments. A case study on a *lignocellulose-to-phenol* biorefinery recently conducted by our group exemplifies the value of this approach, hence it will serve as a guide throughout this section.

3.1 Process design

Flow-sheeting is a common practice in chemical engineering design as it attempts to combine all design functions of individual unit operations in one package to simulate a steady-state model of an existing or future chemical plant. Chemical plant design relies heavily on this method in conceptual (and consecutive) development phases as it is a powerful tool to assess upscale potential of new technologies with low to medium technology readiness level such as reductive catalytic fractionation (RCF) by providing a complete framework of a commercial production process. The process development of any chemical plant consists of three phases. Selecting a series of processing steps and determining proper connections between these steps are conducted during the *process synthesis* phase. This is followed by a *process design* phase in which the appropriate operating conditions for each unit are selected and modeled. Finally, the *process integration* phase involves exploiting the connections and interactions between these processing steps to obtain an environmentally friendly, safe, and profitable process. This systematic approach was employed by our group to demonstrate the potential of RCF as a biomass extraction technology for the sustainable production of two bulk chemicals, phenol and propylene,[4] along with a carbohydrate pulp and phenolic oligomers.[5]

3.1.1 Summary of a case study: A biorefinery for the production of phenol and propylene

The proposed *lignocellulose-to-phenol* biorefinery process consists of three sections: (i) RCF fractionation, (ii) separation and (iii) upgrading. In a first step, woody biomass is catalytically fractionated by batch RCF using methanol solvent and hydrogen gas as a reducing agent, into a carbohydrate pulp and crude lignin oil (CRLO) containing stabilized phenolic monomers and oligomers, and other mainly sugar-derived by-products. Subsequently, the complex mixture is separated into pure and intermediate product streams, comprising phenolic monomers, oligomers, and a carbohydrate pulp, and recycle and waste streams through a series of processing steps of conventional mechanical and chemical unit operations including a hydrocyclone, filtration, distillation, and liquid-liquid extraction (LLE). In a final upgrading stage, the purified phenolic monomer stream is catalytically funneled by gas-phase hydroprocessing using a Ni on SiO_2 catalyst to alkylphenols followed by dealkylation over a tailor-made hierarchical ZSM-5 zeolite in two continuous plug flow reactors to produce phenol and propylene as extensively discussed in Section 2.2.1. Separation of the dealkylation product stream with four distillation columns (including one cryogenic unit) results in (99%) pure streams of phenol and propylene. For this design, an incineration/trigeneration unit is specifically added to combust gaseous waste streams making the biorefinery energy self-sufficient. Liquid waste streams are processed in a wastewater treatment unit before disposal. A process flow diagram (PFD) of the RCF biorefinery in Fig. 13 shows the sequence of processing units and the main process flows between them. Beside the production of 99% pure streams of phenol and propylene, a stream rich in phenolic oligomers and a carbohydrate pulp are generated as by-products. It was demonstrated that the phenolic oligomers can serve as replacement for controversial, fossil-based *p*-nonylphenol in lithographic printing ink. This accomplishes near complete valorization of all biomass components, since the carbohydrate pulp is regarded as a marketable product for, e.g., the production of ethanol with semi-simultaneous saccharification-fermentation process in this scenario.[5,36] These valorization steps are not shown on the PFD.

3.1.2 Process synthesis

The current process synthesis broadly corresponds to an upscale equivalent of a lab-scale production procedure. This structure sets just one example which is evidently prone to change depending on altering (economic or environmental) design criteria and/or new outcomes in research and development.[5,36] It is therefore useful to evaluate this particular design along with state-of-the-art technologies which show potential for application in this biorefinery concept, as not to overlook any promising alternatives in this critical phase of the process development.

3.1.2.1 Reactor: Producing lignin oil

3.1.2.1.1 Case study: Upscaling batch reactors The reactor unit is modeled as an upscale of the 2 L laboratory batch reactor installation in which 24 g of lignin oil is extracted from 150 g of birch wood with RCF.[5] Recently, the feasibility to upscale this batch process has been demonstrated by a 50 L pilot scale study in which 1.3 kg of pine wood was

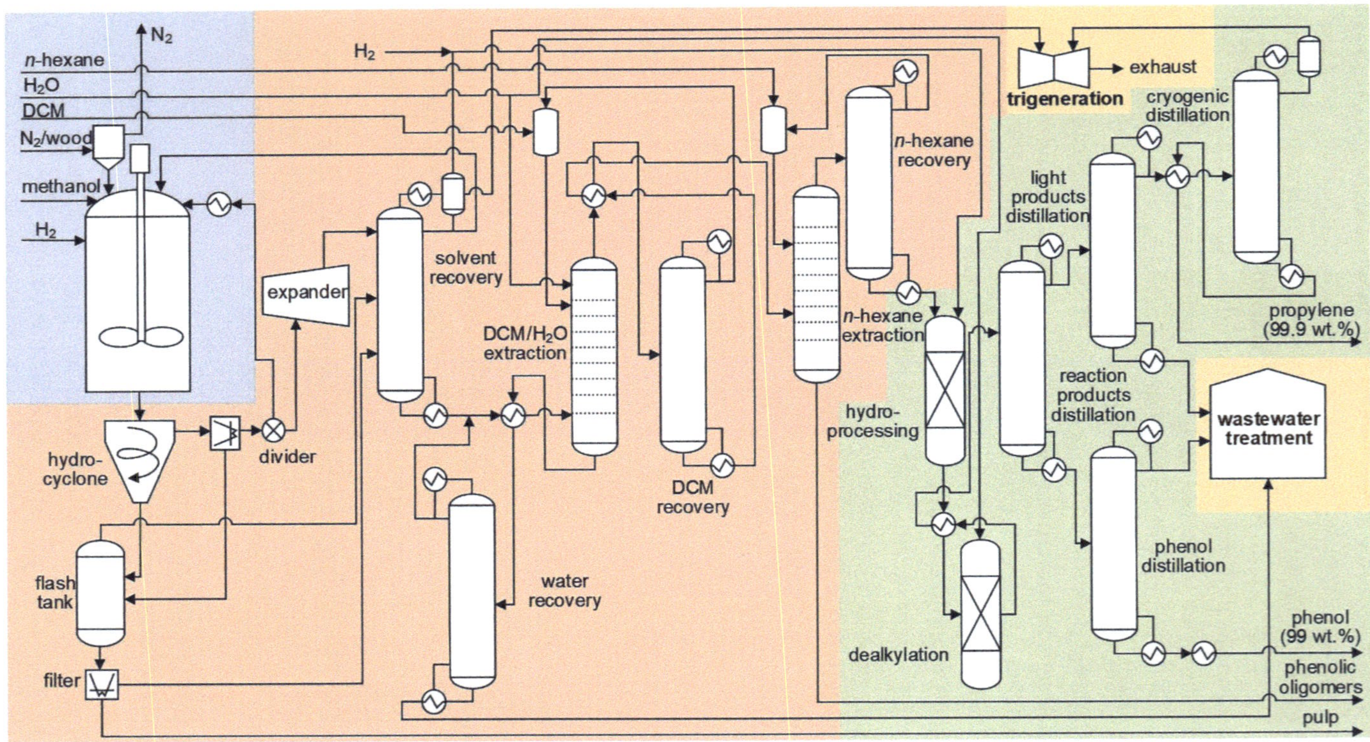

Fig. 13 Process flow diagram of the multiproduct biorefinery. *Adapted from Liao, Y.; Koelewijn, S.-F.; Van den Bossche, G.; et al.* Science **2020**, 367, *1385–1390.*

successfully converted to 190 g of lignin oil.[56] In comparison, our own group processed up to 2 kg of beech wood obtaining 380 g of lignin oil.[30] The batch upscale studies prove that interaction between the catalyst, reactive lignin fragments and hydrogen is sufficient to extract, depolymerize and stabilize lignin oil at larger scales. However, at industrial scale, batch reactors are typically used to produce high value fine and specialty chemicals when the priority is given to the product, rather than the process. For example, high-value active pharmaceutical ingredients are usually produced in small quantities (<1 kt/year) in batch reactor systems where successive reaction and separation steps (e.g., crystallization) are performed in the same vessel. In comparison, low-value, bulk chemicals production is rarely conducted in batch mode as one cannot benefit from the economies of scale (>1 kt/year) and techno-economic optimization of the entire process operation is prevalent.[156]

3.1.2.1.2 From batch to flow-through to continuous reactors The development of an efficient reactor is anticipated to be pivotal to guarantee economic competitiveness of any biorefinery concept opting for RCF as fractionation technology. State of the art reactor designs for RCF operation at lab scale comprise batch and flow-through. Most of the RCF research is performed through experiments with mechanically stirred batch autoclaves as they are easily implemented and allow flexibility in process parameter variation.[30,36,39] Recently, several independent research groups successfully conducted RCF flow-through experiments at lab scale, opening interesting yet unchartered fields of research by overcoming shortcomings of batch-wise operations (e.g., kinetic studies, vide infra).[43,44,47] Current flow-through setups consist of two separated but interconnected beds operating in series. Solvent flowing through a first bed containing the biomass disassembles the lignin. The formed liquor proceeds to the second, catalyst bed where oligomeric fragments further depolymerize, via solvolysis and catalytic hydrogenolysis, and reactive lignin fragments are stabilized. For industrial implementation of RCF in an upscale biorefinery, either design entails certain benefits and drawbacks from multiple perspectives.[23,30]

From a process design perspective, the drawbacks associated with batch reactor system mainly come from its dynamic nature.[30] Specific operational cycles of batch reactors may complicate efficient integration (e.g., heat integration and recovery, recycle networks) of the reactor section with a steady-state continuous downstream separation and/or upgrading sections unless a large number of batch reactors can be alternately operated, essentially functioning as one continuous reactor section. This will undoubtedly lead to large operational expenses and higher capital investments. For example in our case study, it was estimated that 17 batch reactors of 250 m^3 have to operate in parallel to provide the minimum reaction time of 1.5 h in order to produce ±250 kton of monomers annually, making the RCF step one of the highest contributors toward CAPEX (see Section 3.2.2).[5] Another drawback is the need for two additional processing steps for solid isolation and catalyst recuperation to obtain a catalyst-free pulp and a catalyst recycle stream. The latter can be avoided by physically separating biomass and catalyst pellets in the reactor using a catalyst basket or with dedicated reactor designs (e.g., Robinson-Mahoney).[157] Flow-through reactors are preferred over batch for a number of reasons: (i) the continuity of the process simplifies integration of RCF with downstream processing units and recycle streams and allows easier and individual control of the process conditions in the reactors, (ii) less separation steps are expected as the effluent reactor stream is solid-free and (iii) catalyst recuperation for, e.g., regeneration is simpler as biomass and catalyst are physically separated in two beds already. A major drawback using flow-through is the high liquid flow rates making solvent recuperation difficult due to very low product concentrations in the stream fed to the separation section. It must be noted that current flow-through setups are essentially non-continuous systems from a biomass perspective implying the need for at least one parallel biomass bed to allow alternated loading and unloading. The number of biomass beds may be even higher to ensure a flow with constant composition passing over a consecutive catalyst bed as the sizes of the released depolymerized lignin fragments alter throughout the course of the fractionation process. This can affect the continuity of all downstream operations.[47] Finally, also more than one catalyst bed may be required to allow catalyst regeneration when deactivation by inorganic compounds and coking becomes too severe.

A major engineering challenge for successful implementation of RCF technology at industrial scales is thus the development of a customized fully continuous reactor which allows for simple process integration and simultaneously fulfills other requirements to guarantee optimal chemical operation. Custom-made reactor designs developed for other liquid chemical pulping technologies, although operated at different conditions, reinforce this assumption. For example, special continuous reactor systems named Kamyr Digesters were developed for the Kraft pulping process[158] and a recent publication of Ruiz et al. extensively discusses continuous screw conveyor reactors for hydrothermal treatment of lignocellulosic biomass.[159] Continuously operating fluidized bed reactors could straightforwardly be used when the solid biomass completely dissolves during a reaction so that its addition to the reaction mixture can be matched with its degradation kinetics.[160] This is for example possible with (fast-)pyrolysis of solid biomass, where only gas and liquids are formed and leave the reactor.[161] However, for an RCF reactor to operate continuously, a continuous solid-solid separation (i.e., biomass—pulp separation) must be achieved. For this purpose, the expected (small) differences of wood and pulp hydrodynamic properties (e.g., density, particle size, etc.) could potentially enable a continuous hydrodynamic separation of solid pulp product from the fluidized mixture.

3.1.2.2 Separation: Refining lignin oil

3.1.2.2.1 Case study: A combination of conventional mechanical and chemical unit operations RCF research of the past 5 years has mainly covered conceptual and optimization studies of the reactive step to tailor CRLO composition by altering biomass substrates,[36] solvents,[37,50,138] additives,[51] catalysts[33,37,39] and more recently, the reactor set-up.[44,47] However, successful implementation of RCF hinges on the development of effective refining strategies to obtain pure streams which can be marketed or functionalized to valuable products. The proposed separation model of our process simulation is an approximate upscale equivalent of the laboratory purification procedure used before chemical analysis. Firstly, the solid separation is achieved through a hydrocyclone, flash tank and filters. The solvent and hydrogen gas stream are then removed via the overheads of a distillation column for recycle and successive use in the hydroprocessing/dealkylation units, respectively. The CRLO, obtained in the bottoms, is sent through a series of extraction and distillation columns to obtain streams rich in phenolic monomers and oligomers, respectively, and recycle streams of the extraction solvents: dichloromethane (DCM), water, and *n*-hexane (Fig. 13).[5] Equivalent laboratory steps are filtration, rotary evaporation and liquid-liquid extraction.[36] It must be noted that this separation sequence is just one example and that alternative sequences using different techniques and/or other solvents can be explored.

3.1.2.2.2 Liquid-liquid extraction Liquid-liquid extraction (LLE) is used twice in the overall separation strategy of our case study as it has proven to be an efficient method to process crude lignin oils. A first extraction step aims to separate lignin-derived phenolics and sugar depolymerization products after solvent removal. DCM/water extraction is commonly applied by several research groups for this purpose.[5,47,162] Barta and co-workers recently employed ethyl acetate as a chlorine-free and biodegradable[163] alternative for toxic DCM.[38] Renders et al. from our group came up with a system exploiting the temperature-dependent phase behavior of a solvent mixture comprising *n*-butanol and water. At reaction temperatures (160–250 °C), the solvents form one phase in which RCF takes place. Upon cooling (<125 °C), phase separation occurs automatically, resulting in a concentrated *n*-butanol product phase rich in lignin-derived phenolics, and an aqueous solution of depolymerized sugar products (mainly polyols).[37] In comparison with the methanol-based system of our case study, this approach may simplify the downstream processing sequence as at least one distillation and extraction column would become obsolete, likely resulting in reduced operational costs and capital investments. Lignin derived phenolics are separated in a second extractive step using *n*-hexane which becomes enriched with phenolic monomers when brought in contact with refined lignin oil (RLO) comprising monomeric and oligomeric phenolics.[36] Highly volatile *n*-hexane is often chosen due to its ease of separation via distillation and because it can be produced from the (hemi)cellulosic pulp fraction, hereby integrating monomer production with the carbohydrate platform.[96] A few other solvents like diethyl ether[57] and pentane have been used for RLO extraction as well. However, these solvents entail similar hazards as hexane.[164] Van Aelst et al. recently employed a mixture of ethyl acetate and *n*-heptane as a safer alternative for *n*-hexane for softwood-derived lignin oil fractionation.[58]

3.1.2.2.3 Distillation Distillation is probably the most widespread separation technique used for various applications in the chemical process and allied industries. Its use in future biorefineries is therefore expected to be essential. The separation sequence modeled in our biorefinery concept also consists of four distillation units to recover and recycle the solvents used in RCF reaction (methanol) and in LLE (DCM, water, *n*-hexane), respectively.[5] The large difference in volatility between these solvents and the oily products (phenolics, pyranoses) makes it a suitable technique, however, if properly designed, distillation columns can separate components with a low relative volatility, equally well. For example, Koelewijn et al. proposed a separation sequence to purify the monomer-rich hexane-extracted lignin oil to obtain two streams rich in 4-*n*-propylguaiacol (4-PG) and 4-*n*-propylsyringol (4-PS), respectively.[96] The *n*-hexane phase is processed in a first distillation column (consisting of 10 trays) working at atmospheric pressure to recover the solvent via the overheads and to obtain a phenolics stream in the bottoms. The latter is sent through a second distillation column (consisting of 57 trays) operating at a vacuum to accomplish a proper split between 4-PG and 4-PS whose molecular structures differ by one methoxy group attached to the aromatic ring only. Before column design, the feasibility of distilling 4-PG and 4-PS was examined by using the well-known concept of relative volatility (α) and a novel concept named the distillation resistance (Ω) introduced by Lange et al.[165] The relative volatility of the light to the heavy key component, i.e., 4-PG to 4-PS, is 1.458 (calculated at the estimated boiling point of the refined lignin oil: 278.15 °C at 1 atm) hereby exceeding 1.05, which is the critical threshold for an economically feasible distillation.[165] The diffusional resistance is a measure solely based on the crude composition and the atmospheric boiling points of its components and correlates with an estimated heating duty which can be translated to a total cost for separation. It evolves from a moderate 1.5 $°C^{-1}$ (<1.0 $°C^{-1}$, for simple distillations) for the *n*-hexane-extracted oil to 14.4 $°C^{-1}$, for the monomer-rich phase only. The diffusional resistance of the latter is just below the ceiling of 15–17 $°C^{-1}$ which corresponds to a recovery cost of \$100 per ton at a 200–400 $kt_{product}$ per year scale. The recovery cost is therefore expected to reach an amount of \$85–95 per ton, which seems economically feasible. A rigorous separation model of a small-scale industrial continuous distillation unit (85 kt_{feed} per year of phenolic monomers) built in Aspen Plus® modeled the heating duty at 2.3 GJ/ton of feed corresponding to a recovery cost of \$112 per ton of product. These estimations agree reasonably well with the recovery cost predicted by the diffusional resistance, hence proving its value in preliminary estimations.[96]

3.1.2.2.4 Green solvents The separation sequence as introduced above requires organic solvents in LLE for refining CRLO and RLO. Certain solvents form a significant environmental hazard when leaked as volatile organic compounds (VOCs) to the atmosphere.[166] For example, DCM and *n*-hexane are two common VOCs originating from (petro)chemical industries where they are lost to the environment in several ways such as, e.g., the breathing and loading losses from storage tanks, vents from process vessels, leaks from piping and equipment, contaminants of waste water, and others.[166] So-called *green solvents* may provide a solution in that regard as long as they show acceptable performance beyond their intrinsic environmental benefits. Ionic liquids (ILs), deep eutectic solvents (DESs) and switchable hydrophilicity solvents (SHSs) are examples of highly versatile solvent classes. They are often regarded as greener alternatives for conventional organic solvents due to certain intrinsic properties such as a low volatility, non-flammability, biodegradability and low toxicity.[167–169] All have successfully been applied in LLEs recovering phenolic compounds from water[170] and/or oily mixtures of various compositions[171–175] so that, when specially tailored, they may replace DCM and/or *n*-hexane in one of the extractive steps. SHSs are especially interesting as their recuperation is achieved by removal of, e.g., CO_2, which triggers a switch from a hydrophilic to a hydrophobic state, thereby avoiding an energy-intensive distillation step.[168] However, the actual greenness of solvent types must be critically assessed at least by calculating green chemistry metrics (e.g., atom economy, carbon efficiency, etc.). Even more preferable is to conduct an extensive (desirably cradle-to-cradle or cradle-to-grave) LCA covering solvent synthesis, the chemical process in which it is used and the disposal strategy as not only its VOC emissions determine its environmental impact.[176,177] A recent review authored by Hallett's group extensively discusses emerging solvent classes from a technical, environmental and economical perspective serving as an excellent guide for bridging research and development, and commercialization of new solvents.[177]

ILs, DESs and SHSs and other solvent classes (e.g., liquid polymers, switchable solvents) are excellent subjects for academic research given their highly versatile chemical nature, and their wide field of potential applications.[176,177] However, despite having found their way to certain applications,[178] widespread industrial use in chemical production processes remains rather limited. Solvent selection is therefore usually restricted to available organic solvents. For this purpose Pfizer developed a solvent selection tool which employs several criteria such as (i) worker's safety, (ii) process safety and (iii) environmental and regulatory considerations to classify organic solvents in three classes being (a) preferred, (b) usable and (c) undesirable. In addition, they created a replacement table containing suitable solvents as replacements for undesirable ones based on technical and other criteria introduced above. For example, *n*-hexane can be replaced by less toxic *n*-heptane, and DCM by ethyl acetate, toluene, methyl *tert*-butyl ether or 2-Me-THF.[179] Useful tools to verify the replace-ability of a solvent by a greener alternative are Kamlet-Taft plots[176] and Hansen Solubility Parameters.[180] The former is a 2D plane with polarity and basicity as axis-variables. The latter is a 3D space with three types of interactions, namely, (i) dispersive, (ii) hydrogen bonding and (iii) polarity as axis-variables. Coinciding plots of solvents in these variable spaces indicate similar physicochemical properties suggesting that they can be interchanged.

Finally, renewable solvents, i.e., solvents derived from biomass, may partially replace fossil-derived organics. For example, 2-Me-THF, valerolactone, cyclopentyl methyl ether are common examples of renewable solvents derived from starch or (hemi-)cellulose. Similar tools as introduced for conventional organic solvents are suitable in classifying renewable solvents and predicting their applicability in certain processes such as LLE based on their intrinsic physicochemical properties. Despite their bio-based origins, these solvents can still be hazardous and may entail higher societal costs compared to fossil-derived equivalents, especially when originating from biomass which is competitive with food production.[176,177] In that regard, the biorefinery proposed in our work has the advantage of using non-edible lignocellulosic biomass as primary substrates, allowing for a self-sufficient and integrated solvent production.

3.1.2.2.5 Membrane separation The separation sequence proposed in our biorefinery concept employs a cascade of solvent-consuming LLE and energy-intensive distillation steps for refining the liquid mixture formed during RCF of lignocellulosic biomass. In 2019, Rinaldi and Livingstone investigated a two-stage membrane cascade for refining an oily liquor produced via catalytic upstream biorefining (CUB). This RCF-like process employs Raney Ni catalysts, isopropyl alcohol as hydrogen donor and solvent, and water as co-solvent.[41] Separation is achieved by applying a pressure over a polymeric film with certain porosity, i.e., the membrane, allowing low-MW components to pass (the permeate), while higher MW components are retained (the retentate). In a first set of experiments, commercial and in-house ultra- and nanofiltration membranes were tested to up-concentrate high-MW (>1000 Da) components on the retentate side, allowing low-MW phenolics (<1000 Da) and solvents to pass (stage 1). Subsequently, nanofiltration and reverse osmosis membranes with smaller pore sizes were examined to separate the permeate of the first stage, comprising low-MW phenolics and solvents, into an up-concentrated retentate and a recyclable, pure solvent stream (stage 2). An optimized six-stage cascade membrane setup, using the optimal-performing membranes for stages 1 and 2, fulfilled all technical requirements, though was not deemed economically feasible due to low permeate fluxes making the total membrane area demand to exceed 0.1 km^2 in order to process 500 kt of oil liquor in a theoretical up-scale system. An economically optimized, alternative three-unit system using hypothetical membranes with permeances of 5 $Lm^{-2}h^{-1}bar^{-1}$ for both stages 1 and 2 and a low-MW rejection of 95% in stage 2 reduced overall separation costs. Compared to the six-unit cascade, costs dropped from \$8.20 per kg of dry phenolic product to \$0.38 per kg of dry product, rendering the hypothetical system economically feasible, given the phenolic product value of \$1.11 kg^{-1}. This system also turned out to be much cheaper than thermal distillation, for which a separation cost of \$1.34 per kg of phenolic products was roughly estimated. However, rather than strictly comparing membranes versus distillation, the potential of combinatorial systems of LLE, distillation and membrane units to achieve complete fractionation can be concurrently investigated. Lastly, an additional advantage of a membrane system is its ease of scalability: the membrane filtration capacity can simply be expanded or reduced by parallel addition of new filtration units or by-passing existing ones corresponding to the magnitude of the upstream process streams and conditions.[181]

3.1.2.3 Upgrading: Synthesizing valuable chemicals

3.1.2.3.1 Case study: Production of phenol and propylene with tailor-made gas phase technology Section 2 outlines various fundamental concepts and strategies to upgrade monomeric phenolics into (potentially) valuable and marketable chemicals. Most of these studies employ pure components or carefully prepared mixtures thereof, to investigate chemical pathways of reaction, catalyst functioning and reaction conditions to optimize yield and selectivity. However, impurities and trace components may significantly impact optimal operation, hence testing *real* crude mixtures is recommended. This has been done for our case study where phenolic monomers are upgraded in two consecutive reactive steps extensively discussed in Section 2.2.1.

The gas-phase technology is integrated with the upstream sections by recuperating hydrogen of the overheads of the solvent distillation column, which is sent, along with fresh hydrogen and the monomeric stream, to a plug-flow reactor for isothermal hydroprocessing at 300 °C. Next, dealkylation is performed in a fire-heated tube reactor at 400 °C. Product separation following this step is achieved by four distillation columns. A first column separates the light (propylene, hydrogen, methane) from the heavy components (phenol, water and byproduct cresol). A 99% pure propylene stream is obtained in the bottoms of a cryogenic distillation column. Phenol is purified to 99% in the bottoms of the final distillation column (Fig. 13).

3.1.3 *Process design*

Once a suitable process scheme is defined, the operating conditions must be determined along with appropriate equipment sizing for each individual process step in order to meet the design criterion. This modeling step is usually performed with chemical simulation software (ChemCad®, Aspen Plus®) which have pre-defined design functions describing the process kinetics and thermodynamics. Only a limited number of variables must be fixed by the design engineer hereby greatly simplifying this task. Challenges arise when novel compounds or product streams (e.g., RCF lignin oil) are used, or kinetic descriptions are unavailable.

3.1.3.1 Development of a reactor model: Kinetics and thermodynamics

For example, the reactor units of this case study are not described by predefined sets of design equations of the process simulator. Instead, its mass balances, optimized at lab-scale, are up-scaled using mass ratios, and energy balances are computed with the heats of reaction, estimated by the enthalpy of formation of the reagents and products. Group contribution theory (GCT) is used to calculate formation enthalpies when unavailable for unconventional components,[182,183] although its accuracy is not always ensured and experimental validation through reaction calorimetry is necessary.[184] The development of accurate reactor models requires appropriate kinetic expressions representing the time-dependency of substrate consumption and products formation. This is an especially challenging task for RCF given the multitude of kinetic events occurring simultaneously and the potential involvement of mass transfer limitations at the biomass particle and catalyst level. Several studies covering this topic have recently been published by Roman-Leshkov and Beckham et al.: they compared batch and flow-through systems from a kinetic perspective[47] and developed CFD mass transfer models of wood solvolysis including the intrinsic kinetics of lignin extraction.[32] To the best of our knowledge, a complete model covering the entire RCF operation has not been published yet.

3.1.3.2 Modeling separation units: Thermodynamics of complex mixtures

Research and development is expected to play a key role in advancing innovative separation technologies specifically designed for CRLO refining as outlined in Section 3.1.2. It will also be essential to fundamentally understand and simulate conventional refining methods such as distillation and LLE in context of the RCF biorefinery process simulations. Namely, thermodynamic equilibria of the complex mixtures processed by these unit operations are described by parameterized mathematical functions, i.e., equations of state, which are used in chemical process simulation software such as Aspen Plus® and ChemCad®. Experimental data to fit these parameters, e.g., activity coefficients, is often missing for unconventional compounds formed with RCF. In that case, one can solely rely on rational methods such as UNIFAC group contribution to compute activity coefficients of non-ideal mixtures in order to predict vapor-liquid and liquid-liquid equilibria with large uncertainty.[185] This may significantly reduce the reliability of the process simulation. Therefore, it is highly encouraged to explore this new field of research in which RCF overlaps with fundamental thermodynamics.

3.1.4 Process integration

The final phase of a process design involves exploiting and optimizing the connections and interactions between the processing steps. This includes identifying the recycle streams and integrating hot and cold streams by pinch-point analysis to minimize raw materials and energy consumption, respectively.[156] Examples of our case study are the solvent recycle streams originating from the RCF reactor and extraction columns and the integration of excess hydrogen originating from RCF with the hydroprocessing and dealkylation units. Part of the heat integration network is visible on the PFD of Fig. 13 where multiple output streams cool or heat input streams of process units in heat exchangers.

3.2 Techno-economic assessment

3.2.1 TEA methodology

RCF technology and the use of biomass as raw material will only become relevant if it can compete economically with existing processes to produce drop-in chemicals or achieves to create new markets by producing new high-value chemicals. The profitability of a new production process is typically evaluated by TEA, a methodology in which the economics and technical design and operation are concurrently evaluated and optimized. Traditional TEA consists of four steps. Firstly, a market study identifies market-related parameters potentially influencing the ability to commercialize a project. Secondly, a process flow diagram is developed, providing mass and energy balances of process units as demonstrated in Section 3.1 and shown by Fig. 13. Thirdly, an economic assessment estimates the economics feasibility of the project. Finally, a sensitivity analysis of certain technical and economic variables is conducted. The outcome of an economic assessment (step 3) is described by typical economic performance indicators such as the internal rate of return (IRR, Eq. 1) and net present value (NPV, Eq. 2). These metrics estimate the long-term profitability on the capital investments (CAPEX) required to have a new process installed and running.

$$0 = \sum_{i=1}^{20} \frac{CF_i}{(1 + IRR)^i} - CAPEX \tag{1}$$

$$NPV = \sum_{i=1}^{20} \frac{CF_i}{(1 + r_{wacc})^i} - CAPEX \tag{2}$$

IRR, internal rate of return; *NPV*, net present value of an investment (€); *CAPEX*, total capital investments (€); r_{wacc}, weighted average cost of capital; CF_i, cash flow in year i (€); *i*, year *i* of project life time (e.g., 20 years).

The IRR is the percentage return made on a capital investment over a life-time of the project when the NPV of that investment is set to zero, or else, it is the cost of capital which makes the NPV of an investment zero. Therefore, an IRR larger than the weighted average cost of capital (r_{wacc}) will lead to a profitable investment as capital can be "loaned" at a lower rate than it will return. The magnitude of that profit is estimated by the NPV, i.e., the current value of an investment over the envisioned project lifetime. Negative NPVs and IRRs lower than the cost of capital signify losses on investment in a particular project. Only projects with high NPV are likely to be commercialized to limit the payback period of the investment to 3–5 years.

The IRR and NPV are calculated by the capital costs (CAPEX) and annual cash flows (CF_i) from which the latter are determined by the gross profits (i.e., difference of the operational costs (OPEX) and revenues of product sales) minus the tax allowances. Hence, they are indirectly related to the technical design and operation of the refinery. Capital investments mainly involve process equipment along with other minor costs which are normally estimated as factors of the purchase costs of the major equipment with factorial cost estimation methods.[156,186] The process equipment, central in CAPEX estimations, is designed to meet a predefined criterion, i.e., the production capacity at which a plant is expected to operate. OPEX include the costs of raw materials, utilities (steam, cooling water, electricity, air and nitrogen), consumables, waste disposal and the fixed costs of production (labor, maintenance, etc.). Chemical engineering design handbooks (e.g., Chemical Engineering Design of Sinnott & Towler) extensively discuss how to create detailed cost estimations for chemical plants.[156]

3.2.2 *TEA of a* lignocellulose-to-phenol *biorefinery*

3.2.2.1 Case study: Production of phenol and propylene with tailor-made gas phase technology

The design criterion for the *lignocellulose-to-phenol* biorefinery is fixed at an annual phenol production of 100 kt (i.e., average scale for fossil-based phenol production) which corresponds to a birch wood supply of 2590 kt y^{-1}. TEA of the rigorous process simulation of this case study presented in Section 3.1 indicates a profitable investment given a high IRR of 23% and large NPV. The highest contributors to CAPEX are the RCF system consisting of 17 reactors of 250 m^3 and the incinerator/trigeneration unit, whose investment is justified by the reduced energy costs (Fig. 14).

The highest contributors to the OPEX are the feedstock costs of wood (158€ t^{-1}) which have, along with the product pricing of phenol (1300€ t^{-1}), propylene (830€ t^{-1}), crude pulp (400€ t^{-1}) and oligomers (1750€ t^{-1}) the largest economic sensitivity according to a sensitivity analysis as shown in Fig. 15. This underlines the importance of the integrated biorefinery concept in which complete wood valorization to valuable products is achieved. The catalyst cost is negligible assuming it can be sufficiently recycled and reused (Fig. 15).[5]

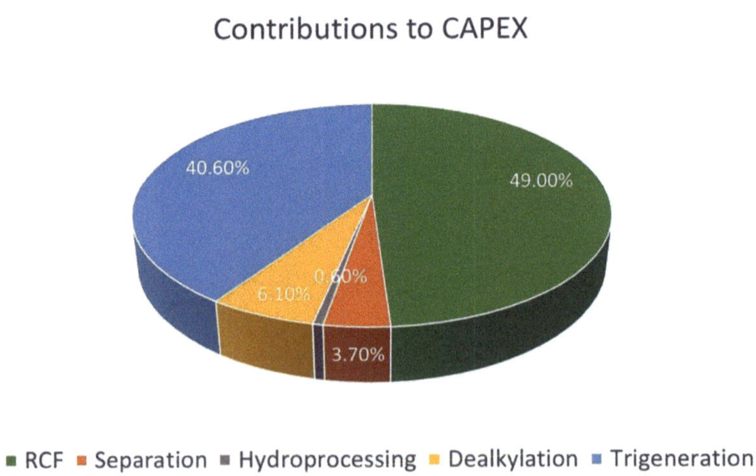

Fig. 14 Contribution of (i) reaction, (ii) separation and (iii) upgrading section (hydroprocessing/dealkylation/trigeneration) to the capital investments of the lignocellulose-to-phenol biorefinery. *Adapted from Liao, Y.; Koelewijn, S.-F.; Van den Bossche, G.; et al.* Science ***2020****,* 367, *1385–1390.*

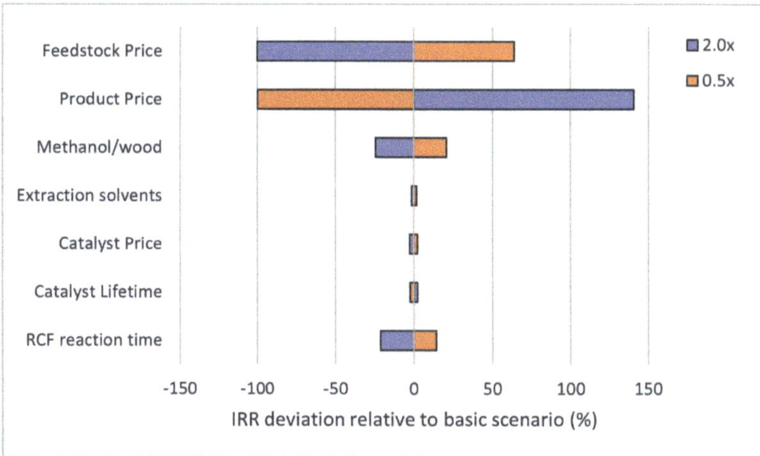

Fig. 15 Tornado diagram illustrating the results from the sensitivity analysis on several design and cost assumptions. The internal rate of return (base case) is 23.3%. *Adapted from Liao, Y.; Koelewijn, S.-F.; Van den Bossche, G.; et al.* Science ***2020***, 367, *1385–1390.*

3.2.3 Integrated TEA of a RCF biorefinery

In a follow-up study, the RCF biorefinery was economically assessed by a fully integrated TEA considering up- and downstream market conditions as well as midstream process-based technical and economic variables.[48] This study aimed to identify the key elements determining the economic feasibility and to create an integrated TEA, adjustable to changing R&D outcomes. The scope of the refinery in this study was limited to the production of phenolic monomers and oligomers and a carbohydrate pulp by removing the hydroprocessing, dealkylation and distillation modules and the tri-generation unit (hence, named RCF biorefinery). The latter was not deemed realistic at smaller scales below 150 kt y^{-1} processed wood considered in this study in comparison with a wood processing capacity of 2590 kt y^{-1} of the case study. Total equipment costs were scaled down using the six-tenth rule (Eq. 3), where C_A, C_B and S_A, S_B are the equipment costs and sizing of projects A and B, respectively, and n is the capital exponent (set to 0.6).

$$\frac{C_A}{C_B} = \left(\frac{S_A}{S_B}\right)^n \tag{3}$$

The details of this study are extensively described in our recent publication.[48] An outline of the results is given below.

Firstly, the importance of the capacity level which resembles the wood supply condition was assessed by calculating the NPV of a 20, 75 and 150 kt y^{-1} scaled biorefinery. Notably, the economies of scale are not always beneficial since the estimated NPV for the 20 kt y^{-1} (−35 M€) was lower than the 75 kt y^{-1} biorefinery (−42 M€). The NPV of the 150 kt y^{-1} was estimated at −23.45 M€, showing that a minimum level scale exists at which the NPV starts to increase. Negative NPVs imply a need for reducing costs per product unit.

Secondly, a Monte-Carlo sensitivity analysis was conducted to identify the high impact variables on the economics of the 150 kt y^{-1} biorefinery. Ten technical and economic variables which relate most sensitively with the NPV were determined. Feedstock prices (−11%), lignin products (+21), and pulp product (+10%) all contribute above 10% to the variance in NPV. Technical variables including lignin (+21%) and pulp product (10%) outputs also significantly contribute to the sensitivity in NPV indicating that improved reactor yields may significantly impact the biorefinery economics. The capacity exponent (*n*, Eq. 3) representing the economies of scale (e.g., new technologies, productivity improvement, learning effects result in larger *n* value) contributes most to the variance in NPV with +22%. Therefore, research should focus on improving and optimizing plant facilities and operations.

Thirdly, the impact of different feedstock on the biorefinery economics were assessed, given its large price and output sensitivity. Mass balances were adapted for poplar, pine and waste wood from the base case birch. The total capacity of the plant is kept constant at 150 kt y^{-1}. Waste wood showed to be the preferred feedstock given its low-price originating from reduced processing costs in comparison with pristine wood (no need for harvesting and forwarding, plus low drying costs) resulting in a positive NPV (59 M€) and high IRR (26%). Adaptation of waste wood in RCF may however be problematic from a technical and consumer perspective due to potential presence of contaminants. Furthermore, it was shown that processing various feedstock is beneficial to absorb drops in individual feedstock supplies which depends on the specific regional availabilities of the wood species (location factor).

Finally, a break-even point (BEP) analysis determined pairs of capacity level and feedstock costs for birch, poplar, pine and waste wood for which the NPV is zero. For a capacity level of 150 kt y^{-1} biorefinery, birch, poplar and pine feedstock costs need to decrease to become profitable, whereas the waste wood price can increase to a certain extent. At current pricing, the capacity level of the refineries should increase for all wood species except for waste wood to become profitable. Further, a BEP analysis of downstream variables showed that product prices need to increase significantly for biorefinery operations with pristine wood species as feedstock to be viable. This can be achieved by increasing R&D efforts to improve the quality of the end-products as to increase product pricing.

3.3 Life cycle assessment

3.3.1 LCA methodology

Beside economic viability, designing a bio-based process only makes sense if a lower environmental impact is attained compared to its fossil-based counterparts. For this purpose, a Life Cycle Assessment (LCA), a well-known technique to assess sustainability impacts of a production systems, should be performed.[187] LCA is an ISO standardized procedure comprising four parts: (i) goal and scope definition, (ii) inventory analysis, (iii) life cycle impact assessment and (iv) life cycle interpretation. Key in performing a good LCA is defining a functional unit and scope or system boundaries (phase 1) and acquiring a complete dataset of input and outputs. This dataset, called the life cycle inventory (LCI), models the product system that fulfills a certain function (e.g., the production of hydrogen via steam reforming) (phase 2). Next, the LCI is converted into multiple environmental performance indicators, namely detailed mid-point (e.g., global warming potential) and generalized end-point (e.g., fossil resource scarcity) categories by an LCA calculation method (phase 3). Multiple methods exist with varying mathematical procedures representing different scenarios regarding time and expectations on how a production system will impact the environment. These are employed to evaluate LCIs in various ways. Interpretation of the LCA occurs throughout all LCA development phases (phase 4). LCA is typically performed with specialist software (e.g., GaBi®, SimaPro®) in which these four phases can be executed simultaneously.

3.3.2 LCA of a lignocellulose-to-phenol *biorefinery*

3.3.2.1 Case study: Production of phenol and propylene with tailor-made gas phase technology

The proposed *lignocellulose-to-phenol* biorefinery produces phenol and propylene offering alternative synthesis pathways for the fossil oil-based Hock process and steam-cracking routes, respectively.[4] The sustainability of these production processes are assessed and compared using LCA. The system boundaries of all three processes span almost all stages in the products' life cycle, starting from the extraction of raw materials, transport of raw materials and finally, the production of chemical products. This *cradle-to-gate* approach suffices as the use-phase and end-of-life of the products are identical for biomass-derived and fossil-derived phenol and propylene, hence the LCA results are not affected by exclusion of these phases. The functional unit of the production process is either "*1 kg of phenol produced*" or "*1 kg of propylene produced*" depending on the studied process (other products are considered as by-products in either scenario). The total emissions of the *lignocellulose-to-phenol* biorefinery are allocated over the products based on their economic value. The LCI is built with the mass and energy balances obtained for the rigorous process simulation as introduced in Section 3.1. LCI of the Hock and steam-cracking processes are present in the software LCI databases. GWP is calculated for the three processes using the CML 2001—Jan 2016 assessment method.[5]

The production of phenol and propylene shows a reduced impact on global warming compared to the fossil-based counterparts. The GWPs for phenol and propylene decrease from 1.73 to 0.74 kg CO_2 equivalent and 1.47 kg CO_2 to 0.47 kg CO_2 equivalent, respectively. Additionally, the GWP for *p*-nonylphenol is significantly higher (>1.58 kg CO_2 equivalent) than the negative GWP of the substitute oligomers (−0.95 kg CO_2 equivalent). The carbohydrate pulp also shows a negative GWP of −0.22 kg CO_2 equivalent signifying carbon capturing effect or net CO_2 consumption for their production. Hydrogen originates from non-renewable steam-reforming process and wood is not sourced via sustainable forest management practices.

The relative contribution of each processing step, i.e., reactor, separation and upgrading unit, to the phenol GWP is determined individually as shown in Fig. 16. Consumption of wood pellets and limited hydrogen use result in a negative GWP for the reactor step in comparison with the large GWP of the hydrogen-consuming upgrading section, comprising hydro processing and dealkylation. The phenol GWP decreases significantly to a negative value when a renewable hydrogen source (e.g., electrolysis) is used instead of non-renewable fossil-based hydrogen. Additionally, wood sourced from sustainable forest management positively affects the phenol GWP and to a larger extent than changing hydrogen source from renewable to non-renewable due to the large wood processing volumes. Combining wood obtained from sustainable forest management and renewable hydrogen gives the largest reduction in phenol GWP to approximately −4.5 kg CO_2 equivalents. Propylene, oligomers and carbohydrate pulp have large negative GWPs in this scenario too.

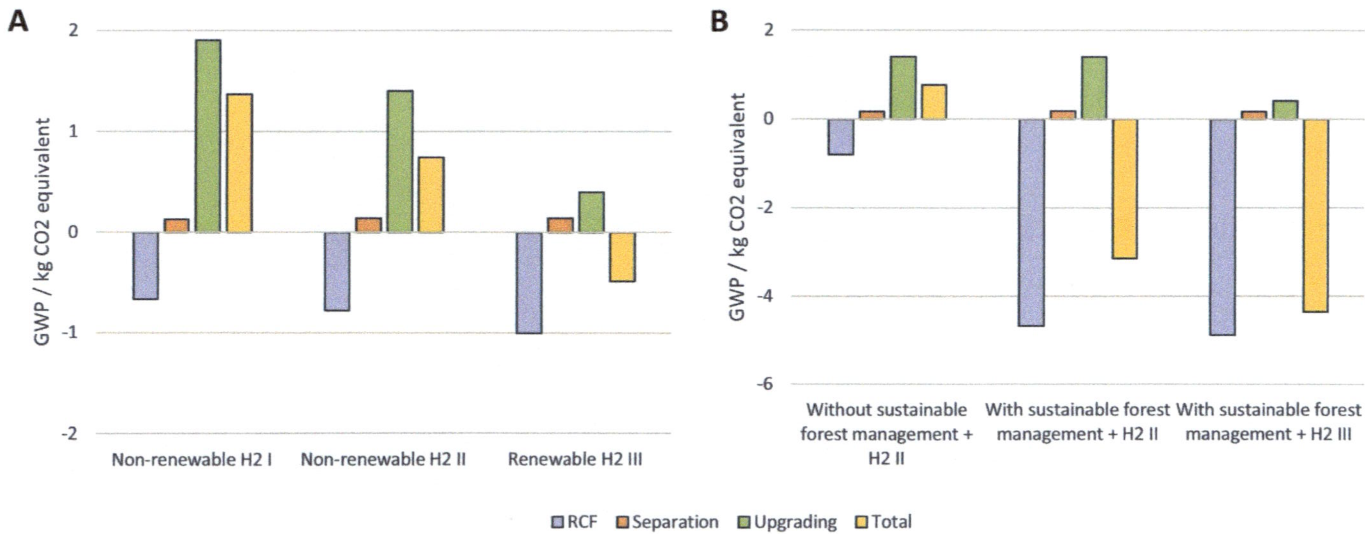

Fig. 16 Contribution of (i) reactor (reductive catalytic fractionation), (ii) separation and (iii) upgrading (hydroprocessing/ dealkylation/distillation) steps to the GWP of phenol production from lignocellulose (birch wood) with different hydrogen sources (A) and without/with sustainable forest management strategies in this integrated biorefinery (B). *Adapted from Liao, Y.; Koelewijn, S.-F.; Van den Bossche, G.; et al.* Science ***2020***, 367, *1385–1390.*

3.4 Critical considerations

It must be noted that the methodology used, and assumptions made in the preliminary process design of the "Introduction" section may impact the TEA and LCA results of Sections 1 and 2. For example, the cycle/residence time of the RCF reactor is set to 1.5 h, whereas current cycle times of pilot scale batch reactors are currently much longer.[30] Given the large contribution of the RCF system to the CAPEX, even a slightly increased cycle time may impact the profitability and environmental impact of the biorefinery as more reactors, utilities and energy are then required to meet the demands of the design criterion. Other assumptions directly affecting TEA and LCA are the catalyst reusability and solvent up-concentration. Despite the flaws that may be introduced herewith, the preliminary TEA and LCA of the *lignocellulose-to-phenol* and TEA of the RCF biorefineries show that a profitable and sustainable production process relying on RCF as biomass extraction technology is achievable. Nevertheless, the technology, still in its infancy, needs further advancement in certain scientific and engineering areas before actual industrial implementation seems possible. Recommendations for this are to:

1. Design a dedicated reactor to lower cycle times, achieve maximum yields and selectivity as to increase profits and reduce environmental burdens. This reactor must work continuously to allow easy integration in and optimization of the entire biorefinery.
2. Develop efficient separation strategies using alternative solvents or innovative methods to achieve higher processing yields, hereby reducing costs and environmental impacts.
3. Investigate the benefits and drawbacks associated with various feedstock as to weaponize the biorefinery for contingent supply discrepancies caused by economic and/or environmental factors. Waste wood may be essential for this.
4. Continue research in product valorization to raise the value of the biorefinery products, making the operation generally more profitable.

To improve the accuracy of the process model, it is recommended to:

5. Develop a kinetic model as part of a larger reactor model to enable mathematical optimization of the reactor individually and within the biorefinery operation.
6. Acquire thermodynamic and other data to better describe the properties of process streams, comprising unconventional compounds of the biorefinery.

4. Conclusions and future perspectives

Lignocellulosic biomass is a promising raw material to replace current fossil fuel feedstocks in the production of new and existing chemicals. *Lignin-first* fractionation strategies are emerging, enabling the valorization of currently underutilized lignin fraction in commercial carbohydrate-focused biorefinery schemes. One of the promising *lignin-first* fractionation techniques is *reductive catalytic fractionation* (RCF). Its most valuable feature is the in situ stabilization of reactive bonds of the depolymerized lignin fragments due to the presence of a reducing hydrogen atmosphere and a redox active catalyst. RCF safeguards the inherent chemical complexity of lignin, thereby encouraging researchers to find novel (synthetic) approaches utilizing the interesting molecules contained within the lignin-derived product streams. The development of RCF has made substantial progress over the past 5 years, nevertheless, many challenges remain. These must be resolved before an RCF-based biorefinery is deemed feasible in the future.

Characterization of the unique chemical structures contained in the lignin product streams is essential for developing valorizing strategies and to unravel the catalytic and solvolytic chemistry enabling the cleavage of the β-O-4 inter unit bonds. However, this field of research is still underexplored due the highly complex lignin oligomer fraction. Nevertheless, some advancements have been made recently, namely by using *NMR techniques* such as ^{1}H–^{13}C HSQC and ^{31}P NMR which enable the identification of inter-phenolic linkages and various hydroxyl chemical functionalities. These advancements are outlined in Section 1.

Whereas valorization of dimeric (and oligomeric) phenolics is still limited partially due to difficult chemo-analytics, a plurality of promising *upgrading strategies* has been developed utilizing various phenolic monomers as reagents. These strategies can be categorized according to one main feature. Strategies under *functionalization* utilize of the intrinsic functional groups to increase the chemical complexity and diversify the product gamma. A first example within this category called dimerization is featured by the coupling of methoxy-functionalized monomers for the synthesis of bisphenol equivalents with unique properties. A second valuable example within this category is the *amination* of phenolic monomers, which allows to incorporate nitrogen into the phenolic monomers. A final category named *defunctionalization* includes all strategies where the intrinsic complexity of the phenolic monomers is reduced to produce drop-in chemicals. Essential for the success of any upgrading strategy is the ability of the utilized catalytic system to cope with the complex product stream and presence of impurities, which is not always consistently investigated. These upgrading strategies and corresponding challenges are discussed in Section 2.

The RCF-based biorefinery must be able to deliver chemical product(s) at a large scale with a process that is profitable and environmentally friendly. An example of such a biorefinery is introduced in part three by means of a case study. In this biorefinery example, lignin is fractionated using RCF, after which the purified phenolic monomers stream is converted to phenol and propylene by means of a catalytic defunctionalization route. This *lignocellulose-to-phenol* case study introduces a potential process design, which is analyzed by a techno-economic and a life cycle assessment. The value of this approach, even in early-stage research, is exemplified by identifying the most sensitive process, economic and environmental variables and certain bottlenecks, which can complicate upscaling. Overall, besides indicating a feasible biorefinery operation using RCF as fractionation strategy, this case study aids in setting targets for future research and development.

Acknowledgments

W.A. acknowledges funding from Internal Funds of KU Leuven (IDN Project-FFASSD), D.R. and K.V.A. acknowledge funding from EOS Excellence of Science program BioFact, and L.T. acknowledges funding from the Catalisti SBO project ARBOREF and BIODENT (C3 KU Leuven) for their doctoral fellowships.

References

1. European Commission. *European Green Deal Investment Plan*; European Commission, 2020.
2. United Nations. *Paris Agreement*; UNFCCC, 2015.
3. Myers, S. L. *Xi Sets 40-Year Target in Ambitious Pledge on Carbon Neutrality*; The New York Times, 2020. September 24.
4. Levi, P. G.; Cullen, J. M. *Environ. Sci. Technol.* **2018**, *52*, 1725–1734.
5. Liao, Y.; Koelewijn, S.-F.; Van den Bossche, G.; et al. *Science* **2020**, *367*, 1385–1390.
6. Schutyser, W.; Renders, T.; Van Den Bosch, S.; et al. *Chem. Soc. Rev.* **2018**, *47*, 852–908.
7. Rinaldi, R.; Jastrzebski, R.; Clough, M. T.; et al. *Angew. Chem. Int. Ed.* **2016**, *55*, 8164–8215.
8. Ragauskas, A. J.; Beckham, G. T.; Biddy, M. J.; et al. *Science* **2014**, *344*, 1246843-1–1246843-10.
9. Li, C.; Zhao, X.; Wang, A.; et al. *Chem. Rev.* **2015**, *115*, 11559–11624.
10. Vanholme, R.; De Meester, B.; Ralph, J.; et al. *Curr. Opin. Biotechnol.* **2019**, *56*, 230–239.
11. Vanholme, R.; Demedts, B.; Morreel, K.; et al. *Plant Physiol.* **2010**, *153*, 895–905.
12. Ralph, J.; Lapierre, C.; Boerjan, W. *Curr. Opin. Biotechnol.* **2019**, *56*, 240–249.

13. Renders, T.; Van Den Bosch, S.; Koelewijn, S.-F.; et al. *Energy Environ. Sci.* **2017**, *10*, 1551–1557.
14. Questell-Santiago, Y. M.; Galkin, M. V.; Barta, K.; et al. *Nat. Rev. Chem.* **2020**, *4*, 311–330.
15. Zijlstra, D. S.; Lahive, C. W.; Analbers, C. A.; et al. *ACS Sustain. Chem. Eng.* **2020**, *8*, 5119–5131.
16. Lancefield, C. S.; Panovic, I.; Deuss, P. J.; et al. *Green Chem.* **2017**, *19*, 202–214.
17. Lan, W.; Amiri, M. T.; Hunston, C. M.; et al. *Angew. Chem. Int. Ed.* **2018**, *57*, 1356–1360.
18. Li, Y.; Shuai, L.; Kim, H.; et al. *Sci. Adv.* **2018**, *4*, 1–10.
19. Shuai, L.; Amiri, M. T.; Questell-Santiago, Y. M.; et al. *Science* **2016**, *354*, 329–333.
20. Lahive, C. W.; Deuss, P. J.; Lancefield, C. S.; et al. *J. Am. Chem. Soc.* **2016**, *138*, 8900–8911.
21. De Santi, A.; Galkin, M. V.; Lahive, C. W.; et al. *ChemSusChem* **2020**, *13*, 4468–4477.
22. Deuss, P. J.; Lancefield, C. S.; Narani, A.; et al. *Green Chem.* **2017**, *19*, 2774–2782.
23. Renders, T.; Van den Bossche, G.; Vangeel, T.; et al. *Curr. Opin. Biotechnol.* **2019**, *56*, 193–201.
24. Godard, H. P.; McCarthy, J. L.; Hibbert, H. *J. Am. Chem. Soc.* **1940**, *62*, 988.
25. Sakakibara, A. In *Methods in Lignin Chemistry*; Lin, S. Y., Dence, C. W., Eds.; Springer Berlin Heidelberg: Berlin, Heidelberg, 1992; pp. 350–368.
26. Koranyi, T. I.; Fridrich, B.; Pineda, A.; Barta, K. *Molecules* **2020**, *25*, 2815.
27. Abu-Omar, M. M.; Barta, K.; Beckham, G. T.; et al. *Energy Environ. Sci.* **2020**, *14*, 262–292.
28. Galkin, M. V.; Samec, J. S. M. *ChemSusChem* **2016**, *9*, 1544–1558.
29. Sun, Z.; Fridrich, B.; de Santi, A.; et al. *Chem. Rev.* **2018**, *118*, 614–678.
30. Cooreman, E.; Vangeel, T.; Van Aelst, K.; et al. *Ind. Eng. Chem. Res.* **2020**, *59*, 17035–17045.
31. Sun, Z.; Cheng, J.; Wang, D.; et al. *ChemSusChem* **2020**, *13*, 5199–5212.
32. Thornburg, N. E.; Pecha, M. B.; Brandner, D. G.; et al. *ChemSusChem* **2020**, *13*, 4495–4509.
33. Van Den Bosch, S.; Renders, T.; Kennis, S.; et al. *Green Chem.* **2017**, *19*, 3313–3326.
34. Li, H.; Song, G. *ACS Catal.* **2020**, *10*, 12229–12238.
35. Li, H.; Song, G. *ACS Catal.* **2019**, *9*, 4054–4064.
36. Van Den Bosch, S.; Schutyser, W.; Vanholme, R.; et al. *Energy Environ. Sci.* **2015**, *8*, 1748–1763.
37. Renders, T.; Cooreman, E.; Van Den Bosch, S.; et al. *Green Chem.* **2018**, *20*, 4607–4619.
38. Elangovan, S.; Afanasenko, A.; Haupenthal, J.; et al. *ACS Cent. Sci.* **2019**, *5*, 1707–1716.
39. Van Den Bosch, S.; Schutyser, W.; Koelewijn, S.-F. F.; et al. *Chem. Commun.* **2015**, *51*, 13158–13161.
40. Rinaldi, R.; Woodward, R. T.; Ferrini, P.; et al. *J. Braz. Chem. Soc.* **2019**, *30*, 479–491.
41. Ferrini, P.; Rinaldi, R. *Angew. Chem. Int. Ed.* **2014**, *53*, 8634–8639.
42. Luo, H.; Klein, I. M.; Jiang, Y.; et al. *ACS Sustain. Chem. Eng.* **2016**, *4*, 2316–2322.
43. Li, Y.; Demir, B.; Vázquez Ramos, L. M.; et al. *Green Chem.* **2019**, *21*, 3561–3572.
44. Anderson, E. M.; Stone, M. L.; Katahira, R.; et al. *Joule* **2017**, *1*, 613–622.
45. Kumaniaev, I.; Subbotina, E.; Sävmarker, J.; et al. *Green Chem.* **2017**, *19*, 5767–5771.
46. Anderson, E. M.; Stone, M. L.; Katahira, R.; et al. *Nat. Commun.* **2019**, *10*, 1–10.
47. Anderson, E. M.; Stone, M. L.; Hülsey, M. J.; et al. *ACS Sustain. Chem. Eng.* **2018**, *6*, 7951–7959.

48. Tschulkow, M.; Compernolle, T.; Van den Bosch, S.; et al. *J. Cleaner Prod.* **2020**, *266*, 1–11.
49. Torr, K. M.; van de Pas, D. J.; Cazeils, E.; et al. *Bioresour. Technol.* **2011**, *102*, 7608–7611.
50. Schutyser, W.; Van Den Bosch, S.; Renders, T.; et al. *Green Chem.* **2015**, *17*, 5035–5045.
51. Renders, T.; Schutyser, W.; Van Den Bosch, S.; et al. *ACS Catal.* **2016**, *6*, 2055–2066.
52. Renders, T.; Van Den Bosch, S.; Vangeel, T.; et al. *ACS Sustain. Chem. Eng.* **2016**, *4*, 6894–6904.
53. Lange, H.; Giannì, P.; Crestini, C. *Lignin Valorization: Emerging Approaches*; The Royal Society of Chemistry, 2018; pp. 413–476.
54. Meng, X.; Crestini, C.; Ben, H.; et al. *Nat. Protoc.* **2019**, *14*, 2627–2647.
55. Feghali, E.; van de Pas, D. J.; Parrott, A. J.; et al. *ACS Macro Lett.* **2020**, *9*, 1155–1160.
56. Feghali, E.; Van De Pas, D. J.; Torr, K. M. *Biomacromolecules* **2020**, *21*, 1548–1559.
57. Van De Pas, D. J.; Torr, K. M.; van de Pas, D. J.; et al. *Biomacromolecules* **2017**, *18*, 2640–2648.
58. Van Aelst, K.; Van Sinay, E.; Vangeel, T.; et al. *Chem. Sci.* **2020**, *11*, 11498–11508.
59. Wen, J. L.; Sun, S. L.; Xue, B. L.; et al. *Materials (Basel)* **2013**, *6*, 359–391.
60. Ralph, J.; Landucci, L. *Lignin and Lignans*; Taylor & Francis, 2010; pp. 137–243.
61. Anderson, E. M.; Katahira, R.; Reed, M.; et al. *ACS Sustain. Chem. Eng.* **2016**, *4*, 6940–6950.
62. Sun, Z.; Bottari, G.; Afanasenko, A.; et al. *Nat. Catal.* **2018**, *1*, 82–92.
63. Ouyang, X.; Huang, X.; Zhu, J.; et al. *ACS Sustain. Chem. Eng.* **2019**, 7, 13764–13773.
64. Huang, Y.; Duan, Y.; Qiu, S.; et al. *Sustain. Energy Fuels* **2018**, *2*, 637–647.
65. Vendamme, R.; Behaghel De Bueren, J.; Gracia-Vitoria, J.; et al. *Biomacromolecules* **2020**, *21*, 4135–4148.
66. Sudarsanam, P.; Ruijten, D.; Liao, Y.; et al. *Trends Chem.* **2020**, *2* (10), 898–913.
67. Anastas, P. T.; Zimmerman, J. B. *Environ. Sci. Technol.* **2003**, *37*, 94A–101A.
68. Czub, P. In *Handbook of Engineering and Speciality Thermoplastics*; Thomas, S., Visakh, P. M., Eds.; John Wiley & Sons, Inc: Hoboken, NJ, USA, 2011; pp. 221–269.
69. *Bisphenol A (BPA) Market—Growth, Trends, and Forecast*; 2020–2025. https://www.researchandmarkets.com/reports/4520075/bisphenol-a-bpa-market-growth-trends-and. Accessed Sep 28, 2020.
70. *Bisphenol A Market Size & Share, Industry Report*; 2018–2025. https://www.grandviewresearch.com/industry-analysis/bisphenol-a-bpa-market. Accessed Sep 28, 2020.
71. PlasticEurope. *Production-and-Demand-Volumes.pdf.* https://bisphenol-a-europe.org/wp-content/uploads/2017/07/Production-and-demand-volumes.pdf. Accessed Sep 28, 2020.
72. Peiffer, D. G. *J. Macromol. Sci. Part B Phys.* **1978**, *15*, 595–611.
73. Pacchierotti, F.; Ranaldi, R.; Eichenlaub-Ritter, U.; et al. *Mutat. Res. Toxicol. Environ. Mutagen.* **2008**, *651*, 64–70.
74. Horan, T. S.; Pulcastro, H.; Lawson, C.; et al. *Curr. Biol.* **2018**, *28*, 2948–2954.
75. Heindel, J. J.; Newbold, R.; Schug, T. T. *Nat. Rev. Endocrinol.* **2015**, *11*, 653–661.
76. Liu, B.; Lehmler, H.-J.; Sun, Y.; et al. *Lancet Planet. Health* **2017**, *1*, 114–122.
77. Alonso-Magdalena, P.; Morimoto, S.; Ripoll, C.; et al. *Environ. Health Perspect.* **2006**, *114*, 106–112.
78. Rehan, M.; Ahmad, E.; Sheikh, I. A.; et al. *PLoS One* **2015**, *10*, 1–18.
79. Doherty, L. F.; Bromer, J. G.; Zhou, Y.; et al. *Horm. Cancer* **2010**, *1*, 146–155.
80. Brotons, J. A.; Olea-Serrano, M. F.; Villalobos, M.; et al. *Environ. Health Perspect.* **1995**, *103*, 5.
81. Olea, N.; Pulgar, R.; Pérez, P.; et al. *Environ. Health Perspect.* **1996**, *104*, 298–305.

82. Vandenberg, L. N.; Hauser, R.; Marcus, M.; et al. *Reprod. Toxicol.* **2007**, *24*, 139–177.
83. Ehlert, K. A.; Beumer, C. W. E.; Groot, M. C. E. *Food Addit. Contam. Part A* **2008**, *25*, 904–910.
84. Brede, C.; Fjeldal, P.; Skjevrak, I.; et al. *Food Addit. Contam.* **2003**, *20*, 684–689.
85. Kang, J.-H.; Kito, K.; Kondo, F. *J. Food Prot.* **2003**, *66*, 1444–1447.
86. Hahladakis, J. N.; Velis, C. A.; Weber, R.; et al. *J. Hazard. Mater.* **2018**, *344*, 179–199.
87. ECHA. *4,4′-Isopropylidenediphenol—Substance Infocard.* https://echa.europa.eu/substance-information/-/substanceinfo/100.001.133. Accessed Sep 28, 2020.
88. ECHA. *Bisphenol A*; https://echa.europa.eu/hot-topics/bisphenol-a (Accessed Sep 28, 2020), 2020.
89. Szafran, A. T.; Stossi, F.; Mancini, M. G.; et al. *PLoS One* **2017**, *12*, 1–19.
90. Kitamura, S. *Toxicol. Sci.* **2005**, *84*, 249–259.
91. Soto, A. M.; Schaeberle, C.; Maier, M. S.; et al. *Environ. Sci. Technol.* **2017**, *51*, 1718–1726.
92. Koelewijn, S. F.; Ruijten, D.; Trullemans, L.; et al. *Green Chem.* **2019**, *21*, 6622–6633.
93. Koelewijn, S.-F. *Safety by Natural Design: Benign Bisphenols From Biomass*; KU Leuven, 2019.
94. Meylemans, H. A.; Groshens, T. J.; Harvey, B. G. *ChemSusChem* **2012**, *5*, 206–210.
95. Koelewijn, S.-F.; Van den Bosch, S.; Renders, T.; et al. *Green Chem.* **2017**, *19*, 2561–2570.
96. Koelewijn, S.; Cooreman, C.; Renders, T.; et al. *Green Chem.* **2018**, *20*, 1050–1058.
97. Ferrini, P.; Koelewijn, S.-F.; Van Aelst, J.; et al. *ChemSusChem* **2017**, *10*, 2249–2257.
98. Reno, K. H.; Stanzione, J. F.; Wool, R. P.; et al. Bisphenol Alternative Derived From Renewable Substituted Phenolics and Their Industrial Application. 2017/0210689 A1, 2017.
99. Hernandez, E. D.; Bassett, A. W.; Sadler, J. M.; et al. *ACS Sustain. Chem. Eng.* **2016**, *4*, 4328–4339.
100. Nicastro, K. H.; Kloxin, C. J.; Epps, T. H. *ACS Sustain. Chem. Eng.* **2018**, *6*, 14812–14819.
101. Eller, K.; Erhard, H.; Rossbacher, R.; et al. *Ullmann's Encyclopedia of Industrial Chemistry*; Wiley-VCH Verlag GmbH & Co. KGaA: Weinheim, 2012; pp. 647–692.
102. Vardanyan, R.; Hruby, V. *Synthesis of Best-Seller Drugs*; Elsevier/AP, 2016.
103. Greene, S. A. In *Sittig's Handbook of Pesticides and Agricultural Chemicals*; Pohanish, R. P., Ed.; 2nd ed.; William Andrew Publishing: Norwich, NY, USA, 2007.
104. Froidevaux, V.; Negrell, C.; Caillol, S.; et al. *Chem. Rev.* **2016**, *116*, 14181–14224.
105. Bruice, P. Y. In *A Primer of Ecology*; Jaworski, A., Ed.; Pearson Education, 2017; p. 1392.
106. Shimizu, K. *Catal. Sci. Technol.* **2015**, *5*, 1412–1427.
107. Pelckmans, M.; Renders, T.; Van de Vyver, S.; et al. *Green Chem.* **2017**, *19*, 5303–5331.
108. Guillena, G.; Ramón, D. J.; Yus, M. *Chem. Rev.* **2010**, *110*, 1611–1641.
109. Verboekend, D.; Liao, Y.; Schutyser, W.; et al. *Green Chem.* **2015**, *18*, 297–306.
110. Pera-Titus, M.; Shi, F. *ChemSusChem* **2014**, 7, 720–722.
111. Liang, G.; Zhou, Y.; Zhao, J.; et al. *ACS Catal.* **2018**, *8*, 11226–11234.
112. Liu, Y.; Afanasenko, A.; Elangovan, S.; et al. *ACS Sustain. Chem. Eng.* **2019**, 7, 11267–11274.
113. Kumaniaev, I.; Samec, J. S. M. *Ind. Eng. Chem. Res.* **2019**, *58*, 6899–6906.
114. Yan, T.; Feringa, B. L.; Barta, K. *Sci. Adv.* **2017**, *3*, 1–7.
115. Kahl, T.; Schröder, K.-W.; Lawrence, F. R.; et al. *Ullmann's Encyclopedia of Industrial Chemistry*; Wiley-VCH Verlag GmbH & Co. KGaA: Weinheim, 2011; pp. 1–58.
116. Natte, K.; Narani, A.; Goyal, V.; et al. *Adv. Synth. Catal.* **2020**, *362*, 5143–5169.

117. Ahrens, S.; Mueller, C.; Paul, A.; et al. Process for the Preparation of Aromatic Amines. US Patent 2012/0071692, 2012.
118. Ono, Y.; Ishida, H. *J. Catal.* **1981**, *72*, 121–128.
119. Chen, Q. *React. Kinet. Catal. Lett.* **2005**, *84*, 215–221.
120. Jiang, R.; Xie, Z.; Zhang, C.; et al. *Catal. Today* **2004**, *93–95*, 359–363.
121. Haese, F.; Wulff-Doering, J.; Koehler, U.; et al. Method for the Continuous Production of an Amine. US Patent 7767855B2, 2010.
122. Chen, Z.; Zeng, H.; Girard, S. A.; et al. *Angew. Chem. Int. Ed.* **2015**, *54*, 14487–14491.
123. Tomkins, P.; Valgaeren, C.; Adriaensen, K.; et al. *J. Catal.* **2019**, *371*, 207–213.
124. Qiu, Z.; Lv, L.; Li, J.; et al. *Chem. Sci.* **2019**, *10*, 4775–4781.
125. Cuypers, T.; Morias, T.; Windels, S.; et al. *Green Chem.* **2020**, *22*, 1884–1893.
126. Tomkins, P.; Valgaeren, C.; Adriaensen, K.; et al. *ChemCatChem* **2018**, *10*, 3689–3693.
127. Chen, Z.; Zeng, H.; Gong, H.; et al. *Chem. Sci.* **2015**, *6*, 4174–4178.
128. Wang, M.; Sjöholm, E.; Li, J. *Holzforschung* **2017**, *71*, 27–34.
129. Wang, B.; Chen, T.-Y.; Wang, H.-M.; et al. *Int. J. Biol. Macromol.* **2018**, *107*, 426–435.
130. Du, X.; Li, J.; Lindström, M. E. *Ind. Crop Prod.* **2014**, *52*, 729–735.
131. Jiao, G.-J.; Peng, P.; Sun, S.-L.; et al. *Int. J. Biol. Macromol.* **2019**, *127*, 544–554.
132. Blondiaux, E.; Bomon, J.; Smoleń, M.; et al. *ACS Sustain. Chem. Eng.* **2019**, *7*, 6906–6916.
133. Fache, M.; Boutevin, B.; Caillol, S. *ACS Sustain. Chem. Eng.* **2016**, *4*, 35–46.
134. Weber, M.; Weber, M.; Weber, V. *Ullmann's Encyclopedia of Industrial Chemistry*; Wiley-VCH Verlag GmbH & Co. KGaA: Weinheim, 2020; pp. 1–20.
135. Fiege, H.; Voges, H.-W.; Hamamoto, T.; et al. *Ullmann's Encyclopedia of Industrial Chemistry*; Wiley-VCH Verlag GmbH & Co. KGaA: Weinheim, 2000; pp. 521–576.
136. Ravichandiran, P.; Gu, Y.; Mubofu, E. B.; et al. *Bio-Based Solvents*; Wiley, 2017; pp. 149–173.
137. Kim, S.; Kwon, E. E.; Kim, Y. T.; et al. *Green Chem.* **2019**, *21*, 3715–3743.
138. Schutyser, W.; Van den Bossche, G.; Raaffels, A.; et al. *ACS Sustain. Chem. Eng.* **2016**, *4*, 5336–5346.
139. Schutyser, W.; Van Den Bosch, S.; Dijkmans, J.; et al. *ChemSusChem* **2015**, *8*, 1805–1818.
140. Saidi, M.; Samimi, F.; Karimipourfard, D.; et al. *Energy Environ. Sci.* **2014**, 7, 103–129.
141. Zhao, C.; Kou, Y.; Lemonidou, A. A.; et al. *Angew. Chem. Int. Ed.* **2009**, *48*, 3987–3990.
142. Sheehan, J. R. *Ullmann's Encyclopedia of Industrial Chemistry*; Wiley-VCH Verlag GmbH & Co. KGaA: Weinheim, 2011; pp. 17–27.
143. James, D. H.; Castor, W. M. *Ullmann's Encyclopedia of Industrial Chemistry*; Wiley-VCH Verlag GmbH & Co. KGaA: Weinheim, 2011; pp. 45–67.
144. Liao, Y.; D'Halluin, M.; Makshina, E.; et al. *Appl. Catal. B Environ.* **2018**, *234*, 117–129.
145. Zhang, J.; Lombardo, L.; Gözaydın, G.; et al. *Chin. J. Catal.* **2018**, *39*, 1445–1452.
146. Ouyang, X.; Huang, X.; Boot, M. D.; et al. *ChemSusChem* **2020**, *13*, 1705–1709.
147. Liao, Y.; Zhong, R.; D'Halluin, M.; et al. *ACS Sustain. Chem. Eng.* **2020**, *8*, 8713–8722.
148. Liao, Y.; Zhong, R.; Makshina, E.; et al. *ACS Catal.* **2018**, *8*, 7861–7878.
149. Settle, A. E.; Berstis, L.; Rorrer, N. A.; et al. *Green Chem.* **2017**, *19*, 3468–3492.
150. Song, S.; Zhang, J.; Gözaydın, G.; et al. *Angew. Chem. Int. Ed.* **2019**, *58*, 4934–4937.
151. Schmidt, R.; Griesbaum, K.; Behr, A.; et al. *Ullmann's Encyclopedia of Industrial Chemistry*; Wiley-VCH Verlag GmbH & Co. KGaA: Weinheim, 2014; pp. 1–74.
152. Xin, Y.; Jing, Y.; Dong, L.; et al. *Chem. Commun.* **2019**, *55*, 9391–9394.
153. Li, L.; Dong, L.; Liu, X.; et al. *Appl. Catal. B Environ.* **2020**, *260*, 118143.

154. Finnveden, G.; Hauschild, M. Z.; Ekvall, T.; et al. *J. Environ. Manage.* **2009**, *91*, 1–21.
155. Patel, M.; Zhang, X.; Kumar, A. *Renew. Sustain. Energy Rev.* **2016**, *53*, 1486–1499.
156. Sinnott, R.; Gavin, T. *Chemical Engineering Design*, 5th ed.; Elsevier, 2009.
157. Lauwaert, J.; Raghuveer, C. S.; Thybaut, J. W. *Chem. Eng. Sci.* **2017**, *170*, 694–704.
158. Richter, J. C. F. C.; Richter, O. J. Continuous Vertical Digester. US Patent 5300195, 1994.
159. Ruiz, H. A.; Conrad, M.; Sun, S. N.; et al. *Bioresour. Technol.* **2020**, *299*, 122685.
160. Atta, A.; Razzak, S. A.; Nigam, K. D. P.; et al. *Ind. Eng. Chem. Res.* **2009**, *48*, 7876–7892.
161. Aho, A.; Kumar, N.; Eränen, K.; et al. *Process Saf. Environ. Prot.* **2007**, *85*, 473–480.
162. Park, J.; Riaz, A.; Verma, D.; et al. *ChemSusChem* **2019**, *12*, 1743–1762.
163. EPA. Action Memorandum: Inert Reassessment—Ethyl Acetate (CAS Reg. No. 141-78-6) and Amyl Acetate (CAS Reg. No. 628-63-7). Washington D.C. 2006, pp 1–24.
164. Desimone, J. M. *Green Chem.* **2002**, *297*, 799–804.
165. Lange, J. *ChemSusChem* **2017**, *10*, 245–252.
166. Khan, F. I.; Ghoshal, A. K. *J. Loss Prev. Process Ind.* **2000**, *13*, 527–545.
167. Smith, E. L.; Abbott, A. P.; Ryder, K. S. *Chem. Rev.* **2014**, *114*, 11060–11082.
168. Jessop, P. G.; Mercer, S. M.; Heldebrant, D. J. *Energy Environ. Sci.* **2012**, *5*, 7240–7253.
169. Dietz, M. L. *Separation Science and Technology*, Vol. 41; Taylor & Francis, 2006; pp. 2047–2063.
170. Sas, O. G.; Domínguez, I.; González, B.; et al. *J. Environ. Manage.* **2018**, *228*, 475–482.
171. Ji, Y.; Hou, Y.; Ren, S.; et al. *Fuel Process. Technol.* **2018**, *171*, 183–191.
172. Poole, C. F.; Poole, S. K. *J. Chromatogr. A* **2010**, *1217*, 2268–2286.
173. Ruesgas-Ramón, M.; Figueroa-Espinoza, M. C.; Durand, E. *J. Agric. Food Chem.* **2017**, *65*, 3591–3601.
174. García, A.; Rodríguez-Juan, E.; Rodríguez-Gutiérrez, G.; et al. *Food Chem.* **2016**, *197*, 554–561.
175. Fu, D.; Farag, S.; Chaouki, J.; et al. *Bioresour. Technol.* **2014**, *154*, 101–108.
176. Jessop, P. G. *Green Chem.* **2011**, *13*, 1391–1398.
177. Clarke, C. J.; Tu, W.; Levers, O.; et al. *Chem. Rev.* **2018**, *118*, 747–800.
178. Plechkova, N. V.; Seddon, K. R. *Chem. Soc. Rev.* **2008**, *37*, 123–150.
179. Alfonsi, K.; Colberg, J.; Dunn, P. J.; et al. *Green Chem.* **2008**, *10*, 31–36.
180. Hansen, C. M. *Hansen Solubility Parameters: A User's Handbook*; CRC Press, 2007.
181. Sultan, Z.; Graça, I.; Li, Y.; et al. *ChemSusChem* **2019**, *12*, 1203–1212.
182. Constantinou, L.; Gani, R. *AIChE J.* **1994**, *40*, 1697–1710.
183. Joback, K. G.; Reid, R. C. *Chem. Eng. Commun.* **1987**, *57*, 233–243.
184. Landau, R. N. *Thermochim. Acta* **1996**, *289*, 101–126.
185. Fredenslund, A.; Jones, R. L.; Prausnitz, J. M. *AIChE J.* **1975**, *21*, 1086–1099.
186. Brennan, D. J.; Golonka, K. A. *Chem. Eng. Res. Des.* **2002**, *80*, 579–586.
187. Rebitzer, G.; Ekvall, T.; Frischknecht, R.; et al. *Environ. Int.* **2004**, *30*, 701–720.

CHAPTER NINE

Enhancing product selectivity in biomass and bioalcohol reactions over Cu-doped porous metal oxides

Jacob A. Barrett*
Department of Chemistry and Biochemistry, University of California, San Diego, CA, United States
*Corresponding author: e-mail address: j1barrett@ucsd.edu

Contents

Abstract

The ever-increasing global fuel demand and environmental consequences of anthropogenic carbon emissions have motivated interest in catalytic disassembly of biomass into liquid fuels and chemicals. Over 1 billion tons of combined residual biomass (i.e., agricultural, forestry, and food waste) are available annually (Langholtz et al., 2016).[1] Humankind is dependent on fossil carbon for much of the energy we consume, the medicine we take, and the clothes we wear. Our society will likely remain dependent on fossil carbon so long as they are the primary chemical feedstock for our medicines, polymers, and fuels. The selective upgrading of biomass and biomass derived alcohols may potentially provide "green" alternatives to commonly used fossil carbon derived fuels, solvents, plasticizers, and a multitude of synthetic precursor chemicals. Cu-doped porous metal oxides were identified as highly active catalysts due to their ability to generate hydrogen from alcohols and subsequently catalyze hydrogenolysis of various carbon-oxygen bonds, a crucial component of biomass macromolecules and derivatives. This chapter outlines recent efforts within and without our laboratories for addressing a vital challenge in catalytic

Advances in Inorganic Chemistry, Volume 77
ISSN 0898-8838
https://doi.org/10.1016/bs.adioch.2020.12.001

biomass disassembly using Cu-doped porous metal oxides (CuPMO) selectivity. Herein we described the innovative strategies used in the context of selective conversion of organosolv poplar lignin, condensation of ethanol, and reaction environment (i.e., temperature, solvent, and substrate) effects on CuPMO structure.

Abbreviations

ATR-IR	attenuated total reflectance—infrared spectroscopy
BuOH	butanol
CuPMO, Cu-doped PMO, Cu_{20}PMO	copper doped porous metal oxide (the 20 denotes that 20% of the Mg was replaced with Cu in the catalyst synthesis)
DMC	dimethyl carbonate
EtOAc	ethyl acetate
EtOH	ethanol
GC–MS	gas chromatography–mass spectrometry
GHG	greenhouse gas
HexOH	hexanol
HTC	hydrotalcite
MeOH	methanol
NMR	nuclear magnetic resonance
OPL	organosolv poplar lignin
RFS	renewable fuels standard
sc	supercritical
v/v	volume by volume
XRD	X-ray diffraction

1. Introduction

1.1 Lignocellulosic biomass

Lignocellulosic biomass (i.e., dry plant matter) is the most abundantly available raw material for fuels, chemicals, and materials.[1] Human beings have utilized biomass derived energy ever since *homo erectus* lit their first fire, and wood remains the largest biomass energy source in the world today. The foundations of our modern lives were built on the applications of lignocellulosic biomass to various aspects of our developing society like architecture, transportation, agriculture, and tools. Biomass was the primary energy source used in the United States up until the replacement of wood and other biofuels with coal during the Industrial Revolution. Wood and wood processing waste accounts for only about 2.5% of the energy used in the United States today, yet biomass remains a crucial fuel source in developing countries. Scientists and researchers have sought to do more than harness its energy via combustion for the last 100 years and have invented processes for producing biochemicals and biomaterials. The first report on the reaction of organosolv hardwood lignin with hydrogen was published in the late 1930s.[2]

Over the past 10 years, several comprehensive reviews of the heterogeneous catalytic disassembly of lignocellulosic biomass have appeared. Pharmaceutical and plastics industries depend on petrochemical feedstocks.[3] Kiloton-scale biorefineries are being developed to produce biomass derived fuels and chemicals as an alternative.[4] However, research advances in catalytic biomass conversion have been focused on the selective conversion of the carbohydrates, whereas lignin is considered waste. The carbohydrate fraction of lignocellulose is readily converted by biotic degradation to various bioalcohols, primarily ethanol. Lignin is the largest potential source of renewable aromatic chemicals; therefore, the development of economically sustainable processes is paramount for our future chemical industry.[5] The intricate, amorphous structure of lignins make them incredibly resistant to degradation. This inherent recalcitrance to degradation has hindered development of large-scale processes.[6] Furthermore, a primary objective for the development of a renewable chemical industry is to maximize atom economy to improve the economic feasibility of kiloton-scale biorefineries.[7]

Lignocellulose is principally composed of cellulose, hemicellulose, and lignin.[8] Cellulose is a biopolymer composed of linear chains of glucose monomers. Hemicellulose is a biopolymer composed of pentoses (i.e., sugars with five carbon atoms) as well as glucose. Lignin is an aromatic biopolymer made from the radical polymerization of three hydroxycinnamyl alcohols; conniferyl alcohol (G), coumaryl alcohol (H), sinapyl alcohol (S).[9] Unlike the carbohydrate polymers, its monomers are linked by various chemical bonding motifs. This heterogeneity imparts natural recalcitrance to biotic and abiotic degradation.[9] Additionally, the techniques employed to extract lignin from lignocellulose typically crosslink benzylic carbons that are considerably strong C—C bonds.[10] The primary lignin linkages are aryl ether carbon-oxygen (C—O) bonds. The more common ones are aryl ether (β-O-4 and α-O-4), phenylcoumaran (β-5), and diaryl ether (4-O-5) linkages (Fig. 1). The abundances of these linkages depend on factors such as the plant species, growth conditions, and lignin isolation technique.[10]

Various lignin disassembly methods utilizing acids, bases, and transition metal based hetero- and homogeneous catalysis have been described.[11–15] Catalytic hydrogenolysis is a promising method to produce aromatic compounds from several types of lignin.[16,17] Hydrogenolysis uses hydrogen (or hydrogen equivalents) to cleave C-X (X=O, S, Cl, and F) bonds.[18] Reductive cleavage of aryl-ether bonds is a potential route to aromatic chemicals from lignin. β-O-4 and α-O-4 linkages generally undergo

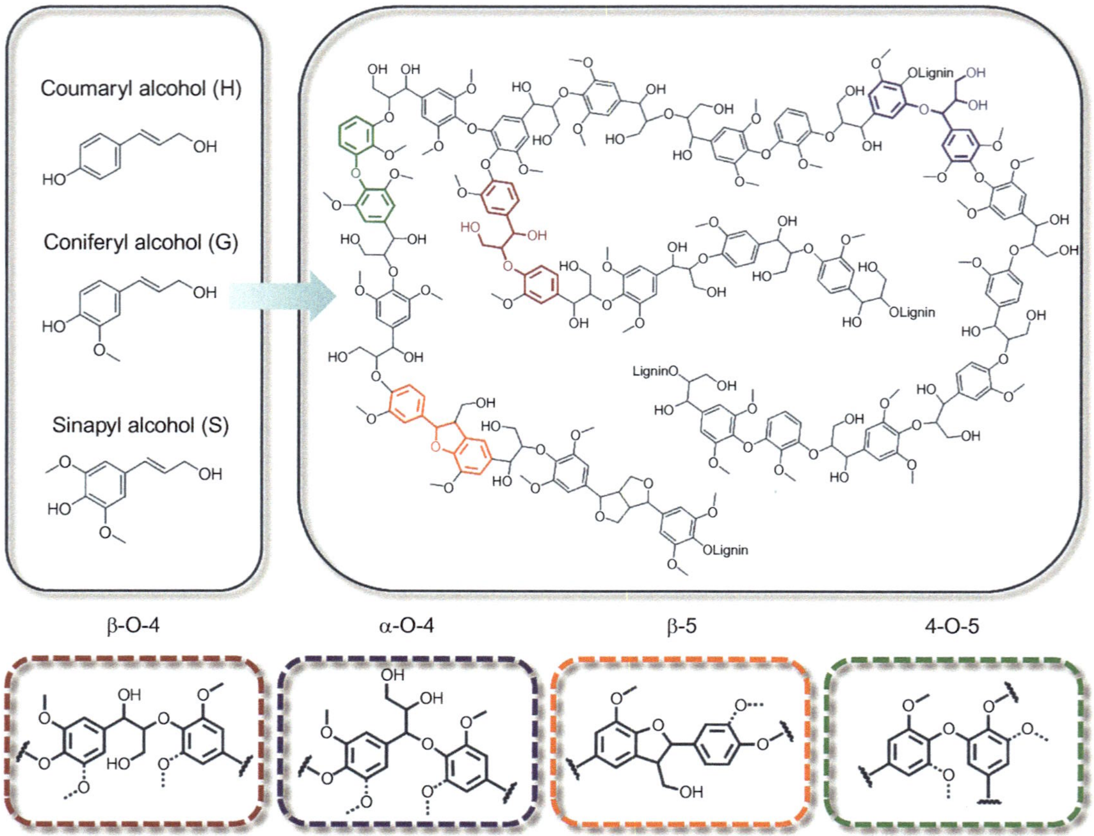

Fig. 1 Visual depiction of lignin with common linkages highlighted in color based on structural data on poplar wood from Sannigrahi et al.[8] The three hydroxycinnamyl alcohols are shown on the top left with four of the more common aryl–ether linkages depicted along the bottom. *Reproduced with permission from Barrett, J. A.; Gao, Y.; Bernt, C. M.; Chui, M.; Tran, A. T.; Foston, M. B.; Ford, P. C.* ACS Sustain. Chem. Eng. ***2016****, 4, 6877–6886. Copyright 2016 American Chemical Society.*

hydrogenolysis more easily than do other types of bonds linking lignin fragments.[19] However, catalysts effective for C—O bond hydrogenolysis also catalyze phenol hydrogenation to give cyclohexyl derivatives.[13] Thus, the desired aromatic compounds are produced and consumed under the same reaction conditions. In this context, one principal objective of the research described in this chapter is the disassembly of lignin through hydrogenolysis to break aryl-ether bonds with minimal aromatic hydrogenation.[17]

1.2 Biomass derived alcohols

The two primarily employed production methods for cellulosic ethanol are cellulases (biotic) and gasification (abiotic). Then former relies heavily on a pretreatment phase to separate the cellulosic fibers from recalcitrant lignin. The resultant cellulosic biomass is chemically or enzymatically hydrolyzed to produce the individual sugar molecules. These sugars are converted to ethanol using bacterial fermentation. Some fermentation processes enable the production of other molecules in addition to ethanol such as acetone, butanol, and/or acetic acid. The U.S. government has supported the production of ethanol, specifically from edible biomass (i.e., corn starch), as a renewable fuel additive to reduce greenhouse gas (GHG) emissions for decades.[20] The Renewable Fuel Standard (RFS) Program was established to combat the economic and environmental expense of widespread fossil fuel consumption and is continually updated.[21] The program's mission is to facilitate US energy independence and security as well as reduce GHG emissions. The RFS program is partially responsible for driving research in the production of lignocellulosic biofuels, like cellulosic ethanol.

Currently, a major portion of biofuels produced in the US and globally is ethanol.[22,23] Virtually all gasoline we use today is blended with at least 10% ethanol.[24–26] Life-cycle analysis of cellulosic ethanol suggests we cannot simply replace gasoline with ethanol to decrease our GHG emissions without increasing overall environmental degradation.[27] Suh and coworkers showed that corn-derived ethanol as a gasoline replacement leads to shifted environmental impact like increased eutrophication and local water scarcity in place of GHG emissions.[27] Gasoline with or without ethanol will inevitably become as obsolete as the horse drawn carriage, as our society continues its transition to electric vehicles. The high water solubility and corrosiveness of EtOH presents challenges in fuel transportation and structural integrity of the vehicle's engine.[28] The U.S. Government Accountability Office recently reported that the RFS program is unlikely to meet the current GHG emissions reduction goals for 2022 due to the infrastructure investment costs needed to overcome fuel transportation and storage complications for ethanol blended gasoline.[29] In this context, there is current motivation to develop chemical processes that use cellulosic biomass derived ethanol as feedstock for the production of chemicals or of biofuels for advanced heavy-duty vehicles. This account will present some of our research on developing a catalytic process to produce value-added chemicals and/or fuels from ethanol over Cu-doped porous metal oxide catalysts.

The Guerbet ethanol condensation is one method for the production of n-butanol (n-BuOH) and industrially relevant higher alcohols.[30–32] Marcel Guerbet described this condensation reaction in the late 1800s.[33] Galadima and Muraza recently examined decades worth of research and attributed the lack of industrial application of the Guerbet condensation to poor selectivity in most catalytic systems.[34] The aforementioned reaction is a cascade of steps; (1) dehydrogenation of an alcohol to an aldehyde, (2) aldol condensation, (3) dehydration to give an α,β-unsaturated aldehyde, and (4) hydrogenation of the α,β-unsaturated aldehyde to form the longer carbon chain alcohol. For ethanol, the alcohol is dehydrogenated to form acetaldehyde. Under basic conditions acetaldehyde is deprotonated ($pK_a = 13.57$ at 25 °C), and then condenses with a second equivalent of acetaldehyde to form crotonaldehyde (Scheme 1). Hydrogenation of crotonaldehyde yields n-BuOH, the longer carbon chain alcohol (Scheme 1).[30,34,35]

Part of the reason for poor selectivity in this reaction is that n-BuOH undergoes further Guerbet condensations and transesterifications that lead to branched alcohols (2-ethyl-butanol or 2-ethyl-hexanol), linear longer-chain alcohols (1-hexanol and 1-octanol), and esters. These chemicals have potential industrial applications; however, a highly selective product stream is desirable to avoid product separations. Additionally, some EtOH is likely present as ethoxide under basic conditions in most catalytic systems. Ethoxide may react with acetaldehyde formed from dehydrogenation to

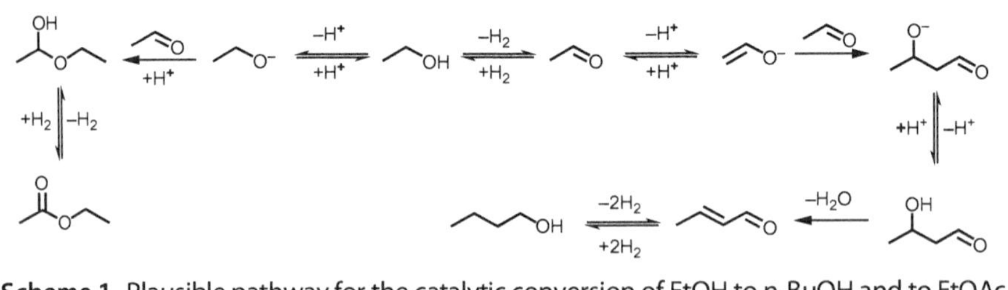

Scheme 1 Plausible pathway for the catalytic conversion of EtOH to n-BuOH and to EtOAc showing plausible intermediates with transition metal doped PMOs.[30,34,35] *Adapted with permission from Barrett, J. A.; Jones, Z. R.; Stickelmaier, C.; Schopp, N.; Ford, P. C. ACS* Sustainable Chem. Eng. ***2018****, 6, 15119–15126. Copyright 2018 American Chemical Society.*

form the hemiacetal 1-ethoxyethan-1-ol (Scheme 1). Dehydrogenation of the hemiacetal forms ethyl acetate (Scheme 1). Another side reaction in EtOH condensation via Guerbet chemistry is the Tishchenko reaction. Acetaldehyde disproportionates in the presence of an alkoxide under elevated temperature.[36] Aldehydes produced under the Guerbet reaction conditions, acetaldehyde and crotonaldehyde undergo reactions which lead to various organic esters. Altogether these reactions lead to poor selectivity toward a single chemical product in most Guerbet condensation systems.[33]

The potential use of n-BuOH as a chemical feedstock should be considered to evaluate the long-term viability of biomass based chemical production. The primary industrial production process for 1-butanol is hydroformylation of propylene.[37,38] Other methods employed are carbonylation of propylene and *Clostridium* bacterial biomass fermentation.[39] The former uses a fossil carbon feedstock and the latter uses cellulose for the so-called Acetone-Butanol-Ethanol (ABE) fermentation. This fermentation process has garnered renewed attention as research interest in biomass based chemical production has increased.[39,40] Surface coatings such as wood varnishes often contain n-butanol as a solvent or curing reagent.[37] Butyl acrylate is one such chemical derived from n-butanol used in latex paints. Butyl acetate and butyl butyrate, produced from n-butanol transesterification, are commonly used in perfume and flavor mixtures.[38] Overall, the longer chain alcohol and ester by-products in the Guerbet condensation reaction have industrial uses as chemical synthesis feedstocks, flavors, coatings, and solvents.[36–38,41] Controlling the product selectivity of a process enables alternative pathways toward upgrading ethanol.[42,43] High selectivity in Guerbet condensation reactions is advantageous for industrial implementation to avoid the high energy costs associated with separations.[44,45] The selective conversion of bio-renewable ethanol to larger alcohols like butanol, hexanol, and octanol has potential in a future advanced biofuel and/or biochemical industry.

As a potential "green" fuel additive, butanol (n-BuOH) has higher energy density (Fig. 2) and lower water solubility than cellulosic EtOH.[26] Spark-ignition engines showed the largest GHG emissions reduction with butanol-gasoline blends when compared to, isobutanol, ethanol, methanol, and acetone-gasoline blends.[47] These characteristics make butanol an attractive advanced biofuel to help reduce light duty vehicle GHG emissions.

Olarte et al. examined the fuel characteristics of ethanol condensation reaction mixtures that consisted primarily of n-butanol and other Guerbet alcohols.[48] The motor octane numbers for the mixtures were similar to gasoline. The production of a chemical feedstock requires high selectivity, but

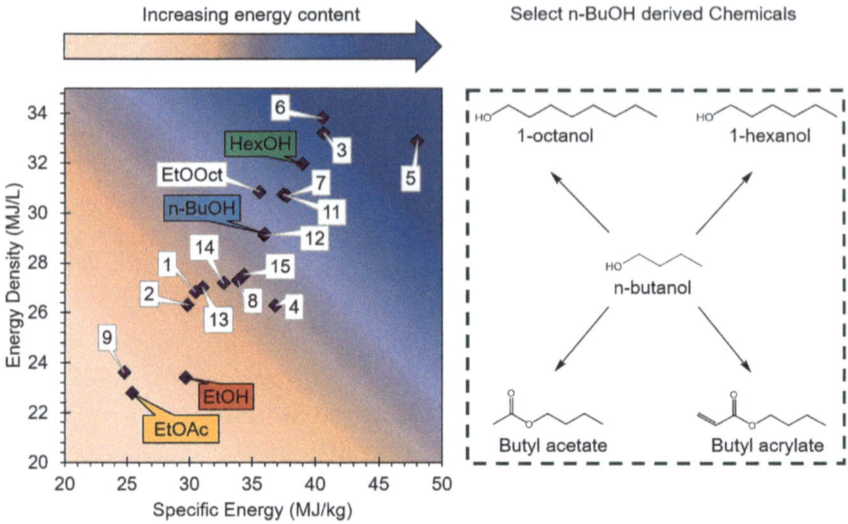

Fig. 2 *Left*: Energy density (MJ/L) and specific energy (MJ/kg) of ethanol and ethanol condensation products calculated using NIST condensed phase thermochemical data.[46] *Right*: Selected chemicals synthesized from n-BuOH as discussed in the account. EtOAc = ethyl acetate, EtOH = ethanol, n-BuOH = n-butanol, HexOH = 1-hexanol. Numbered entries are minor products from Cu_{20}PMO reaction: 1 = ethyl butanoate, 2 = butyl acetate, 3 = 1-octanol, 4 = diethyl ether, 5 = heptane, 6 = 2-ethyl-1-hexanol, 7 = 2-pentanol, 8 = 2-butanone, 9 = butanoic acid, 10 = ethyl octanoate, 11 = 3-hexanone, 12 = 2-pentanone, 13 = butyl butanoate, 14 = 1,1-diethoxyethane, 15 = butyraldehyde. *Modifed with permission from Barrett, J. A.; Jones, Z. R.; Stickelmaier, C.; Schopp, N.; Ford, P. C.* ACS Sustainable Chem. Eng. ***2018****, 6, 15119–15126. Copyright 2018 American Chemical Society.*

combustion fuels are typically mixtures of compounds with varying molecular weights, thus Guerbet alcohol mixtures may potentially circumvent major separation issues. Furthermore, the relative energy content of compounds from transition metal doped porous metal oxide catalyzed ethanol condensation shows these compounds are up to 30% more energy dense than EtOH (Fig. 2). The energy obtained by combustion of 1 kg of EtOH is equivalent to 80% of the energy from the equivalent amount of n-butanol and 74% from that of 1-hexanol. Catalytic systems focused on generating biofuels using Guerbet condensation of EtOH should aim to reduce ethyl acetate production.

Various homogeneous[45,49,50] and heterogenous[34,43,51] catalyst systems for Guerbet chemistry have been explored.[52] Homogenous transition metal catalysts (Group 7, 8, and 9) yield 90–99% selectivity for n-BuOH at low temperatures ($\leq$100 °C) with a base additive and long reaction time (6–24 h).[53,54] Homogeneous catalysts are superior to heterogeneous catalysts in terms of selectivity, however, catalyst separation is typically difficult. Additionally, when a base additive is used, materials that are stable at elevated temperature under basic conditions are required for the reactor design. Heterogeneous catalysts are advantageous in the context of catalyst separation from products. Transition metal doped porous metal oxides (PMOs) derived by calcining doped Mg^{2+}/Al^{3+} hydrotalcites are among some of the most active heterogeneous catalysts for the Guerbet reaction.[42,43,55–59] However, ethyl acetate (EtOAc) is a major byproduct of the ethanol condensation catalyzed by such PMOs.[59]

Cu-doped PMOs readily catalyze the Guerbet reaction of ethanol. Sun and coworkers showed EtOH converts to n-BuOH over Cu_xNi_y-PMOs.[60] Cu_{20}PMO, has nearly equivalent selectivity for EtOAc and n-BuOH. This chapter will detail how poisoning of Cu_{20}PMO with chloride increases activity and selectivity toward n-BuOH.[61] Some structural characteristics of the chloride-modified Cu_{20}PMO are also discussed.

1.3 Porous metal oxide catalysts in biomass catalysis

Ford and coworkers used supercritical methanol and a Cu-doped porous metal oxide catalyst to convert lignocellulose (i.e., sawdust, poplar wood chips, and cellulose) to cyclohexyl alcohol mixtures with little to no char.[62] Porous metal oxides are prepared by calcination of synthetic hydrotalcites. Magnesium-aluminum hydrotalcite is a layered double hydroxide mineral composed of Mg^{2+} and Al^{3+} cations in a 3:1 ratio with interstitial CO_3^{2-} and H_2O. The calcination of a synthetic copper-doped layered double hydroxide where 20% of the Mg^{2+} is replaced with Cu^{2+} yields a Cu-doped PMO (Cu_{20}PMO), an effective catalyst for alcohol dehydrogenation and aryl-ether hydrogenolysis.[63] Barta et al. used ^{1}H and ^{2}H NMR spectroscopy to determine that the aromatic ethers of organosolv lignin readily undergo hydrogenolysis and hydrogenation to form cyclohexyl alcohols in sc-CD_3OD.[64]

Cu_{20}PMO catalyzes MeOH reformation and water gas shift reactions under these conditions, which generates the reducing equivalents (H_2) needed (Eqs. 1 and 2).[65] In addition to hydrogenolysis of C—O bonds, Cu_{20}PMO can use the H_2 produced via Eq. (1) and (2) to catalyze hydrogenation of alkenes and arenes. Subsequent studies have shown Cu_{20}PMO and related catalysts disassemble various types of lignin[66] and dehydrogenate other alcohols (like ethanol).[67]

$$CH_3OH \rightleftarrows 2\,H_2 + CO \quad (1)$$

$$CO + H_2O \rightleftarrows CO_2 + H_2 \quad (2)$$

Catalyst stability is crucial for biomass conversion processes to be economically and environmentally sustainable compared to fossil carbon processes. Biomass feedstocks need to be free of impurities that can potentially poison catalysts. There are several such impurities in biomass that are detrimental to catalysts. Lignocellulose contains more oxygen in its chemical structure than fossil carbon which leads to water in hydrogenolysis based catalytic systems. Additionally, trace elements in woody biomass include alkali metals (K and Na), alkaline-earth metals (Mg and Ca), chalcogens (O and S), transition metals (Fe, Mo, Mn, Cu and Zn), pnictogens (N and P), and halogens (Cl and Br).[68] Chlorine is a well-known poison for copper catalysts in amounts as low 100 ppm Cl.[69] Lignocellulosic biomass contains various amounts of chlorine. For example, poplar wood has low chlorine content (200 ppm) in contrast to wheat straw (2600 ppm) and fescues/grasses (3400 ppm).[70] Woody biomass solids, like those from poplar and eucalyptus, have chlorine contents between 30 and 1900 ppm.[68,70,71] Furthermore the organosolv extraction can use mineral acids like HCl which contribute to the chlorine content of the substrate. The chlorine abundance in biomass is a concern in the deactivation of many transition metal based heterogeneous catalysts with limited regeneration options.[72]

The commonly described deactivation pathways for heterogeneous metal catalysts are (1) metal atom sintering to form large metal nanoparticles, (2) carbonaceous solid (i.e., coke or tar) deposition, and (3) formation of strongly bound species to catalyst active sites.[72] Catalyst preparation methods and supports have been researched to limit sintering. Coke and tar can sometimes be removed by facile calcination. Unfortunately, some poisons are so difficult to remove that catalyst regeneration is impractical if not impossible.[72] In petrochemical processes, potential catalyst poisons are removed from feedstocks first. For example, sulfur in fossil carbon feedstocks (like naphtha) is removed by hydrodesulfurization to avert poisoning of noble metal catalysts.[73] Fossil carbon and lignocellulose feedstocks have some similarites,[73] however, the impurities and potential catalyst poisons are different and should be considered in catalytic biomass systems.[74] Specifically, chlorine was hypothesized to increase copper atom mobility for Cu distributed on ZnO(0001)–Zn which facilitates formation of thermodynamically favored large spherical particles.[75] In this context, the nature of structural and reactivity changes of Cu_{20}PMO are described.

2. Understanding and manipulating lignin model compound conversion

Bernt and coworkers showed that operating temperature (280–320 °C)[a] and catalyst contact time[b] can be used to enhance product selectivity toward aromatics for the lignin model compounds phenethyl phenyl ether (PPE), benzyl phenyl ether (BPE), dihydrobenzofuran (DHBF), and phenol.[76] Phenol has been previously shown to greatly contribute to product proliferation, typically resulted in various cyclohexyl alcohols. It is typically used as a simple model of the *p*-hydroxyphenyl (coumaryl alcohol) lignin monomer. The apparent activation energies for PPE and BPE hydrogenolysis (83 and 91 kJ/mol) were lower than for phenol hydrogenation (228 kJ/mol), the primary side reaction leading to lower aromatic product yield. Temperature tuning is thus a potential tool to enhance aromatic product yield as shown by the more aromatic products maintained the reaction of aryl ether model compounds at lower temperatures. Additionally, H_2 overproduction from methanol reforming over Cu_{20}PMO likely contributes to product proliferation in batch reactions. One potential tool for tuning the rates of methanol reforming over Cu_{20}PMO could be selective poisoning to decrease the dehydrogenation and hydrogenation active surface area. Furthermore, the *O*-methylation product of phenol, was previously shown to be relatively stable under these reaction conditions. Thus, another method for preserving aromaticity in the products is to increase the rate O-alkylation of phenol by adding a known alkylating agent.

Dimethyl carbonate (DMC) is a non-toxic and green methylating reagent that was shown to increase the aromatic product yield from both lignin model compounds and organosolv lignin. Bernt and coworkers initially demonstrated increased aromatic product yield from benzyl phenyl ether (α-O-4 lignin model) hydrogenolysis and subsequent phenol methylation.[77] The reactivity of Cu_{20}PMO with BPE in sc-MeOH normally yields primarily toluene and cyclohexanols (from phenol hydrogenation). Under otherwise identical conditions in 2:1 MeOH/DMC, BPE is converted to toluene and mainly anisole (from phenol *O*-methylation).[77] Hydrogenolysis of the benzyl ether bond is rapid in both solvent systems. Toluene is relatively

[a] The supercritical point for MeOH is about 239.5 °C and 79.8 atm. At these temperature (280–320°C), the pressure of the reactor typically reaches above 100 atm.

[b] These experiments were done using a Swagelok stainless steel mini reactor with an internal volume of approximately 10 mL. Typically, there is 3 mL of solvent, 1 mmol of a model compound, and 50–100 mg of Cu_{20}PMO added.

unreactive and is unchanged for a minimum of 12-h residence time. In 1 h at 310 °C in sc-MeOH with Cu_{20}PMO, 85% of the BPE is converted to toluene (99% yield) and phenol (90% yield). After 3 h only 24% of the phenol remains with the major products as cyclohexyl alcohols (cyclohexanol and methylcyclohexanol) and the minor products as aromatics (cresol and anisole).[77] After 6 h a 72% combined yield of cyclohexyl alcohols was observed and no phenol was present in the product mixture. In contrast, phenol largely undergoes *O*-methylation to anisole (methoxybenzene) in the 2/1 (v/v) MeOH/ DMC co-solvent. After 1 h at 300 °C with Cu_{20}PMO in this solvent mixture, 89% of the BPE was converted. Phenol (32% yield) and anisole (28% yield) were the major aromatic products observed. After 6 h, phenol is completely consumed with anisole (61% yield) and cresols as the major products. Bernt and coworkers previously determined hydrogenolysis reactions are first order with respect to the substrate and independent of H_2 pressure between operating temperatures of 280–330 °C in sc-MeOH.[76] The authors hypothesized that the fast hydrogen production rate of CuPMO from methanol reforming saturates the catalyst active sites for hydrogenolysis.[76] In the 2/1 MeOH/DMC co-solvent system, both solvents are in much higher concentration than the substrates and thus the reactions are under pseudo first order conditions.

Phenol *O*-methylation was directly observed in 2/1 MeOH/DMC under the same reaction conditions.[78] The reactivity network for data collected from 6 identical reactors that were subjected to the same treatment for increasing 1-h time intervals (i.e., reaction times of 1, 2, 3, 4, 5, 6 h) is shown in Scheme 2. Briefly, 87% of the phenol is converted to anisole in 6 h at 300 °C. Cyclohexanol and methylcyclohexanol account for about 5% of the phenol mass balance (Scheme 2), and the rest are methylated aromatics (methylanisole, *O*-cresol, and xylenol). Hydrodeoxygenation of anisole to form benzene is a minor and slow reaction pathway. *O*-cresol is the major ring methylation product observed expected as the -OH group on phenol is an activating ortho-para directing group. Limited quantities of para- and meta-cresols were also observed. Xylenol, methylcyclohexanol, and methylanisole were previously shown to derive from o-cresol in sc-MeOH.[77]

One outstanding question is: Why is phenol so much more reactive with Cu_{20}PMO than anisole? One plausible explanation is that phenol has a greater affinity for the catalyst surface than does anisole. Calcination of synthetic transition metal hydrotalcites typically generate porous metal oxides that act as solid bases. PMOs have been previously shown to act as solid base catalysts in transesterification of triglycerides.[79] Acidic substrates (like phenol)

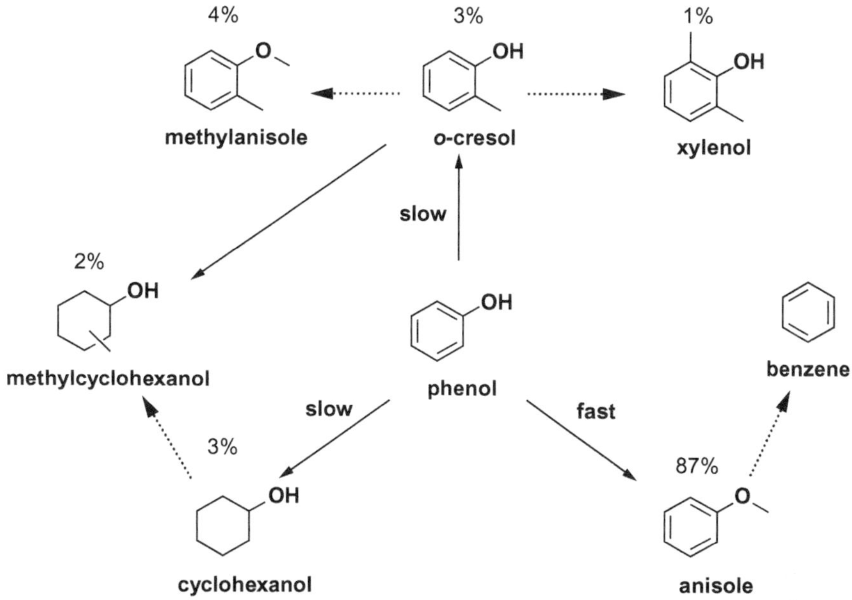

Scheme 2 Reaction pathways for phenol and its derivatives with $Cu_{20}PMO$ in 2/1 MeOH/DMC at 300 °C. Values shown are mole % yield. *Reproduced with permission from Barrett, J. A.; Gao, Y.; Bernt, C. M.; Chui, M.; Tran, A. T.; Foston, M. B.; Ford, P. C.* ACS Sustain. Chem. Eng. ***2016***, 4, *6877–6886. Copyright 2016 American Chemical Society.*

should adsorb more easily to the surface than neutral substrates (like anisole). Molecules like anisole, toluene, and benzene may react relatively slowly under these conditions because they do not strongly adsorb to the $Cu_{20}PMO$ surface. Studies on SiO_2-Al_2O_3 surfaces show a much lower molar adsorptivity for anisole than for phenol at 400 °C,[80] however, to the best of our knowledge similar studies have not been performed on PMOs.

Phenol is a very simple model for the phenolic species generated in lignin disassembly. A more complex model is 2-methoxy-4-propylphenol (MPP) which has the propyl chain and methoxy (-OCH_3) group that would be present in lignin derived phenolic species (Scheme 3). The two monolignols that contain *ortho*-OCH_3 groups are coniferyl alcohol (G or guaiacyl with one -OCH_3 groups) and sinapyl alcohol (S or syringyl with two-OCH_3 groups) (Fig. 1). Approximately 50% of MPP is converted to a broad distribution of products after 6 h with $Cu_{20}PMO$ in sc-MeOH at 300 °C.[78] The largest component (21% of products) was 1,2-dimethoxy-4-propylbenzene (i.e., O-methylated MPP), in 21% of the products. The reactivity of MPP with in 2/1 MeOH/DMC at 300 °C results in primarily aromatic products with

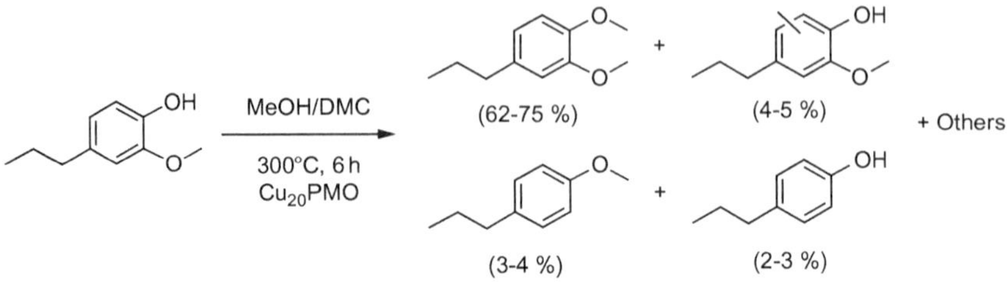

Scheme 3 Major products from the reaction of MPP with $Cu_{20}PMO$ in 2/1 MeOH/DMC at 300 °C. Values shown are mole % yield. *Reproduced with permission from Barrett, J. A.; Gao, Y.; Bernt, C. M.; Chui, M.; Tran, A. T.; Foston, M. B.; Ford, P. C.* ACS Sustain. Chem. Eng. ***2016****, 4, 6877–6886. Copyright 2016 American Chemical Society.*

the same major product as in sc-MeOH (Scheme 3). In the 2/1 MeOH/DMC co-solvent, high conversion (95%) of MPP to products was observed. A 67% yield of 1,2-dimethoxy-4-propylbenzene and small amounts of 4-propylphenol (2%) and 4-propylanisole (3%) were also observed (Scheme 3).

The reaction of BPE in 2/1 MeOH/DMC with $Cu_{20}PMO$ at 300 °C was examined in more detail to develop a more comprehensive reaction network. The temporal distribution of products from the reaction of BPE over $Cu_{20}PMO$ is shown in Fig. 3. Dynafit is a free-to-use (for academics) software that concurrently produces growth and decay curves for species in a theoretical reaction network proposed by the user using an algorithm of first-order ordinary differential equations.[81] The rate constants for a given reaction pathway are optimized using a non-linear least squares approach to best fit the experimental data. For instance, the ordinary differential equation for BPE in the fit model for Fig. 3 was $d[BPE]/dt = -k_1[BPE] - k_{s1}[BPE]$. The rate of BPE disappearance was k_1 ($1.7\ h^{-1}$) and the rate constant k_{s1} ($\sim 0\ h^{-1}$) is an unidentified product sink to compensate for mass balance. Low sink rate constants indicate that all the products are accounted for by the model of the reaction network.

Hydrogenolysis of BPE is nearly quantitative in 2 h (Fig. 4). Toluene is unreactive and remains in the reaction solution once produced. There are three known pathways shown for phenol. Under these conditions (2/1 MeOH/DMC) the minor pathways are aromatic ring hydrogenation and ring methylation. *O*-cresol is primarily hydrogenated to 2-methylcyclohexanol ($k = 0.42\ h^{-1}$). The predominant reaction pathway for phenol in MeOH/DMC is *O*-methylation. The anisole yield after 2 h was 71% and decreases very slightly over time. Anisole reacts slowly to form benzene primarily ($k = 0.04\ h^{-1}$). When a smaller fraction of DMC was used in the

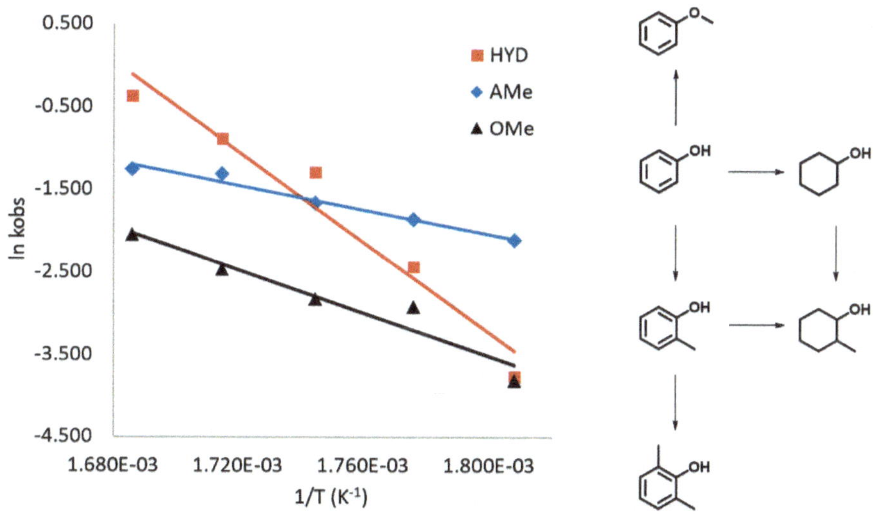

Fig. 3 Arrhenius plots for the competing phenol hydrogenation (HYD), aromatic methylation (AMe), and *O*-methylation (OMe) in sc-MeOH with $Cu_{20}PMO$ from 280 to 320 °C for up to 1.5 h. Apparent activation energies are $E_a(HYD) = 228 \pm 30$ kJ/mol, $E_a(AMe) = 61 \pm 6$ kJ/mol, and $E_a(OMe) = 108 \pm 17$ kJ/mol. The observed reaction network for phenol is also shown. *Adapted with permission from Bernt, C. M.; Manesewan, H.; Chui, M; Boscolo, M.; Ford, P.C.* ACS Sustain. Chem. Eng. ***2018***, 6, *2, 2510–2516. Copyright 2018 American Chemical Society.*

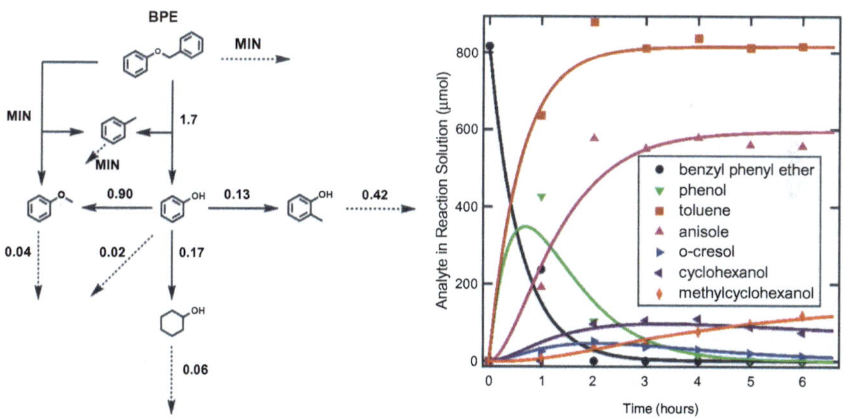

Fig. 4 *Left*: Reaction pathways for BPE with $Cu_{20}PMO$ in 2/1 MeOH/DMC at 300 °C. Units for rate constants (k_{obs}) are h^{-1}. MIN = ~0. *Right*: Temporal product evolution and for BPE (initial amount = 800 µmol). Minor products can be found in the supporting information of Ref. 78. All progress curve fits were generated using Dynafit.[42] *Reproduced with permission from Barrett, J. A.; Gao, Y.; Bernt, C. M.; Chui, M.; Tran, A. T.; Foston, M. B.; Ford, P. C.* ACS Sustain. Chem. Eng. ***2016***, 4, *6877–6886. Copyright 2016 American Chemical Society.*

co-solvent system (2.5/0.5, v/v MeOH/DMC), a 42% lower anisole yield and accordingly higher yield of methylcyclohexanols were found. This indicated that the *O*-methylation of phenol is performed by DMC and not MeOH.

The heat and pressure of sc-MeOH likely plays an important role in solvolysis of lignin to smaller fragments.[82,83] Model compounds like BPE can react in sc-MeOH and MeOH/DMC without catalyst under similar conditions. BPE potentially undergoes hydrolysis from residual water in reaction solvents or substrates in a similar mechanism to acidolysis proposed by Pelzer et al.[84] About 23% BPE conversion after 3 h in sc-MeOH (89% material balance) at 300 °C and 51% conversion after 6 h (79% material balance). The easily quantifiable products were mostly toluene (49%) and phenol (38%). As the reaction time increased the material balance decreased due to the formation of insoluble materials consistent with the reaction mechanisms in literature.[84] Note that reaction of BPE with Cu_{20}PMO under the same conditions had 74% conversion after 1 h and near quantitative conversion after 3 h with almost 100% material balance. The Cu_{20}PMO catalyst accelerates the rate of aromatic-ether hydrogenolysis and prevents formation of insoluble materials (like char and tar). The reaction of BPE in the 2/1 MeOH/DMC without catalyst showed ~43% conversion after 6 h with toluene (34%) and phenol (37%) as the major products. Anisole (24%) was also observed as expected. Porous metal oxide catalyst derived from calcining commercially available synthetic 3:1 Mg-Al hydrotalcite has been previously shown to catalyze *O*-methylation of phenol by Talawar, Jyothi, and coworkers.[85,86] However, this PMO does not catalyze aryl-ether hydrogenolysis as well as Cu_{20}PMO, presumably because it does not reform methanol to produce hydrogen. The reaction of BPE reaction in 2/1 MeOH/DMC at 300 °C with copper-free PMO showed 32% conversion after 6 h with similar product distributions as the reactions without catalyst.

The α-O-4 model compound, benzyl phenyl ether, is very useful as a tool for quickly evaluating the efficacy of a catalyst system because it is commercially available and breaks down to very easily measurable products. The β-O-4 aryl-ether bond is the most abundant linkage in most lignins, so model compounds like 2-phenoxy-1-phenylethan-1-ol (PPE) generate products and intermediates that are more like those that may be formed in lignin hydrogenolysis. Hydrogenolysis of the β-O-4 bond in PPE is nearly quantitative (98% conversion) in 1 h (Fig. 5). The expected products are 1-phenylethanol and phenol. The former rapidly undergoes hydrodeoxygenation to ethylbenzene ($k = 2.8\ h^{-1}$). Propylbenzene ($k = 0.3\ h^{-1}$)

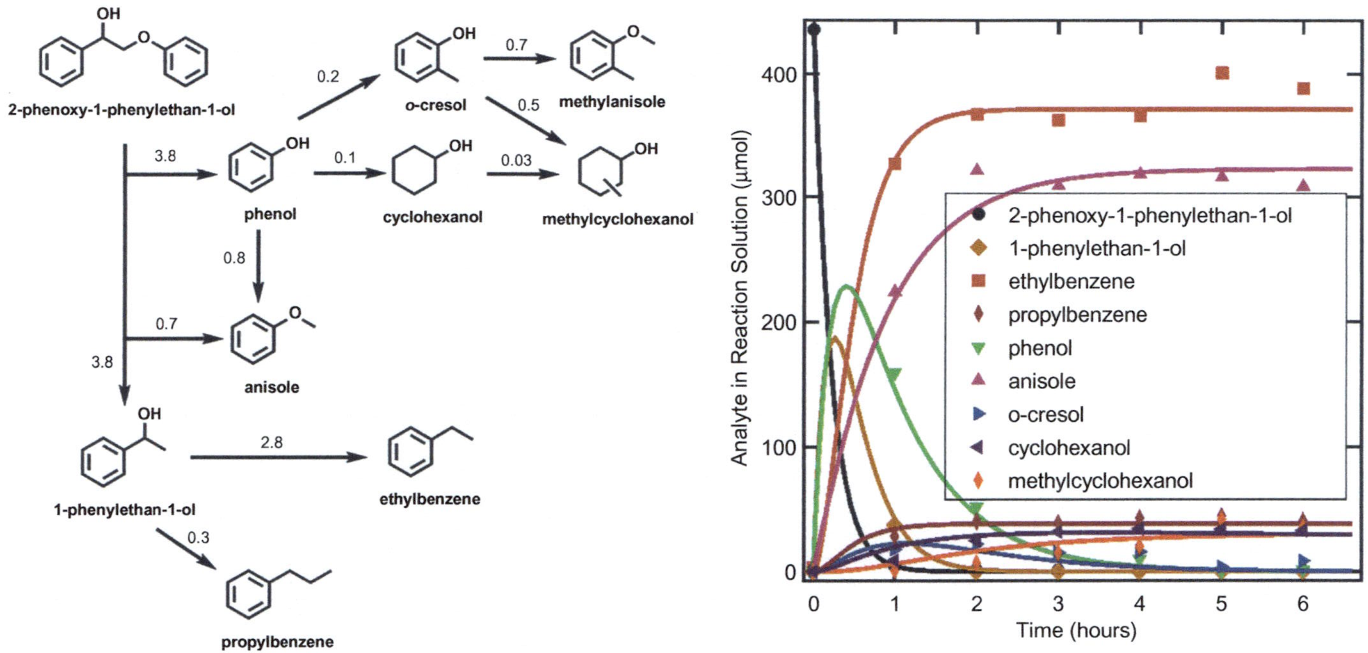

Fig. 5 *Left*: Reactivity network from the reaction of PPE in MeOH/DMC with $Cu_{20}PMO$. The units for the calculated rate constants (k_{obs}) are h^{-1}. *Right*: Temporal product distribution from the reaction of 2-phenoxy-1-phenylethan-1-ol (~500 μmol) in MeOH/DMC with $Cu_{20}PMO$ (50 mg) at 300 °C. *Reproduced with permission from Barrett, J. A.; Gao, Y.; Bernt, C. M.; Chui, M.; Tran, A. T.; Foston, M. B.; Ford, P. C.* ACS Sustain. Chem. Eng. ***2016****, 4, 6877–6886. Copyright 2016 American Chemical Society.*

was a minor product from β-carbon methylation of acetophenone, the 1-phenylethanol dehydrogenation intermediate (Scheme 4).[87] The anisole yield after 2 h was 74%. Aromatic products account for ~91% of the theoretical product yield and remain constant for at least 6 h with some redistribution to other aromatic species. Cyclohexanols and cyclohexanes make up the aliphatic products. In terms of phenol derived products, aromatics were 89% at 3 h and 83% at 6 h with material balances of ~100% for each time point.

In contrast, the total yield of aromatic products from PPE was only 69% after 6 h in sc-MeOH with $Cu_{20}PMO$. These products were mostly ethylbenzene and propylbenzene which are derived from 1-phenylethanol. The aromatic product yield from phenol was only 33%. Similar experimental data and fits were previously reported for sc-MeOH when key reactions (i.e., aryl-ether hydrogenolysis) are compared in Table 1.[77,78]

As shown in Fig. 1, lignin has several aryl-ether binding motifs. The 4-O-5 and β-5 bonds are modeled by diphenyl ether (DPE) and 2,3-dihydrobenzofuran (DHBF), respectively. In both softwoods and hardwoods

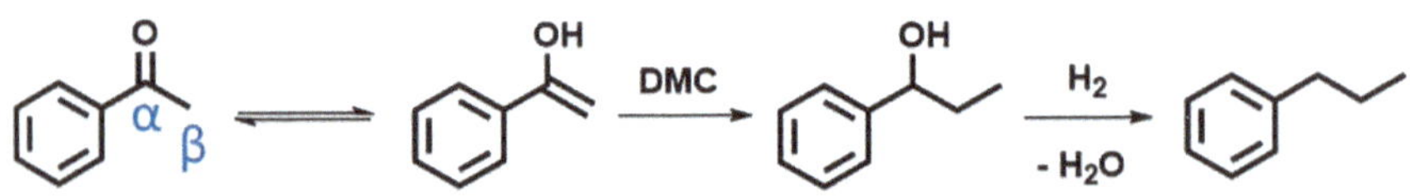

Scheme 4 Plausible mechanism for the β-carbon methylation of 1-phenylethanol to form propylbenzene. Acetophenone tautomerizes to the enol 1-phenylethen-1-ol. DMC = dimethyl carbonate. *Reproduced with permission from Barrett, J. A. Part 1: Investigations of Reaction Pathways for Catalytic Biomass Conversion, Doctoral Dissertation, University of California-Santa Barbara, 2019. Copyright 2019 Jacob A. Barrett.*

Table 1 Comparison of relative k_{obs} for key reactions in sc-MeOH (from Ref. 77) or MeOH/DMC.[77,78]

Reaction	Solvent	k_{obs} (h^{-1})	Qualitative rate
β-O-4/α-O-4 ether hydrogenolysis	Sc-MeOH	>1.0	Fastest
	MeOH/DMC	>1.0	Fastest
Phenol Hydrogenation	Sc-MeOH	0.2–0.75	Fast
	MeOH/DMC	0.10	Slow
Phenol O-methylation	Sc-MeOH	0.05–0.10	Slow
	MeOH/DMC	0.8–1.2	Fast

only about 2% of aryl-ether bonds are 4-O-5 and 3–12% are β-5.[8–10] The 4-O-5 and β-5 aryl-ether bonds are more difficult to break than the α-O-4 and β-O-4 linkages, and the conversion in $Cu_{20}PMO$ system in both sc-MeOH and MeOH/DMC are low.[19] Notably, a larger percentage of aromatic products (73%) was observed in MeOH/DMC than in sc-MeOH (29%). This is likely because H_2 pressure is lower in 2/1 MeOH/DMC.

Barrett, Jones, and co-workers sought to modulate the activity for $Cu_{20}PMO$ using parts-per-million quantities of a known copper catalyst poison, chloride. When $Cu_{20}PMO$ was pretreated in sc-MeOH with 15 ppm chloride (added as $MgCl_2{\cdot}H_2O$) for 72 h, the phenol hydrogenation activity was greatly decreased. BPE was still completely converted in 2 h but the phenol yield was 59% which was almost double the yield observed by Bernt et al.[78] Furthermore, the phenol derived products are mainly aromatics (20%) with some cyclohexyl alcohols (5%). We hypothesized that the added chloride facilitates sintering of the Cu atoms to generate Cu nanoparticles which would decrease the active surface area for phenol hydrogenation. This explanation was supported by measurements of Cu particle size using N_2O temperature programmed desorption and estimation of domain sizes from the crystalline Bragg reflections (51° and 75° for Cu metal) observed in the XRD pattern (Fig. 6) In short, Cu(0) nanoparticles between 5 and 10 nm seem to be active for aryl-ether hydrogenolysis but less active for phenol hydrogenation.

Phenolic compounds are the typical intermediates formed in reductive lignin disassembly process and likely to be the source of product proliferation and consequently decreased selectivity. Trapping phenolic intermediates by selective conversion to more stable products (like *O*-methylation to methoxyarenes) is a potential strategy that can be applied to lignin. The MeOH/DMC solvent limits the aliphatic product yield (cyclohexyl alcohols) by selective *O*-methylation shown by the observed rate constants in Table 1. In this case DMC was used in excess of the model substrates. A future goal for such a system would be the use stoichiometric amounts of methylating agent. Reaction temperature and solvent system provide two handles to tune for increased selectivity. Alternatively, selectively tuning the catalyst structure may enable similar *O*-methylation activity but reduced hydrogenation activity. As described above, selective catalyst poisoning is one approach toward limiting specific activities to increase selectivity. Chui and co-workers showed that the addition of Sm^{3+}, as $Sm(OTF)_3$, or formic acid drastically decreased the hydrogenation activity of phenol derived

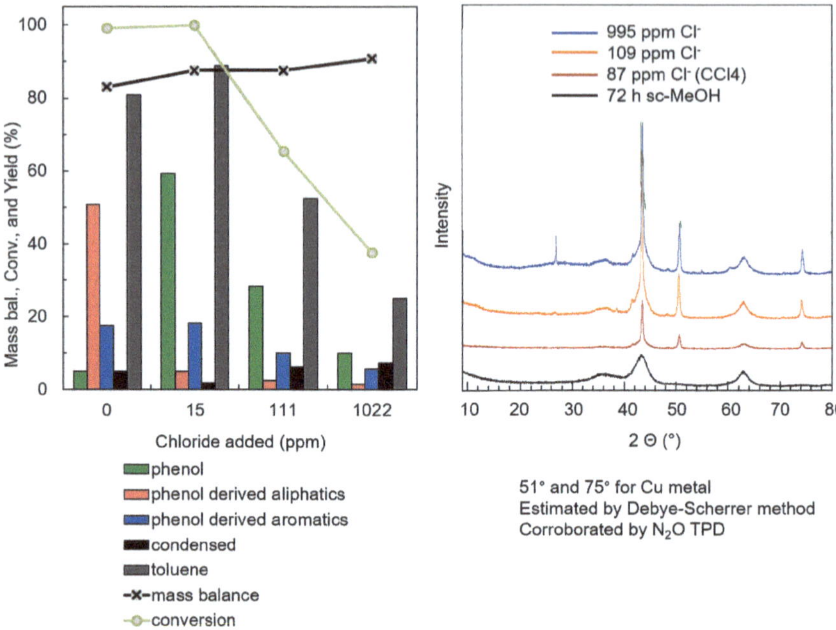

Fig. 6 *Left*: Product yields, BPE conversion, and material balance for the reaction of BPE with $Cu_{20}PMO$ at 300 °C. The catalyst was calcined at 600 °C and heated at 300 °C for 72 h in sc-MeOH. Bars = product yields. Lines w/markers = mass balance and BPE conversion. "Condensed" products are unidentified but show *m/z* over 180 in GC–MS. *Right*: Powder X-ray diffraction patterns of $Cu_{20}PMO$ treated with sc-MeOH for 72 h that contained varying amounts and sources of chloride. Unless otherwise noted, chloride was derived from $MgCl_2$. *Adapted with permission from Barrett, J. A. Part 1: Investigations of Reaction Pathways for Catalytic Biomass Conversion, Doctoral Dissertation, University of California-Santa Barbara,* ***2019*** *and Jones, Z. R. Elucidating Structure-Activity Relationships through X-ray Absorption Fine Structure Analyses: from Site-Isolated Systems, to Nanoclusters, and to Supported Nanoparticle Catalysts, Doctoral Dissertation, University of California-Santa Barbara,* ***2019****. Copyright 2019 Jacob Alexander Barrett and Copyright 2019 Zachary Jones.*

from BPE hydrogenolysis with $Cu_{20}PMO$.[88] There was no apparent explanation for why Sm^{3+} addition resulted in decreased phenol hydrogenation/increased aromatic product yield. The formic acid likely decreases phenol adsorption to the surface by neutralizing the basic sites of $Cu_{20}PMO$ thus decreasing all activities with acidic substrates (like phenol).

3. Organosolv poplar lignin conversion

The conversion of woody biomass would lead to a complex product mixture from the cellulosic fraction as well as lignin. In order to study the reactivity of lignin with $Cu_{20}PMO$, several research groups have employed as a model substrate organosolv lignin. This is lignin isolated from woody biomass using an acidified organic solvent reflux for long periods.[c] The Ford group has typically used organosolv poplar lignin (OPL) to assess how the observations and lessons learned from the aforementioned model compound reactions can be applied to lignin and lignocellulosic solids.[d] Organosolv poplar lignin disassembly in sc-MeOH or in MeOH/DMC at 300 °C, likely proceeds via solvolysis to give smaller fragments,[83] which can then adsorb onto the catalyst and undergo catalytic aryl-ether bond cleavage (Scheme 5).

The organosolv poplar lignin is converted to a brown liquid mixture after reaction with Cu_{20}PMO under supercritical conditions. The unprocessed samples contain a mixture of organic compounds and the remaining solvent (these are typically referred to as "unprocessed"). "Unprocessed" liquid samples are the raw product mixture after Cu_{20}PMO was removed by filtration. Solvent removal also removes some of the more volatile organic products as well (cyclohexyl alcohols). Thus "non-volatile" liquid samples are the

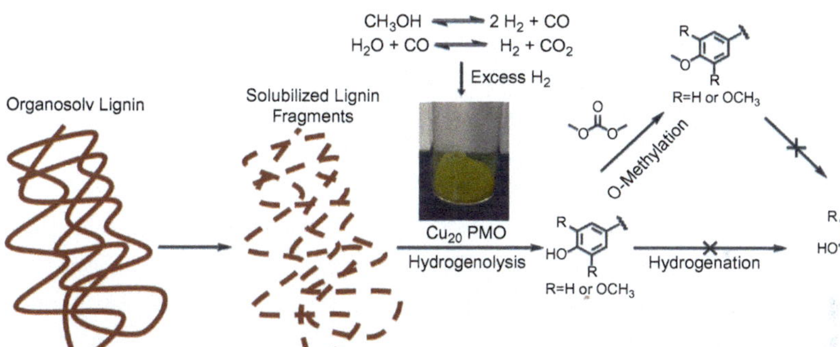

Scheme 5 Proposed step-wise mechanism of reductive organosolv lignin disassembly in MeOH/DMC with Cu_{20}PMO. Solvolysis breaks lignin into smaller fragments and subsequent or synchronous hydrogenolysis further converts the fragments to phenolic intermediates. *O*-methylation by DMC yields methoxybenzenes. *Reproduced with permission from Barrett, J. A.; Gao, Y.; Bernt, C. M.; Chui, M.; Tran, A. T.; Foston, M. B.; Ford, P. C.* ACS Sustain. Chem. Eng. ***2016***, 4, *6877–6886. Copyright 2016 American Chemical Society.*

[c] Mineral acids (i.e. H_2SO_4, HCl, HNO_3, H_3PO_4) are typically used in the organosolv process. The Ford laboratory uses HCl. Notably, there is an increase in chlorine content in the resultant organosolv poplar lignin even after successive washing with deionized water.

[d] These experiments were done using a Swagelok stainless steel mini reactor with an internal volume of approximately 10 mL. Typically, there is 3 mL of solvent, 50–100 mg of organosolv poplar lignin, sawdust, or poplar shavings, and 50–100 mg of Cu_{20}PMO added.

filtered product mixtures after solvent evaporation (vacuum at −20 °C).[78] The non-volatile liquid mixtures from OPL reacted with $Cu_{20}PMO$ in sc-MeOH was 45 wt% and in MeOH/DMC was 64 wt% (based on the initial 100 mg of OPL added). When no catalyst was used, large amounts of char were isolated in both solvent systems (35 and 32 wt%, respectively), whereas no char was observed when $Cu_{20}PMO$ was present. Untreated OPL, unprocessed and non-volatile product mixtures from reaction in sc-MeOH and 2/1 MeOH/DMC with and without $Cu_{20}PMO$ for 3 h were analyzed by Gel-Permeation Chromatography (GPC). The molecular weight distributions are nearly unchanged between "unprocessed" and "non-volatile" mixtures. All samples had a high polydispersity index (PDI) which indicates a broad distribution of molecular weights. Notably, the M_n for each liquid sample (~310–390 g/mol) is ~3–4 times lower than that of untreated OPL (~1290 g/mol). Thus, the lignin macrostructure is largely disassembled to monomeric (~200 g/mol), dimeric (~400 g/mol), and/or trimeric (~600 g/mol) species within 3 h. The *O*-methylation by DMC contributes 15 g/mol to the molecular weight of products, however, the experimental uncertainty in GPC is too large to accurately observe this increase. Nonetheless, M_n and M_w for products from reaction with $Cu_{20}PMO$ in 2/1 MeOH/DMC are slightly larger than their sc-MeOH counterparts.[78]

^1H NMR analysis of OPL and liquid product mixtures was previously employed by Barta et al.[64] Three different proton environments, aromatic, aliphatic, and oxygenated are integrated and holistically compared. Solvolytic fragmentation and catalytic hydrogenolysis of the lignin aryl-ether bonds is supported by ^1H NMR data (Fig. 7). Untreated OPL is characterized by 76% O-aliphatic protons, 15% aromatic protons, and 9% aliphatic protons. The extent of aryl-ether bond cleavage and hydrodeoxygenation is evident in the decrease in the 3.0–5.0 ppm range. A decrease in the proton chemical shifts for the β- and γ-carbon (oxygenated) of various aryl-ether bonds is observed for products from sc-MeOH and MeOH/DMC with $Cu_{20}PMO$. Consequently, a large increase in proton chemical shifts associated with alkyl carbons (aliphatic) attached to aromatic rings. The aromatic proton chemical shift integration remains very similar for product mixtures which highlights a shortcoming in ^1H NMR analysis for lignin and product mixture analysis. There are more oxygenated protons observed in 2/1 MeOH/DMC likely due to *O*-methylation of phenolic products. However, it is not entirely clear from ^1H NMR analysis that aromatic products were maintained in MeOH/DMC. Solvent evaporation likely removes some volatile aliphatic products, as shown by GC–MS analysis of the liquid nitrogen-trapped solvent and volatile

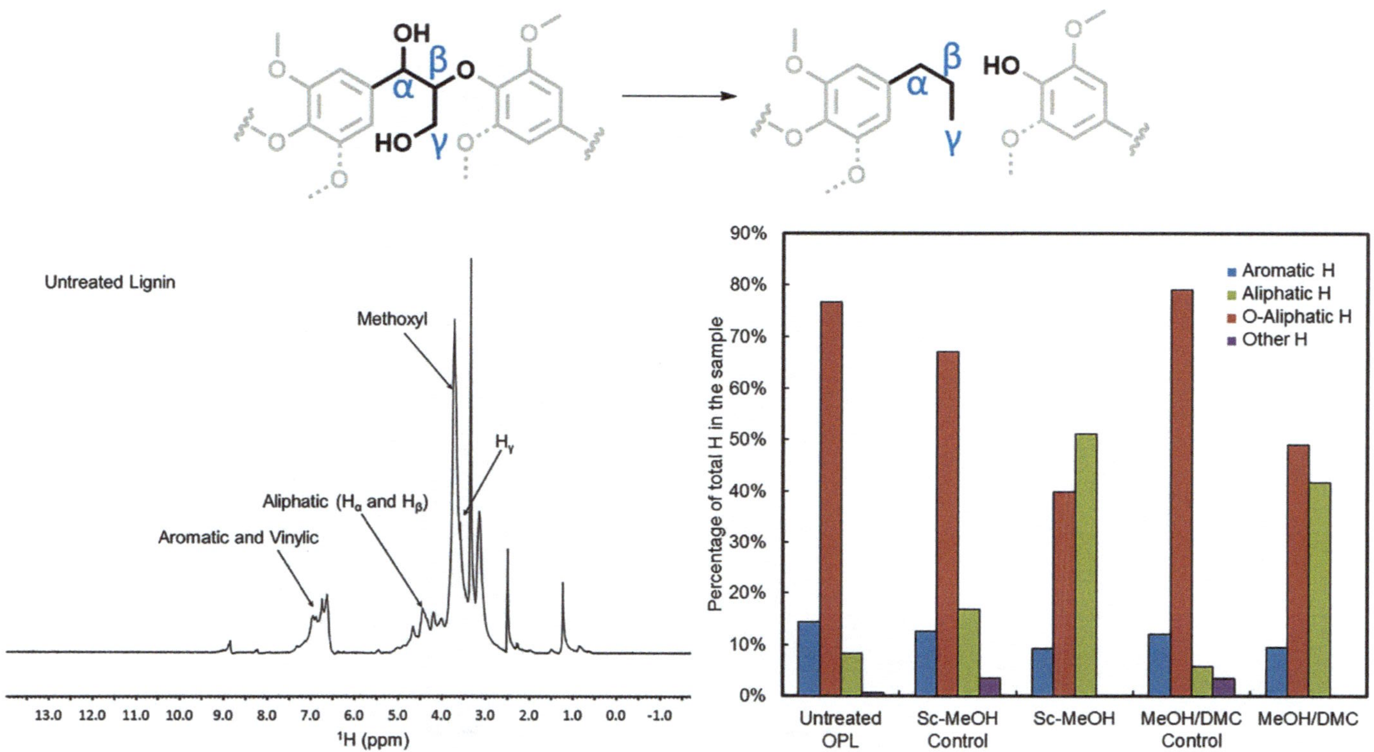

Fig. 7 *Top*: Depiction of hydrogenolysis and hydrodeoxygenation of β-O-4 bond showing the α-, β-, and γ-carbon positions represented in the aryl-ether (left) and the aryl-propyl (right) species. *Bottom*: ^{1}H NMR spectrum for untreated organosolv poplar lignin (OPL). Proton percentage (%) for untreated OPL, non-volatile liquid product mixtures in sc-MeOH, and non-volatile liquid product mixtures in MeOH/DMC. Control = without Cu_{20}PMO. *Reproduced with permission from Barrett, J. A.; Gao, Y.; Bernt, C. M.; Chui, M.; Tran, A. T.; Foston, M. B.; Ford, P. C.* ACS Sustain. Chem. Eng. ***2016***, 4, *6877–6886. Copyright 2016 American Chemical Society.*

product components.[78] Thus, NMR analyses only observes low volatility products (e.g., oligomers or low vapor pressure chemicals).

Carbon (^{13}C) NMR analysis is another powerful technique to examine complex liquid product mixtures from lignin disassembly. The ^{13}C NMR spectrum of untreated OPL in DMSO-d_6 solutions is shown in Fig. 8. The chemical shift areas are assigned by the functional group carbons which was previously determined by Ben et al.[89] Integration of the chemical shifts enables quantification of the carbon environments (i.e., aromatic C—H, aromatic C—O, aliphatic C—H, etc.). The aliphatic C—C or C—H accounts for 63% of carbon in the non-volatile products from the OPL reaction in sc-MeOH for 6 h (Fig. 8). The peaks from 55 to 14 ppm are indicative of both non-volatile aliphatic products as well as the alkyl chain on aromatic products. The untreated OPL has 13% abundance of aliphatic carbons adjacent to an oxygen atom (aliphatic C—O). The abundance aliphatic C—O (from 96 to 61 ppm) decreased to ~1% after reaction in sc-MeOH or MeOH/DMC with $Cu_{20}PMO$. The untreated OPL has 3% abundance of aliphatic carbons (aliphatic C—C or C—H, noted as "other aliphatic C" in Fig. 8). The abundance of aliphatic C—C or C—H (from 55 to 14 ppm) increased to 35% after 3 h and 63% after 6 h reaction in sc-MeOH which indicates that much of the substrate aromaticity was reduced (Fig. 8). In contrast, the abundance of aliphatic C—C or C—H increased to 24% after 3 h and 26% after 6 h in 2/1 MeOH/DMC which indicates that much of the substrate aromaticity is retained (Fig. 8).

The change in abundance of aromatic carbons (aromatic C—O from 167 to 142 ppm, C—H from 125 to 96 ppm, and C—C from 142 to 125 ppm) also implies that aromaticity is maintained in the product mixture after reaction with MeOH/DMC. The untreated OPL has 52% abundance of aromatic carbons and this decreased to ~42% after 3 h reaction in both sc-MeOH and MeOH/DMC. After 6 h in sc-MeOH, the aromatic carbon further decreased to ~23%. After 6 h in MeOH/DMC the abundance of aromatic carbon is unchanged with a slight change in distribution. We concluded that reductive disassembly accompanied by some arene hydrogenation occurs in both sc-MeOH and MeOH/DMC, however, DMC maintains the aromaticity of the product mixture. Based on the model studies with phenol, BPE, and PPE, we hypothesized that DMC would selectively *O*-methylate phenolic species. This is indicated by the relatively constant abundance of aromatic bound methoxyl carbons (~32%).

A more complete picture of the product distribution is developed by combining GC–MS analyses of the unprocessed products mixtures and

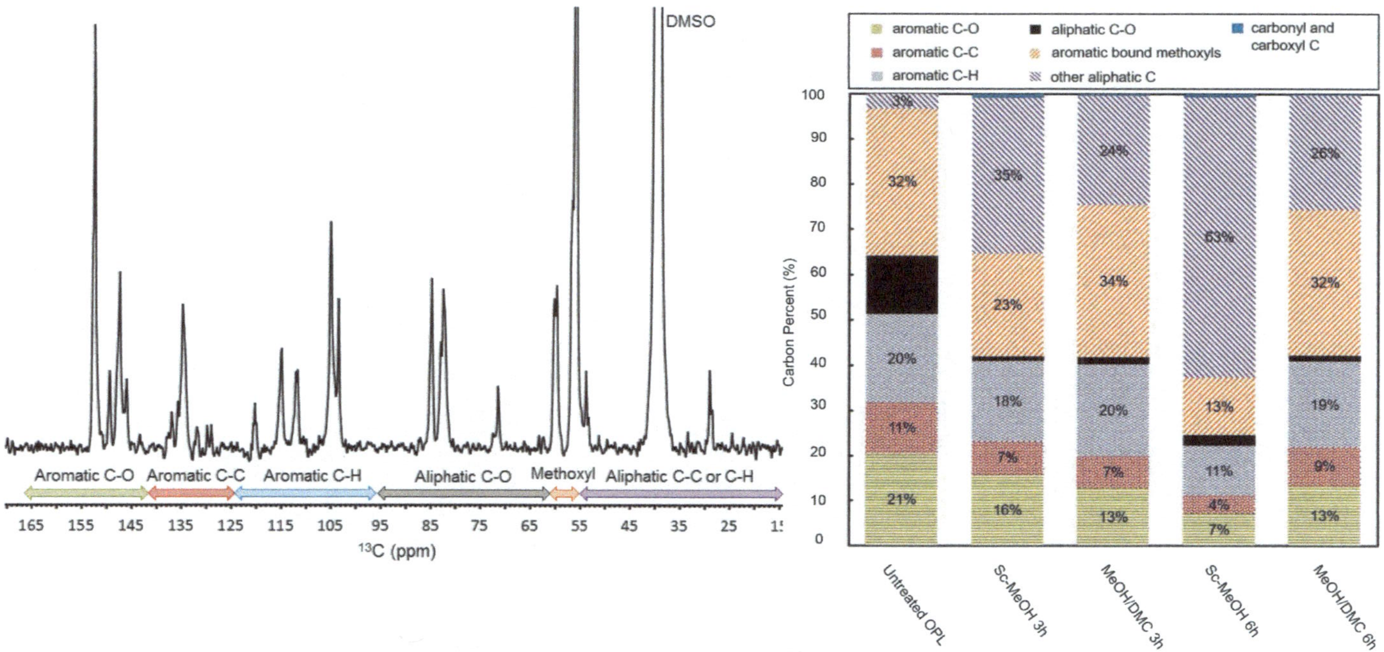

Fig. 8 *Left*: ^{13}C NMR spectrum in DMSO-d_6 of untreated organosolv poplar lignin with peak assignments. *Right*: Percent of carbon environments based on ^{13}C NMR data for untreated OPL and for the products of disassembled OPL in sc-MeOH, and in MeOH/DMC. *Reproduced with permission from Barrett, J. A.; Gao, Y.; Bernt, C. M.; Chui, M.; Tran, A. T.; Foston, M. B.; Ford, P. C.* ACS Sustain. Chem. Eng. ***2016***, 4, *6877–6886. Copyright 2016 American Chemical Society.*

NMR analyses of the non-volatile product mixtures. The mass spectra are often convoluted for OPL product mixtures; however, factor analysis can be used to group products based on chemical features. Mass spec "factors" are established using characteristic *m/z* peaks for that specific chemical. For example, Gao and co-workers observed *m/z* peaks at 39, 50, 65, 74, 93 for some non-polar aromatics and *m/z* peaks at 39, 65, 79, 91, 107, 119, 135, 151 for dimethoxybenzenes.[90] This enables the chemical assignment of peaks in the chromatogram to individual factors even if they are not cataloged in the GC–MS library. Fig. 9 shows the chromatograms from GC–MS analysis of product mixtures from the 6 h reaction of OPL in MeOH/DMC and sc-MeOH with Cu_{20}PMO. There are 2 major peaks and about 10 minor peaks in the MeOH/DMC reaction mixture. The largest peak is 1,2,3-trimethoxy-5-propylbenzene and the second largest peak is 1,2-dimethoxy-4-propylbenzene. The minor peaks were other various methoxybenzenes and methoxyphenols. In short, the products generated in reductive OPL disassembly by Cu_{20}PMO in sc-MeOH are susceptible to hydrogenation and other secondary reactions that lead to significant product proliferation and low yields of aromatic products. We showed that by using the co-solvent mixture MeOH/DMC (previously shown to selectively *O*-methylate phenol from aryl ether hydrogenolysis), a narrower product distribution can be obtained, which is largely methoxybenzene derivatives.

In any ideal catalytic system, the catalyst remains active for many turnovers or is easily regenerated after deactivation. Cu_{20}PMO exhibited reproducible activity with OPL for three reaction cycles, however after the fourth consecutive reaction with the same 100 mg of catalyst activity greatly decreased. These recycles were run at 300 °C for 3 h in 3 mL fresh 2/1 MeOH/DMC with 100 mg OPL in 10 mL Swagelok mini-reactors in each case, so that the major products would be easy to identify. Analysis of the two major products 1,2,3-trimethoxy-5-propylbenzene and 1,2-dimethoxy-4-propylbenzene by GC–MS showed the same yields for the first three consecutive reactions but about 50% decreased yield in the fourth reaction and 25% decreased yield in the fifth. Studies on the structural and reactivity changes of Cu_{20}PMO are discussed further in a later section.[87,91]

Solvolysis of lignin aggregates in alcohol solvents has been hypothesized as a crucial step in reductive cleavage.[83] Cheng, Truong, and coworkers showed that a mixture of ethanol and isopropanol was able to solubilize and disassemble organosolv poplar lignin. After heating the OPL in a pressure reactor at 270 °C for 4 h, an oil was recovered (70 wt% of initial OPL)

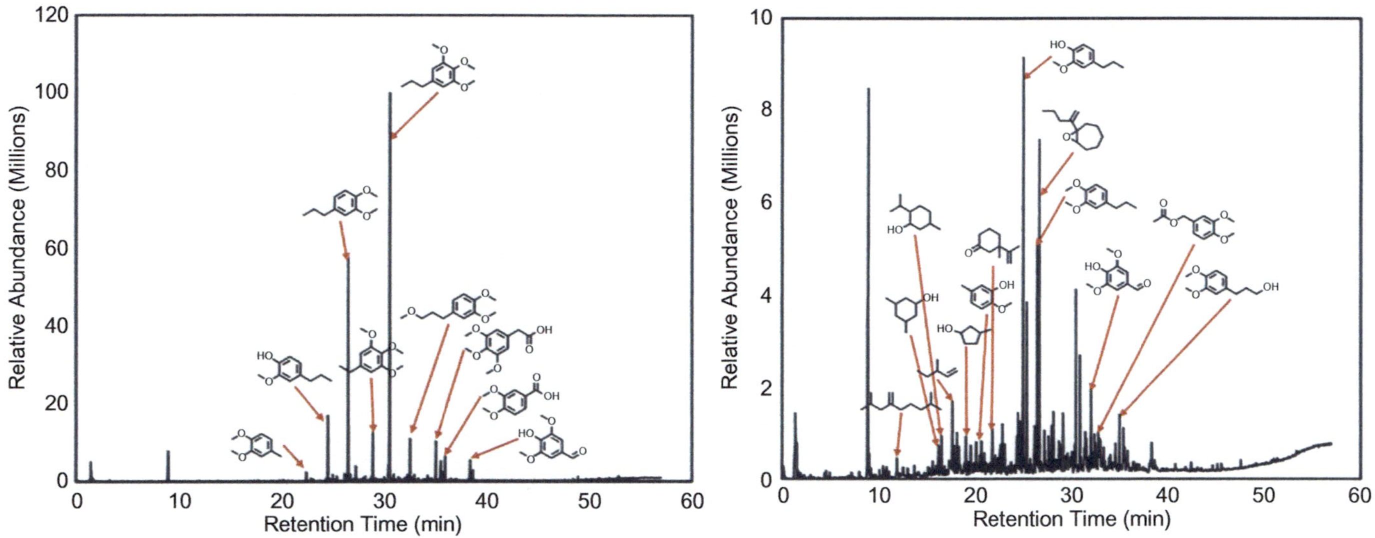

Fig. 9 GC–MS chromatograms of liquid product mixture from OPL disassembly for 6 h in sc-MeOH (right) and in MeOH/DMC (left) with $Cu_{20}PMO$. The structures that could be easily assigned using the MS library are drawn. *Reproduced with permission from Barrett, J. A.; Gao, Y.; Bernt, C. M.; Chui, M.; Tran, A. T.; Foston, M. B.; Ford, P. C.* ACS Sustain. Chem. Eng. ***2016****, 4, 6877–6886. Copyright 2016 American Chemical Society.*

that contain almost 50% two aromatic compounds (*E*)-4-propenyl syringol and isoeugenol.[92] This is a remarkable result compared to the experiments in sc-MeOH and MeOH/DMC which resulted in 45 and 64 wt% liquid yield of over a dozen products.[78] Exactly how the mixed EtOH/*i*-PrOH system was able to produce such a selective product stream without any catalyst remains unexplained.

4. Guerbet chemistry over Cu-doped porous metal oxides

The reactivity of BPE with Cu_{20}PMO that had been treated with ppm quantities of chloride (in the form of magnesium chloride) was briefly described in the prior section. The researchers concluded that the chloride facilitates significant structural change in Cu_{20}PMO, primarily through the action of sintering. In-line GC–MS measurements showed that the chloride treated Cu_{20}PMO had decreased methanol reforming activity, which may have led to less phenol hydrogenation. However, undesirable condensation products of phenol were also observed. The reactivity of ethyl alcohol (i.e., ethanol) with chloride treated Cu_{20}PMO was explored to gauge how the concentration of chloride affected the catalyst's activity for alcohol dehydrogenation and aldehyde condensation. This experiment is what led to our studies to control the selectivity in the Guerbet condensation of alcohols over copper porous metal oxides.[61]

Marcel Guerbet observed that ethyl, isobutyl, and isopentyl alcohols underwent self-condensation at temperatures above 200 °C under basic conditions.[33] Renewed interest in the Guerbet condensation of ethanol has led to research to better understand the reaction over metal oxide catalysts.[93,94] Specifically, the selective condensation of ethanol to n-butanol would provide a renewable chemical feedstock for n-butanol, which is typically produced via hydroformylation of propylene followed by hydrogenation of the butyraldehyde product. Propylene and hydrogen are produced from petroleum or natural gas processing. Transition metal catalysts (Mn, Fe, Ru, Rh, and Ir) are capable of highly selective condensation of EtOH to n-BuOH at relatively mild conditions.[45] Homogeneous metal catalysts are difficult to separate from the products after catalysis and homogeneous bases are often corrosive to standard reactor materials at high temperatures.[30] In contrast, heterogeneous catalysts are easily separated from the products. Furthermore, metal doped PMOs are some of the most active heterogeneous catalysts for Guerbet condensation.[58]

Ethyl acetate (EtOAc) and n-butanol (n-BuOH) are the major products of EtOH condensation catalyzed by PMOs.[59] Sun and coworkers found that Cu_{20}PMO was equally selective for EtOAc and n-BuOH. Small quantities of chloride (ppm) enhances the selectivity of EtOH condensation with Cu_{20}PMO toward n-BuOH (Fig. 10).[61] We initially hypothesized that chloride facilitated Cu atom sintering to Cu(0) nanoparticles decreases overall catalyst basicity by replacing Cu—O sites with Cu—Cu sites. These metal-oxide pairs likely act as Brønsted Acid-Base sites for EtOH chemisorption and deprotonation to ethoxide.[87] Jones previously showed that sintering results in more Cu—Cu pairs and less Cu—O pairs using Cu K-edge extended X-ray absorption fine structure of Cu_{20}PMO heated in sc-MeOH.[91] Thus, decreased catalyst basicity due to chloride poisoning of Cu_{20}PMO would lead to decreased EtOH dehydrogenation but increased aldol condensation.

Ethanol conversion was 43% for the reaction of 100% EtOH with Cu_{20}PMO at 320 °C for 6 h.[e] When a substrate solution containing 0.3 mg $MgCl_2 \cdot 6H_2O$ per 1 mL of EtOH was used the conversion increased to 59% under otherwise identical conditions (Fig. 10). When no chloride was present, the primary products were 16% EtOAc (45% selectivity) and 10% n-BuOH (29% selectivity) while secondary products identified were butyl acetate (2%) and ethyl butanoate (3%), results similar to what Sun and coworkers reported.[60] The n-butanol yield was significantly greater (28%) when 140 ppm Cl^- was present in the EtOH substrate (53% selectivity). Furthermore, ethyl acetate yield was only 5% (8% selectivity) and the most abundant secondary product was 1-hexanol. In this case, the space time yield of n-butanol was 748 $g_{n\text{-BuOH}}\ kg_{cat}^{-1}\ h^{-1}$, which is similar to Cu_xNi_y-PMO.[60] A single exposure of Cu_{20}PMO to 72–297 ppm Cl^- increased overall EtOH conversion and n-BuOH selectivity, however above 464 ppm Cl^-, overall activity decreased (Fig. 10). Ethanol conversion was ~60% and n-butanol selectivity was ~55% for the reactions with 72–297 ppm Cl^-. The largest minor product was hexanol (~15% selectivity) highest. The reactions with 464 and 630 ppm Cl^- resulted in decreased conversion but similarly high n-BuOH selectivity (~51%).

When CCl_4 was added to EtOH to give similar Cl atom concentration (150 ppm Cl), EtOH conversion and n-BuOH selectivity increased, similar to observations with added $MgCl_2 \cdot 6H_2O$. It is likely that Cu_{20}PMO

---

[e] In a similar fashion to the lignin reactions, these experiments were performed in Swagelok stainless steel mini reactors containing 3 mL of EtOH and 25–100 mg of Cu_{20}PMO. Stock solutions of ethanol containing various amounts of chloride were used when noted that a reaction had "chloride added".

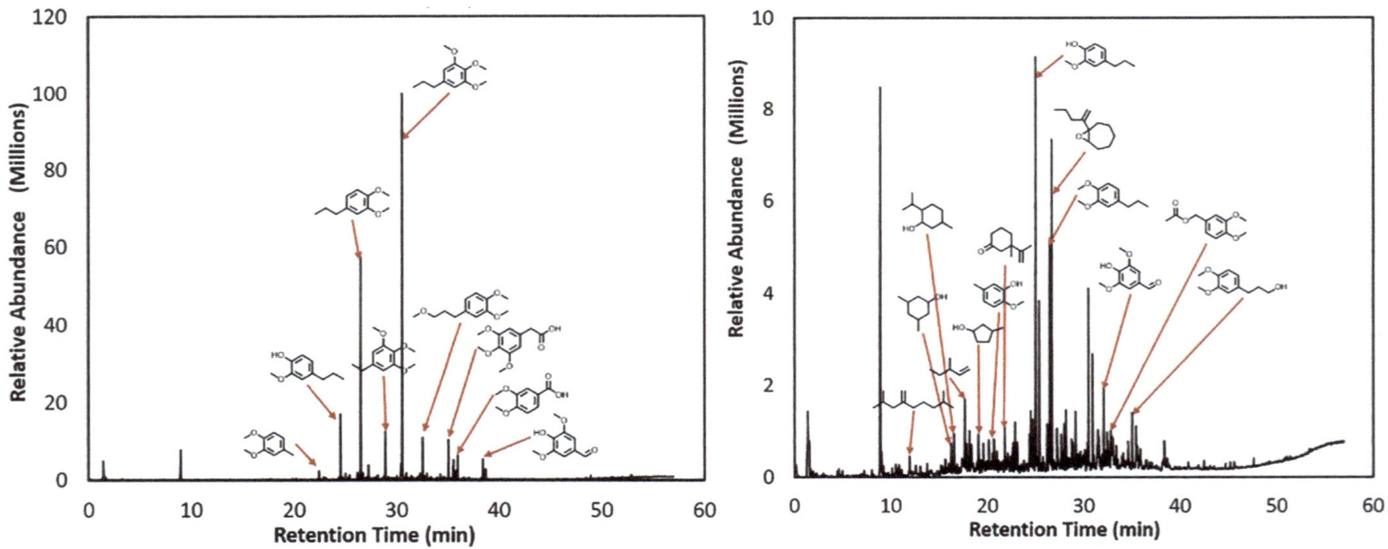

Fig. 10 *Top*: Proposed reaction scheme for the formation of n-BuOH and EtOAc from EtOH condensation over PMOs. *Left*: Ethanol conversion (red rhombus) and product yield after EtOH reaction with $Cu_{20}PMO$ at 320 °C for 6 h. Catalyst loading was 5 wt% (100 mg). "Cl^- Added" = ~140 ppm Cl^- EtOH (added as $MgCl_2 \cdot 6H_2O$). *Right*: Ethanol conversion (red rhombus), and product distribution as a function of added chloride (ppm Cl^-) "Other" products are primarily C6+ alcohols and alkyl esters. A full list of can be found in Ref. 61. *Adapted with permission from Barrett, J. A.; Jones, Z. R.; Stickelmaier, C.; Schopp, N.; Ford, P. C.* ACS Sustainable Chem. Eng. ***2018****, 6, 15119–15126. Copyright 2018 American Chemical Society.*

catalyzes hydrodechlorination as this reaction has been previous studied on copper catalysts.[95] We wanted to know if the chloride was permanently changing the catalyst or simply acting as a co-catalyst. Recycle experiments of the Cu_{20}PMO were conducted where the first reaction was in 140 ppm Cl^- EtOH but the subsequent reactions were with EtOH that contained no chloride (Fig. 11). Ethanol conversion, EtOAc selectivity, n-BuOH selectivity, and 1-hexanol selectivity for 18 recycles are shown in Fig. 11. The reactivity is relatively consistent with EtOH conversion between 41% and 59%, n-BuOH yield between 18% and 26%, HexOH yield between 3% and 6%, and EtOAc yield between 1% and 3%. There are some unexplained fluctuations, but this result indicated that chloride from the first reaction permanently changed the catalyst. When chloride was included in each subsequent reaction activity continually decreased. EtOH conversion was 40–50% and product selectivity was consistent with the results in Fig. 10

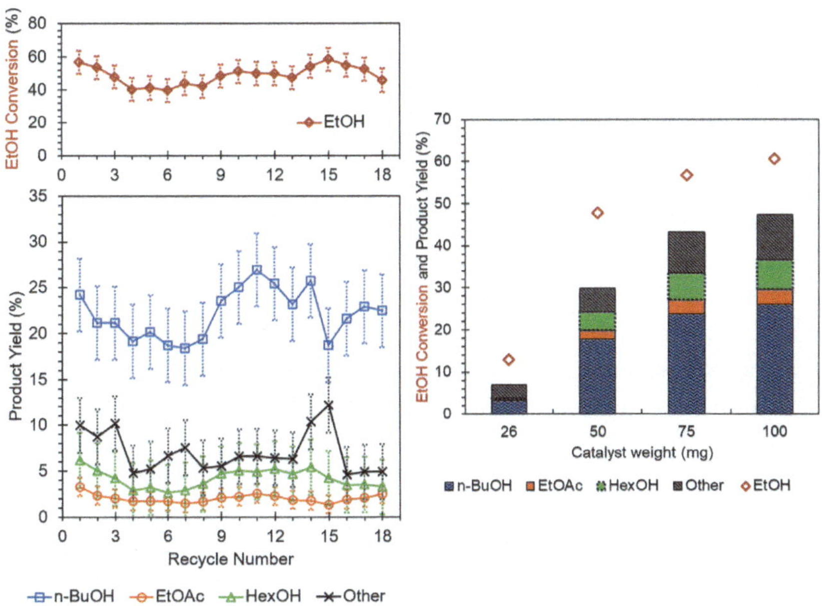

Fig. 11 *Left*: Product yield and ethanol conversion (red) versus recycle number. For the first reaction, the substrate was 2.5 mL of 140 ppm Cl^- EtOH and subsequent reactions contain 100% EtOH with no added chloride. Markers = mean of parallel duplicate or triplicate experiments. Error bars = one standard deviation. Other products found in Ref. 61. *Right*: Product yield and ethanol conversion (%) as a function of catalyst loading. Reaction conditions = 6 h at 320 °C with Cu_{20}PMO. *Adapted with permission from Barrett, J. A.; Jones, Z. R.; Stickelmaier, C.; Schopp, N.; Ford, P. C.* ACS Sustainable Chem. Eng. ***2018****, 6, 15119–15126. Copyright 2018 American Chemical Society.*

(~15% EtOAc yield and ~10% n-BuOH yield) for sequential reactions of 100% EtOH without chloride with Cu_{20}PMO.

Changes in catalyst loading (26–100 mg) resulted in only modest changes in product selectivity (Fig. 11). However, EtOH conversion was much lower at 26 mg loading compared to 100 mg. The highest n-BuOH selectivity observed was 59% for 140 ppm Cl^- EtOH with 50 mg of Cu_{20}PMO. Catalyst loadings of 75 and 100 mg gave 55%. As the catalyst loading increased, yield of C6 + products increased (i.e., butyl acetate and ethyl butanoate). The lower catalyst loading has a higher Cu:Cl ratio that likely decreased catalyst activity in a similar fashion to high ppm Cl^- shown in Fig. 10. The limited change in product distribution and conversion between the 6 and 8 h reactions at 300 °C indicated that the theoretical maximum yield of n-BuOH under these conditions is ~30%.[87] One potential strategy to improve n-BuOH selectivity over chloride treated Cu_{20}PMO would be to employ flow reactors. The decreased substrate-catalyst contact time in a flow reactor could also reduce the proliferation to larger alcohols from the desired product (i.e., butanol). The recycles shown in Fig. 11 indicate that chloride treated Cu_{20}PMO is stable enough to be kept in a catalyst bed for up to 108 h with relatively little change to activity.

As previously stated, our initial hypothesis was that the added chloride results in more Cu—Cu pairs and less Cu—O pairs due to sintering, and this would lead to decreased catalyst basicity, decreased EtOH dehydrogenation, and increased aldol condensation. Basicity has been previously researched for PMOs for the Guerbet condensation of EtOH.[56] The initial reaction steps for ethanol over PMOs are dehydrogenation to form acetaldehyde or deprotonation to form ethoxide (Fig. 10). The formation of n-BuOH depends on the rate of deprotonation of acetaldehyde to form the corresponding enolate that condenses with another acetaldehyde to form 3-hydroxybutanal (Fig. 10). The formation of EtOAc depends on the rate of deprotonation of EtOH to ethoxide and the condensation of ethoxide with acetaldehyde to form the hemiacetal (Fig. 10). We hypothesized that the chloride induced structural changes of Cu_{20}PMO change the catalyst basicity. Notably, acetaldehyde is several orders of magnitude more acidic than is ethanol,[96] thus changes in catalyst basicity should directly impact reactivity. The Hammett indicator method is a simple way to evaluate solid basicity.[97] The Hammett function (H_-) describes the ability of a solid to ionize a weak acid indicator (like neutral red ($pK_a = 6.8$), bromothymol blue ($pK_a = 7.2$), phenolphthalein ($pK_a = 9.3$), Nile blue ($pK_a = 10.1$), and 2,4-dinitroaniline ($pK_a = 15$).). The H_- of the solid is qualitatively determined to be between

Table 2 Hammett function (H_-) of calcined $Cu_{20}PMO$ catalysts before and after reaction with EtOH for 6 h at 320 °C.

Sample conditions	Base strength	Total basicity (mmol/g)
Calcined 600 °C	$10.1 < H_- < 15$	1.07 ± 0.10
No chloride	$10.1 < H_- < 15$	1.34 ± 0.26
140 ppm Cl^-	$9.3 < H_- < 10.1$	1.90 ± 0.20

Reproduced with permission from Bernt, C. M.; Manesewan, H.; Chui, M; Boscolo, M.; Ford, P.C. *ACS Sustainable Chem. Eng.* **2018**, *6*, 15119–15126.

the pK_a of an indicator it can ionize and the highest pK_a indicator it cannot ionize. The H_- determined for calcined $Cu_{20}PMO$ and for catalyst recovered from 6-h reactions with EtOH at 320 °C are summarized in Table 2. Notably, $Cu_{20}PMO$ recovered after reaction with 140 ppm Cl^- displays the lowest H_- (9.3–10.1), but a higher total basicity (determined by titration with benzoic acid). Overall, the strength of the base sites decreased but the concentration of base sites increased. This may result in a decreased ability to deprotonate ethanol to form ethoxide but relatively equivalent or increased ability to deprotonate acetaldehyde to form the enolate. Therefore, chloride treated $Cu_{20}PMO$ would have increased selectivity for the reaction leading to n-BuOH over EtOAc.

5. $Cu_{20}PMO$ structure and activity changes

Porous metal oxide catalysts are typically prepared from the corresponding synthetic layered double hydroxides. Hydrotalcite is specifically used to refer to layered double hydroxides that have a 3 to 1 Mg:Al composition. Synthetic hydrotalcites can often substitute divalent or trivalent metal anions in the lattice and retain the hydrotalcite structure. The theoretical molecular formula for the $Cu_{20}PMO$ synthetic precursor ($Cu_{20}HTC$) is $Cu_{1.2}Mg_{4.8}Al_2CO_3(OH)_{16}{\cdot}4(H_2O)$. The overall hydrotalcite ($Mg_6Al_2CO_3(OH)_{16}{\cdot}4(H_2O)$) structure with less crystallinity as evidenced by powder XRD (Fig. 12). The observed mass ratios of Cu, Mg, and Al are also measured by inductively coupled plasma-optical emission spectroscopy (ICP-OES) and are consistent expected ratio of Cu:Mg:Al as 0.6:2.5:1.0.[60,61,64,66,76–78,88,90] Surface area of the calcined $Cu_{20}HTC$ ($Cu_{20}PMO$) is typically between 100 and 300 m^2/g as determined by Brunauer–Emmett–Teller (BET) method.[87,91,99] The XRD pattern of $Cu_{20}PMO$ shows broad features which index to MgO.[91,98] Typically, the

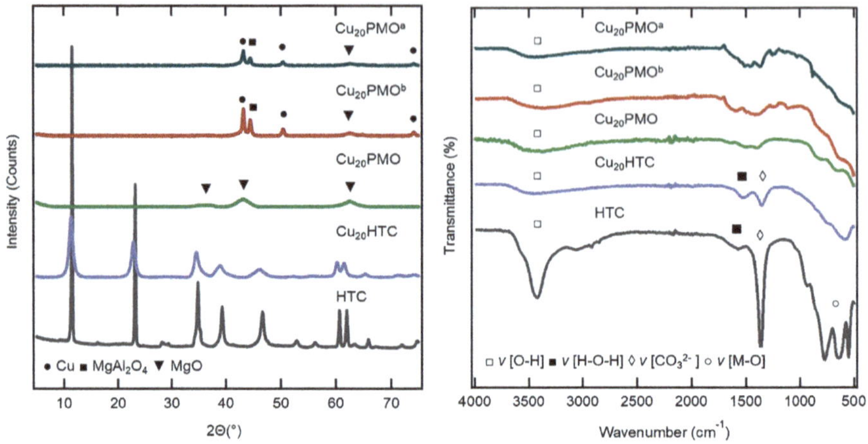

Fig. 12 XRD patterns (*left*) and ATR-IR (*right*) for HTC and PMO materials. [a]MeOH/DMC reaction and [b]sc-MeOH reaction for 6 h at 300 °C. HTC = hydrotalcite. XRD database indexes found in Refs. 61, 98. *Adapted with permission from Barrett, J. A. Part 1: Investigations of Reaction Pathways for Catalytic Biomass Conversion, Doctoral Dissertation, University of California-Santa Barbara, 2019. Copyright 2019 Jacob Alexander Barrett.*

Cu, Mg, and Al content for $Cu_{20}PMO$ catalysts reactions in sc-MeOH or MeOH/DMC are unchanged.[77,78] Reactions with EtOH or acidic MeOH often lead to a leaching of Mg and Al cations into the solution.[88,99] The XRD patterns of $Cu_{20}PMO$ after reaction in sc-MeOH or MeOH/DMC exhibit features that index to metallic Cu(0) (Fig. 12). The reflection at 44° 2Θ is consistent with a Mg—Al spinel phase; however, the low crystallinity of the sample prevents definitive assignment.[91,98]

The ATR-IR spectra of the HTC, $Cu_{20}HTC$, and $Cu_{20}PMO$ are also shown in Fig. 12. The large peak at 1450 cm^{-1} is consistent with the antisymmetric stretching mode of carbonate ($CO_3{}^{2-}$).[100–102] The smaller peak at 1550 cm^{-1} is often assigned to interlayer water molecules but there is evidence this peak is bicarbonate ($HCO_3{}^{-}$) due to the high degree of hydrogen bonding in the interlayer region of HTC.[101] The interlayer carbonate and water are removed by calcination (heating above 460 °C) to form PMOs. These materials absorb CO_2 from air, so a small peak at 1450 cm^{-1} is still apparent.[101,103] Small and broad stretching frequencies between 1400 and 1600 cm^{-1} appear after reaction in sc-MeOH or MeOH/DMC. These peaks are not well enough defined to assign but may represent organics adsorbates with C=C and C—H bonds.

Scanning electron microscopy (SEM) images provide further evidence that spherical Cu particles (around 10 nm) are formed after reactions in supercritical alcohols or under flowing hydrogen.[87,91,99] In short, SEM images and XRD patterns indicate the loss of characteristic HTC reflections after calcination and appearance of Cu metal reflections which IR shows near quantitative loss of interlayer ions. These Cu metal nanoparticles are the proposed active sites for hydrogenolysis, dehydrogenation, and hydrogenation.[91] The Cu_{20}PMO surface likely reaches hydrogen surface coverage saturation after 15 min at or above the supercritical point of MeOH (240 °C and 81 bar).[76] We hypothesize that Cu(0) nanoparticles (<1 nm) start forming during the first 15 min of the reaction. The formation of larger copper nanoparticles (30 nm) was previously observed by transmission electron microscopy (TEM) for the reactions of carbohydrate bio-oils with Cu_{20}PMO in supercritical alcohols[99] and of ethanol condensation catalyzed by Cu_xNi_yPMO.[60]

A proposed mechanism for the aryl-ether cleavage of PPE with DMC is shown in Scheme 6. The initial step is adsorption onto the Cu nanoparticle (Scheme 6A). This is straightforward for model compounds but lignin would need to be solubilized to interact with any part of the Cu_{20}PMO surface.[83] Step B shows hydrogen transfer from surface Cu-hydrides to break the aryl-ether C—O bond. The acidic phenol is likely stabilized via

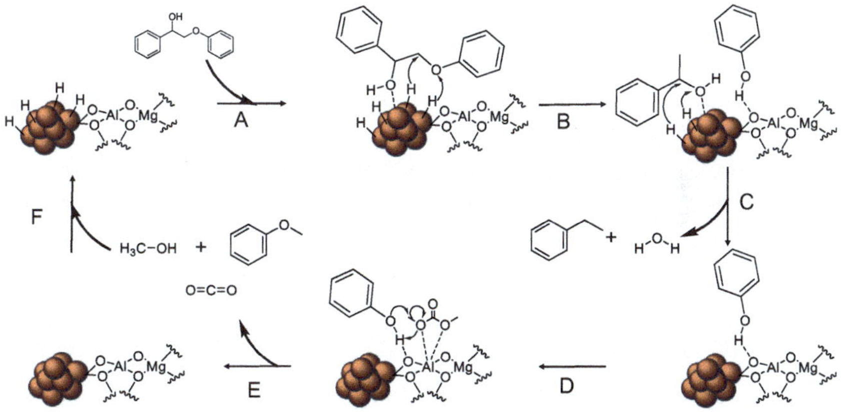

Scheme 6 Proposed mechanism of the reaction of aryl-ether (PPE) in MeOH/DMC with Cu(0) nanoparticle and Lewis acid-base pair active sites. Chemisorption (A), hydrogenolysis (B), hydrodeoxygenation (C), DMC adsorption (D), *O*-methylation (E), and methanol reforming (F) occur over the Cu nanoparticle (spheres) distributed on the metal-oxide pairs. Note: some O atoms are omitted for clarity. *Reproduced with permission from Barrett, J. A. Part 1: Investigations of Reaction Pathways for Catalytic Biomass Conversion, Doctoral Dissertation, University of California-Santa Barbara, 2019. Copyright 2019 Jacob Alexander Barrett.*

hydrogen bonding to a surface oxygen atom. Methyl transfer from DMC to the phenolic O atom is facilitated by an adjacent Lewis base (the O atom bound to Al^{3+}).[104] The products (anisole and CO_2) can desorb from the catalyst surface and the catalyst is regenerated by methanol reforming (Scheme 6F).

The basicity of surface O atoms as well as the exposure of the Cu atoms likely plays a vital role in the extent and specificity of Cu_{20}PMO activity. Modification of the catalyst by acid-treatment, Lewis acid dopants, and selective poisoning should be accompanied by surface characterization techniques, i.e., XRD, TPD, BET, IR, and Hammett indicator titrations. This would enable a standardized set of comparable factors to evaluate new methods for tuning the reactivity of PMOs.

The ATR-infrared spectrum of Cu_{20}PMO shows only broad weak peaks at ~3400 and ~1500 cm^{-1} associated with water and atmospheric CO_2 adsorption, respectively (Fig. 13). All of the IR spectra of Cu_{20}PMO after 6 h reactions with chloride-containing EtOH at 320 °C have small sharp peaks at 2950, 2900 and 2850 cm^{-1} consistent with C—H stretches of organic molecules that remain on the catalyst and an increased broad peak at

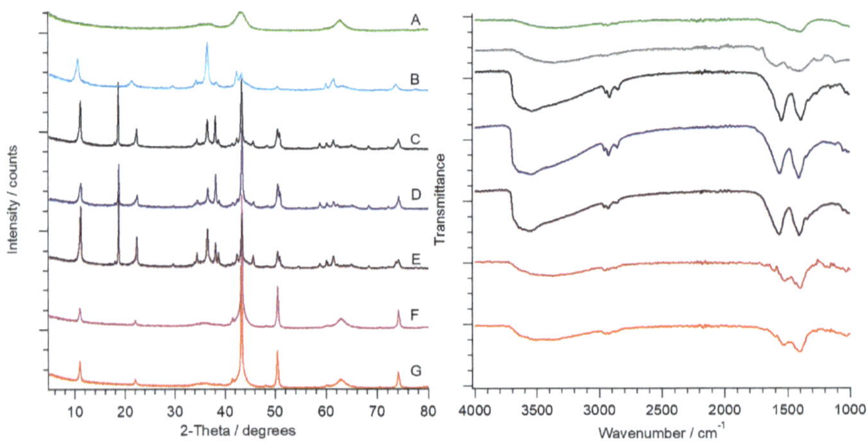

Fig. 13 XRD patterns (*left*) and ATR-IR (*right*) of Cu_{20}PMO after calcination at 600 °C (A, green) and reaction with EtOH for 6 h at 320 °C with (B) no chloride added, (C) 72 ppm Cl^-, (D) 140 ppm Cl^-, (E) 297 ppm Cl^-, (F) 464 ppm Cl^- and (G) 630 ppm Cl^-. Data analysis was done using Malvern Panalytical HighScore Software with two structural databases; Powder Diffraction Files and Inorganic Crystal Structure Database. Database indexes can also be found in Refs. 61, 91, 98. *Reproduced with permission from Barrett, J. A. Part 1: Investigations of Reaction Pathways for Catalytic Biomass Conversion, Doctoral Dissertation, University of California-Santa Barbara, 2019. Copyright 2019 Jacob Alexander Barrett.*

~3600 cm^{-1} which is likely from water. There are also two new peaks at ~1575 and ~1411 cm^{-1}. Interlayer water (or bicarbonate) in hydrotalcites is typically observed at ~1640 cm^{-1}.[100,105] Interlayer carbonate appears at 1365/ 1400 cm^{-1} in hydrotalcites.[106] One possibility is acetate, which appears at 1554 and 1428 cm^{-1} on a ZnO surface.[107] Acetate becoming trapped on the catalyst might explain mass balance losses. The most plausible explanation with the given data is that the peaks correspond to bicarbonate, since the reaction with ethanol produces water, the chloride is added as a hydrate salt, and PMOs are known to adsorb CO_2 from the atmosphere.

We previously observed reactivity changes for Cu_{20}PMO toward BPE and EtOH when ppm quantities of chloride were introduced in tandem to the substrate. These chloride-induced structural changes were examined via X-ray diffraction (XRD) patterns of the post-reaction Cu_{20}PMO (Fig. 13). The XRD pattern for Cu_{20}PMO (trace A, Fig. 13) shows only broad Bragg reflections consistent with some MgO domains (at 36°, 43°, and 63° 2Θ).[98] Bragg reflections consistent with the reformation of a HTC-like structure (11°, 22°, 35°, 39°, 60°, and 61°) appear after 6 h at 320 °C with EtOH (trace B). We hypothesize that H_2O formed as a byproduct of the ethanol condensation and the Cl^- contribute to reformation of a layered phase. Alternatively, the reflections at 35° and 39° may correspond to CuO (cupric oxide). Bragg reflections at 36°, 42°, 51°, 62° and 73° correspond to Cu_2O (cuprous oxide); Many of these reflections overlap with MgO domains and thus assignment of the reacted catalyst to a specific phase or oxidation state is not possible using only XRD.

The Cu_{20}PMO reacted with 72–297 ppm Cl^- in EtOH (trace C-E) contained reflection that index to brucite ($Mg(OH)_2$, 19° and 38°). These reflections are also characteristic of Mg/Al hydrotalcite, however it is unlikely that the parent HTC is reformed. If these reflections are brucite then gibbsite, $Al(OH)_3$ and/or Al(O)OH, are also possibly formed, but not all the Bragg reflections for Al-containing phases are observed. Cupric oxide (35° and 39°), cuprous oxide (36°, 42°, 51°, 62° and 73°) and copper metal (44°, 51°, and 75°) reflections are all observed. These Cu(0) Bragg reflections are consistent with the analyses for Cu_{20}PMO treated with magnesium chloride in sc-MeOH described earlier (Fig. 6). The XRD patterns for Cu_{20}PMO after reactions with 464 or 630 ppm Cl EtOH contained significantly larger Cu(0) reflections (traces F & G) compared to 72–297 ppm Cl^-. Increased chloride concentration lead to increased Cu sintering.

The Scherer equation (Particle Size $= 0.9\lambda/(L \cos \Theta)$) can be used to estimate crystalline domain sizes using the full width at half maximum

(FWHM) of the Bragg reflection. We used the equation to estimate the Cu(0) domain sizes using the Bragg reflection at 44° 2Θ.[108] In the Scherer equation, the shape factor is 0.9, the X-ray instrument wavelength λ has Cu Kα =1.5405 Å), the FWHM of the reflection is *L*, and Θ is the Bragg angle. The Panalytical HighScore software package, X'PERT, was used to determine the particle sizes in the same way for each pattern. Ideally, the particle sizes are estimated from every indexed reflection for the crystallite; however, Bragg reflection overlap often makes this impractical. The Cu(0) particle size for the patterns shown in Fig. 13 were 24 nm for $Cu_{20}PMO$ reacted with 72 ppm Cl^- EtOH and 25 nm for $Cu_{20}PMO$ reacted with 140 and 297 ppm Cl^- EtOH. The higher chloride content reactions (464 or 630 ppm Cl^-) resulted in a larger Cu(0) crystallite size (29 nm). The Bragg reflection at 44° for $Cu_{20}PMO$ recovered after reaction with EtOH (no chloride) was too small to extract a meaningful estimation. The XRD pattern of $Cu_{20}PMO$ after a single reaction with 140 ppm Cl^- in EtOH was remarkably similar to the XRD pattern of $Cu_{20}PMO$ from the recyclability experiments. Two differences are the estimated Cu(0) particle size (27 nm) and the absence of the Bragg reflections for a HTC-like structure (11°, 22°, and 35°) in the $Cu_{20}PMO$ after 18 consecutive reactions. Overall, examination of the structural changes in $Cu_{20}PMO$ enable us to rationalize reactivity changes seen with aryl-ether model compounds, alcohols, and ultimately lignin.

6. Concluding remarks and future perspectives

Lignocellulosic biomass and bio-derived alcohols are increasingly finding industrial uses in materials.[109–114] Apart from vanillin production there is very little use of lignin as an aromatic commodity chemical feedstock, which is an incredibly large missed opportunity. Nature has concentrated a resource and we have not been able to fully tap into its usefulness. The sustainable chemistry research with dimethyl carbonate[115] and other green alkylating agents provides a potential tool for generating value-added chemicals from biomass. We have specifically shown that dimethyl carbonate can be used to enhance the aromatic product yield from reductive lignin disassembly. The resulting alkylmethoxybenzenes might serve as potential synthetic precursors. Optimization of these green methodologies will further enhance the economic sustainability of future biorefineries.

Cu-doped porous metal oxides are robust to polar high temperature environments. Furthermore, modification of these catalysts via selective halogen poisoning can be used to tune the product stream for lignin model compounds and bio-derived alcohols (i.e., ethanol). However, the effects of large amounts of chloride present an important issue for catalytic biomass conversion using metal catalysts. While there appears to be an amount of chloride (10–100 ppm) which can beneficially decrease aromatic hydrogenation, the chloride amounts found in lignocellulosic biomass much greater (100–1000 ppm).[71] A sample of the OPL used in these studies was analyzed and contained 998 mg chloride per kg of OPL (~0.1%). The chloride concentrations in organosolv poplar lignin are likely influenced by the HCl extraction used. The small inorganic component of biomass may play a larger role in irreversible catalyst deactivation than high temperature and pressure. Nonetheless, limited chloride poisoning proved to be useful for ethanol condensation and may also be beneficial for increased selectivity in other organic transformations such as the condensation and alkylation of furfurals.

Catalytic conversion of biorefinery-derived organosolv lignin via (selective reductive disassembly) and of ethanol (via Guerbet condensation) enables mankind to utilize renewable feedstocks in place of non-renewable fossil carbon.[116] This account shows that chloride can tune the selectivity of transition metal PMOs for these types of biomass reactions. The enhancement in selectivity and activity for the Guerbet condensation was an important step toward producing renewable n-butanol, which is both a valuable synthetic precursor and fuel additive. The drastically increased production of alkylmethoxybenzenes from organosolv lignin was an impressive advance for using lignin as a renewable aromatic chemical feedstock. The efficient production of these chemicals from biomass and biomass-derived ethanol could enhance the economic viability and sustainability of bio-refineries.

Acknowledgments

The author acknowledges the University of San Diego Chancellor's Postdoctoral Research Fellowship for its financial and professional support. The author also acknowledges the NSF Graduate Research Fellowship program for funding his graduate studies while at UC Santa Barbara. Studies at UC Santa Barbara were supported by the Center for the Sustainable Use of Renewable Feedstocks (CenSURF), a NSF Center for Chemical Innovation (NSF CHE-1240194).

References

1. Langholtz M. H.; Stokes B. J.; Eaton L. M., U.S. Department of Energy, 2016. 2016 Billion-Ton Report, https://doi.org/10.2172/1271651.
2. Harris, E. E.; D'lanni, J.; Adkins, H. *J. Am. Chem. Soc.* **1938**, *60* (6), 1467–1470.
3. Wittcoff, H. A.; Reuben, B. G.; Plotkin, J. S; Chapter 4, *Industrial Organic Chemicals*, John Wiley & Sons, Inc: New York, 2013; pp. 93–138.

4. Pu, Y. Q.; Zhang, D. C.; Singh, P. M.; Ragauskas, A. J. *Biofuels Bioprod. Biorefin.* **2008**, *2* (1), 58–73.
5. Sun, Z.; De Santi, A.; Elangovan, S.; Barta, K. *Chem. Rev.* **2018**, *118*, 614–618.
6. Tuck, C. O.; Perez, E.; Horvath, I. T.; Sheldon, R. A.; Poliakoff, M. *Science* **2012**, *337* (6095), 695–699.
7. Ragauskas, A. J.; Williams, C. K.; Davison, B. H.; Britovsek, G.; Cairney, J.; Eckert, C. A.; Frederick, W. J., Jr.; Hallett, J. P.; Leak, D. J.; Liotta, C. L.; Mielenz, J. R.; Murphy, R.; Templer, R.; Tschaplinski, T. *Science* **2006**, *311* (5760), 484–489.
8. Sannigrahi, P.; Ragauskas, A. J.; Tuskan, G. A. *Biofuels Bioprod. Biorefin.* **2010**, *4* (2), 209–226.
9. Ragauskas, A. J.; Beckham, G. T.; Biddy, M. J.; Chandra, R.; Chen, F.; Davis, M. F.; Davison, B. H.; Dixon, R. A.; Gilna, P.; Keller, M.; Langan, P.; Naskar, A. K.; Saddler, J. N.; Tschaplinski, T. J.; Tuskan, G. A.; Wyman, C. E. *Science* **2014**, *344*, 1246843-1–1246843-10.
10. Rinaldi, R.; Jastrzebski, R.; Clough, M. T.; Ralph, J.; Kennema, M.; Bruijnincx, P. C. A.; Weckhuysen, B. M. *Angew. Chem. Int. Ed.* **2016**, *55* (29), 8164–8215.
11. Zakzeski, J.; Bruijnincx, P. C. A.; Jongerius, A. L.; Weckhuysen, B. M. *Chem. Rev.* **2010**, *110*, 3552–3599.
12. Zakzeski, J.; Jongerius, A. L.; Bruijnincx, P. C. A.; Weckhuysen, B. M. *ChemSusChem* **2012**, *5*, 1602–1609.
13. Wang, X.; Rinaldi, R. *Angew. Chem. Int. Ed.* **2013**, *52*, 11499–11503.
14. Deuss, P. J.; Barta, K. *Coord. Chem. Rev.* **2015**, *306*, 510–532.
15. Okuda, K.; Ohara, S.; Umetsu, M.; Takami, S.; Adschiri, T. *Bioresour. Technol.* **2008**, *99* (6), 1846–1852.
16. Zaheer, M.; Kempe, R. *ACS Catal.* **2015**, *5*, 1675–1684.
17. Cheng, C.; Shen, D.; Gu, S.; Luo, K. H. *Catal. Sci. Technol.* **2018**, *8*, 6275–6296.
18. Sinfelt, J. H. *Catal. Lett.* **1991**, *9*, 159–172.
19. Kim, S.; Chmely, S. C.; Nimlos, M. R.; Bomble, Y. J.; Foust, T. D.; Paton, R. S.; Beckham, G. T. *J. Phys. Chem. Lett.* **2011**, *2*, 2846–2852.
20. U.S. Environmental Protection Agency. *Fed. Regist.* **2011**, *76* (17), 4662–4683.
21. U.S. Environmental Protection Agency. *Fed. Regist.* **2017**, *82* (237), 58486–58527.
22. BP. *BP Statistical Review of World Energy 2017*; British Petroleum, 2017.
23. U. S. Energy Information Administration. *October Monthly Energy Review*; U.S. EIA, 2020.
24. Agarwal, A. K. *Prog. Energy Combust. Sci.* **2007**, *33* (3), 233–271.
25. Zapata-Mina, J.; Restrepo, A.; Romero, C.; Quintero, H. *Sustainable Energy Technol. Assess.* **2020**, *42*, 100803.
26. Alleman, T. L.; McCormick, R. L.; Yanowitz, J. *Energy Fuels* **2015**, *29* (8), 5095–5102.
27. Yang, Y.; Bae, J.; Kim, J.; Suh, S. *Environ. Sci. Technol.* **2012**, *46* (7), 3671–3678.
28. Jin, C.; Yao, M.; Liu, H.; Lee, C. F. F.; Ji, J. *Renew. Sustain. Energy Rev.* **2011**, *15* (8), 4080–4106.
29. U.S. GAO. *RFS Program Unlikely to Meet Its Targets for Reducing Greenhouse Gas Emissions, GAO-17-94*; Washington, D.C, 2016.
30. Gabriëls, D.; Hernández, W. Y.; Sels, B.; Van Der Voort, P.; Verberckmoes, A. *Catal. Sci. Technol.* **2015**, *5*, 3876–3902.
31. Angelici, C.; Weckhuysen, B. M.; Bruijnincx, P. C. A. *ChemSusChem* **2013**, *6* (9), 1595–1614.
32. Falbe, J.; Bahrman, H.; Lipps, W.; Mayer, D.; Frey, G. D. *Ullmann's Encyclopedia of Industrial Chemistry*; Wiley-VCH Verlag GmbH & Co. KGaA, 2013.
33. Galadima, A.; Muraza, O. *Ind. Eng. Chem. Res.* **2015**, *54* (29), 7181–7194.
34. Kozlowski, J. T.; Davis, R. J. *ACS Catal.* **2013**, *3* (7), 1588–1600.
35. Veibel, S.; Nielsen, J. I. *Tetrahedron* **1967**, *23* (4), 1723–1733.

36. Riemenschneider, W.; Bolt, H. M. *Ullmann's Encyclopedia of Industrial Chemistry*; Wiley-VCH Verlag GmbH & Co. KGaA, 2012.
37. Hahn, H.-D.; Dambkes, G.; Rupprich, N.; Bahl, H.; Frey, G. D. *Ullmann's Encyclopedia of Industrial Chemistry*; Wiley-VCH Verlag GmbH & Co. KGaA, 2013.
38. Panten, J.; Surburg, H. *Ullmann's Encyclopedia of Industrial Chemistry*; Wiley-VCH Verlag GmbH & Co. KGaA, 2015.
39. Xue, C.; Zhao, J.; Chen, L.; Yang, S.-T.; Bai, F. *Biotechnol. Adv.* **2017**, *35* (2), 310–322.
40. Li, Y.; Tang, W.; Chen, Y.; Liu, J.; Lee, C.-F. F. *Fuel* **2019**, *242*, 673–686.
41. Bahrmann, H.; Hahn, H. D.; Mayer, D.; Frey, G. D. *Ullmann's Encyclopedia of Industrial Chemistry*; Wiley-VCH Verlag GmbH & Co. KGaA, 2013.
42. Di Cosimo, J. I.; Dı´ez, V. K.; Xu, M.; Iglesia, E.; Apesteguı´a, C. R. *J. Catal.* **1998**, *178* (2), 499–510.
43. Leon, M.; Diaz, E.; Ordonez, S. *Catal. Today* **2011**, *164* (1), 436–442.
44. de Haan, A. B.; Bosch, H. *Industrial Separation Processes. Fundamentals*; De Gruyter: Berlin, Boston, 2013.
45. Aitchison, H.; Wingad, R. L.; Wass, D. F. *ACS Catal.* **2016**, *6* (10), 7125–7132.
46. Linstrom, P. J.; Mallard, W. G., Eds. *NIST Chemistry WebBook, NIST Standard Reference Database Number 69*; National Institute of Standards and Technology: Gaithersburg MD, 2017; p. 20899.
47. Elfasakhany, A. *Renew. Sustain. Energy Rev.* **2017**, *71*, 404–413.
48. Olarte, M. V.; Albrecht, K. O.; Bays, J. T.; Polikarpov, E.; Maddi, B.; Linehan, J. C.; Hagan, M. J. O.; Gaspar, D. J. *Fuel* **2019**, *238*, 493–506.
49. Fu, S.; Shao, Z.; Wang, Y.; Liu, Q. *J. Am. Chem. Soc.* **2017**, *139*, 11941–11948.
50. King, A. M.; Sparkes, H. A.; Wingad, R. L.; Wass, D. F. *Organometallics* **2020**, *39* (21), 3873–3878.
51. Neumann, C. N.; Rozeveld, S. J.; Yu, M.; Rieth, A. J.; Dincă, M. *J. Am. Chem. Soc.* **2019**, *141* (44), 17477–17481.
52. Guerbet, M. *C. R. Hebd. Séances Acad. Sci.* **1899**, *14*, 1002–1004.
53. Wingad, R. L.; Gates, P. J.; Street, S. T. G.; Wass, D. F. *ACS Catal.* **2015**, *5* (10), 5822–5826.
54. Chakraborty, S.; Piszel, P. E.; Hayes, C. E.; Baker, R. T.; Jones, W. D. *J. Am. Chem. Soc.* **2015**, *137* (45), 14264–14267.
55. Di Cosimo, J. I.; Apesteguıa, C. R.; Ginés, M. J. L.; Iglesia, E. *J. Catal.* **2000**, *190* (2), 261–275.
56. Marcu, I. C.; Tichit, D.; Fajula, F.; Tanchoux, N. *Catal. Today* **2009**, *147* (3–4), 231–238.
57. León, M.; Díaz, E.; Vega, A.; Ordóñez, S.; Auroux, A. *Appl. Catal. B Environ.* **2011**, *102* (3–4), 590–599.
58. Marcu, I. C.; Tanchoux, N.; Fajula, F.; Tichit, D. *Catal. Lett.* **2013**, *143* (1), 23–30.
59. Bravo-Suárez, J. J.; Subramaniam, B.; Chaudhari, R. V. *Appl. Catal. A Gen.* **2013**, *455*, 234–246.
60. Sun, Z.; Couto Vasconcelos, A.; Bottari, G.; Stuart, M. C. A.; Bonura, G.; Cannilla, C.; Frusteri, F.; Barta, K. *ACS Sustain. Chem. Eng.* **2016**, *5* (2), 1738–1746.
61. Barrett, J. A.; Jones, Z. R.; Stickelmaier, C.; Schopp, N.; Ford, P. C. *ACS Sustainable Chem. Eng.* **2018**, *6*, 15119–15126.
62. Matson, T. D.; Barta, K.; Iretskii, A. V.; Ford, P. C. *J. Am. Chem. Soc.* **2011**, *133* (35), 14090–14097.
63. Macala, G. S.; Matson, T. D.; Johnson, C. L.; Lewis, R. S.; Iretskii, A. V.; Ford, P. C. *ChemSusChem* **2009**, *2* (3), 215–217.
64. Barta, K.; Matson, T. D.; Fettig, M. L.; Scott, S. L.; Iretskii, A. V.; Ford, P. C. *Green Chem.* **2010**, *12* (9), 1640.

65. Barta, K.; Ford, P. C. *Acc. Chem. Res.* **2014**, *47*, 1503–1512.
66. Barta, K.; Warner, G. R.; Beach, E. S.; Anastas, P. T. *Green Chem.* **2014**, *16*, 191–196.
67. Huang, X.; Korányi, T. I.; Boot, M. D.; Hensen, E. J. M. *ChemSusChem* **2014**, 7, 2276–2288.
68. Vassilev, S. V.; Baxter, D.; Andersen, L. K.; Vassileva, C. G. *Fuel* **2013**, *105*, 40–76.
69. Heldal, J. A.; Mørk, P. C. *J. Am. Oil Chem. Soc.* **1982**, *59* (9), 396–398.
70. Da Silva Perez, D.; Dupont, C.; Guillemain, A.; Jacob, S.; Labalette, F.; Briand, S.; Marsac, S.; Guerrini, O.; Broust, F.; Commandre, J. M. *Waste Biomass Valoriz.* **2015**, *6* (4), 515–526.
71. Thy, P.; Yu, C.; Jenkins, B. M.; Lesher, C. E. *Energy Fuels* **2013**, *27* (7), 3969–3987.
72. Argyle, M.; Bartholomew, C. *Catalysts* **2015**, *5* (1), 145–269.
73. Huber, G. W.; Corma, A. *Angew. Chem. Int. Ed.* **2007**, *46* (38), 7184–7201.
74. Likun, P. K. W.; Zhang, H.; Vitidsant, T.; Reubroycharoen, P.; Xiao, R. *Energy Fuels* **2017**, *31* (6), 6120–6131.
75. Grant, A.; Ranney, J.; Campbell, C. *Catal. Lett.* **2000**, *65*, 159–168.
76. Bernt, C. M.; Manesewan, H.; Chui, M.; Boscolo, M.; Ford, P. C. *ACS Sustain. Chem. Eng.* **2018**, *6* (2), 2510–2516.
77. Bernt, C. M.; Bottari, G.; Barrett, J. A.; Scott, S. L.; Barta, K. *Catal. Sci. Technol.* **2016**, *6*, 2984–2994.
78. Barrett, J. A.; Gao, Y.; Bernt, C. M.; Chui, M.; Tran, A. T.; Foston, M. B.; Ford, P. C. *ACS Sustainable Chem. Eng.* **2016**, *4*, 6877–6886.
79. Macala, G. S.; Robertson, A. W.; Johnson, C. L.; Day, Z. B.; Lewis, R. S.; White, M. G.; Iretskii, A. V.; Ford, P. C. *Catal. Lett.* **2008**, *122* (3–4), 205–209.
80. Popov, A.; Kondratieva, E.; Goupil, J. M.; Mariey, L.; Bazin, P.; Gilson, J. P.; Travert, A.; Maugé, F. *J. Phys. Chem. C* **2010**, *114*, 15661–15670.
81. Kuzmic, P. *Anal. Biochem.* **1996**, *237*, 260–273.
82. Song, Q.; Wang, F.; Xu, J. *Chem. Commun.* **2012**, *48*, 7019–7021.
83. Song, Q.; Wang, F.; Cai, J.; Wang, Y.; Zhang, J.; Yu, W.; Xu, J. *Energy Environ. Sci.* **2013**, *6*, 994–1007.
84. Pelzer, A. W.; Sturgeon, M. R.; Yanez, A. J.; Chupka, G.; O'Brien, M. H.; Katahira, R.; Cortright, R. D.; Woods, L.; Beckham, G. T.; Broadbelt, L. J. *ACS Sustain. Chem. Eng.* **2015**, *3*, 1339–1347.
85. Talawar, M. B.; Jyothi, T. M.; Raja, T.; Rao, B. S.; Sawant, P. D. *Green Chem.* **2000**, *2*, 266–268.
86. Jyothi, T. M.; Raja, T.; Talawar, M. B.; Sugunan, S.; Rao, B. S.; Energy, H. *Synth. Commun.* **2000**, *30*, 3929–3934.
87. Barrett, J. A; Doctoral Dissertation, *Part 1: Investigations of Reaction Pathways for Catalytic Biomass Conversion*, University of California-Santa Barbara, 2019.
88. Chui, M.; Metzker, G.; Bernt, C. M.; Tran, A. T.; Burtoloso, A. C. B.; Ford, P. C. *ACS Sustain. Chem. Eng.* **2017**, *5* (4), 3158–3169.
89. Ben, H.; Ragauskas, A. J. *Energy Fuels* **2011**, *25*, 2322–2332.
90. Gao, Y.; Walker, M. J.; Barrett, J. A.; Hosseinaei, O.; Harper, D. P.; Ford, P. C.; Williams, B. J.; Foston, M. B. *Green Chem.* **2018**, *20*, 4366–4377.
91. Jones, Z. R; Doctoral Dissertation, *Elucidating Structure-Activity Relationships through X-Ray Absorption Fine Structure Analyses: From Site-Isolated Systems, to Nanoclusters, and to Supported Nanoparticle Catalysts*, University of California-Santa Barbara, 2019.
92. Cheng, C.; Truong, J.; Barrett, J. A.; Shen, D.; Abu-Omar, M. M.; Ford, P. C. *ACS Sustain. Chem. Eng.* **2020**, *8* (2), 1023–1030.
93. Lopez-Olmos, C.; Morales, M. V.; Guerrero-Ruiz, A.; Rodríguez-Ramos, I. *Ind. Eng. Chem. Res.* **2020**, *59* (38), 16626–16636.
94. Nezam, I.; Zak, J.; Miller, D. J. *Ind. Eng. Chem. Res.* **2020**, *59* (31), 13906–13915.
95. Huang, C. C.; Lo, S. L.; Lien, H. L. *Chem. Eng. J.* **2012**, *203*, 95–100.

96. Serjeant, E. P.; Dempsey, B. *IUPAC Chemical Data Series*; Pergamon Press: Oxford; New York, 1979; p. 989.
97. Busca, G. *Chem. Rev.* **2010**, *110* (4), 2217–2249.
98. Downs, R. T.; Hall-Wallace, M. *Am. Mineral.* **2003**, *88*, 247–250.
99. Yin, W.; Venderbosch, R. H.; Bottari, G.; Krawzcyk, K. K.; Barta, K.; Heeres, H. J. *Appl. Catal. B Environ.* **2015**, *166–167*, 56–65.
100. Bitetto, A. D.; Kervern, G.; André, E.; Durand, P.; Carteret, C. *J. Phys. Chem. C* **2017**, *121* (11), 6104–6112.
101. Li, H.; Tran, T.-N.; Lee, B.-J.; Zhang, C.; Park, T.-H.; Yu, J.-S. *ACS Appl. Mater. Interfaces* **2017**, *9* (24), 20294–20298.
102. Sasai, R.; Sato, H.; Sugata, M.; Fujimara, T.; Ishihara, S.; Deguchi, K.; Ohki, S.; Tansho, M.; Shimizu, T.; Oita, N.; Numoto, M.; Fujii, Y.; Kawaguchi, S.; Matsuoka, Y.; Hagura, K.; Abe, T.; Moriyoshi, C. *Inorg. Chem.* **2019**, *58*, 10928–10935.
103. Yavuz, C. T.; Shinall, B. D.; Iretskii, A. V.; White, M. G.; Golden, T.; Atilhan, M.; Ford, P. C.; Stucky, G. D. *Chem. Mater.* **2009**, *21* (15), 3473–3475.
104. Debecker, D. P.; Gaigneaux, E. M.; Busca, G. *Chem. A Eur. J.* **2009**, *15* (16), 3920–3935.
105. Frost, R. L.; Kloprogge, J. T. *Spectrochim. Acta A Mol. Biomol. Spectrosc.* **1999**, *55* (11), 2195–2205.
106. Loring, J. S.; Thompson, C. J.; Zhang, C.; Wang, Z.; Schaef, H. T.; Rosso, K. M. *J. Phys. Chem. A* **2012**, *116* (19), 4768–4777.
107. Kandare, E.; Hossenlopp, J. M. *Inorg. Chem.* **2006**, *45* (9), 3766–3773.
108. Patterson, A. L. *Phys. Rev.* **1939**, *56* (10), 978–982.
109. Sadeghifar, H.; Cui, C.; Argyropoulos, D. S. *Ind. Eng. Chem. Res.* **2012**, *51*, 16713–16720.
110. Zhao, S.; Abu-omar, M. M. *ACS Sustain. Chem. Eng.* **2016**, *4*, 6082–6089.
111. Oroumei, A.; Fox, B.; Naebe, M. *ACS Sustain. Chem. Eng.* **2015**, *3*, 758–769.
112. Mahajan, J. S.; O'Dea, R. M.; Norris, J. B.; Korley, L. T. J.; Epps, T. H., III. *ACS Sustain. Chem. Eng.* **2020**, *8* (40), 15072–15096.
113. Fang, Z.; Nikafshar, S.; Hegg, E. L.; Nejad, M. *ACS Sustain. Chem. Eng.* **2020**, *8* (24), 9095–9103.
114. Feghali, E.; van de Pas, D. J.; Torr, K. M. *Biomacromolecules* **2020**, *21* (4), 1548–1559.
115. Fiorani, G.; Perosa, A.; Selva, M. *Green Chem.* **2018**, *20*, 288–322.
116. Horvath, I. T.; Cséfalvay, E.; Mika, L. *ACS Sustain. Chem. Eng.* **2017**, *5*, 2734–2740.

CHAPTER TEN

Heterogeneous catalytic ethanol transformation into chemicals: Some Brazilian contributions

Gustavo Metzker[a,*], Jorge Andrés Mora Vargas[a,b], and Mauricio Boscolo[a,*]

[a]Sao Paulo State University (UNESP), Chemistry and Environmental Sciences Department (IBILCE), Sao Jose do Rio Preto, SP, Brazil

[b]Universidad Santiago de Cali (USC), Facultad de Ciencias Basicas, Santiago de Cali, Colombia

[*]Corresponding authors: e-mail address: g.metzker@unesp.br; mauricio.boscolo@unesp.br

Contents

Abstract

Described here are the ethanol heterogeneous catalytic transformations to higher value-added products such as hydrogen (H_2), ethylene (C_2H_4), Guerbet condensation products (mostly 1-butanol), the Lebedev dehydration product 1,3-butadiene, and other relevant products, focusing on catalysts based on non-noble metals. Ethanol catalytic transformation can be a crucial technology in producing organic chemicals and building blocks since it is renewable and fits low-carbon economy requirements. Although promising, much effort is still needed to develop new catalysts, reaction methodologies, and up-scaling, which open new perspectives for intense research and development. Due to the low-cost and high-production of ethanol in Brazil, as well as the energy surplus in the sugarcane mills, research groups in this country are dedicating efforts to develop new ethanol catalytic transformations. However, there remains a need for ethanol valorization into more complex condensed products.

Advances in Inorganic Chemistry, Volume 77
ISSN 0898-8838
https://doi.org/10.1016/bs.adioch.2021.02.005

Abbreviations

BTX benzene, toluene, and xylenes
E100 100% ethanol fuel
E25 25% ethanol and 75% gasoline (v/v) fuel
E27 27% ethanol and 73% gasoline (v/v) fuel
ED ethanol dehydration
ESR ethanol steam reforming
GR Guerbet reaction
HAP hydroxyapatite
LDH layered double hydroxide
LR Lebedev reaction
MMO mixed metal oxides
MSR methane steam reforming
OSR oxidative steam reforming
PE polyethylene
POX partial oxidation reaction
WGSR water-gas shift reaction

1. Introduction

1.1 Ethanol in Brazil

Brazil is now the second world producer of ethanol, with ca. 30% of the produced volume, with sugar cane (*Saccharum officinarum*) being the raw material. However, ethanol production in Brazil during the first decades of the 20th century was marginal since the sugarcane industry was almost entirely dedicated to producing and exporting sugar and ethanol production for fuel applications was practically nonexistent.[1] This panorama started to change with the petroleum supply crisis in the 1970s. The Brazilian government encouraged the production of cars powered exclusively by ethanol (E100) through an effective national program called "ProAlcool" (Brazilian Alcohol Program, launched in 1975) that included rural producers, agriculture industry of heavy machinery, and the automobile industry.[2,3] In 1979, ca. 3000 of ethanol vehicles were produced, and in 1986, a record of ca. 700,000 units was reached. Since then, the ethanol industry in Brazil has been well established in its entire production chain. Due to several political and economic factors, including the scarcity of ethanol in off-season periods, production decelerated to 41,000 vehicles in 1995, reaching practically zero in the following years. Consequently, a massive decrease in ethanol production was observed.

In 2003, with the advent of flex-fuel technology, which adjusts car engine operation for any proportion of hydrated ethanol/gasoline, ethanol utilization as fuel grew again. Currently, in Brazil, cars exclusive powered by gasoline are no longer manufactured. Ethanol from sugarcane is offered to consumers as a biofuel throughout the country at a low price, pure or mixed with gasoline (E27), and consumers can choose any ethanol/gasoline ratio. The hydrated ethanol sold at the gas stations contains 5% water since the storage and handling of anhydrous ethanol in large quantities are impractical due to its hygroscopicity. Since the consumers may want to fill the vehicle tank with variable amounts of each fuel, the Brazilian gasoline must be produced with at least 18% anhydrous ethanol to avoid liquid-phase separation.[4] The addition of ethanol in gasoline also helps Brazil to reduce the gasoline importation-dependence from other countries.

Brazilian authorities require gasoline to be produced with an anhydrous ethanol content between 18% and 27.5%, depending on the market. In 2015, the ethanol content was fixed at 25% (E25) for premium gasoline and 27% for regular gasoline (E27). This demand was fundamental for establishing about 400 sugarcane mills located basically on the northeast coast and in the center-south, of which half produce only ethanol instead of sugar. In the 2019/20 harvest, this industry processed 590 million tons of sugarcane, 65% used for ethanol production.[5] However, the sugarcane mills usually change the ethanol production volume in response to ethanol and sugar global prices.

The Brazilian light vehicles fleet has more than 40 million units (cars and motorcycles), which annually demand ca. 33 billion liters of anhydrous ethanol (installed capacity for 46 billion liters) and 23 billion liters of hydrated ethanol (installed capacity to 85 billion liters). Usually, the price of hydrated ethanol (E100) at gas stations does not exceed 70% of the gasoline (E27). This price ratio is essential to the consumers since ethanol has a lower mileage yield per liter than gasoline (ca. 30% higher for gasoline). Thus, if the ethanol price is more than 70% of the gasoline price, there is no economic advantage. In 2020, the average selling price of ethanol was USD 0.35/L at the sugarcane mills.[6]

In addition to the low cost, another great advantage of sugarcane ethanol is its energy output/input ratio for its production, which is ca. 9:1, much higher if compared to corn ethanol (ca. 1:1) or sugar beet ethanol (ca. 2:1).[7,8] This advantage comes from the possibility of energy-generation from sugarcane harvesting residues. Each ton of sugarcane processed results in ca. 250 kg of bagasse and 200 kg of straw. These are burned in sugarcane mills furnaces to heat the steam boilers, generating 15,000 MJ/ton of sugarcane harvested.[9] Usually, 40% of this energy is consumed by the sugarcane mills, and the surplus is sold to the national electric power transmission grid, although, only 55% of sugarcane mills have the implemented system to sell energy surplus during the harvest season. Thus, the sugarcane industry contributes to biomass-generated electrical energy in Brazil, reaching in 2018 a total of 21.5 TWh.[9] It is estimated that ca. 6.0 TWh still could be generated in 150 sugarcane mills.

Thinking forward, sugarcane mills, are not only fuel and sugar producers, but also are possible bio-refinery plants to convert ethanol into more value-added products. The energy surplus at these sugarcane mills is a key advantage available in loco at low cost. Ethanol conversion into other chemicals demands high amounts of thermal energy due to its endothermic nature. Thus, the energy surplus at the sugarcane mills can enhance the economic viability of ethanol-based bio-refineries.

Ethanol-based bio-refineries preceded the Brazilian petrochemical industry (initiated in 1953) by almost 40 years, producing ethyl chloride, diethyl ether, and acetic acid.[5] However, due to low petroleum prices until 1970s, these bio-refineries were shut down. During the petroleum crisis, the "ProAlcool" program implemented economic subsidies to re-build ethanol-based bio-refineries to produce organic chemicals such as ethylene, acetaldehyde, *n*-butanol, and 1,3-butadiene. Due to petroleum price decreases during the 1980s and technical drawbacks such as the catalyst development (high conversion and selectivity issues and long-term stability) and process up-scaling, the subsidy was canceled in 1984.

Ethanol-based bio-refineries start to gain attention again due to the efforts to find alternative routes for producing base-chemicals in order to reduce the petroleum environmental impact. Ethanol is considered a renewable carbon source since a considerable fraction of CO_2 emissions during the production process (sugars fermentation, heavy machinery based on diesel fuel, bagasse burning for steam generation, vinasse decomposition, among others) can be reincorporated during sugarcane harvesting.[2,10–12] Also, the energy-environmental efficiency of the sugarcane mills should be considered on CO_2 emissions, since there are variations of production conditions thought Brazilian territory.[13]

However, taking into account ethanol as fuel, the CO_2 balance is less favorable than its utilization as carbon source for chemicals synthesis. Also, the increasing ethanol demand as fuel impacts on the CO_2 emissions related to land utilization. For example, the deforestation aiming to increase the sugarcane production may change the CO_2 balance for negative in comparison to gasoline beyond other environmental impacts.[14–16] Although it is still a matter of discussion in literature, ethanol utilization as fuel may reduce CO_2 emissions compared to gasoline up to 45%, depending on the prediction model employed.[17,18] Thus, ethanol utilization as renewable carbon source should be encouraged, not simply as a fuel.

Despite the urgency for developing and utilizing greener industrial processes, the economic viability will always be linked to the price of the produced chemicals by green routes versus the ones based on petroleum. Thus, ethanol-based bio-refineries need to be environmental-friendly, profitable, and competitive with the petroleum industry. From this perspective, sugarcane mills bring together all the factors needed to operate as a bio-refinery: low-cost raw material (ethanol), energy (from harvesting waste), and the possibility to integrate in loco other industrial plants to produced high value-added chemicals.[19,20]

Thus, ethanol-based bio-refineries may represent an excellent alternative to produce basic chemicals. Much effort needs to be done to develop new catalysts, new chemical reactions and synthetic methodologies, and up-scaling the process.[21,22] In the next topics, we will discuss the most relevant ethanol-based products and how they may be produced using heterogeneous catalysts and focusing on non-noble metals. Also, Brazilian research group contributions to the development of the ethanol-bio-refinery will be highlighted.

2. Ethanol steam reforming and dehydration reactions

Ethanol transformation into high value-added chemicals can be achieved along distinct reaction pathways. Considering the total conversion of ethanol into gaseous products (H_2 and CO_2), three preferential ethanol reforming reactions were proposed: steam reforming (ESR, Eq. 1), partial oxidation (POX, Eq. 2), and oxidative steam reforming (OSR, Eq. 3).[23–26] On the other hand, the conversion of ethanol into long-chain carbon products (alcohols, aldehydes, ketones, ethers, esters, and aromatics) occurs through the Guerbet condensation reaction (GR).[27,28] In both reaction mechanisms, the reactivity of ethanol and the catalyst must be tuned to give the desired products. The reactions involved in the ethanol reforming can be summarized as follows:

$$\textbf{ESR}\; C_2H_5OH + 3H_2O \xrightarrow{\text{catayst}} 2CO_2 + 6H_2 \tag{1}$$

$$\textbf{POX}\ C_2H_5OH + 1.5O_2 \xrightarrow{\text{catayst}} 2CO_2 + 3H_2 \quad (2)$$

$$\textbf{OSR}\ C_2H_5OH + (3-2x)H_2O + xO_2 \xrightarrow[0<x<0.5]{\text{catayst}} 2CO_2 + (6\text{-}2x)H_2 \quad (3)$$

During ethanol reforming a series of non-desired reactions can occur, giving rise to gaseous products such as methane (CH_4) and carbon monoxide (CO), as depicted in Eq. (4). These products can undergo a series of consecutive and/or parallel reaction pathways such as methane steam reforming (MSR, Eq. 5) and water-gas shift reaction (WGSR, Eq. 6). Ethanol can be dehydrogenated into acetaldehyde (C_2H_4O, Eq. 7), which further decomposed into CH_4 and CO (Eq. 8). Finally, ethanol dehydration reaction yields ethylene (C_2H_4, Eq. 9), an essential building block for the syntheses of other compounds including materials such as polyethylene. However, it is well established that ethylene decomposition can be one of the major pathways to catalyst poisoning due to coke formation (Eq. 10).[29,30]

$$C_2H_5OH \xrightarrow{\text{catalyst}} CH_4 + CO + H_2 \quad (4)$$

$$CH_4 + H_2O \xrightarrow{\text{catalyst}} CO + 3H_2 \quad (5)$$

$$CO + H_2O \xrightarrow{\text{catalyst}} CO_2 + H_2 \quad (6)$$

$$C_2H_5OH \xrightarrow{\text{catalyst}} C_2H_4O + H_2 \quad (7)$$

$$C_2H_4O \xrightarrow{\text{catalyst}} CH_4 + CO \quad (8)$$

$$C_2H_5OH \xrightarrow{\text{catalyst}} C_2H_4 + H_2O \quad (9)$$

$$C_2H_4 + H_2O \xrightarrow{\text{catalyst}} C_{coke} + H_2 \quad (10)$$

Following the pathways described above, catalyst development is a key-factor for the fine-tuning of ethanol reactions. Thus, depending on the desired outcome, the catalyst can be tailored to enhance and/or to inhibit one or more reaction pathways.

Considering the concept of the low-carbon economy,[31,32] which needs to be developed in response to the already felt effects of global warming, two gaseous products hydrogen (H_2) and ethylene (C_2H_4) from ethanol are of special interest.

2.1 Hydrogen production

Nowadays, H_2 production can be split into two major routes: natural gas steam reforming (95%) and seawater electrolysis (5%).[33] The main uses of

H_2 are in order of utilization: Haber-Bosch process (ammonia production),[34] methanol production (carbon monoxide reduction),[35] chemical synthesis,[34] and as fuel (direct combustion and fuel cell applications).[36,37] The hydrogen production is still heavily based on fossil hydrocarbons technology, which must be substituted due to the high levels of CO_2 emissions.

Ethanol is emerging as a green alternative to H_2 production through ethanol reforming reactions,[23–26] and many Brazilian research groups have been exploring these reactions as a strategic technology for the future. The research focused on developing new catalysts based on non-noble transition metals, such as Ni (Table 1) and Co (Table 2), with fewer examples with Rh, which was chosen for the chemical similarity to Co. Some examples also exploited catalyst doping with small percentages of Pt and Pd to enhance the ethanol reactivity for H_2 production. Ni and Co are among the preferential choice for catalysts development since it is well established that these metals can break the C—C bond of ethanol, which is a crucial step for the reforming reactions.[34–36]

Besides the choice of transition metal, the support employed plays a pivotal role in the catalyst activity for ESR, POX and OSR. A variety of supports have been utilized, such as mixed metal oxides derived from hydrotalcite[37,46] and perovskite[52]; magnesium,[38] aluminum,[39,42,43,45] silicon,[44,51,56] zirconium,[47] lanthanum,[51] cerium,[40,47,48,51,55,56] titanium[56] and zinc[56] oxides; cerium-zirconium oxides[47,53]; magnesium-aluminum oxides[50]; lanthanum-cerium-silicon oxides[51]; cerium-gadolinium[41]; activated carbon[49]; carbon nanofibers[49]; carbon nanotubes,[49] and metallic nanoparticles supported on magnesium-aluminum oxide,[54] among others. The supports determine the catalyst surface area, degree of dispersion of the transition metals, and, most important, the acid-base (Lewis and Brønsted-Lowry) properties of the catalysts. Supports with high acidity enhance the ethanol dehydration, yielding ethylene instead of dehydrogenation. Thus, the fine-tuning of the catalyst's acid-basic properties is a controlling factor.

The most important pathway leading to catalyst deactivation is coke deposition over the surface. This process is a result of ethylene dehydrogenation during ESR, POX, and OSR reactions. Since the coke prevents ethanol access to the catalyst reaction sites, the H_2 yield usually decreases during the reaction time. To circumvent this drawback, several alternatives have been proposed. (i) Catalyst tailoring (acidity control) aiming to reduce the ethanol dehydration reaction to prevent ethylene formation; (ii) Co-feeding water with the ethanol, which produces OH^- species that can remove the deposited coke over the catalyst surface; (iii) Co-feeding

Table 1 Ni catalysts for hydrogen production from ethanol.

Metal[a] (%)	Metal[a] (%)	Support	Temp.[b] (°C)	Ethanol conv. (%)	χ_p H_2[c] (%)	Selec. H_2[d] (%)	Ref.
Ni (8.9)	–	MgAl[e]	550	78.5	98	–	37
Ni (9.3)	La (2.9)	MgAl[e]	550	81.1	85	–	
Ni (8.9)	Ce (3.4)	MgAl[e]	550	95.5	35	–	
Ni (8.9)	–	MgAl[e]	650	98.7	75	–	
Ni (9.3)	La (2.9)	MgAl[e]	650	98.8	65	–	
Ni (8.9)	Ce (3.4)	MgAl[e]	650	99.3	92	–	
Ni (6.7)	–	MgO	500	95	62	–	38
Ni (6.6)	–	MgO	500	65	65	–	
Ni (7.2)	–	MgO	500	50	65	–	
Ni (15.0)	–	Al_2O_4	500	82	62	–	39
Ni (15.0)	Zn	Al_2O_4	500	100	65	–	
Ni (15.0)	ZnO	Al_2O_4	500	100	72	–	
Ni (10.0)	–	CeO_2	300	58	50	–	40
Ni (10.0)	Pt (1.0)	CeO_2	300	62	45	–	
Pt (1.0)	–	CeO_2	300	55	42	–	
–	–	CeGd	500	50	58	–	41
Ni (56.3)	–	CeGd	500	90	62	–	
Ni (49.7)	–	CeGd	500	40	60	–	
Ni (56.3)	–	CeGd	500	45	60	–	
Ni (5.0)[f]	–	Al_2O_3	500	99	–	1.05	42
Ni (5.0)[f]	Ca (0.5)[f]	Al_2O_3	500	99	–	1.00	
Ni (5.0)[f]	Ca (2.5)[f]	Al_2O_3	500	99	–	1.23	
Ni (5.0)[f]	Ca (5.0)[f]	Al_2O_3	500	99	–	1.79	
Ni (5.0)[g]	–	Al_2O_3	500	99	–	1.11	
Ni (5.0)[g]	Ca (0.5)[g]	Al_2O_3	500	95	–	0.97	
Ni (5.0)[g]	Ca (2.5)[g]	Al_2O_3	500	95	–	1.28	
Ni (5.0)[g]	Ca (5.0)[g]	Al_2O_3	500	95	–	1.40	

Table 1 Ni catalysts for hydrogen production from ethanol.—cont'd

Metal[a] (%)	Metal[a] (%)	Support	Temp.[b] (°C)	Ethanol conv. (%)	χ_p H_2[c] (%)	Selec. H_2[d] (%)	Ref.
Ni (8.2)[h]	–	α-Al_2O_3	540	19	19	–	43
Ni (8.2)[i]	–	α-Al_2O_3	540	22	68	–	
Ni (7.3)[h]	–	α-Al_2O_3	540	38	34	–	
Ni (7.3)[i]	–	α-Al_2O_3	540	25	70	–	
Ni (9.3)	–	SiO_2	600	100	60	–	44
Ni (9.5)	Pd (1.1)	SiO_2	600	100	60	–	
Ni (15.0)		γ-Al_2O_3	500	72	63	–	45
Ni (15.0)		CeO_2/Al_2O_3	500	70	60	–	
Ni (15.0)		$CeZrO_2/Al_2O_3$	500	72	60	–	
Ni (15.0)		$CeZrO_2$	500	68	60	–	

[a]Mean mol%.
[b]Reaction temperature.
[c]Molar fraction of H_2 in the gaseous mixture.
[d]Selectivity calculated as (n mols of produced H_2/n mols of converted EtOH).
[e]Hydrotalcite-like catalyst precursor.
[f]Synthesis by impregnation.
[g]Synthesis by co-precipitation.
[h]Ethanol partial oxidation reaction.
[i]Ethanol steam reforming reaction.

Table 2 Co and Rh catalysts for hydrogen production from ethanol.

Metal[a] (%)	Metal[a] (%)	Support	Temp.[b] (°C)	Ethanol conv. (%)	χ_p H_2[c] (%)	Selec. H_2[d] (%)	Ref.
Co (0.1)	–	MgAl[e]	550	97	75	–	46
Co (0.2)	La (0.04)	MgAl[e]	550	84	68	–	
Co (0.2)	Ce (0.04)	MgAl[e]	500	63	75	–	
Co (10.0)	–	CeO_2	600	–	–	3.3	47
Co (10.0)	–	ZrO_2	600	–	–	3.8	
Co (10.0)	–	CeO_2-ZrO_2[f]	600	–	–	2.5	
Co (10.0)	–	CeO_2-ZrO_2[g]	600	–	–	3.0	
Co (10.0)	–	CeO_2-ZrO_2[h]	600	–	–	2.4	
Co (10.0)	–	CeO_2-ZrO_2[i]	600	–	–	3.4	

Continued

Table 2 Co and Rh catalysts for hydrogen production from ethanol.—cont'd

Metal[a] (%)	Metal[a] (%)	Support	Temp.[b] (°C)	Ethanol conv. (%)	χ_p H_2[c] (%)	Selec. H_2[d] (%)	Ref.
Co (5.0)	–	CeO_2	500	94	–	1.5	48
Co (10.0)	–	CeO_2	500	98	–	1.5	
Co (20.0)	–	CeO_2	500	98	–	1.9	
Co (1.0)	–	C[j]	500	60	48	–	49
Co (1.0)	–	CNF[k]	500	60	62	–	
Co (1.0)	–	CNT[l]	500	38	50	–	
–	Ni (8)	$MgAl_2O_4$	500	100	58	–	50
–	Ni (15)	$MgAl_2O_4$	500	100	62	–	
Co (8.0)	–	$MgAl_2O_4$	500	100	62	–	
Co (15.0)	–	$MgAl_2O_4$	500	88	38	–	
Co (4.0)	Ni (4)	$MgAl_2O_4$	500	70	30	–	
Co (4.0)	Ni (11)	$MgAl_2O_4$	500	95	60	–	
Co (7.5)	Ni (7.5)	$MgAl_2O_4$	500	95	42	–	
Co (11.0)	Ni (4)	$MgAl_2O_4$	500	100	55	–	
–	–	LaCeSi[m]	500	31		–	51
Co (10.0)	–	SiO_2	500	87		1.0	
Co (10.0)	–	CeO_2	500	93		1.4	
Co (10.0)	–	La_2O_3	500	57		1.4	
Co (10.0)	–	LaCeSi[m]	500	97	–	1.5	
Co (10.0)	–	LaCeSi[n]	500	86	–	1.6	
Co (10.0)	–	LaCeSi[o]	500	73	–	1.5	
Co (1.0.0)	–	La-perov.[p]	600	100	68	–	52
Co (1.0)	–	La-perov.	800	100	68	–	
Co (0.8)	Cu (0.2)	La-perov.	600	100	68	–	
Co (0.8)	Cu (0.2)	La-perov.	800	58	62	–	
Co (9.6)	–	CeO_2-ZrO_2	400	46	60	–	53
Co (9.5)	K (2.1)	CeO_2-ZrO_2	400	98	62	–	
Co (8.1)	Rh (1.9)	CeO_2-ZrO_2	375	100	38	–	

Table 2 Co and Rh catalysts for hydrogen production from ethanol.—cont'd

Metal[a] (%)	Metal[a] (%)	Support	Temp.[b] (°C)	Ethanol conv. (%)	χ_p H_2[c] (%)	Selec. H_2[d] (%)	Ref.
Co (7.6)	K (1.6) Rh (3.1)	CeO_2-ZrO_2	375	100	35	–	
Co (8.1)	K (1.4) Rh (0.2)	CeO_2-ZrO_2	375	100	40	–	
Co (12)	–	$MgAl_2O_4$	550	100	70	–	54
Co (12)	Cu (3.0)	$MgAl_2O_4$	550	100	68	–	
Rh (1.0)	–	CeO_2-LS[q]	500	100	62	–	55
Rh (1.0)	–	CeO_2-HS[ryyy]	500	100	60	–	
Rh (0.02)	–	CeO_2	550	100	40	–	56
Rh (0.02)	–	TiO_2	550	100	20	–	
Rh (0.02)	–	SiO_2	550	100	22	–	
Rh (0.5)	–	ZnO	550	100	10	–	
Rh (0.04)	–	Al_2O_3	550	58	10	–	

[a]Mean mol%.
[b]Reaction temperature.
[c]Molar fraction of H_2 on the gaseous mixture.
[d]Selectivity calculated as (n mols of produced H_2/n mols of converted EtOH) and values higher than 3.0 may be due to parallel or side reactions.
[e]Hydrotalcite-like catalyst precursor.
[f]20% ceria in zirconia.
[g]40% ceria in zirconia.
[h]60% ceria in zirconia.
[i]80% ceria in zirconia.
[j]Activated carbon.
[k]Carbon nanofibers.
[l]Carbon nanotubes.
[m]$La_{2\%}Ce_{48\%}Si_{50\%}$.
[n]$La_{4\%}Ce_{46\%}Si_{50\%}$.
[o]$La_{6\%}Ce_{441\%}Si_{50\%}$.
[p]perovskite-type oxide.
[q]Low surface area.
[r]High surface area.

molecular oxygen with ethanol or with ethanol and water. Oxygen can oxidize coke into CO, being removed from the reactor as gaseous product or by reacting with water, yielding H_2 and CO_2 via WGSR.[23–26]

Thus, due to the coke deposition, the ethanol reforming reaction is almost always conducted with water and oxygen co-feeding. Usually, ESR, POX and OSR reactions are conducted with 3:1 water:ethanol mixture (v/v) or higher. However, even with water co-feeding, the catalysts usually undergo deactivation during the time. It was observed that the deactivation strongly depends on the support composition, transition metal ion, and flow reactor feeding parameters.

The catalysts containing Ni showed ethanol conversion ranging from 19% to 100% with a wide range of selectivity and the molar fraction H_2 in the product composition.[37–45] Table 1 summarizes the data obtained for Ni catalysts.

Better conversions and selectivity were achieved by doping the Ni-based catalysts with transition metals (Zn, Pt, and Pd) or lanthanides (La and Ce). The reaction temperature ranged from 300 to 650 °C, and higher temperatures did not favor the ESR, POX and OSR reactions due to coke deposition. The Ni catalysts doped with noble metals allowed decreasing the reaction temperature, but they also decreased the ethanol conversion rates, thus making the further utilization of these catalysts less attractive for up-scaling.

Regarding the temporal behavior of ethanol conversion and H_2 production, the Ni catalysts presented three distinct scenarios: (i) drop in both ethanol conversion and H_2 production; (ii) drop in ethanol conversion and H_2 production almost constant; (iii) ethanol conversion and H_2 production constant from 6 to 30 h. Considering the behavior described in (ii), it is reasonable to suppose that the H_2 production remained constant due to ethanol side reactions at the beginning of the catalytic process.

The Ni catalysts for which the ethanol conversion and H_2 production remained constant, were the ones containing ZnO over Al_2O_4 (30 h of reaction),[39] Ni and Pd over SiO_2 (30 h of reaction)[44] and Ni over CeO_2/Al_2O_3 (6 h of reaction).[45] Considering the ethanol conversion, H_2 molar fraction or selectivity, the catalyst containing Ni and ZnO over Al_2O_4 seems to be the more advantageous since Zn is more abundant and cheaper than Pd. The authors claim that ZnO addition in the synthetic process leads to a Ni–Zn bimetallic alloy, providing the remarkable stability and activity observed, since coke deposition was not observed due to structural and electronic effects.[45]

It is established in the literature that H_2 selectivity during ethanol reforming follows the order Pt > Rh > Ni > Co regarding the transition metal incorporated into the catalyst support. Ni has higher selectivity for H_2 than does Co, but it is also more susceptible to coke deposition.[34–36] Thus, taking into account the long-term stability, a series of Co and Rh catalysts were synthesized and tested for ESR, POX and OSR. Table 2 summarizes the data for Co and Rh catalysts.[46–56]

The Co-based catalysts converted ethanol in the range of 38–100%, at temperatures varying from 375 to 800 °C. However, the temperature range of 500–600 °C proved to be the better choice for catalytic reactions.

The doping of Co catalysts with Rh did not lead to significant improvements in the H_2 selectivity.[53] However, the addition of K (as KNO_3) to the Co-catalyst on a CeO_2-ZrO_2 during the catalyst synthesis, proved to be more efficient in enhancing the H_2 selectivity than was Rh.[53] Furthermore, the addition of La, Ce, and Cu as dopants in metal oxides derived from Co-hydrotalcites, perovskites, and $MgAl_2O_4$ as supports, enhanced the H_2 selectivity when compared with non-doped Co catalysts.[46,47]

Regarding the temporal stability of Co catalysts, it was observed that as a general trend, these catalysts demonstrated more extended periods of constant activity compared to the Ni ones. Usually, the Co presented a drop in the ethanol conversion over the first 2 h, then stabilized for as long as 30 h of reaction. The initial drop in the ethanol conversion was attributed to catalyst surface modifications at the beginning of the reaction, or redox reactions involving Co. The H_2 selectivity was almost constant during the reaction time for all Co catalysts, suggesting that the coke deposition was lower or that the deposited coke was more easily oxidized to CO. It is important to recall that Co–Ni bimetallic catalysts were unsuccessful on enhancing the ethanol conversion and H_2 selectivity, indicating no synergic effect of the two metals.

Analyzing the temporal behavior of Rh-based catalysts (ethanol consumption and H_2 selectivity), a remarkable stability (24 h of reaction) was observed for CeO_2 as support.[55,56] Rh catalysts did not show a drop in ethanol conversion, indicating the absence of surface modifications during the reaction. The authors claimed that the long-term catalyst stability was due to coke oxidation enhancement in a synergic effect between Rh and CeO_2.

The choice of non-noble metals for catalyst tailoring is still a key point in developing the H_2 economy based on ethanol. In the literature, heterogeneous catalysts containing noble metals such as Ir, Pt, Pd, Rh, and Ru are active for ESR, POX and OSR. However, these metals have a high market price due to their low abundance and extraction issues. Even using small quantities of these metals would increase the final cost of H_2 produced, thus, making this pathway, although greener, less attractive economically than the well-established petrochemical routes.

The utilization of Ni and Co catalysts for ESR, POX or OSR seems feasible for the production of greener H_2 from ethanol. However, these methods have not yet been up-scaled, which must be the next step for this technology. The economics of large-scale application of H_2 production are very sensitive to the cost of ethanol. Since Brazilian ethanol has a competitive price due to the low costs of the sugarcane-based process, Brazil is a candidate to implement this technology in the future.

2.2 Ethylene production

Ethylene is a key industrial raw material, being the organic compound with the highest production globally (ca. 207 million tons in 2019), since it is the starting material for large-scale chemical processes. Some examples of ethylene utilization are the synthesis of polyethylene (PE), acetaldehyde (Wacker process), acetic acid (acetaldehyde oxidation), ethylene oxide (key intermediate in addition reactions), vinyl chloride (and further polyvinylchloride), styrene (Friedel-Crafts reaction of benzene and ethylene and further polystyrene), among other organic building blocks.[57,58] Since ethylene is almost entirely obtained from petroleum (hydrocarbon cracking), new chemical routes to obtain it are required.

Ethanol dehydration (ED, Eq. 9) attracted considerable attention in the first decades of the 20th century as a crucial technology to produce ethylene,[59] and many ED industrial plants were built, despite high ethanol production costs, mostly in Europe and the USA. However, once petrochemical routes to produce ethylene were developed, more profitable due to the low cost and abundant raw material, the ED plants were shut down.

The requirement of a greener route to produce ethylene has made the ED industrial plants attractive again. Taking into account the raw material cost and logistic issues, only sugarcane produced Brazilian and Indian ethanol has the competitive price, compared to the USA (from corn) and European (from sugar beets) ethanol, to fulfill ethylene production cost requirements for utilization in ED plants.[59–64]

It is well established that ED is performed under an acid catalyst since it requires removal of the hydroxyl group of ethanol.[65–71] Thus, acid catalysts such as H_3PO_4 supported on silica, Al_2O_3, zeolites, transition metal oxides, heteropoly acids, were employed to achieve this objective.[65–71] The reaction is typically conducted in a temperature range of 300–500 °C, although zeolite catalysts operate at lower reaction temperatures but with lower selectivity to ethylene. Table 3 summarizes the catalysts that have been employed for ED and the advantages, and disadvantages of each.

Several ED reactor designs have been proposed and studied, but the continuous flow reactor is the best choice. Among these are isothermal,

Table 3 Catalysts properties for ethanol dehydration (ED).

Catalyst	Advantages	Disadvantages	Ref.
H_3PO_4-SiO_2	• Minor side products	• Coke deposition • Difficult to regenerate	[65]
γ-Al_2O_3	• High chemical stability • Low coke deposition • High ethylene production • Low cost	• High reaction temp. (450–500 °C)	[66,67]
Zeolites	• High ethylene production • Low reaction temp. (200–300 °C)	• Low chemical stability • Coke deposition • High acidity • High cost	[68,69]
Heteropolyacids	• Low reaction temp. (220–250 °C)	• Chemical stability not fully tested • Requires feeding with high % of water • High cost • Low ethanol conv. rate	[70,71]

adiabatic, and fluidized bed reactors; the first two options were already tested (and some cases operated) for ED on an industrial scale. Ethanol conversions as high as 95% with selectivity for C_2H_4 reaching 97–99% have been achieved.[63]

Despite the higher selectivities reported, ED gives rise to other undesirable side products, such as acetaldehyde and diethyl ether (DEE). Other by-products such as higher alcohols, hydrocarbons (butene, propene), acids (acetic and butanoic), and gaseous (CO and CO_2) are also found in minor concentrations. Although ED side products generally are found in low percentages relative to ethylene, their presence requires careful purification, since polyethylene (PE) production requires a high purity grade monomer (>99.9% pure) to fulfill quality requirements. Also, catalyst poisoning by the ED side products can be an issue.

A triangular reaction network for ED was proposed,[72–74] as shown in Fig. 1. For ethylene production, it is expected that only the dehydration reaction should occur on the catalyst surface. However, two ethanol molecules can also dehydrate to give rise to DEE. Ethanol can also be dehydrogenated into acetaldehyde that can react in the GR mechanism to furnish long-chain carbon alcohols (butanol, hexanol, octanol) or the

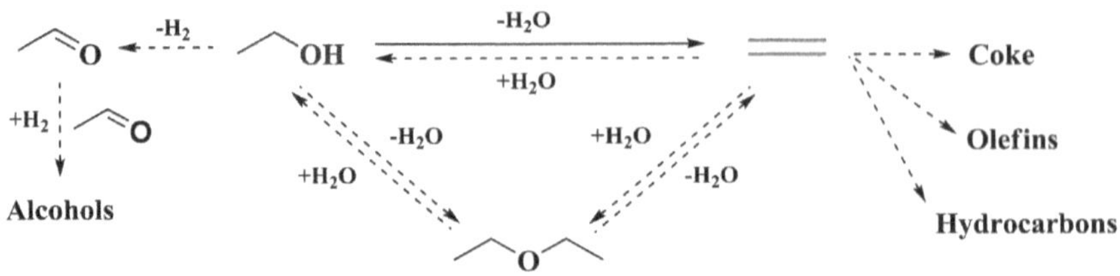

Fig. 1 Reaction network proposed for ethanol conversion into ethylene and most common side products. The dashed lines indicate the undesired reaction pathways. *Adapted from Phillips, C. B.; Datta, R.* Ind. Eng. Chem. Res. ***1997***, 36, *4466–4475; Nguyen, T. M.; Mao, R. L. V.* Appl. Catal. ***1990***, 58, *119–129.*

respective aldehydes (see below). Ethylene can also react further during the catalytic process generating long-chain olefins, hydrocarbons, and coke. The prevention of coke formation is strongly desirable since its deposition on the catalyst surface is the main pathway for deactivation of the catalyst.

Due to large and economic competitivity of ethanol production in Brazil, ED to ethylene is an economic and logistic feasible process. In 2010, Braskem (Brazilian Petrochemical Industry) started to produce ethylene from ED in an industrial plant in Rio Grande do Sul State.[74] The aim of this process was to produce ethylene in a high purity grade (>99.9%) for an *in loco* polymerization industrial plant to produce PE, which was named by the company as "I'm green,"™ with an annual production of 200,000 tons. A review about Braskem history and ED process development can be found in a publication by Do Carmo et al.[74]

The Braskem ED process is a continuous flow operation in an adiabatic reactor, using alumina as a catalyst. Although the company's detailed industrial process is undisclosed, the overall reactions and separation processes are shown in Fig. 2.

The reactor is fed with hydrated ethanol (95%, v/v) since it helps to prevent coke formation and catalyst poisoning. The catalytic process is optimized to produce a lower amount of undesired products by the reaction pathways presented. However, the reactor effluent contains variable amounts of DEE, hydrocarbons, oxygenates, gases (CO, CO_2, H_2), and organic acids. Thus, a set of ethylene purification processes must be conducted.

At the end of the purification, ethylene with a purity higher than 99.9%, is conducted to the polymerization process. Braskem produces high-density polyethylene, linear low-density polyethylene, and low-density polyethylene. These PE families cover almost all the range of industrial and commercial applications.

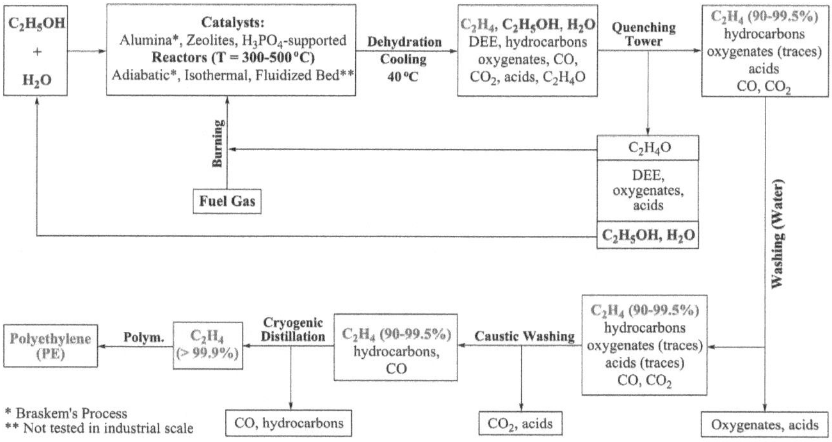

Fig. 2 Block diagram of ethanol conversion into ethylene and polyethylene-based on the Braskem process.

3. Guerbet and Lebedev reactions (GR and LR)

3.1 1-Butanol

n-Butanol is an important alcohol employed as a raw material in different fields of the chemical industry. Its isomers based on the position of the OH group (*tert*-butanol, *sec*-butanol, isobutanol, and 1-butanol) are all widely used as solvents, plasticizers, and building blocks for the synthesis of commercial compounds (Table 4).[75–78]

In recent years, 1-butanol has been proposed as a fuel additive and also considered as an alternative fuel due to its physiochemical characteristics, such as hydrophobicity, high calorific value, and high-octane rating.[79–81] Due to its wide applications, different chemical and biological methodologies to obtain 1-butanol have been developed. Among the main processes studied are the chemo-catalytic routes from ethanol and from sugar fermentation.[82–86]

Ethanol has been employed as a carbon source for the synthesis of 1-butanol by the GR (chemo-catalytic route). This reaction consists of the condensation of two molecules of ethanol (self-condensation) by several endothermic processes into the final "Guerbet" alcohol and loss of one molecule of water. Other longer-chain alcohols (C_6–C_{10}) can also be obtained from the GR of ethanol.[87–90] Importantly, the condensation requires that the alcohol contains a methylene group adjacent to the carbinol carbon.

Table 4 Physicochemical properties and main application of butanol isomers.

Isomers	Density ($g\ mL^{-1}$)	RON[a]	Boiling point (°C)	$\Delta H_{vap.}$[b] (kJ/kg)	Main applications
1-Butanol	0.808	92	117.7	582	Solvents, plasticizers, chemical intermediate, cosmetic, and gasoline additive
sec-butanol	0.806	101	99.5	557	Solvent, chemical intermediate, perfumes, domestic cleaner agent
Isobutanol	0.789	105	82.4	527	Solvent, gasoline additive
tert-butanol	0.802	113	108.0	566	Solvent, oxygenate intermediate, industrial cleaners, etc.

[a]Research octane number.
[b]Vaporization enthalpy at boiling point.

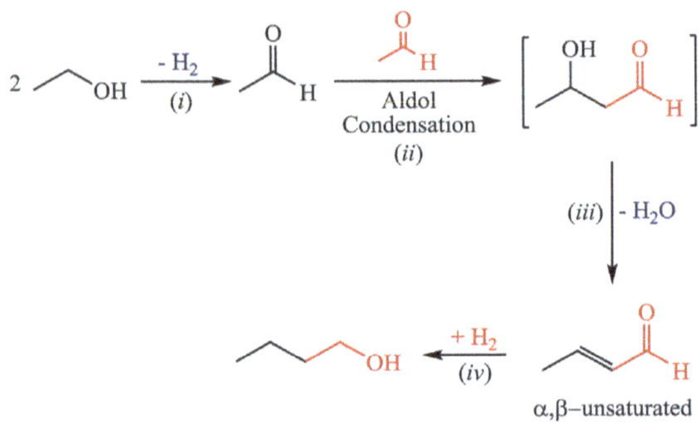

Fig. 3 Guerbet reaction (GR) for ethanol conversion into 1-butanol.

Mechanistically, the GR (Fig. 3) involves dehydrogenation of the ethanol to give acetaldehyde (i). Subsequently, this species undergoes an α-proton abstraction followed by C—C bond formation via nucleophilic attack at carbonyl affording an enolate intermediate (ii), which by successive dehydration (to form α,β-unsaturated carbonyl compound), (iii) and hydrogenation (iv) reactions providing 1-butanol.

Due to the multiple reaction steps required in the GR, the utilization of bifunctional catalysts that possess acid/basic and/or redox sites for dehydrogenation and hydrogenation reactions is necessary.[87] Therefore, different homogeneous and heterogeneous catalysts have been studied to produce long-chain alcohols by GR. Both classes of catalysts provided different yields and selectivity for 1-butanol and distinct by-products.[91–93] However, to make

Table 5 Hydroxyapatites as catalysts in the ethanol Guerbet reaction, ethanol conversion, and butanol selectivity.

Catalysts	Temp. (°C)	Reaction time (h)	Ethanol conv. (%)	1-Butanol selec. (%)	Ref.
$Ca_{10}(PO_4)_6(OH)_2$	300	3	7.1[a]	74.5[b]	94
$Ca_{10}(VO_4)_6(OH)_2$	300	3	6.6[a]	21.9[b]	
$Sr_{10}(PO_4)_6(OH)_2$	300	3	7.6[a]	81.2[b]	
$Sr_{10}(VO_4)_6(OH)_2$	300	3	5.8[a]	8.1[b]	
HAP (Ca/P ratio 1.61)	300	4.7×10^{-4} [e]	14.7[c]	76.3[d]	95
	350	4.7×10^{-4} [e]	26.1[c]	68.8[d]	
	450	4.7×10^{-4} [e]	57.4[c]	44.8[d]	
	450	4.7×10^{-4} [e]	95.3[c]	6.0[d]	
Stoichiometric HAP	340	1	6.6[e]	75[f]	95

[a]Determined by (ethanol conversion%) = [(1-C mol unreacted ethanol/C mol of liquid and gas products and unreacted ethanol) × 100].
[b]Determined by (selectivity%) = [(product yield/ethanol conversion) × 100].
[c]Determined by (ethanol conversion %) = [(1-C mol of unreacted ethanol/ C mol of total outlet gases) × 100].
[d]Determined by (selectivity%) = [(C mol of product/ C mol of reacted ethanol) × 100].
[e]Contact time.

the GR economically viable for its industrial application, heterogeneous catalysts are preferable, since they are more attractive in terms of separation and reusability.

Hydroxyapatites (HAP, the stoichiometric form is $Ca_{10}(PO_4)_6(OH)_2$ with a molar ratio of $Ca/P = 1.67$) have been employed in different fields of the chemical industry as catalysts and also in the gas phase reaction of ethanol due to their acid/basic properties (Table 5).[94–98]

For example, HAPs synthesized in different pH conditions and Ca/P molar ratios (between 1.59 and 1.67) have been reported to be catalysts for GR.[96] The authors reported ethanol conversion of 50% and the enhancement of the selectivity for Guerbet alcohols (C_4, C_6, C_8, and C_{10}) with a higher Ca/P ratio (1.67) reaching a maximum of 70%. The authors propose that the reaction in the gas phase of ethanol into 1-butanol over HAP starts by the dissociative adsorption of ethanol (i) (Fig. 4), in which an ethoxide intermediate (ii) is adsorbed on Lewis acid sites. Then, the proton-like hydrogen is adsorbed on the Brønsted-Lowry basic sites of the HAP and subsequently dissociates into an aldehyde intermediate

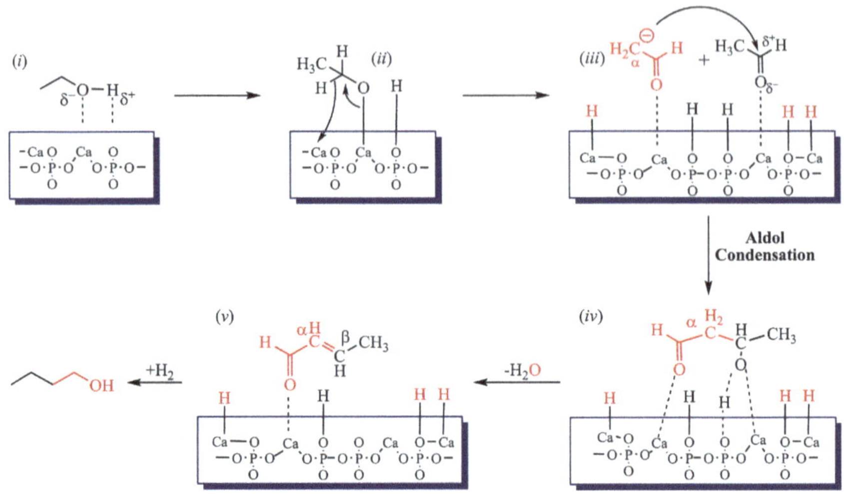

Fig. 4 Mechanism for the conversion of ethanol into 1-butanol over HAP. *Adapted from Tsuchida, T.; Kubo, J.; Yoshioka, T.; Sakuma, S.; Takeguchi, T.; Ueda, W.* J. Catal. ***2008***, 259, *183.*

(iii). The intermediate (iii) reacts with a second aldehyde intermediate, giving intermediate (iv) by aldol condensation. In this step, one of the two aldehydes is converted to a carbanion intermediate by adsorption of proton-like hydrogen on a Brønsted basic site. Subsequently, the aldol intermediate (iv) is dehydrated to form the unsaturated aldehyde (v) followed by hydride reduction providing 1-butanol, utilizing the hydrogen generated by dissociative adsorption during the first steps.

Additionally, several compounds such as acetaldehyde, crotonaldehyde, butyraldehyde, mixtures of acetaldehyde/1-propanol were utilized to rationalize the ethanol GR mechanism over HAP.[99] The authors showed a conversion of $\geq$90% of these compounds into Guerbet intermediates and 1-butanol. For example, acetaldehyde conversion into crotonaldehyde by aldol condensation was observed, being a crucial intermediate in GR. Interestingly, the authors proposed that the enolate formation is the rate-limiting step in the formation of 1-butanol via GR over HAP.

The selectivities for formation of 1-butanol, saturated and high-carbon alcohols, were studied related to the HPA acidic/basic properties and Ca/P ratios higher than the stoichiometric values.[100] The increase of acid/base ratio from 1.3 (carbonate-rich apatite) to 4.8 (carbonated apatite) also enhanced the selectivity for Guerbet alcohols to ca. 88%. However, increasing the acid/base ratio between 4.8 and 23.1 did not provide further improvements, leading to ethanol conversion of 40% and a total Guerbet alcohol yield of 30%. Another example of HAP as catalyst in the gas phase conversion of ethanol into 1-butanol, was carried out using strontium phosphate hydroxyapatite ($Sr_{10}(PO_4)_6(OH)_2$) and a fixed-bed continuous flow reactor at 300 °C under atmospheric pressure.[94] Conversions of ethanol between 1.0% and 24.0%, with 1-butanol selectivity as high as 81% were obtained.

Mixed metal oxides (MMOs) derived from hydrotalcites, also known as layered double hydroxides (LDHs), have also been employed as catalysts for GR. LDHs have the general formula $[M_{1-x}{}^{2+}M_x{}^{3+}(OH)_2]^{x+}A_{x/m}{}^{m-}\bullet nH_2O$, where M^{2+} and M^{3+} are di- and trivalent cations, respectively. These cations are octahedrally coordinated to six groups of hydroxyls and A is an interlamellar charge-compensating anion, such as Cl^-, F^-, $CO_3{}^{2-}$, NO_3^-, OH^-, among others.[100–102]

The LDHs itself did not present relevant catalytic performance for GR, but the calcination of LDHs between 450 and 600 °C forms the MMOs by dehydration, dehydroxylation, and decarboxylation reactions. The MMOs are widely employed as catalysts in different chemical applications, such as reforming methane and ethanol,[103] oxidation reaction,[104] oxidative dehydrogenation,[105] hydrogenation reaction,[106] transesterification, and etherification reactions,[107,108] biomass disassembly,[109–115] as well as the GR.[116,117]

It is well established that MMOs possess Lewis and Brønsted-Lowry acidic/basic sites. These sites play a pivotal role in the catalytic properties of MMOs in the reactions mentioned before. The presence of weak Brønsted-Lowry basic sites is due to the presence of surface hydroxyl groups remaining after heat treatment, and medium-strength Lewis sites are obtained from Mg^{2+}-O^{2-} and Al^{3+}-O^{2-} acid-base pairs. On the other hand, low coordinated O^{2-} species can provide strong Lewis basic sites.[117–120]

Mechanistically, the acidic/basic sites on the MMOs play an important role in 1-butanol synthesis (Fig. 5). Initially, the acetaldehyde formation starts with the adsorption of ethanol in the gas phase onto the Mg^{2+}-O^{2-} pair (i) followed by hydrogen abstraction giving an ethoxy intermediate, dissociating into an aldehyde intermediate and hydride-like hydrogen (ii). Subsequently, the two acetaldehydes undergo aldol condensation needing strong Lewis basic sites and acid-base pairs (iii). In this case, the mechanism starts with the abstraction of an α-hydrogen from a first acetaldehyde molecule (in this step it is necessary the use strong basic sites is necessary due to the low acidity of this hydrogen), followed by an electrophilic attack of the

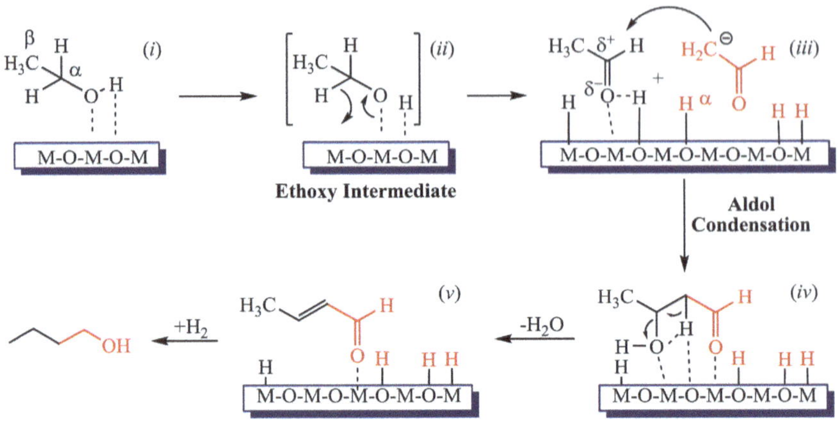

Fig. 5 Mechanism for the conversion of ethanol into 1-butanol over MMO. *Adapted from León, M.; Díaz, E.; Ordóñez, S.* Catal. Today ***2011***, 164, *436–442.*

carbanion formed to the second carbonyl group of the second acetaldehyde molecule providing an aldol intermediate (iv), which by dehydration leads to an α,β-unsaturated aldehyde (v). Subsequently, 1-butanol is formed by successive hydrogenations of 2-butenal.[121–123]

The catalytic properties of MMOs can be tuned by substituting transition or lanthanide metal cations for Mg^{2+} or Al^{3+} ions on the LDHs and subsequently on MMOs structure. This allows for the synthesis of modified catalysts with different properties, such as acidic/basic sites ratio, thermal stability, phase segregation control, redox properties, among others.[124–126]

Therefore, several MMOs with different substituents have been employed as catalysts in ethanol GR for 1-butanol synthesis as summarized in Table 6.

The addition of transition metals such as Cu and Ni to the MMOs structure showed higher ethanol conversion and 1-butanol selectivity than MMOs modified with noble metals. Interestingly, the addition of Pd, Sm, and Yb on the MMOs showed a higher 1-butanol selectivity but a low ethanol conversion ($\leq$3.8%).[117,127–129] Moreover, additives such as $MgCl_2$ and CCl_4, in a small percentage in the reaction medium provided an increase in the ethanol conversion and 1-butanol selectivity of up to 37% and 28%, respectively.[130]

In Brazil, the Boscolo research group has been working on the synthesis, characterization of different MMOs derived from LDHs, and their subsequent use as catalysts in the ethanol GR. Formation of 1-butanol using MMOs modified with Cu, Pd and In were studied.[93] In these cases, the

Table 6 1-Butanol obtained by Guerbet reaction over MMOs.

Metal (%)	Support (%)	Temp. (°C)	Reaction time (h)	Ethanol conv. (%)	1-Butanol selec. (%)	Ref.
Cu (1.0)	Mg (69.0); Al (30.0)	215	12	27.8	52.8	127
Cu (1.0)	Mg (69.0); Al (30.0)	230	12	32.4	52.1	
Cu (1.0)	Mg (63.7); Al (35.3)	260	2	24.7	3.1	117
Cu (1.01)	Mg (76.4); Al (22.5)	260	2	43.9	6.7	
–	Mg (62.5); Al (37.5)	400	2	12.4	4.2	128
–	Mg (81.0); Al (18.9)	400	2	23.5	9.6	
–	Mg (87.8); Al (12.1)	400	2	11.5	1.6	
Pd (5.0)	Mg (71.5); Al (23.7)	200	5	3.8	72.7	129
Ag (5.0)	Mg (71.5); Al (23.7)	200	5	1.5	38.8	
Mn (5.0)	Mg (71.5); Al (23.7)	200	5	0.5	53.3	
Sm (5.0)	Mg (71.5); Al (23.7)	200	5	1.3	66.3	
Yb (5.0)	Mg (71.5); Al (23.7)	200	5	1.2	53.0	
Cu (20.0)	Mg (55.0); Al (25.0)	320	6	43.0	29.0	130
Ni (20.0)	Mg (55.0); Al (25.0)	310	6	30.3	73.0	93
Ni (7.0) Cu (13.0)	Mg (55.0); Al (25.0)	310	6	62.4	66.0	
Cu (7.0) Ni (13.0)	Mg (55.0); Al (25.0)	310	6	51.2	72.0	
Cu (10.0) Ni (10.0)	Mg (55.0); Al (25.0)	310	6	47.9	72.0	

reaction was carried out employing 3.0 mL of ethanol and 100 mg of MMOs in a 10 mL stainless steel autoclave equipped with a withdrawing valve and using different temperatures during 5 h of reaction. The authors reported an increase in ethanol conversion and 1-butanol selectivity with increasing reaction temperature using all MMOs synthesized (Table 7). Additionally, the use of catalysts modified with Pd and Cu can provide an increase in the ethanol dehydrogenation step to give acetaldehyde (a key intermediate in the 1-butanol formation) with yields between 1.0% and 3.0% at low temperatures (from 130 to 210 °C) compared to a catalyst modified only with Cu. Interestingly, the authors observed 1-butanol

Table 7 1-Butanol yields obtained from the ethanol Guerbet reaction over MMOs employed by the Boscolo group.

Metal (%)	Metal (%)	Support (%)	Temp. (°C)	Ethanol conv.[a] (%)	1-Butanol selec.[b] (%)	Ref.
Cu (10.0)	–	Mg (65); Al (25)	300	46.5	62.2	131
Cu (10.0)	Pd (0.5)	Mg (64.5); Al (25)	260	20.9	45.5	
Cu (10.0)	In (0.5)	Mg (65); Al (24.5)	260	26.6	60.0	
Cu (10.0)	–	Mg (65); Al (25)	350	79.6	32.0	132

[a]Determined by (%) = [((*n* ethanol initial − *n* ethanol unconverted)/*n* ethanol initial) × 100].
[b]Determined by (%) = [(*n* desired product/*n* ethanol converted) × 100].

formation at temperatures between 130 and 190 °C using MMOs modified with Cu–Pd or Cu–In. For the MMO modified only with Cu the authors only observed 1-butanol formation using temperatures above 210 °C.

The formation of 1-butanol by ethanol GR over MMOs substituted with 10 mol% of Cu was also reported.[131] The reaction was carried out using an experimental design employing different variables: temperature, reaction time, and catalyst load. The optimal conditions yielded gaseous and condensate products of 69% and 31%, respectively. The major products obtained were 1-butanol and ethyl acetate with 32% and 9.6% of selectivity in the condensed phase, respectively, during 5 h of reaction at 350 °C.

3.2 1,3-Butadiene

1,3-butadiene (C_4H_6) is an important building block for the syntheses of commercial compounds and is widely used for many different applications, such as polymer chemistry.[132] This compound can be obtained as a by-product in the production of ethylene by steam cracking. However, its purification procedure can be expensive and complicated.[133,134] Oxidative dehydrogenation of *n*-butane and *n*-butene can also be used as the production method of 1,3-butadiene.[135–137]

Ethanol can be used as a carbon source in the formation of 1,3-butadiene by the Lebedev reaction (LR, Eq. 11). This reaction consists of the direct conversion in the gaseous ethanol phase to 1,3-butadiene over different catalysts at temperatures between 400 and 450 °C.

The LR mechanism (Fig. 6) starts with the dehydrogenation of ethanol yielding acetaldehyde (i), followed by its condensation into acetaldol (ii) and subsequent dehydration to crotonaldehyde (iii). A second dehydration reaction furnishes 1,3-butadiene (iv).[138–140] For the LR, it is necessary to use an

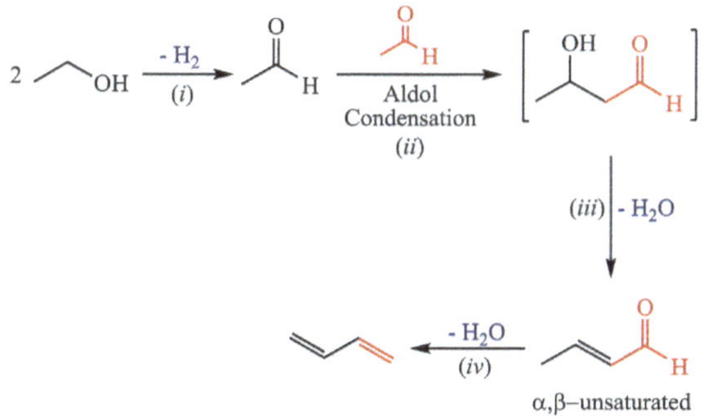

Fig. 6 Lebedev reaction (LR) for ethanol dehydration into 1,3-butadiene.

amphoteric bifunctional catalyst with dehydrogenation/dehydration activities.[141,142] Several catalysts have been studied for this reaction, such basic oxides,[140] zeolites,[143,144] and MMO.[145,146]

The LR and GR reaction mechanisms have similarities and also common intermediates, such as acetaldehyde and crotonaldehyde. The key-step that differentiates both mechanisms is related to the crotonaldehyde reactivity. In the GR, crotonaldehyde is reduced by H_2 (both the olefin and aldehyde groups), yielding 1-butanol, while in LR the crotonaldehyde is dehydrated. Thus, the catalyst must have strong Lewis acid sites to enhance the dehydration reaction and favor LR. However, the dehydrogenation and aldol condensation require Brønsted-Lowry and Lewis basic sites. Thus, the fine-tuning between acidic/basic sites ratio and its strengths on the catalyst is crucial for achieving the Lebedev products instead of Guerbet products.

Silicon-magnesium oxides (SiO_2-MgO) have been studied as effective catalysts for the ethanol to 1,3-butadiene conversion by LR.[147–149] The SiO_2-MgO synthetic methodology and the acid/basic sites on the catalytic obtainment of 1,3-butadiene were investigated.[150] The authors showed that the catalysts synthesized by the co-precipitated method have a higher number of strong acid and basic sites than the catalysts prepared by the wet-kneading method. The co-precipitated SiO_2-MgO showed ethanol conversion between 46% and 49% but low 1,3-butadiene yields (between 1.2% and 6.4%). On the other hand, the SiO_2-MgO synthesized by wet-kneading provided 1,3-butadiene yields of up to 16.3%. Additionally, the utilization of SiO_2-MgO modified with Cu was studied, yielding up to 43% of 1,3-butadiene and ethanol conversion between 75% and 80%.[151]

Some transition metal catalysts have been employed in the Lebedev process.[152–155] Zeolites doped with Ag, Cu, and Zn were used as catalysts in the

LR.[153] In this case, the use of these metals provided an ethanol conversion and 1,3-butadiene selectivity of up to 87% and 72%, respectively. Another example of metal catalyst was reported using M^{2+}:M'^{2} system impregnated on silica[154] (M^{2+}: Co, Mn, Cu, Zn, Zr) and reported ethanol conversion and 1,3-butadiene selectivity of up to 46% and 38%, respectively. Interestingly, the addition of 0.5 wt% of Zn and 1.5 wt% of Zr in the system provides the best conversion and selectivity results.

Additionally, silica modified with Zn and Hf was studied as a catalyst for the LR.[155,156] The use of Cu, Zn, and Hf supported on silica as catalysts was reported to provide ethanol conversion between 34% and 98% using 0.5 and 10 h reaction time at 360 °C, followed by 3 h reaction at 300 °C.[155] The authors reported that the use of Cu-Zn-Zr catalyst could provide an ethanol conversion between 79% and 98% and a selectivity of 66% to 1,3-butadiene. However, changing Hf for Zr improved ethanol conversion to 96–99% and 1,3-butadiene selectivity of up to 72%.[156]

3.3 Other relevant products indirectly obtained from ethanol

Although ethanol can be used as starting material for several chemical reactions aiming to produce high value-added chemicals, these products can also be employed as reagents to obtain organic compounds that are not accessible directly from ethanol. Among the GR derived products from ethanol, 1-butanol and its isomers deserve attention as carbon sources to produce valuable chemicals.[157–159] For example, synthesis of aromatic compounds from isobutanol by successive dehydration and aromatization reaction using zeolites (Si/Al = 13.4–34.3) as catalysts were studied.[160] The authors reported BTX compounds (benzene, toluene, and xylene) with yields ranging between 33% and 35% using a pure acid zeolite. Additionally, it was shown that the insertion of Zn on the zeolitic (2.3–5.1% wt) increased BTX yields by up to 52%.

Hydrocarbons obtained from 1-butanol using SiO_2-Al_2O_3 as catalysts were also reported.[161] The reaction was conducted under the vapor phase using a continuous down-flow fixed-bed reactor at atmospheric pressure and N_2 as a carrier gas. The results showed an excellent 1-butanol conversion (up to 96%) and C_3 and C_4 hydrocarbons selectivity (between 63% and 81%) at 250 °C. Under this condition, C_5–C_{12} hydrocarbons can be obtained with selectivity between 16% and 35%. Also, aromatic compounds were obtained in this reaction. At 300 °C, the authors reported C_9 aromatic compounds production with selectivity between 25% and 35%.

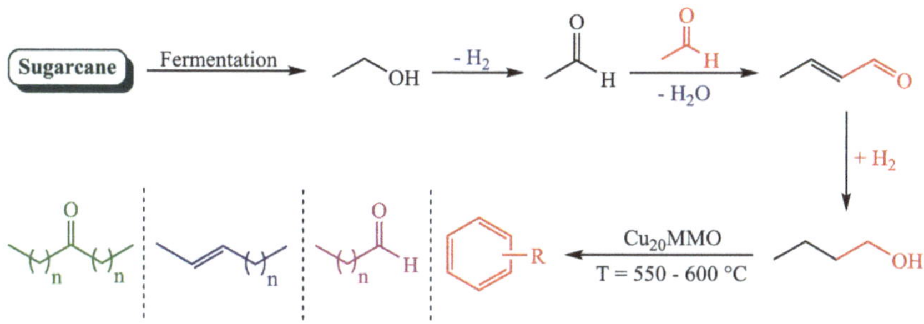

Fig. 7 Conversion of 1-butanol into high value-added chemicals. *Adapted from Vargas, J. A. M.; Morelato, L. H. T.; Ortega, J. O.; Boscolo, M.; Metzker, G.* Green Chem. ***2020***, 22, *2365–2369.*

However, the selectivity for aromatics decreased with increasing reaction temperature (350 and 400 °C). BTX compounds can be obtained with a selectivity of up to 75% at 400 °C. At low temperatures (300 and 350 °C) BTX compounds were obtained with 50% and 65% of selectivity, respectively.

The utilization of 1-butanol as a carbon source yielded unsaturated, carbonyl, and aromatic products (Fig. 7) using copper-magnesium-aluminum MMO catalyts.[162] In this case, the reaction was carried out in a fixed-bed flow reactor at temperatures between 450 and 600 °C. A good 1-butanol conversion (≥88%) was reported, obtained at temperatures above 500 °C. Notably, the products obtained in this reaction can be split into two different classes: condensed (alkenes, aldehydes, ketones, and aromatic) and gaseous products (mainly H_2, CO_2, and CO). Interestingly, the aromatic compounds can only be obtained at 600 °C. The main products found in this reaction were BTX, butyraldehyde, 4-heptanone, and 3-heptene.

4. Conclusions and future perspectives

Ethanol production in Brazil is a well-established technology and has a deep impact on the Brazilian economy. However, ethanol is mainly employed as fuel for light vehicles, limiting the great potential of this raw material. A new strategic vision for ethanol utilization must be developed, facing it as a renewable carbon source that can, in the future, supply the demand for chemicals that are today exclusively obtained from petroleum.

Considering the ethanol reforming reactions (ESR, POX and OSR) and ethanol dehydration for ethylene production, Brazil has advanced considerably in both research fields, culminating as the first country with a renewable and profitable ethylene industrial plant (Braskem process). However, the condensed product obtained from Guerbet and Lebedev reactions is still almost an unexplored field for Brazilian researchers. This fact can be both an excellent research opportunity and an alert sign since the development of this technology can be a key point for survival of the Brazilian sugarcane industry. Since ethanol is almost entirely employed as a fuel for light-transportation, the emerging of electrical-based cars may drastically reduce ethanol demand in the next 10–15 years.

The research focus should be given for catalyst development for ethanol conversion into high-added-value products. Catalyst selectivity, poisoning, long-term stability, and non-noble transition metals utilization, are sensitive points. The obtained results indicate that heterogeneous catalysts based on Co, Ni, and Cu are promising for ESR, POX, OSR, ED, GR, and LR, making these transition metals the basis for new achievements. New reactions and methodologies must be developed to utilize further ethanol-derived chemicals, such as 1-butanol, which proved to be a valuable starting material for obtaining aromatics, carbonyl, and unsaturated organic building blocks.

Acknowledgments

G.M. acknowledges the National Council for the Improvement of Higher Education (CAPES)—National Post Doctorate Program (PNPD) for the post-doctoral fellowship (Grant number: 1760327). J.A.M.V. acknowledge Sao Paulo Research Foundation—FAPESP (Grant number: 2017/09520-8) and USC (Convocatoria Interna DGI number 016-2019: Joven Investigador Santiaguino) for financial support. M.B. acknowledges FAPESP (Grant number: 2017/13230-5) for financial support.

References

1. Lopes, M. L.; Paulillo, S. C. L.; Godoy, A.; Cherubin, R. A.; Lorenzi, M. S.; Giometti, F. H. C.; Bernardino, C. D.; Amorim Neto, H. B.; Amorim, H. V. *Braz. J. Microbiol.* **2016**, *47*, 64–76.
2. Lago, A. C.; Bonomi, A.; Cavalett, O.; Cunha, M. P.; Lima, M. A. P. *Biomass Bioenergy* **2012**, *46*, 5–12.
3. Rosillo-Calle, F.; Cortez, L. A. B. *Biomass Bioenergy* **1998**, *14*, 115–124.
4. Melo, T. C. C.; Machado, G. B.; Belchior, C. R. P.; Colaço, M. J.; Barros, J. E. M.; Oliveira, E. J.; Oliveira, D. G. *Fuel* **2012**, *97*, 796–804.
5. Brazilian Sugarcane Industry Association., 2020. http://english.unica.com.br(Accessed 17 September 2020).
6. Global Petrol Prices., 2020. https://www.globalpetrolprices.com/Brazil/ethanol_prices/(Accessed 17 September 2020).
7. Stewart, G. G. In *Brewing and Distilling Yeasts*; Stewart, G. G., Ed.; Springer: New York, 2017; pp. 167–187.
8. Von Blottnitz, H.; Curran, M. A. *J. Clean Prod.* **2007**, *15*, 607–619.

9. Rabelo, S. C.; Carrere, H.; Maciel Filho, R.; Costa, A. C. *Bioresour. Technol.* **2011**, *102*, 7887–7889.
10. Wang, L.; Quiceno, R.; Price, C.; Malpas, R.; Woods, J. *Renew. Sust. Energ. Rev.* **2014**, *40*, 571–582.
11. Jaiswal, D.; Souza, A. P.; Larsen, S.; LeBauer, D. S.; Miguez, F. E.; Sparovek, G.; Bollero, G.; Buckeridge, M. S.; Long, S. P. *Nat. Clim. Change* **2017**, *7*, 788–792.
12. Macedo, I. C.; Seabra, J. E. A. In *Sugarcane Ethanol: Contributions to Climate Change Mitigation and the Environment*; Zuurbier, P., van Vooren, J., Eds.; Wageningen Academic Publishers: Wageningen, 2008; pp. 95–110.
13. Lima, N. D. S.; Naas, I. A.; Reis, J. G. M.; Silva, R. B. T. R. *Energies* **2020**, *13*, 2067.
14. Cavalett, O.; Chagas, M. F.; Seabra, J. E. A.; Bonomi, A. *Int. J. Life Cycle Assess* **2013**, *18*, 647–658.
15. Tsao, C.-C.; Campbell, J. E.; Mena-Carrasco, M.; Spak, S. N.; Carmichael, G. R.; Chen, Y. *Nat. Clim. Change* **2012**, *2*, 53–57.
16. Oliveira, M. D. In *Biofuels, Solar and Wind as Renewable Energy Systems*; Pimentel, D., Ed.; Springer: Dordrecht, 2008; pp. 215–230.
17. Oliveira, M. E. D.; Vaughan, B. E.; Rykiel, E. J., Jr. *Bioscience* **2005**, *55*, 593–602.
18. DeCicco, J. M. *Clim. Change* **2013**, *121*, 801–814.
19. Albarelli, J. Q.; Onorati, S.; Caliandro, P.; Peduzzi, E.; Meireles, M. A. A.; Marechal, F.; Ensinas, A. V. *Energy* **2017**, *138*, 1281–1290.
20. Khatiwada, D.; Leduc, S.; Silveira, S.; McCallum, I. *Renewable Energy* **2016**, *85*, 371–386.
21. Gallo, J. M. R.; Bueno, J. M. C.; Schuchardt, U. *J. Braz. Chem. Soc.* **2014**, *25*, 2229–2243.
22. Dagle, R. A.; Winkelman, A. D.; Ramasamy, K. K.; Dagle, V. L.; Weber, R. S. *Ind. Eng. Chem. Res.* **2020**, *59*, 4843–4853.
23. Rossetti, I.; Tripoli, A.; Ramis, G. *Int. J. Hydrogen Energy* **2020**, *45*, 10292–10303.
24. Mosca, L.; Jimenez, J. A. M.; Wassie, S. A.; Gallucci, F.; Palo, E.; Colazzi, M.; Taraschi, S.; Galdieri, G. *Int. J. Hydrogen Energy* **2020**, *45*, 7266–7277.
25. Mattos, L. V.; Jacobs, G.; Davis, B. H.; Noronha, F. B. *Chem. Rev.* **2012**, *112*, 4094–4123.
26. Zanchet, D.; Santos, J. B. O.; Damyanova, S.; Gallo, J. M. R.; Bueno, J. M. C. *ACS Catal.* **2015**, *5*, 3841–3863.
27. Hernandez, W. Y.; De Vlieger, K.; Van Der Voort, P.; Verberckmoes, A. *ChemSusChem* **2016**, *9*, 3196–3205.
28. Gabriels, D.; Hernandez, W. Y.; Sels, B.; Van Der Voort, P.; Verberckmoes, A. *Catal. Sci. Technol.* **2015**, *5*, 3876–3902.
29. Montero, C.; Remiro, A.; Valle, B.; Oar-Arteta, L.; Bilbao, J.; Gayubo, A. G. *Ind. Eng. Chem. Res.* **2019**, *58*, 14736–14751.
30. Masiran, N.; Vo, D.-V. N.; Salam, M. A.; Abdullah, B. *Procedia Eng.* **2016**, *148*, 1289–1294.
31. Dou, X. *Environ. Dev. Sustain.* **2015**, *17*, 123–136.
32. Stahel, W. R. *Nature* **2016**, *531*, 435–438.
33. Aramouni, N. A. K.; Touma, J. G.; Tarboush, B. A.; Zeaiter, J.; Ahmad, M. N. *Renew. Sustain. Energy Rev.* **2018**, *82*, 2570–2585.
34. Bepari, S.; Kuila, D. *Int. J. Hydrogen Energy* **2020**, *45*, 18090–18113.
35. Phung, T. K.; Pham, T. L. M.; Nguyen, A.-N. T.; Vu, K. B.; Giang, H. N.; Nguyen, T.-A.; Huynh, T. C.; Pham, H. D. *Chem. Eng. Technol.* **2020**, *43*, 672–688.
36. Ogo, S.; Sekine, Y. *Fuel Process. Technol.* **2020**, *199*, 106238.
37. Lucredio, A. F.; Bellido, J. D. A.; Assaf, E. M. *Appl. Catal. A* **2010**, *388*, 77–85.
38. Wurzler, G. T.; Rabelo-Neto, R. C.; Mattos, L. V.; Fraga, M. A.; Noronha, F. B. *Appl. Catal. A* **2016**, *518*, 115–128.

39. Anjaneyulu, C. A.; Costa, L. O. O.; Ribeiro, M. C.; Rabelo-Neto, R. C.; Mattos, L. C.; Venugopal, A.; Noronha, F. B. *Appl. Catal. A* **2016**, *519*, 85–98.
40. Moraes, T. M.; Rabelo-Neto, R. C.; Ribeiro, M. C.; Mattos, L. V.; Kourtelesis, M.; Ladas, S.; Verykios, X.; Noronha, F. B. *Appl. Catal. B* **2016**, *181*, 754–768.
41. Costa, L. O. O.; Silva, A. M.; Noronha, F. B.; Mattos, L. V. *Int. J. Hydrogen Energy* **2012**, *37*, 5930–5939.
42. Elias, K. F. M.; Lucredio, A. F.; Assaf, E. M. *Int. J. Hydrogen Energy* **2013**, *38*, 4407–4417.
43. Palm, M. O.; Silva Junior, M. E.; Cardoso, L. R.; Duarte, D. A.; Cataplan, R. C. *Energy Fuels* **2020**, *34*, 2205–2213.
44. Chagas, C. A.; Manfro, R. L.; Toniolo, F. S. *Catal. Lett.* **2020**, *150*, 3424–3436.
45. Dantas, S. C.; Resende, K. A.; Avila-Neto, C. N.; Noronha, F. B.; Bueno, J. M. C.; Hori, C. E. *Int. J. Hydrogen Energy* **2016**, *41*, 3399–3413.
46. Lucredio, A. F.; Bellido, J. D. A.; Assaf, E. M. *Catal. Commun.* **2011**, *12*, 1286–1290.
47. Maia, T. A.; Assaf, J. M.; Assaf, E. M. *Mater. Chem. Phys.* **2012**, *132*, 1029–1034.
48. Carvalho, F. L. S.; Asencios, Y. J. O.; Bellido, J. D. A.; Assaf, E. M. *Fuel Process. Technol.* **2016**, *142*, 182–191.
49. Augusto, B. L.; Ribeiro, M. C.; Aires, F. J. C. S.; Silva, V. T.; Noronha, F. B. *Catal. Today* **2020**, *344*, 66–74.
50. Braga, A. H.; Ribeiro, M. C.; Noronha, F. B.; Galante, D.; Bueno, J. M. C.; Santos, J. B. O. *Energy Fuels* **2018**, *32*, 12814–12825.
51. Marcos, F. C. F.; Lucredio, A. F.; Assaf, E. M. *RSC Adv.* **2014**, *4*, 43839–43849.
52. Silva, A. A. A.; Ribeiro, M. C.; Cronauer, D. C.; Kropf, A. J.; Marshall, C. L.; Gao, P.; Jacobs, G.; Davis, B. H.; Noronha, F. B.; Mattos, L. V. *Top. Catal.* **2014**, *57*, 637–655.
53. Pereira, E. B.; de la Piscina, P. R.; Marti, S.; Homs, N. *Energy Environ. Sci.* **2010**, *3*, 487–493.
54. Avila-Neto, C.; Zanchet, D.; Hori, C. E.; Ribeiro, R. U.; Bueno, J. M. C. *J. Catal.* **2013**, *307*, 222–237.
55. Silva, A. M.; Souza, K. R.; Jacobs, G.; Graham, U. M.; Davis, B. H.; Mattos, L. V.; Noronha, F. B. *Appl. Catal. B* **2011**, *102*, 94–109.
56. Rodrigues, T. S.; Silva, F. A.; Candido, E. G.; Silva, A. G. M.; Geonmonond, R. S.; Camargo, P. H. C.; Linardi, M.; Fonseca, F. C. J. *Mater. Sci.* **2019**, *54*, 11400–11416.
57. Zhang, M.; Yu, Y. *Ind. Eng. Chem. Res.* **2013**, *52*, 9505–9514.
58. Fan, D.; Dai, D.-J.; Wu, H.-S. *Materials* **2013**, *6*, 101–115.
59. Morschbacker, A. *J. Macromol. Sci. Part C* **2009**, *49*, 79–84.
60. Haro, P.; Ollero, P.; Trippe, F. *Fuel Process. Technol.* **2013**, *114*, 35–48.
61. Tsiropoulos, I.; Faaij, A. P. C.; Lundquist, L.; Schenker, U.; Briois, J. F.; Patel, M. K. *J. Clean. Prod.* **2015**, *90*, 114–127.
62. Alvarenga, R. A. F.; Dewulf, J. *Renewable Energy* **2013**, *59*, 49–52.
63. Oliveira, C. C. N.; Rochedo, P. R. R.; Bhardwaj, R.; Worrell, E.; Szklo, A. *Biofuels, Bioprod. Biorefin.* **2020**, *14*, 286–300.
64. Mohsenzadeh, A.; Zamani, A.; Taherzadeh, M. J. *ChemBioEng Rev.* **2017**, *4*, 75–91.
65. Pearson, D. E.; Tanner, R. D.; Picciotto, I. D.; Sawyer, J. S.; Cleveland, J. H., Jr. *Prod. Res. Dev.* **1981**, *20*, 734–740.
66. Adkins, H.; Perkins, P. P. *J. Am. Chem. Soc.* **1925**, *47*, 1163–1167.
67. Pines, H.; Haag, W. O. *J. Am. Chem. Soc.* **1960**, *82*, 2471–2483.
68. Pan, L. R.; Li, H. X. *Petrochem. Technol.* **1985**, *14*, 154–157.
69. Styskalik, A.; Vykoukal, V.; Fusaro, L.; Aprile, C.; Debecker, D. P. *Appl. Catal. B* **2020**, *271*, 118926.
70. Varislia, D.; Dogua, T.; Dogu, G. *Chem. Eng. Sci.* **2007**, *62*, 5349–5352.
71. Vaźquez, P.; Pizzio, L.; Caćeres, C. *J. Mol. Catal. A: Chem.* **2000**, *161*, 223–232.
72. Phillips, C. B.; Datta, R. *Ind. Eng. Chem. Res.* **1997**, *36*, 4466–4475.

73. Nguyen, T. M.; Mao, R. L. V. *Appl. Catal.* **1990**, *58*, 119–129.
74. Coutinho, P. L. A.; Morita, A. T.; Cassinelli, L. F.; Morschbacker, A.; Do Carmo, R. W. In *Catalytic Process Development for Renewable Materials*; Imhof, P., van de Waal, J. C., Eds.; Wiley-VCH Verlag GmbH & Co: Weinheim, 2013; pp. 149–164.
75. Wallmer, T.; Miers, S. A.; McConnell, S. *J. Eng. Gas Turbines Power* **2009**, *131*, 1–9.
76. Jin, C.; Yao, M.; Liu, H.; Lee, C. F.; Ji, J. *Renew. Sustain. Energy Rev.* **2011**, *15*, 4080–4106.
77. Bepari, S.; Basu, S.; Pradhan, N. C.; Dalai, A. K. *Catal. Today* **2017**, *291*, 47–57.
78. Trindade, W. R. S.; Santos, R. G. *Renew. Sustain. Energy Rev.* **2017**, *69*, 642–651.
79. Yusri, I. M.; Mamat, R.; Akasyah, M. K.; Jamlos, M. F.; Yusop, A. F. *Appl. Therm. Eng.* **2019**, *156*, 209–219.
80. Kumar, S.; Cho, J. H.; Park, J.; Moon, I. *Renew. Sustain. Energy Rev.* **2013**, *22*, 46–72.
81. Wang, H.; Reitz, R. D.; Yao, M.; Yang, B.; Jiao, Q.; Qiu, L. *Combust. Flame* **2013**, *160*, 504–519.
82. Berzin, V.; Tyurin, M.; Kiriukhin, M. *Appl. Biochem. Biotechnol.* **2013**, *169*, 950–959.
83. Badr, H. R.; Toledo, R.; Hamdy, M. K. *Biomass Bioenergy* **2011**, *20*, 119–132.
84. Ezeji, T. C.; Qureshi, N.; Blaschek, H. P. *Curr. Opin. Biotechnol.* **2007**, *18*, 220–227.
85. Keitaro, K.; Matsu-ura, T.; Obora, Y.; Ishi, Y. *Chem. Lett.* **2009**, *38*, 838–839.
86. Birgen, C.; Durre, P.; Preisig, H. A.; Wentzel, A. *Biotechnol. Biofuels* **2019**, *12*, 167.
87. Lopez-Olmos, C.; Morales, M. V.; Guerrero-Ruiz, A.; Rodríguez-Ramos, I. *Ind. Eng. Chem. Res.* **2020**, *59*, 16626–16636.
88. Burk, P. L.; Pruett, R. L.; Campo, K. S. *J. Mol. Catal.* **1985**, *33*, 1–14.
89. Gines, M. J. L.; Iglesia, E. *J. Catal.* **1998**, *176*, 155–172.
90. Gutierrez, A.; Karinen, R.; Airaksinen, S.; Kaila, R.; Krause, A. O. I. *Int. J. Hydrogen Energy* **2011**, *36*, 8967–8977.
91. Aitchison, H.; Wingad, R. L.; Wass, D. F. *ACS Catal.* **2016**, *6*, 7125–7132.
92. Dowson, G. R. M.; Haddow, M. F.; Lee, J.; Wingad, R. L.; Wass, D. F. *Angew. Chem., Int. Ed.* **2013**, *52*, 9005–9008.
93. Sun, Z.; Vasconcelos, A. C.; Bottari, G.; Stuart, M. C. A.; Bonura, G.; Cannilla, C.; Frusteri, F.; Barta, K. *ACS Sustain. Chem. Eng.* **2017**, *5*, 1738–1746.
94. Ogo, S.; Onda, A.; Yanagisawa, K. Appl. Catal. A. **2011**, *402*, 188–195.
95. Hanspal, S.; Young, Z. D.; Shou, H.; Davis, R. J. *ACS Catal.* **2015**, *5*, 1737–1746.
96. Tsuchida, T.; Kubo, J.; Yoshioka, T.; Sakuma, S.; Takeguchi, T.; Ueda, W. *J. Catal.* **2008**, *259*, 183.
97. Silvester, L.; Lamonier, J. F.; Faye, J.; Capron, M.; Vannier, R. N.; Lamonier, C.; Dubois, J. L.; Couturier, J. L.; Calais, C.; Dumeignil, F. *Catal. Sci. Technol.* **2015**, *5*, 2994–3006.
98. Tsuchida, T.; Sakuma, S.; Takeguchi, T.; Ueda, W. *Ind. Eng. Chem. Res.* **2006**, *45*, 8634–8642.
99. Ho, C. R.; Shylesh, S.; Bell, A. T. *ACS Catal.* **2016**, *6*, 939–948.
100. Coelho, A.; Perrone, O. M.; Gomes, E.; Da-Silva, R.; Thoméo, J. C.; Boscolo, M. *Appl. Catal. A* **2017**, *532*, 32–39.
101. Frost, R. L.; Musumeci, A. W.; Kloprogge, J. T.; Adebajo, M. O.; Martens, W. N. *J. Raman Spectrosc.* **2006**, *37*, 733–741.
102. Miyata, S. *Clays Clay Miner.* **1975**, *23*, 369–375.
103. Daza, C. E.; Gallego, J.; Mondragón, F.; Moreno, S.; Molina, R. *Fuel* **2010**, *89*, 592–603.
104. Chmielarz, L.; Węgrzyn, A.; Wojciechowska, M.; Witkowski, S.; Michalik, M. *Catal. Lett.* **2011**, *141*, 1345–1354.
105. Venugopal, A. K.; Venugopalan, A. T.; Kaliyappan, P.; Raja, T. *Green Chem.* **2013**, *15*, 3259–3267.
106. Monzón, A.; Romeo, E.; Royo, C.; Trujillano, R.; Labajos, F. M.; Rives, V. *Appl. Catal. A* **1999**, *185*, 53–63.

107. Guerrero-Urbaneja, P.; García-Sancho, C.; Moreno-Tost, R.; Mérida-Robles, J.; Santamaría-González, J.; Jiménez-López, A.; Maireles-Torres, P. *Appl. Catal. A* **2014**, *470*, 199–207.
108. Tantirungrotechai, J.; Chotmongkolsap, P.; Pohmakotr, M. *Materials* **2010**, *128*, 41–47.
109. Barta, K.; Warner, G. R.; Beach, E. S.; Anastas, P. T. *Green Chem.* **2014**, *16*, 191–196.
110. Warner, G.; Hansen, T. S.; Riisager, A.; Beach, E. S.; Barta, K.; Anastas, P. T. *Bioresour. Technol.* **2014**, *161*, 78–83.
111. Matson, T. D.; Barta, K.; Iretskii, A. V.; Ford, P. C. *J. Am. Chem. Soc.* **2011**, *133*, 14090–14097.
112. Bernt, C. M.; Bottari, G.; Barrett, J. A.; Scott, S. L.; Barta, K.; Ford, P. C. *Catal. Sci. Technol.* **2016**, *6*, 2984–2994.
113. Chui, M.; Metzker, G.; Bernt, C. M.; Tran, A. T.; Burtoloso, A. C. B.; Ford, P. C. *ACS Sustain. Chem. Eng.* **2017**, *5*, 3158–3169.
114. Sun, Z.; Fridrich, B.; de Santi, A.; Elangovan, S.; Barta, K. *Chem. Rev.* **2018**, *118*, 614–678.
115. Sun, Z.; Bottari, G.; Afanasenko, A.; Stuart, M. C. A.; Deuss, P. J.; Fridrich, B.; Barta, K. *Nat. Catal.* **2018**, *1*, 82–92.
116. Carlini, C.; Marchionna, M.; Noviello, M.; Raspolli Galletti, A. M.; Sbrana, G.; Basile, F.; Vaccari, A. *J. Mol. Catal. A: Chem.* **2005**, *232*, 13–20.
117. Cheng, F.; Hou, B.; Guo, H.; Cui, J.; Li, D. *J. Fuel Chem. Technol.* **2018**, *46*, 1472–1481.
118. Balsamo, N. F.; Sapag, K.; Oliva, M. I.; Pecchi, G. A.; Eimer, G. A.; Crivello, M. E. *Catal. Today* **2017**, *279*, 209–216.
119. Kuśtrowski, P.; Chmielarz, L.; Bozek, E.; Sawalha, M.; Roessner, F. *Mater. Res. Bull.* **2004**, *39*, 263–281.
120. Jiang, B.; Xi, Z.; Lu, F.; Huang, Z.; Yang, Y.; Sun, J.; Liao, Z.; Wang, J.; Yang, Y. *Catal. Sci. Technol.* **2019**, *9*, 6335–6344.
121. Barta, K.; Ford, P. C. *Acc. Chem. Res.* **2014**, *47*, 1503–1512.
122. Scheid, A. J.; Barbosa-Coutinho, E.; Schwaab, M.; Salau, N. P. G. *Ind. Eng. Chem. Res.* **2019**, *58*, 12981–12995.
123. León, M.; Díaz, E.; Ordóñez, S. *Catal. Today* **2011**, *164*, 436–442.
124. Di Cosimo, J. I.; Díez, V. K.; Xu, M.; Iglesia, E.; Apesteguía, C. R. *J. Catal.* **1998**, *178*, 499.
125. Pavel, O. D.; Tichit, D.; Marcu, I. C. *Appl. Clay Sci.* **2012**, *61*, 52–58.
126. Bîrjega, R.; Pavel, O. D.; Costentin, G.; Che, M.; Angelescu, E. *Appl. Catal. A* **2005**, *288*, 185–193.
127. Benito, P.; Vaccari, A.; Antonetti, C.; Licursi, D.; Schiarioli, N.; Rodriguez-Castellon, E.; Raspolli Galletti, A. M. *J. Clean. Prod.* **2019**, *209*, 1614–1623.
128. Stošić, D.; Hosoglu, F.; Bennici, S.; Travert, A.; Capron, M.; Dumeignil, F.; Couturier, J. L.; Dubois, J. L.; Auroux, A. *Catal. Commun.* **2017**, *89*, 14–18.
129. Marcu, I. C.; Tanchoux, N.; Fajula, F.; Tichit, D. *Catal. Lett.* **2012**, *143*, 23–30.
130. Barret, J. A.; Jones, Z. R.; Stickelmaier, C.; Schopp, N.; Ford, P. C. *ACS Sustain. Chem. Eng.* **2018**, *6*, 15119–15126.
131. Perrone, O. M.; Lobefaro, F.; Aresta, M.; Nocito, F.; Boscolo, M.; Dibenedetto, A. *Fuel Process. Technol.* **2018**, *177*, 353–357.
132. Siqueira, M. R.; Perrone, O. M.; Metzker, G.; Lisboa, D. C. O.; Thoméo, J. C.; Boscolo, M. *Mol. Catal.* **2019**, *476*, 110516.
133. Gunther, W. R.; Michaelis, V. K.; Griffin, R. G.; Román-Leshkov, Y. *J. Phys. Chem. C* **2016**, *120*, 28533–28544.
134. White, W. C. *Chem.-Biol. Interact.* **2007**, *166*, 10–14.
135. Abghari, S. Z.; Sadi, M. *J. Taiwan Inst. Chem. Eng.* **2013**, *44*, 365–376.

136. Lee, H.; Jung, J. C.; Kim, H.; Chung, Y. M.; Kim, T. J.; Lee, S. J.; Oh, S. H.; Kim, Y. S.; Song, I. K. *Catal. Lett.* **2008**, *124*, 364–368.
137. Lee, H.; Jung, J. C.; Kim, H.; Chung, Y. M.; Kim, T. J.; Lee, S. J.; Oh, S. H.; Kim, Y. S.; Son, I. K. *Catal. Commun.* **2008**, *9*, 1137–1142.
138. Madeira, L. M.; Portela, M. F. *Catal. Rev.* **2002**, *44*, 247–286.
139. Taifan, W. E.; Bučko, T.; Baltrusaitis, J. *J. Catal.* **2017**, *346*, 78–91.
140. Chieregato, A.; Velasquez Ochoa, J.; Bandinelli, C.; Fornasari, G.; Cavani, F.; Mella, M. *ChemSusChem* **2014**, 7, 1–13.
141. Angelici, C.; Weckhuysen, B. M.; Bruijnincx, P. C. A. *ChemSusChem* **2013**, *6*, 1595–1614.
142. Yan, T.; Dai, W.; Wu, G.; Lang, S.; Hunger, M.; Guan, N.; Li, L. *ACS Catal.* **2018**, *8*, 2760–2773.
143. Ordomsky, V. V.; Sushkevich, V. L.; Ivanova, I. I. *J. Mol. Catal. A Chem.* **2010**, *333*, 85–93.
144. Sekiguchi, Y.; Akiyama, S.; Urakama, W.; Koyama, T.; Miyaji, A.; Motokura, K.; Baba, T. *Catal. Commun.* **2015**, *68*, 20–24.
145. Tsuchida, T.; Yoshioka, T.; Sakuma, S.; Takeguchi, T.; Ueda, W. *Ind. Eng. Chem. Res.* **2008**, *47*, 1443–1452.
146. Baylon, R. A. L.; Sun, J.; Wang, Y. *Catal. Today* **2016**, *259*, 446–452.
147. Ohnishi, R.; Akimoto, T.; Tanabe, K. *Chem. Commun.* **1985**, 1613–1614.
148. Kvisle, S.; Aguero, A.; Sneeden, R. P. A. *Appl. Catal.* **1988**, *43*, 117–131.
149. Chung, S. H.; Angelici, C.; Hinterding, S. O. M.; Weingarth, M.; Baldus, M.; Houben, K.; Weckhuysen, B. M.; Bruijnincx, P. C. A. *ACS Catal.* **2016**, *6*, 4034–4045.
150. Angelici, C.; Velthoen, M. E. Z.; Weckhuysen, B. M.; Bruijnincx, P. C. A. *Catal. Sci. Technol.* **2015**, *5*, 2869–2879.
151. Angelici, C.; Meirer, F.; van der Eerden, A. M. J.; Schaink, H. L.; Goryachev, A.; Hofmann, J. P.; Hensen, E. J. M.; Weckhuysen, B. M.; Bruijnincx, P. C. A. *ACS Catal.* **2015**, *5*, 6005–6015.
152. Sushkevich, V. L.; Ivanova, I. I.; Taarning, E. *Green Chem.* **2015**, *17*, 2552–2559.
153. Kyriienko, P. I.; Larina, O. V.; Soloviev, S. O.; Orlyk, S. M.; Calers, C.; Dzwigaj, S. *ACS Sustain. Chem. Eng.* **2017**, *5*, 2075–2083.
154. Jones, M.; Keir, C.; Iulio, C.; Robertson, R.; Williams, C.; Apperley, D. *Catal. Sci. Technol.* **2011**, *1*, 267.
155. Sushkevich, V. L.; Ivanova, I. I.; Ordomsky, V. V.; Taarning, E. *ChemSusChem* **2014**, 7, 2527–2536.
156. De Baerdemaeker, T.; Feyen, M.; Müller, U.; Yilmaz, B.; Xiao, F. S.; Zhang, W.; Yokoi, T.; Bao, X.; Gies, H.; De Vos, D. E. *ACS Catal.* **2015**, *5*, 3393–3397.
157. Gunst, D.; Alexopoulos, K.; Van Der Borght, K.; John, M.; Galvita, V.; Reyniers, M.-F.; Verberckmoes, A. *Appl. Catal. A* **2017**, *539*, 1–12.
158. Raizada, V. K.; Tripathi, V. S.; Lal, D.; Singh, G. S.; Dwivedi, C. D.; Sen, A. K. *J. Chem. Technol. Biotechnol.* **1993**, *56*, 265–270.
159. Jiang, B.-S.; Chang, R.; Hou, Y.-C.; Lin, Y.-C. *Ind. Eng. Chem. Res.* **2012**, *51*, 13993–13998.
160. Yu, L.; Huang, S.; Zhang, S.; Liu, Z.; Xin, W.; Xie, S.; Xu, L. *ACS Catal.* **2016**, *2*, 1203–1210.
161. Palla, V. C. S.; Shee, D.; Maity, S. K. *Appl. Catal. A* **2016**, *526*, 28–36.
162. Vargas, J. A. M.; Morelato, L. H. T.; Ortega, J. O.; Boscolo, M.; Metzker, G. *Green Chem.* **2020**, *22*, 2365–2369.

Index

Note: Page numbers followed by "*f*" indicate figures, "*t*" indicate tables, and "*s*" indicate schemes.

N

O

P

R

S

FSC
www.fsc.org
MIX
Papier | Fördert gute Waldnutzung
FSC® C083411

Kolibri 360 GmbH
Ferdinand-Jühlke-Straße 7
99095 Erfurt, Deutschland
produktsicherheit@kolibri360.de